ANALOGIES BETWEEN ANALOGIES

The mathematical reports of S.M. Ulam
and his Los Alamos collaborators

S.M. Ulam

ANALOGIES BETWEEN ANALOGIES

The mathematical reports of S.M. Ulam
and his Los Alamos collaborators

Edited by
A.R. Bednarek and Françoise Ulam

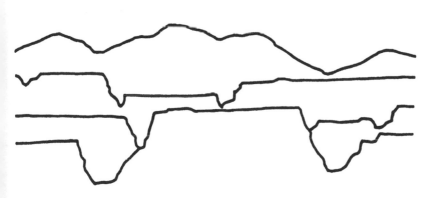

UNIVERSITY OF CALIFORNIA PRESS
Berkeley • Los Angeles • Oxford

University of California Press
Berkeley and Los Angeles, California

University of California Press, Ltd.
Oxford, England

Library of Congress Cataloging-in-Publication Data

Analogies between Analogies; the mathematical reports of
S. M. Ulam and his Los Alamos collaborators/S. M. Ulam;
A. R. Bednarek and Françoise Ulam, editors.
(Los Alamos series in basic and applied sciences)
ISBN 978-0-520-30230-3 (paper)
Includes Index.
1. Stochastic processes. 2. Nonlinear theories.
3. Mathematical physics. 4. Ulam, Stanislaw M.
I. Ulam, Stanislaw M. II. Bednarek, A. R.
III. Ulam, Françoise. IV. Series.
QA274.A53 1990
519.2–dc20   89-20275
                                                              CIP

CONTENTS

The Los Alamos Laboratory in 1944 and some forty years later.

FOREWORD

*"Good mathematicians see analogies
between theorems or theories,
the very best ones see
analogies between analogies."*
Stefan Banach

Stanislaw Ulam's affiliation with Los Alamos National Laboratory spanned over two-thirds of his professional life. There was no aspect of its mathematical activity during this period in which he was not involved, either centrally or tangentially.

His catholic view of the role of mathematics vis-à-vis other sciences extended far beyond into the mathematical and scientific community at large, as did his genius for problem formulation and for applying the most abstract ideas from the foundations of mathematics to computing, physics, and biology. In addition he possessed the ability to excite others—many of them not trained as mathematicians—and involve them in his researches. The impact of his work is still felt both at the Laboratory and in those larger communities, for he liked to disseminate his ideas orally in an ever widening round of lectures and seminars from where they took on a life of their own. The Monte Carlo method which he originated with von Neumann in order to study neutron scattering and other nuclear problems at Los Alamos is one such example. Its offshoots are now so universal that they are even applied to regulate traffic lights!

His influence, along with that of John von Neumann, the brilliant Hungarian mathematician, contributed to the establishment at the Laboratory of an atmosphere and a tradition that fostered and supported an exceptional—if not unique—interaction between mathematics and science. Extensive testimony and documentation concerning the integral role that Stan Ulam played in this interaction can be found in "From Cardinals to Chaos, Reflexions on the Life and Legacy of Stanislaw Ulam" published by Cambridge University Press in 1989.

From 1944 until his death in 1984, while connected with Los Alamos in a variety of ways, from staff member, to group leader, to research advisor, to 'no-fee consultant'—one of his favorite expressions—he wrote Laboratory reports (many with the help of trusted and talented collaborators) that show a breadth of scientific interests unusual for a mathematician. They cover pioneering work, in the horse-and-buggy days of computing, on mathematical modeling of physical processes, nuclear rocketry, space travel, and biomathematics. (Another eleven, weapons-related reports, with Evans, Everett, Fermi, Metropolis, von Neumann, Richtmyer, Teller, Tuck, and others are still classified and unavailable for publication.)

Mathematically speaking, three motifs run through Ulam's theoretical and applied work: the iteration or composition of functions, or relations; the use of evolving computer capability in the exploration of analytically intractable problems; and the introduction of probabilistic approaches—while knowing that most practical applications are made in the presence of uncertainty. The fusion of these themes is characteristic of the central contributions of this collection.

As to the quotation from which the title of this book is derived, one must remember that Ulam held Banach, along with von Neumann and Fermi, "as one of the three great men whose intellects impressed me the most." Stefan Banach was an outstandingly original Polish mathematician and one of the founders of the now famous Lwów school of mathematics (see Chapter 20, Preface to "The Scottish Book.") Banach was Ulam's friend and mentor in Poland before World War II. His influence on Ulam was profound and Ulam liked to quote his comment on the ability of some mathematicians to see "analogies between analogies." There is no question that in the practice of his craft and his art, Ulam was guided by this principle, and that he, in turn, epitomized its application. In addition, to Ulam the idea of analogy was itself amenable to mathematical discussion.

In 1983, when D. Sharp and M. Simmons, editors of this Los Alamos Science series, asked Ulam to gather his unclassified—and declassified—reports for publication in one volume, they intended to omit a few that had appeared elsewhere. After his death in 1984, it was decided to publish them all as many represent preliminary studies of subjects that were subsequently expanded elsewhere, leading, in several instances, to the development of new and extensive theories.

Ulam had dictated brief introductory notes and a sketch for a preface which he intended to develop. Rather than put words in his mouth, it was thought more appropriate to reproduce his notes in their short and unpolished form. It was also decided to leave the style and substance of the reports untouched, as evidence that scientific advances do not usually arise in their final, definite form. More often than not they are the product of sequences of tentative, sometimes repetitive, and even at times inaccurate steps. Two appendices complete the volume: a list of Ulam's publications and a brief biographical chronology. (More detailed biographical material can be found in his autobiography, "Adventures of a Mathematician," as well as in "From Cardinals to Chaos.")

This collection represents an important complement to the selection of papers and problems, mostly in pure mathematics, published by MIT Press in 1974 as "Sets, Numbers, Universes," and to a volume of essays, "Science, Computers, and People," published by Birkhäuser in 1985. And, whereas these two books are composed of papers readily available albeit scattered in the scientific literature, the Los Alamos Reports have been for the most part difficult of access and little known. Their historic value is therefore very real.

As mentioned earlier, many of these reports and much of Ulam's work was done in collaboration. He liked to stress the importance of working with collaborators, with whom, he said, the nature of their "shared ideas and techniques" depended "on the personality and experience of the individuals." We add a few words about these colleagues as well as about the work they and Ulam engaged in.

As early as 1944, when he joined the Manhattan Project, Ulam and David Hawkins (Chapter 1) were "playing" with the notion of branching processes, or multiplicative systems, as they called them, motivated by their application to atomic physics.

David Hawkins, a philosopher of science by profession and mathematician by inclination, was, in Ulam's words "the best amateur mathematician I know," and they became fast friends. Hawkins is presently professor emeritus at the University of Colorado.

The work in the Ulam-Hawkins report was subsequently developed with Everett in the three extensive reports grouped in Chapter 3, which are reproduced here for the first time. While clearly motivated by the need to understand neutron multiplication in fission processes, the reports lay the foundations for—in their own words—a "formalism general enough to include as special cases the multiplication of bacteria, radioactive decay, cosmic ray showers, diffusion theory and the theory of trajectories in mechanical systems."

C. J. Everett, who died in 1987, was a mathematician with whom Ulam worked on a conceptual as well as technical level in Wisconsin and at Los Alamos, where he became a member of Ulam's group. An eccentric, shy, and witty man, he was quite probably the only person who ever opted for bus transportation to come to Los Alamos for a hiring interview, and he was known for having turned in a monthly progress report—in which staff members were supposed to describe their research—which said tersely "progress was made on last month's progress report."

The first written proposals for the Monte Carlo method put together in a 1946 "report" called "Statistical Methods in Neutron Diffusion" appear in Chapter 2. This method of approaching precise but intractable problems through the introduction of random processes and probabilistic experimentation, has found wide application not only in areas close to those motivating its origin but others more removed, such as operations research, and combinatorics. In fact, the "report" —of which only eight copies were made—consists of two letters and handwritten calculations photographed and stapled together. Its cover specifies that the "work" was "done" by Ulam and von Neumann and "written" by von Neumann and Robert Richtmyer—then head of the Laboratory's Theoretical Division. Its informality attests to the casual manner in which information was disseminated through the Laboratory at the time.

The long term professional and personal rapports between von Neumann and Ulam need not be recounted here—references to them can be found in the books already mentioned. Suffice it to say that though there exist few papers and abstracts under their joint names, von Neumann's extensive correspondence with Ulam attests to their interacting interests in pure mathematics, in pioneering computer technology and techniques, and in cellular automata and the brain. (The correspondence is now stored in the archives of the Philosophical Society in Philadelphia.)

Ulam's collaboration with Enrico Fermi initiated the computer simulations of nonlinear dynamical systems that lead to the evolution of a major field of research popularly labeled "chaos theory." Fermi called this work, which was developed with the programming assistance of John Pasta, "a minor discovery," a modest understatement given the seminal character of this investigation (Chapter 5.) Chapters 10 and 11, with P. R. Stein, address the subject of nonlinear transformations in greater detail.

Fermi, with whom Ulam became acquainted in Los Alamos during the war, was a man of simple tastes and life style. The Ulams had an opportunity to sample this while motoring together across France one summer. Feeling ill at ease during a lunch in a recommended temple of gastronomy, Fermi decreed he would select the night's lodgings. Meandering through a picturesque valley he chose a modest inn by a babbling brook where, after dinner, sitting under the stars they discussed physics and new mathematical problems to experiment with after the vibrating string calculations. However the night's encounter with fleas, bedbugs, and mosquitoes made him admit the next morning that the higher-class hostelry next door that Ulam had eyed, might perhaps have provided a more restful night.

In the area of space technology, Ulam investigated schemes for nuclear rocketry with Everett and with Conrad Longmire, a physicist from the mountains of Tennessee who played a mean banjo. Chapter 7 describes a way to propel very large space vehicles by a series of small external nuclear explosions which later developed into Project Orion. Chapter 9 deals with the propulsion of space vehicles by extraction of gravitational energy from planets. Schemes based on a similar idea are now used in "flyby" missions to the outer planets and to provide part of the energy for spacecrafts going beyond the planets.

A study of patterns of growth, with Robert Schrandt (Chapter 12) investigates how simple recursively defined codes can give rise to complex objects. Such studies have become a growth industry of their own in the improved computer graphics world of today.

Several other reports are devoted to biomathematical questions. Their findings have opened new fields of biomathematical research. Abstract schemata of mathematics are applied to pattern recognition with the help of computers investigating, for example, the way visual pictures are recognized. Using metrics in molecular biology shows how a new mathematical concept of distance between finite sequences or objects can be applied to reconstruct the evolutionary history of biological organisms.

Closely involved with Ulam in this work was Paul Stein, a physicist turned mathematician—under Ulam's influence—who became an invaluable collaborator able to implement and develop the gist of Ulam's directions. William Beyer, a gifted fellow mathematician, also collaborated on a conceptual and technical level in the biological investigations.

John Pasta, Mary Tsingou-Menzel, Robert Schrandt, and Myron Stein lent their talents to creative programming, at a time when the art was in its infancy and pre-microchip-era machines with names like "Eniac," "Maniac," "Johnniac," presented storage and timing constraints. Pasta, who died in 1980, was a self-made man of Italian descent. He had furthered his education and became a physicist while working on the New York city police force.

The mathematician Al Bednarek, one of the editors of this volume and coauthor of this foreword, also collaborated with Ulam on problems of parallel computation (Chapter 18). He is a former chairman of the mathematics department at the University of Florida.

Shortly after his arrival at Los Alamos in 1944, Ulam was asked by a colleague what it was that he was doing. Since at the time he was a very pure mathematician and had not yet familiarized himself with the nature of the work, his Socratic answer was "I supply the necessary don't know how!" Stan Ulam's "necessary don't know how" as well as his modestly unenunciated "know how" are sorely missed by all who were privileged to have known him or worked with him.

Last but not least, the editors wish especially to thank Peggy Atencio, Ben Atencio, Janet Holmes, Debi Erpenbeck, Gary Benson, Chuck

Calef and Gloria Sharp, among other members of Los Alamos Information Services Division, and above all Chris West and Pat Byrnes, for their herculean efforts in transposing into print and formatting these extremely difficult reports, and also Patricia Metropolis for her invaluable informal advice and help. The editors assume full responsibility for any existing discrepancies or inaccuracies. They also gratefully acknowledge the permission granted by *Rozprawy Matematyczne* to reprint their edition of the report "Non linear transformations studies on electronic computers." The preparation of this book was done under the generous auspices of the Los Alamos National Laboratory.

A. R. Bednarek,
Gainesville, Florida
Françoise Ulam,
Santa Fe, New Mexico

PREFACE

The collection of these reports, which appeared over the considerable span of years that I spent at Los Alamos, concerns a great variety of topics. Its very heterogeneous nature illustrates the diversity of the programs and of the areas of research that interested the laboratory.

Before World War II it was almost exclusively in the universities, in the graduate schools of the larger institutions that scientific research used to take place. The Bureau of Standards and a very few large industrial companies such as Bell Telephone, General Electric, and some pharmaceutical firms were the exception to the rule.

This little collection may bear witness, in a very modest way, to the wide-ranging changes, which are still going on in the organization and practice of research in this country and abroad. Because of the novel problems which confronted its scientists during the wartime establishment of Los Alamos, the need arose for research and ideas in domains contiguous to its central purpose. This trend continues unabated to the present.

Problems of a complexity surpassing anything that had ever existed in technology rendered imperative the development of electronic computing machines and the invention of new theoretical computing methods. There, consultants like von Neumann played an important role in helping enlarge the horizon of the innovations, which required the most abstract ideas derived from the foundations of mathematics as well as from theoretical physics. They were and still are invested in new, fruitful ways.

An enormous number of technological and theoretical innovations were initiated at this laboratory during these forty years. To mention but a few, besides the advances in computing, one can name research on nuclear propulsion of rockets and space vehicles, in molecular biology, and on the technology of separating cells.

The growing importance of research laboratories such as this one became a not exclusively American phenomenon. For instance the aspect of academic research has changed almost beyond recognition in France. What used to be, before World War II, almost exclusively the province of universities, has now shifted to the French National Center of Research (Centre National de Recherche.)

The growing importance of research laboratories such as this one became a not exclusively American phenomenon. For instance the aspect of academic research has changed almost beyond recognition in France. What used to be, before World War II, almost exclusively the province of universities, has now shifted to the French National Center of Research (Centre National de Recherche.)

This collection of Los Alamos Reports ranges over almost four decades and may illustrate, I hope, how a mathematical turn of mind, a mathematical habit of thinking, a way of looking at problems in different subfields of physics, astronomy, or biology can suggest general insights and not just offer the mere use of techniques. Ideas derived from even very pure mathematical fields can provide more than mere "service work," they may help provide true conceptual contributions from the very beginning.

The period in question has seen the origin and development of the art of computing on a scale which vastly surpasses the breadth and depth of the numerical work of the past. In at least two different and separate ways the availability of computing machines has enlarged the scope of mathematical research. It has enabled us to attempt to gather, through heuristic experiments, impressions of the morphological nature of various mathematical concepts such as the behavior of solutions of certain nonlinear transformations, the properties of some combinatorial systems, and some topological curiosities of seemingly general behaviors. It has also enabled us to throw light on the behavior of solutions of many problems concerning complicated systems, by allowing numerical computations of very elaborate special physical problems, using both Monte Carlo type experiments and extensive but "intelligently chosen" brute force approaches, in hydrodynamics for example.

A number of such experiments have revealed, surprisingly, a nonclassical ergodic behavior of several dynamical systems. They have showed unexpected regularities in certain flows of dynamical systems, in the mechanics of many-body problems, and in continuum mechanics. Recently they have been applied to the study of elementary particle physics set-ups and interactions.

And now there appear some most exciting vistas in the applications of mathematics to biology that deal with both the construction and the evolution of living systems, including problems of the codes, which seem to define the basic properties of organisms and ultimately may provide us with a partial understanding of the working and evolution of the nervous system and some of the powers of the brain itself.

In addition these reports show the varying involvments of my collaborators and myself. I particularly want to stress the importance of the role of collaborators. An ever increasing number of publications of mathematical research is proof of the advantages derived when two or more authors share ideas and techniques. The nature of this exchange varies from case to case, depending on the personality and experience of the individuals.

These few very sketchy remarks are merely intended to emphasize how the necessity of defense work at the frontiers of science has continued to this day to stimulate research in a multitude of directions.

S. M. Ulam
Santa Fe,
February 1984

LA-171

November 14, 1944

This document contains **22** pages

THEORY OF MULTIPLICATIVE PROCESSES. 1.

WORK DONE BY	REPORT WRITTEN BY
D. Hawkins	D. Hawkins
S. Ulam	S. Ulam

Faded cover of the original 1944 report after it was released from classification in 1956. Note its low number.

1

THEORY OF MULTIPLICATIVE PROCESSES

With David Hawkins
(LA-171, November 14, 1944)

This report treats branching processes, of neutron proliferation for instance, through a mathematical theory involving compositions of generating functions.

It is a precursor of the studies of multiplicative systems in several variables written with C. J. Everett in 1948 (LA-683, -690 and -707) which develop an elaborate and basic theory of "multiplicative" (branching) processes. See for example a book by T. E. Harris: The Theory of Branching Processes, published by Springer in 1963. (Author's note.)

ABSTRACT

General properties of statistics of multiplicative systems are discussed together with the study of fluctuations in the number of particles in such systems. A general method is indicated through which one may study the fluctuations in the case where one takes into account the factors of geometry and time-dependence of constants.

The statistical theory of multiplicative chain processes does not compare in completeness to date with the corresponding theory of additive processes. The present paper is intended primarily as an exposition of a simple theory of the statistics of multiplication, permitting application to a variety of special problems.

The simplest (the "Bernoullian") case may be described as follows: A particle can produce, with probabilities p_0, p_1, $p_2 \ldots$, $p_n, \ldots$ a number 0, 1, 2, 3, $\ldots$, $n, \ldots$ of similar particles in one generation. We assume that each particle produced has again the same probabilities of producing n offspring. We also assume that each particle dies at procreation. Required is the probability law $p_k(n)$ for any generation k.

We remark parenthetically that this formulation makes the multiplicative process essentially discrete and finite. The statistics of neutron multiplication involves a continuous process as well, namely a random distribution in energy, space, and time. We disregard this aspect initially. Later we shall show that the admission of such continuity leads to a generalization of the methods described below. There are, in the meantime, two physically accurate interpretations of a discrete series: (1) one can represent the chain process as a graph; the n particles in the kth generation are the n lines connecting the kth — and the $k + 1$st — branch points in a chain or set of chains; (2) the n particles are those in existence at the kth unit of time, where the probability law $p_1(n)$ is the distribution one unit of time after the introduction of a single particle. If a time unit be chosen equal to the average time between fissions, the distinction is in many cases not crucial. Frankel, and later Feynman, studied the continuous process. We shall show later that their differential equations of the random process correspond to the infinitesimal transformations of the group in which our iteration (see Theorem I) may be imbedded.

1. The first problem to consider is this: we are given an amount and arrangement of active material. In this system a neutron produces *on the average n* neutrons with probability $p(n)$; $\sum_{n=0}^{\infty} p(n) = 1$. $p(0)$ is the average probability of leakage or absorption, without subsequent production of neutrons. $p(n)$ normalized for $n > 0$ is a nuclear constant, so far purely empirical, known as to its first moment and less accurately as to its second moment. Required is the probability of having n neutrons after k generations (or units of time). This problem is solved, in principle, by:

Theorem I. Let $f(x)$ be the generating function of the distribution of the number of offspring, i.e., $f(x) = \sum_{n=0}^{\infty} p(n)x^n$. Then the generating function for the kth generation $f_k(x) = f^k(x)$, the kth *iterate* of $f(x)$. [The kth iterate is defined as follows: $f^1(x) = f(x)$, $f^k(x) = f(f^{k-1}(x))$. The theorem asserts that the probability $p_k(n)$ is given as the coefficient of x^n in the ascending polynomial or power series expression of $f^k(x)$. The physical multiplication of the random variable is reflected in the iterated substitution by which $x \rightarrow f(x)]$.

Proof: Starting with one neutron in the 0th generation we obtain, with probability $p_k(n)$, n neutrons in the kth generation. Beginning with r neutrons, denote the corresponding probability by $p_k^{(r)}(n)$. Now assume that a chain is started by one neutron. We have

$$p_k(n) = \sum_{r=0}^{\infty} p_{k-1}(r) p_1^{(r)}(n) .$$

Now if $f(x)$ is the generating function of the distribution $p_1(n)$, the generating function of the distribution $p_1^{(r)}(n)$ is $[f(x)]^r$. This follows from the assumption that contemporary neutrons are independent in procreative powers, and from the theorem (of Laplace) that the generating function of a sum of independent random variables is the product of their generating functions. The above proposition may also be verified for $r = 0$, since $p_1^{(0)}(n) = 0$ for all $n > 0$. Substituting generating function for probability in the above equation, we have:

$$f_k(x) = p_{k-1}(0)[f(x)]^0 + p_{k-1}(1)[f(x)]^1 + p_{k-1}(2)[f(x)]^2 + \dots$$
$$| \dots = f_{k-1}[f(x)] = f^k(x) .$$

Two remarks may be made at this point. (a) The simple proof above sustains a more general theorem if the distribution generated by $f(x)$ is not constant, but time- or generation-dependent. Instead of the iterate $ffff \dots (f(x))$, we will have some $fgh \dots (q(x))$. By the mode of argument established, the chain process may be analyzed one step further. Let $g(y) = ay + b$ be the generating function for the probabilities b of loss or absorption of a single neutron and a of producing fission, with $a + b = 1$. Let $h(x) = c_1 x + c_2 x^2 + c_3 x^3 + \dots$ be the generating function of distribution of neutrons per fission. Then if the two are combined by the transformation $y \rightarrow h(x)$, we have that the distribution of neutrons per neutron is generated by $f(x) = g[h(x)]$. (b) If on the other hand we start from a single fission, and wish to know the distribution for the number of first-generation fissions, this is given by $F(y) = h(g(y))$. The iterates of $f(x)$ and $F(y)$ are connected by simple and evident relations.

There remains the practical problem of determining coefficients and other properties of $f^k(x)$, given $f(x)$. To this end we first shall establish some general properties of iteration.

2. Let $f(x)$ be a monotone function. Assume, for example, $f(x)$ increasing, i.e., if $x < y$, $f(x) < f(y)$. A *fixed point* for $f(x)$ is a value x_0 such that $f(x_0) = x_0$. The set of fixed points for a continuous function is closed, i.e., the points which are not fixed form a collection of disjoint intervals, whose endpoints are fixed points. If we form the sequence $f^k(x)$ for a given x we obtain a sequence of points converging to a fixed point x_0 which forms the endpoint of the interval in which x is situated. In fact, there are two cases possible, either $f(x) < x$ or $f(x) > x$. From the monotone character of $f(x)$ it follows that we shall have correspondingly either $f^k(x) < f^{k-1}(x)$ or $f^k(x) > f^{k-1}(x)$ for all k. Unless these sequences tend to $-\infty$ or $+\infty$, they will have limit points. If now $\lim_{k\to\infty} f^k(x) = x_0$, we must have $f(x_0) = x_0$. In fact $\lim_{k\to\infty} f(f^k(x_0)) = f(x_0) = x_0$. In addition, it is easy to see that x_0 is the *next* fixed point to x (on the left or right depending on whether $f(x) < x$ or $f(x) > x$. This follows from the fact that if $f(x)$ is monotone and $f(x_0) = x_0$, $f(x_1) = x_1$, then for all x such that $x_0 < x < x_1$, we have $f(x_0) = x_0 < f(x) < f(x_1) = x_1$.

In our case $f(x)$ is a power series with all coefficients non-negative, $f(0) \geq 0$, $f(1) = 1$. This function is certainly monotone and increasing for all non-negative x. Let x_0 be the first (non-negative) fixed point, x_0 certainly exists, the set of fixed points being closed. From these conditions it follows that $\lim_{k\to\infty} f^k(0) = x_0$. But if the variable in a generating function is set $= 0$, the value of the function is the probability that the random variable takes the value 0. Hence x_0 gives us the limit of the probability of mortality in the system. The probability of immortality is therefore simply $1 - x_0$, where x_0 is the smallest non-negative root of the equation $f(x) = x$. It is easy to see that if, as in our case, all the coefficients in the expansion of $f(x)$ are non-negative and $f(1) = 1$, then from $f'(1) > 1$ it follows that there is a root, and only one root x_0, which is non-negative and < 1. If $f'(1) < 1$, $x_0 = 1$ is the smallest positive root. We obtain immediately therefore the familiar fact that neutrons in a subcritical gadget without source will, with probability 1, die out in a finite time. For the supercritical gadget the probability of indefinite production can be obtained by solving the equation $f(x) = x$.

The kth iterate of a function can be obtained by a simple graphical or mechanical method which is based on the fact that along the diagonal, $f(x) = x$. Thus we may for given x replace this argument by $f(x)$, getting $f^2(x)$ graphically, then repeating $f^4(x)$ and so forth. In the case of the generating function under discussion this shows that $f^k(x)$ very rapidly approaches its asymptotic form: for the critical or subcritical case the asymptote in the interval $0 \leq x < 1$ is $\lim_{k\to\infty} f^k(x) \equiv x_1$; for the supercritical case in the interval $0 \leq x < 1$ the asymptote is

4

$\lim_{k\to\infty} f^k(x) \equiv x_0$. This implies that for all positive powers of x in $f^k(x)$ the coefficients approach 0 uniformly, i.e., the mass of probability is either absorbed altogether into the zero region (subcritical case), or is spread out in an infinitely long tail (supercritical case). In the region of criticality the distribution has an infinitely long tail with mass approaching zero as the probability of mortality approaches one.

3. One of the important properties of generating functions is that they permit the calculation of moments. Thus if p_n is the distribution itself, $f(x) = \sum_{n=0}^{\infty} p_n x^n$ its generating function, we have, because obviously $f(1) = 1$, the first moment or expected value of the random variable $= \sum_{n=0}^{\infty} np_n = f'(1) =$ the first derivative of $f(x)$ at $x = 1$. Similarly the second moment of the number of neutrons can be found if we know the *second* derivative. In fact

$$\sum_{n=0}^{\infty} n^2 p_n = \sum_{n=0}^{\infty} n(n-1)p_n + \sum_{n=0}^{\infty} np_n = f''(1) + f'(1) \, .$$

Similarly the rth moment can be found easily from the values of the first r derivatives of $f(x)$ at $x = 1$. (The rth derivative at $x = 1$ is sometimes called the rth *combinatorial* moment.)

Our generating function is the kth iterate $f^k(x)$. *It turns out that its first m derivatives depend only on the first m derivatives of $f(x)$ itself in a rather simple way.* We have, in fact:

Theorem II. (a) $[f^k(x)]'_{x=1} = f'(1) = \overline{\nu}^n$ (i.e., the proof of the intuitively obvious result that the expected number of neutrons after n generations is $\overline{\nu}^n$.) (b) $[f^k(x)]''_{x=1} = f''(x)_{x=1} \cdot [(f'(1))^k + (f'(1))^{k+1} + \ldots (f'(1))^{2k-1}])$.

The proof is immediate by induction:

$$[f(f^{k-1}(x))]' = f'(f^{k-1}(x)) \cdot [f^{k-1}(x)]' \, .$$

But for $x = 1$, $f^{k-1}(x) = x = 1$; therefore since by assumption $[f^{k-1}(x)]'_{x=1} = [f'(1)]^{k-1}$ we obtain our formula (a). By differentiating twice we obtain (b). Somewhat more complicated formulae hold for higher derivatives:

Their derivation is through recursive relations as follows: by differentiating the identical equation

$$f^k(x) = f(f^{k-1}(x))$$

repeatedly, and in all places substituting x_0 for $f^{k-1}(x_0)$, we obtain a sequence of linear first-order difference equations. Representing $d^r/dx^r \cdot f^k(x) = M_{k,r}(M_{1,r} = M_r)$ we obtain

$$M_{k,1} = M_1 \cdot M_{k-1,1}$$
$$M_{k,2} = M_2 \cdot M_{k-1}^2 + M_1 \cdot M_{k-1,2}$$
$$M_{k,3} = M_3 \cdot M_{k-1,1}^3 + 3M_2 M_{k-1,1} \cdot M_{k-1,2} + M_1 M_{k-1,3}$$

$$- - - -$$
$$- - - -$$

each is of the form

$$x_k = A_{k-1} + M_1 x_{k-1}$$

whose general solution is

$$x_k = \sum_{s=2}^{k} M_1^{k-s} A_{s-1} + M_1^{k-1} x_1 \ .$$

Solutions for the first three derivatives are

$$M_{k,1} = M_1^k$$
$$M_{k,2} = M_2 \cdot M_1^{k-1} \left[\frac{1 - M_1^k}{1 - M_1} \right]$$
$$M_{k,3} = M_3 M_1^{k-1} \left[\frac{1 - M_1^{2k}}{1 - M_1^2} \right] + \frac{3 M_2^2 M_1^{k-1}}{(1 - M_1)} \left[\frac{1 - M_1^{k-1}}{1 - M_1} \right]$$
$$- \frac{3 M_2^2}{(1 - M_1)} M_1^k \left[\frac{1 - M_1^{2k-2}}{1 - M_1^2} \right] \ .$$

Since in the function under discussion $x_0 = 1$ is a fixed point, these derivatives are the combinatorial moments of the distribution. We may now consider the three cases where $M_1(= \bar{\nu}$ of common use) is $> 1, < 1$, or $= 1$.

a) In the supercritical case where $M_1 > 1$, it is clear from the method of deriving these factorial moments that if the random variable n is measured in units of M_1, all moments approach a finite asymptotic form. Computation of moments for this asymptotic distribution may be greatly simplified as follows: Let us define a function $\phi(x) = x^{1/M_1}$, the kth iterate being $\phi^k(x) = x^{(1/M_1)^k}$. The generating function $f^k(\phi^k(x))$ if expanded in powers of $x^{(1/M_1)^k}$ has the same coefficients as $f^k(x)$ but these are now probabilities associated with the number of particles measured as fractions of the expected number. This is to say that the distribution is scaled in units of $M_1^k = \bar{\nu}^k$, and its first moment $= 1$. Since for the supercritical case all moments approach a constant value as $k \to \infty$ when scaled in this way, and since the generating function is monotonic in the region $(0, \infty)$, there exists a common limiting value, $g(x)$ of both $f^k[\phi^k(x)]$ and $f^{k-1}[\phi^{k-1}(x)]$. Since $f^k[\phi^k(x)] = f[\phi^{k-1}[\phi^k(\phi(x))]]$, we may write in the limit: $g(x) = f[g[\phi(x)]]$, $\phi(x) = x^{1/M_1}$, $f(x)$ given, and from this functional equation for g, its moments may be obtained from the second, third, etc., derivatives of g by solving only linear algebraic equations.

b) In the exactly critical case, $M_1 = 1$, the moments are

$$M_{k,1} = 1$$
$$M_{k,2} = k \cdot M_2$$
$$M_{k,3} = kM_3 + 3\binom{k}{2}M_2^2 \ .$$

This is a distribution in which $P_{k,0} \simeq 1 - 1/kM_2$, and such that if the system has not died in the kth generation, the expected number of neutrons is $\simeq kM_2$.

c) In the subcritical case all moments converge to zero, but are approximately proportional to the first moment.

4.   We may consider here briefly a simple special case, in which the iteration problem may be solved exactly.
     Let $f(x) = (ax+b)/(cx+d)$; we have here a three-parameter family of functions (one of the four constants a, b, c, d, is immaterial). We can adjust them so that $f(1) = 1$, and $f'(1) = \bar{\nu}$. We can then impose another condition, either on $f''(1)$, or so that $f(x_0) = x_0$, where x_0 is the "true" probability of mortality. Functions of the above sort form a group under substitution. This can be verified directly by substituting. (They form the so-called projective group of the line.) *A fortiori* the iterated function

$$f^k(x) = \frac{(a_k x + b_k)}{(c_k x + d_k)} \ .$$

By expanding $f^k(x)$ in a power series in x, we obtain the exact solution of our problem in this fairly general case. We determine the constants by the following three relations:

(1) Because $f(1) = 1$, we have for every $k : f^k(1) = 1$ which gives $a_k + b_k = c_k + d_k$.

(2) Similarly, for the second fixed point x_0 of $f(x)$, i.e., the root $x_0 \neq 1$ of $f(x) = x$, we have $f(x_0) = x_0$, and therefore for all k : $f^k(x_0) = x_0$ or $a_k x_0 + b_k = c_k x_0 + d_k$ and $x_0 = -b/c$, from $a x_0 + b = x_0(c x_0 + d)$.

(3) From the results of section **3**, we know that

$$[f^k(x)]'_{x=1} = [f'(1)]^k = \overline{v}^k .$$

This gives

$$\frac{a(cx + d) - (ax + b)}{(cx + d)^2} \Big|_{x=1} = \overline{v}$$

or taking account of (1)

$$\frac{(a - c)}{(c + d)} = \overline{v}$$

and therefore for all k

$$\frac{(a_k - c_k)}{(c_k - d_k)} = \overline{v}^k .$$

From the above three relations it is easy to calculate the constants a_k, b_k, c_k, d_k in terms of $\overline{v}$ and one arbitrary parameter. By eliminating a_k, b_k and developing into a power series, we get, noting that $c_k/d_k = 1/\overline{v}^k - 1$, assuming, e.g., $\overline{v} > 1$, the result in the form

$$f^k(x) = [(ax + b)/\overline{v}^{k+1}]\{1 + (1 - 1/v^k)x + (1 - 1/v^k)^2 x^2 + \dots (1 - 1/v^k)^n x^n + \dots .$$

This constitutes a complete solution of our problem. It is interesting to note that the probability of having n neutrons decreases geometrically with n; the ratio of the successive terms is in the case $\overline{v} > 1$, k large extremely close to 1. The distribution has the form of an *exponential*, decreasing very slowly. Asymptotically the probability of having exactly n neutrons is independent of n. This result shows also the possibility of enormous fluctuations in multiplicative systems. The "law of large numbers" in its ordinary formulation is not true for multiplicative processes. In fact the probability of having more (or less) than ℓ *times* the expected value of neutrons tends to a *positive* constant (dependent on ℓ). The following form of the law of large numbers is valid, as the examination of the distribution shows at once:

Theorem III. Given an $\epsilon > 0$, there exists an N such that for all $k > N$, the probability of the number n of neutrons in the kth generation being such that $(\bar{\nu} - \epsilon)^k < n < (\bar{\nu} + \epsilon)^k$ is greater than $1 - \epsilon$:

$$P\{(\bar{\nu} - \epsilon)^k < n < (\bar{\nu} + \epsilon)^k\} > 1 - \epsilon .$$

It remains to discuss the most general form of the distribution. We hope to do this later through two methods, one consisting of the consideration of functions of the form $hfh^{-1}(x)$, where f is of the projective linear form discussed above, and $h(x)$ is an arbitrary monotonic function. The kth iterate then is simply $hf^kh^{-1}(x)$. The function $h(x)$ will give us more arbitrary parameters for our real distribution. The second method consists in developing $f(x)$ into a series of functions whose terms have the "projective" form.

Finally it may be remarked that the limiting distribution obtained above is formally identical to those obtained by Frankel[1] and Feynman who used a continuous time parameter instead of our discrete-generations model. Their physical model is somewhat different and leads to the finding of the infinitesimal transformation of the continuous, abelian, one-parameter group into which the group of iterates of a function can be imbedded.

5. There are many other problems besides the question of the probable number of neutrons after k generations which can be solved by operational methods.

The first we shall consider is that of a subcritical system ($\bar{\nu} < 1$) with a source. We suppose that the distribution of neutrons entering the system in a given generation has the generating function $\phi(x)$, $f(x)$ being the generating function of the system itself as before, we shall have

Theorem IV. The generating functions in the zero, first, second generations are the functions:

$$\phi(x), \quad \phi(x) \cdot \phi[f(x)], \quad \phi(x) \cdot \phi[f(x)] \cdot \phi[f^2(x)] .$$

Proof is completely analogous to that of Theorem I.

In general, letting $F_k(x)$ represent the distribution in the kth generation
a) $F_k(x) = \phi(x) \cdot F_{k-1}[f(x)]$.
If the system is subcritical, but sustained at a definite level by the source, we shall have the limiting distribution—or its limiting generating function—as a nonsingular function of x : $\lim_{k \to \infty} F_k(x) = F(x)$, $F(1) = 1$. Passing to the limit on both sides of our equation a) we get

b) $F(x) = \phi(x) \cdot F[f(x)]$ where $\phi(x)$, $f(x)$ are given.

One has to determine $F(x)$ from this functional equation. *Even without doing it* one can obtain at once useful statistical information, for example the moments of $F(x)$, by differentiating b). Thus:

$$F'(1) = \frac{\phi'}{(1 - f')}$$

$$F''(1) = \frac{\phi''}{(1 - f'^2)} + \frac{2\phi' f' + f'' \phi'}{(1 - f'^2)(1 - f')}$$

giving us a way to compute standard deviations, and similarly, more complicated expressions for the higher derivatives and moments. The first derivative—the expected value—being inversely proportional to the degree of subcriticality becomes infinite if $f'(1)$ approaches 1.

6. We come now to the probability distribution of the sum of all neutrons in the system from the first to the kth generation. We have established previously that if $f(x) = \sum_{n=0}^{\infty} p_n x^n$ is the generating function for the probabilities of n particles in the first generation then the generating function of the kth generation is given by the kth iterate $f^k(x)$.

If we want the generating function for probabilities of having the *total* of n particles from the first to the kth generation, we shall proceed as follows.

The total of n particles can be obtained by any one of the following mutually exclusive cases: we can have 1 in the first generation and $n-1$ in the remaining $k - 1$, or 2 in the first generation and $n - 2$ in the remaining $k - 1$; in general we can have r in the first and $n - r$ in the remaining $k - 1$ generations. The required probability is therefore the sum of

$$q(n) = \sum_r p_r \cdot p_{n-r}^{k-1}(n) \ .$$

Here $p_{n-r}^{k-1}(n)$ denotes the probability that, starting from r in the first generation, we shall attain from these r a total of $n - r$ in $k - 1$ generations. But the r particles are independent of each other. The probability of getting the total of $n - r$ from them is therefore the probability of $n - r$ in the sum of these r variables. The generating function for the sum of the independent variables is the product of the generating functions corresponding to each of them. In our case it is the rth power of $f(x)$. We are looking for the coefficient of x^{n-r} in $[f^{k-1}(x)]^r$. Our required probability q_k equals therefore the sum with respect to r of coefficients of x^{n-r} in $[f^{k-1}(x)]^r$, or the sum of the coefficients of x^n in $\sum_r p_n x^r \cdot [f^{k-1}(x)]^r$.

But the coefficient of x^n in $\sum_r p_n x^r [f^{k-1}(x)]^r$ is the same as this coefficient in $f(x \cdot f^{k-1}(x))$. This is true for all n. Therefore the generating function for q_n is $f(x f^{k-1}(x))$. Since n here is arbitrary we get:

Theorem V. The generating function for the time sum is:

$$u^k(x) = f[xu^{k-1}(x)] .$$

If we "count" the original particle, this multiplies the generating function by x; expressing this slightly modified form recursively, we obtain the more convenient expression:

$$u^k(x) = xf[u^{k-1}(x)] .$$

As we know we have, in general, a relation between *moments* of the nth order of a distribution function and the nth derivative of the generating function. We shall now show how one can compute the derivatives of $u^k(x)$ for any k in an explicit manner.

Since, as was shown above,

$$u^k(x) = xf[u^{k-1}(x)] ,$$

we may obtain the desired results by repeated differentiations, and by solving the resulting finite difference equations. But if k is allowed to approach infinity, and if the system is subcritical,

$$\lim_{k \to \infty} u^k(x) = \lim_{k \to \infty} u^{k-1}(x) = u(x) .$$

Hence for the distribution of the total number produced, we have

$$u(x) = x \cdot f[u(x)] ,$$

differentiating, we obtain:

$$u'(1) = \frac{1}{(1 - f'(1))}$$

$$u''(1) = \frac{[f'' + f'(1 - f')]}{(1 - f')^3} .$$

These examples show how moments of the distributions can be computed for various problems in our discrete model. Otto Frisch has shown how, from the form of these moments, one can write their correct form for the continuous model, *without* having to solve the partial differential equations of the problem. This correspondence between the two models will be taken up later. It may be said that a generality of method has been established by the foregoing results, which demonstrate that the iteration of suitable operators corresponds to various physical observables connected with chain processes. For example it may be mentioned that the transformation $x \to (1/x)f(x)$ gives us the probability-distribution for differences between the number of neutrons in a generation and the number in the next generation. Thus $f^{k-1}[(1/x) \cdot f(x)]$ generates probabilities of this kind. The mathematical description of a multiplicative chain process is seen to involve the iteration of a functional operator U. These operators U act on the domain of all monotone functions $g(x)$, $g(1) = 1$. To summarize again just a few examples:

(1) $U(g) = f(g)$, f here is a given monotone function, g represents any function of the domain on which U operates, i.e., $g(x)$ is monotonic, $g(1) = 1$. This operator U is the only one that has been studied extensively in the literature. Its iteration leads to the simple iteration process:

$$g(x), \ f(g(x)), \ f[f(g(x))] \ldots f^k(g(x)) \ldots \ .$$

(2) $U(g) = f(x \cdot g)$, f a given function. The domain of the operator, i.e., the admissible g are the same, but there seems to be very little known about the iterates of this operator. This operator is tied to the probability law of the total number of particles produced.

(3) $U(g) = \phi(x) \cdot g(f(x))$; $\phi(x)$, $f(x)$ are given. The iterates of this operator give us the distribution of the number of particles produced when a source with given distribution $\phi(x)$ is acting constantly.

(4) $U(g) = f[(1/x) \cdot g]$.

This operator relates to the probability distribution of the *difference* of the number of particles in successive generations. The study of conjugates, fixed points, etc., for such operators seems to be important. We hope to undertake this study later.

We turn now to a more complex version of the problem. Hitherto it has been assumed that the generating function was independent of temporal and geometrical factors. However, our methods are extensible beyond these limitations.

7. The calculation of the probability distributions in the general case of heterogeneous particles will now be considered. So far we have assumed that the probability of generating n neutrons is the same independently of the parent neutron. If one takes the real situation where the system of the active material is of finite extent, then obviously the probability of leakage and absorption is a function of position of the parent nucleus. It is obvious that in general chemical or nuclear chain-reaction processes one has to deal with several kinds or even a continuous variety of the elementary generating functions.

In order to explain our methods of iteration of functional operators for this general case we shall take the simplest case of two kinds of particles. If we divide, for the first approximation, the sphere of the active material into two parts, an inner sphere and the outer shell, we shall characterize the neutrons generated in the one part by the subscript x, the others by subscript y. An x-particle can generate either x-particles again or penetrating to the outer shell y-particles, or, of course, leak out or be absorbed; the same, though with different probabilities, applies to the y-particles. In reality we should consider a one-dimensional variety of kinds of particles corresponding to all values of their distance r from the center of the sphere or even a two-dimensional one if we want to take into account different velocities. To simplify the presentation we shall limit ourselves here to just two kinds (x and y).

We assume that the following elementary probabilities are given by the nuclear constants and by the integrals of the geometry involved.

An x-particle can produce $n(> 0)$ x-particles with the probabilities p_n and $n(> 0)$ y-particles with probabilities q_n. The probability of dying out—absorption or leakage—will be denoted by p_0.

For the y-particles the corresponding probabilities will be denoted by $\bar{p}_n, \bar{q}_n$, and $\bar{p}_0$. It is because of the geometry of the system that $\bar{p}_0$ and p_0 are certainly different.

We now write the two functions of two variables each:

$$f(x,y) = p_0 + p_1 x + \ldots p_n x^n + \ldots + q_1 y + \ldots q_n y^n + \ldots$$
$$g(x,y) = \bar{p}_0 + \bar{p}_1 x + \ldots \bar{p}_n x^n + \ldots + \bar{q}_1 y + \ldots \bar{q}_n y^n + \ldots .$$

The coefficients of $f(x,y)$ give the probabilities of having in the first generation a given number of x- or y-particles starting with one x-neutron. Those of $g(x,y)$, if we start with a y-neutron.

Required are the probabilities of finding in the next generation a given number of x- and y-particles.

Let us form the function

$$f_2(x,y) = f[f(x,y) \cdot g(x,y)] .$$

13

By reasoning exactly as in the proof of Theorem I (or Theorem III) we see that the probability of having n x-particles and m y-particles is given by the coefficient of $x^n y^m$ in $f_2(x, y)$. If we started in 0th generation with a y-particle we will get these probabilities as the coefficients of $x^n y^m$ in $g[f(x, y), g(x, y)]$. By an obvious induction we obtain:

Theorem VI. The probabilities of having n x-particles and m y-particles in the kth generation are given by the coefficient of $x^n y^m$ in $f[T^{k-1}(r)]$ or $g[T^{k-1}(r)]$ (depending on whether we started from an x- or from a y-particle). $T^k(p)$ is a transformation of the plane (x, y) into itself defined as follows: if $p = (x, y)$ then $T'(p) = T(p) = [f(x, y), g(x, y)]; T^k(r) = T(T^{k-1}(p))$.

Without going into the details of the proof or actual computations of moments we wish to conclude by the following remarks:

(1) In the case of 3 or any finite number r of different kinds of particles, the formalism necessary to obtain the generating function for the kth generation is the same. It consists of iterating a given set of r functions or a *transformation in r dimensions* (variables $x_1, x_2 \ldots x_r$).

(2) One fairly general case where the coefficients of the mixed powers of the variables $x_1^{a_1} x_2^{a_2} \ldots x_r^{a_r}$ can be computed explicitly in a closed form for any number k of generations is when the given transformation is the r dimensional generalization of our projective transformations on the line, i.e., $p = (x_1, x_2 \ldots x_r)$; $p' = T(r) = (x_1' x_2' \ldots x_r')$ where $x_1' = f_1(x_1 \ldots x_r) = (a_{11}x_1 + \ldots a_{1r}x_r + b_1)/(c_{11}x_1 + \ldots c_{1r}x_r + d_1) \ldots \ldots \ldots \ldots \ldots \ldots \ldots \ldots \ldots \ldots \ldots \ldots \ldots \ldots \ldots$
$x_r' = f_r(x_1 \ldots x_r) = (a_{r1}x_1 + \ldots a_{rr}x_r + b_r)/(c_{r1}x_1 + \ldots c_{rr}x_r + d_r)$.

(3) The computation of moments of the distribution in the most general case does not involve the explicit knowledge of $T^k(r)$, but can be obtained through the knowledge of the moments of the r given elementary functions

$$f_1(x_1 \ldots x_r) \ldots f_r(x_1 \ldots x_r) .$$

The role of the numerical multiplication of moments is here taken over by matrix multiplication.

(4) The other operators corresponding, e.g., to $U(g) = f(x \cdot g)$, etc., have not been so far investigated in the r-dimensional case.

Conclusions Regarding Applications

The *expected* value of the number of neutrons per fission "ν" is known with fair accuracy. The critical mass and the *expected* number of neutrons in a gadget depend on this constant alone. Very little seems to be known, however, about the distribution function of the number of neutrons or even only about its second moment. The great fluctuations in multiplicative systems discussed above are of some practical interest for the following reasons:

1. The correct timing of the initiation of the gadget is vital for high efficiency. Even with good sources there will be an uncertainty of several generations time—due to fluctuations in multiplication.

2. The fluctuations of multiplication are of interest in all "integral" experiments.

3. For gadgets large in comparison with the mean free path for fission, the spatial fluctuations may destroy the initial spherical symmetry.

In dealing with such problems it is useful to develop a uniform technique for describing the statistics of multiplicative phenomena. This paper constitutes a first step consisting essentially in the observation that the iterated substitution (of a function, or more generally of a functional operation) represents exactly the statistical laws of multiplicative processes. In the sequel, it is hoped to apply this technique to the study of the problems of geometrical- and time-dependence of the process.

Reference

1. Stanley P. Frankel, The Statistics of the Hypercritical Gadget, LAMS-36, January 8, 1944.

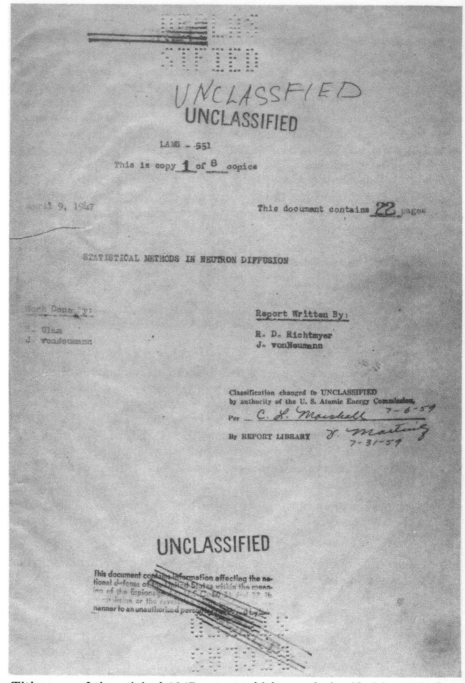

UNCLASSIFIED

LAMS - 551

This is copy 1 of 8 copies

April 9, 1947

This document contains 22 pages

STATISTICAL METHODS IN NEUTRON DIFFUSION

Work Done By:

L. Ulam
J. vonNeumann

Report Written By:

R. D. Richtmyer
J. vonNeumann

Classification changed to UNCLASSIFIED
by authority of the U. S. Atomic Energy Commission,

Per _C. S. Marshall_ 7-6-59

By REPORT LIBRARY _J. Martin_ 7-31-59

UNCLASSIFIED

This document contains information affecting the national defense of the United States within the meaning of the Espionage Act, U.S.C. 50:31 and 32. Its transmission or the revelation of its contents in any manner to an unauthorized person is prohibited by law.

Title page of the original 1947 report which was declassified in 1959. Los Alamos documents used to list the persons who did the work as well as those who wrote the reports. This practice was abandoned soon after the end of the war.

2

STATISTICAL METHODS IN NEUTRON DIFFUSION

With J. von Neumann and R. D. Richtmyer
(LAMS-551, April 9, 1947)

This report, written in 1947 by J. von Neumann and R. D. Richtmyer, is about work done by J. von Neumann and myself—as the title page indicates. It gives the first published ideas and proposals for the Monte Carlo Method. (Author's note.)

ABSTRACT

There is reproduced here some correspondence on a method of solving neutron diffusion problems in which data are chosen at random to represent a number of neutrons in a chain-reacting system. The history of these neutrons and their progeny is determined by detailed calculations of the motions and collisions of these neutrons, randomly chosen variables being introduced at certain points in such a way as to represent the occurrence of various processes with the correct probabilities. If the history is followed far enough, the chain reaction thus represented may be regarded as a representative sample of a chain reaction in the system in question. The results may be analyzed statistically to obtain various average quantities of interest for comparison with experiments or for design problems.

This method is designed to deal with problems of a more complicated nature than conventional methods based, for example, on the Boltzmann equation. For example, it is not necessary to restrict neutron energies to a single value or even to a finite number of values, and one can study the distribution of neutrons or of collisions of any specified type not only with respect to space variables but with respect to other variables, such as neutron velocity, direction of motion, time. Furthermore, the data can be used for the study of fluctuations and other statistical phenomena.

THE INSTITUTE FOR ADVANCED STUDY
Founded by Louis Bamberger and Mrs. Felix Fuld
Princeton, New Jersey
School of Mathematics

March 11, 1947

VIA AIRMAIL: REGISTERED

Mr. R. Richtmyer
Post Office Box 1663
Santa Fe, New Mexico

Dear Bob:

This is the letter I promised you in the course of our telephone conversation on Friday, March 7th.

I have been thinking a good deal about the possibility of using statistical methods to solve neutron diffusion and multiplication problems, in accordance with the principle suggested by Stan Ulam. The more I think about this, the more I become convinced that the idea has great merit. My present conclusions and expectations can be summarized as follows:

(1) The statistical approach is very well suited to a digital treatment. I worked out the details of a criticality discussion under the following conditions:

  (a) Spherically symmetric geometry.
  (b) Variable (if desired, continuously variable) composition along the radius of active material (25 or 49), tamper material (28 or Be or WC), and slower-down material (H in some form).
  (c) Isotropic generation of neutrons by all processes of (b).
  (d) Appropriate velocity spectrum of neutrons emerging from the collision processes of (b), and appropriate description of the

18

cross-sections of all processes of (b) as functions of the neutron velocity; i.e., an infinitely many (continuously distributed) neutron velocity group treatment.

(e) Appropriate account of the statistical character of fissions, as being able to produce (with specified probabilities), say 2 or 3 or 4 neutrons.

This is still a treatment of "inert" criticality: It does not allow for the hydrodynamics caused by the energy and momentum exchanges and production of the processes of (b), and for the displacements, and hence changes of material distribution, caused by hydrodynamics; i.e., it is not a theory of efficiency. I do know, however, how to expand it into such a theory (cf. (5) below).

The details enumerated in (a)-(e) were chosen by me somewhat at will. It seems to me that they represent a reasonable model, but it would be easy to make them either more or less elaborate, as desired. If you have any definite desiderata in this respect, please let me know, so that we may analyze their effects on the set-up.

(2) I am fairly certain that the problem of (1), in its digital form, is well suited for the ENIAC. I will have a more specific estimate on this subject shortly. My present (preliminary) estimate is this: Assume that one criticality problem requires following 100 primary neutrons through 100 collisions (of the primary neutron or its descendants) per primary neutron. Then solving one criticality problem should take about 5 hours. It may be, however, that these figures (100 × 100) are unnecessarily high. A statistical study of the first solutions obtained will clear this up. If they can be lowered, the time will be shortened proportionately.

A common set-up of the ENIAC will do for all criticality problems. In changing over from one problem of this category to another one, only a few numerical constants will have to be set anew on one of the "function table" organs of the ENIAC.

(3) Certain preliminary explorations of the statistical-digital method could be and should be carried out manually. I will say somewhat more subsequently.

(4) It is not quite impossible that a manual-graphical approach (with a small amount of low-precision digital work interspersed) is feasible. It would require a not inconsiderable number of computers for several days per criticality problem, but it may be possible, and

it may perhaps deserve consideration until and unless the ENIAC becomes available. This manual-graphical procedure has actually some similarity with a statistical-graphical procedure with which solutions of a bombing problem were obtained during the war, by a group working under S. Wilks (Princeton University and Applied Mathematics Panel, NDRC). I will look into this matter further, and possibly get Wilks' opinion on the mathematical aspects.

(5) If and when the problem of (1) will have been satisfactorily handled in a reasonable number of special cases, it will be time to investigate the more general case, where hydrodynamics also comes into play; i.e., efficiency calculations, as suggested at the end of (1). I think that I know how to set up this problem, too: One has to follow, say, 100 neutrons through a short time interval Δt; get their momentum and energy transfer and generation in the ambient matter; calculate from this the displacement of matter; recalculate the history of the 100 neutrons by assuming that matter is in the middle position between its original (unperturbed) state and the above displaced (perturbed) state; recalculate the displacement of matter due to this (corrected) neutron history; recalculate the neutron history due to this (corrected) displacement of matter, etc., etc., iterating in this manner until a "self-consistent" system of neutron history and displacement of matter is reached. This is the treatment of the first time interval Δt. When it is completed, it will serve as a basis for a similar treatment of the second time interval Δt; this, in turn, similarly for the third time interval Δt; etc., etc. In this set-up there will be no serious difficulty in allowing for the role of light, too. If a discrimination according to wavelength is not necessary, i.e., if the radiation can be treated at every point as isotropic and black, and its mean free path is relatively short, then light can be treated by the usual "diffusion" methods, and this is clearly only a very minor complication. If it turns out that the above idealizations are improper, then the photons, too, may have to be treated "individually" and statistically, on the same footing as the neutrons. This is, of course, a non-trivial complication, but it can hardly consume much more time and instructions than the corresponding neutronic part. It seems to me, therefore, that this approach will gradually lead to a completely satisfactory theory of efficiency, and ultimately permit prediction of the behavior of all possible arrangements, the simple ones as well as the sophisticated ones.

(6) The program of (5) will, of course, require the ENIAC at least, if not more. I have no doubt whatever that it will be perfectly

tractable with the post-ENIAC device which we are building. After a certain amount of exploring (1), say with the ENIAC, will have taken place, it will be possible to judge how serious the complexities of (5) are likely to be.

Regarding the actual, physical state of the ENIAC my information is this: It is in Aberdeen, and it is being put together there. The official date for its completion is still April 1st. Various people give various subjective estimates as to the actual date of completion, ranging from mid-April to late May. It seems as if the late May estimate were rather safe.

I will inquire more into this matter, and also into the possibility of getting some of its time subsequently. The indications that I have had so far on the latter score are encouraging.

In what follows, I will give a more precise description of the approach outlined in (1); i.e., of the simplest way I can now see to handle this group of problems.

Consider a spherically symmetric geometry. Let r be the distance from the origin. Describe the inhomogeneity of this system by assuming N concentric, homogeneous (spherical shell) zones, enumerated by an index $i = 1, \ldots, N$. Zone No. i is defined by $r_{i-1} \leq r \leq r_i$, the $r_0, r_1, r_2, \ldots, r_{N-1}, r_N$ being given:

$$0 = r_0 < r_1 < r_2 < \ldots < r_{N-1} < r_N = R ,$$

where R is the outer radius of the entire system.

Let the system consist of the three components discussed in (1), (b), to be denoted A, T, S, respectively. Describe the composition of each zone in terms of its content of each of A, T, S. Specify these for each zone in relative volume fractions. Let these be in zones Nos. i x_i, y_i, z_i, respectively.

Introduce the cross sections per cm^3 of pure material, multiplied by $^{10}\log e = .43 \ldots$, and as functions of the neutron velocity v, as follows:

Absorption in A, T, S: $\sum_{aA}(v), \sum_{aT}(v), \sum_{aS}(v)$.
Scattering in A, T, S: $\sum_{sA}(v), \sum_{sT}(v), \sum_{sS}(v)$.
Fission in A, with production of 2, 3, 4 neutrons:

$$\sum_{fA}^{(2)}(v), \sum_{fA}^{(3)}(v), \sum_{fA}^{(4)}(v) .$$

Scattering as well as fission are assumed to produce isotropically distributed neutrons, with the following velocity distributions:

If the incident neutron has the velocity v, then the scattered neutrons' velocity statistics are described for A, T, S, by the relations

$$v' = v\varphi_A(\nu), \ v' = v\varphi_T(\nu), \ v' = v\varphi_S(\nu) \ .$$

Here v' is the velocity of the scattered neutron, $\varphi_A(\nu), \varphi_T(\nu), \varphi_S(\nu)$ are known functions, characteristic of the three substances A, T, S (they vary all from 1 to 0), and ν is a random variable, statistically equidistributed in the interval 0, 1.

Every fission neutron has the velocity v_0.

I suppose that this picture either gives a model or at least provides a prototype for essentially all those phenomena about which we have relevant observational information at present, and actually for somewhat more. It may be expected to provide a reasonable vehicle for the additional relevant observational material that is likely to arise in the near future.

Do you agree with this?

In this model the state of a neutron is characterized by its position r, its velocity v, and the angle θ between its direction of motion and the radius. It is more convenient to replace θ by $s = r\cos\theta$, so that $\sqrt{r^2 - s^2}$ is the "perihelion distance" of its (linearly extrapolated) path.

Note that if a neutron is produced isotropically, i.e., if its direction "at birth" is equidistributed, then (because space is three-dimensional) $\cos\theta$ will be equidistributed in the interval $-1, 1$, i.e., s in the interval $-r, r$.

It is convenient to add to the characterization of a neutron explicitly the No. i of the zone in which it is found, i.e., with $r_{i-1} \leq r \leq r_i$. It is furthermore advisable to keep track of the time t to which the specifications refer.

Consequently, a neutron is characterized by these data:

$$i, r, s, v, t \ .$$

Now consider the subsequent history of such a neutron. Unless it suffers a collision in zone No. i, it will leave this zone along its straight path, and pass into zones Nos. $i + 1$ or $i - 1$. It is desirable to start a "new" neutron whenever the neutron under consideration has suffered a collision (absorbing, scattering, or fissioning—in the last-mentioned case several "new" neutrons will, of course, have to be started), or whenever it passes into another zone (without having collided).

Consider first, whether the neutron's linearly extrapolated path goes forward from zone No. i into zone No. $i + 1$ or $i - 1$. Denote these two possibilities by I and II.

If the neutron moves outward, i.e., if $s \geq 0$, then we have certainly I. If the neutron moves inward, i.e., if $s < 0$, then we have either I or II, the latter if, and only if, the path penetrates at all into the sphere $r_i - 1$. It is easily seen that the latter is equivalent to $s^2 \gg r^2 - r_{i-1}^2$. So we have:

$$s \geq 0. \therefore A$$

$$s < 0. \therefore B \begin{cases} r_{i-1}^2 + s^2 - r^2 \leq 0 & \therefore B'\} & \therefore \text{I}. \\ r_{i-1}^2 + s^2 - r^2 > 0 & \therefore B''\} & \therefore \text{II}. \end{cases}$$

The exit from zone No. i will therefore occur at

$$r^* \begin{cases} = r_i & \text{for I}. \\ = r_{i-1} & \text{for II}. \end{cases}$$

It is easy to calculate that the distance from the neutron's original position to the exit position is $d = s^* - s$, where

$$s^* = \pm\sqrt{r^{*2} + s^2 - r^2}, \quad \left.\begin{matrix} +\text{for I} \\ -\text{for II} \end{matrix}\right| .$$

The probability that the neutron will travel a distance d^1 without suffering a collision is 10^{-fd^1}, where

$$f = \left(\sum_{aA}(v) + \sum_{sA}(v) + \sum_{fA}^{(2)}(v) + \sum_{fA}^{(3)}(v) + \sum_{fA}^{(4)}(v) \right) x_i$$

$$+ \left(\sum_{aT}(v) + \sum_{sT}(v) \right) y_i + \left(\sum_{aS}(v) + \sum_{aS}(v) \right) z_i .$$

It is at this point that the statistical character of the method comes into evidence. In order to determine the actual fate of the neutron, one has to provide now the calculation with a value λ, belonging to a random variable, statistically equidistributed in the interval 0, 1; i.e., λ is to be picked at random from a population that is statistically equidistributed in the interval 0, 1. Then it is decreed that 10^{-fd^1} has turned out to be λ; i.e.,

$$d^1 = \frac{-^{10}\log\lambda}{f} \quad * \quad .$$

From here on, the further procedure is clear.

---

* Today this older notation would read $d^1 = -\log_{10}\lambda/f$. (Eds.)

If $d^1 \geq d$, then the neutron is ruled to have reached the neighboring zone No. $i \pm 1$ ($^{+\ \text{for I}}_{-\ \text{for II}}$) without having suffered a collision. The "new" neutron (i.e., the original one, but viewed at the interzone boundary, and heading into the new zone), has characteristics which are easily determined: i is replaced by $i \pm 1$, r by r^*, s is easily seen to go over into s^*, v is unchanged, t goes over into $t^* = t + d/v$. Hence, the "new" characteristics are

$$i \pm 1, r^*, s^*, v, t^* .$$

If, on the other hand, $d^1 < d$, then the neutron is ruled to have suffered a collision while still within zone No. i, after a travel d^1. The position at this stage is now

$$r^* = \sqrt{r^2 + 2sd^1 + (d^1)^2} ,$$

and the time

$$t^* = t + \frac{d^1}{v} .$$

The characteristic contains, accordingly, at any rate i, r^*, and t^* in place of i, r, and t. It remains to determine what becomes of s and v.

As pointed out before, the "new" s will be equidistributed in the interval $-r^*, r^*$. It is therefore only necessary to provide the calculation with a further value ρ', belonging to a random variable, statistically equidistributed in the interval $0, 1$. Then one can rule that s has the value

$$s' = r^*(2\rho' - 1) .$$

As to the "new" v, it is necessary to determine first the character of the collection: Absorption (by any one of A, T, S); scattering by A, or by T, or by S; fission (by A) producing 2, or 3, or 4 neutrons. These seven alternatives have the relative probabilities

$$f_1 = \sum_{aA}(v)x_i + \sum_{aT}(v)y_i + \sum_{aS}(v)z_i ,$$

$$f_2 - f_1 = \sum_{sA}(v)x_i ,$$

$$f_3 - f_2 = \sum_{sT}(v)y_i ,$$

$$f_4 - f_3 = \sum_{sS}(v)z_i ,$$

$$f_5 - f_4 = \sum_{fA}^{(2)}(v)x_i ,$$

$$f_6 - f_5 = \sum_{fA}^{(3)}(v)x_i ,$$

$$f - f_6 = \sum_{fA}^{(4)}(v)x_i .$$

We can therefore now determine the character of the collision by a statistical procedure like the preceding ones: Provide the calculation with a value μ belonging to a random variable, statistically equidistributed in the interval 0, 1. Form $\bar{\mu} = \mu f$; this is then equidistributed in the interval $0, f$. Let the 7 above cases correspond to the 7 intervals $0, f_1; f_1, f_2; f_2, f_3; f_3, f_4; f_4, f_5; f_5, f_6; f_6, f$, respectively. Rule, that one of those 7 cases holds in whose interval $\bar{\mu}$ actually turns out to lie.

Now the value of v can be specified. Let us consider the 7 available cases in succession.

Absorption: The neutron has disappeared. It is simplest to characterize this situation by replacing v by 0.

Scattering by A: Provide the calculation with a value v belonging to a random variable, statistically equidistributed in the interval 0, 1. Replace v by

$$v' = v\varphi_A(\nu) \ .$$

Scattering by T: Same as above, but

$$v' = v\varphi_T(\nu) \ .$$

Scattering by S: Same as above, but

$$v' = v\varphi_S(\nu) \ .$$

Fission: In this case replace v by v_0. According to whether the case in question is that one corresponding to the production of 2, 3, or 4 neutrons, repeat this 2, 3, or 4 times, respectively. This means that, in addition to the p', s' discussed above, the further $\rho'', s''; \rho''', s'''; \rho'''', s''''$ may be needed.

This completes the mathematical description of the procedure. The computational execution would be something like this:

Each neutron is represented by a card C which carries its characteristics

$$i, r, s, v, t,$$

and also the necessary random values

$$\lambda, \mu, \nu, \rho', \rho'', \rho''', \rho''''.$$

I can see no point in giving more than, say, 7 places for each one of the 5 characteristics, or more than, say, 5 places for each of the 7 random variables. In fact, I would judge that these numbers of places are already higher than necessary. At any rate, even in this way only 70 entries are consumed, and so the ordinary 80-entry punchcard will have 10 entries left over for any additional indexings, etc., that one may desire.

The computational process should then be so arranged as to produce the card C' of the "new" neutron, or rather 1 to 4 such cards C', C'', C''', C'''' (depending on the neutrons' actual history, cf. above). Each card, however, need only be provided with the 5 characteristics of its neutron. The 7 random variables can be inserted in a subsequent operation, and the cards with $v = 0$ (i.e., corresponding to neutrons that were absorbed within the assembly) as well as those with $i = N+1$ (i.e., corresponding to neutrons that escaped from the assembly) may be sorted out.

The manner in which this material can then be used for all kinds of neutron statistic investigations is obvious.

I append a tentative "computing sheet" for the calculation above. It is, of course, neither an actual "computing sheet" for a (human) computer group, nor a set-up for the ENIAC, but I think that it is well suited to serve as a basis for either. It should give a reasonably immediate idea of the amount of work that is involved in the procedure in question.

I cannot assert this with certainty yet, but it seems to me very likely that the instructions given on this "computing sheet" do not exceed the "logical" capacity of the ENIAC. I doubt that the processing of 100 "neutrons" will take much longer than the reading punching, and (once) sorting time of 100 cards, i.e., about 3 minutes. Hence, taking 100 "neutrons" through 100 of these stages should take about 300 minutes, i.e., 5 hours.

Please let me know what you and Stan think of these things. Does the approach and the formulation and generality of the criticality problem seem reasonable to you, or would you prefer some other variant?

Would you consider coming East some time to discuss matters further? When could this be?

With best regards;
Very truly yours,

John von Neumann

TENTATIVE COMPUTING SHEET

Data:

(1) r_i, x_i, y_i, z_i

as functions of $i = 1, \ldots, N^1$. $(r_0 = 0.)$

(2) $\sum_{aA}(v), \sum_{aT}(v), \sum_{aS}(v), \sum_{sA}(v), \sum_{sT}(v),$
 $\sum_{sS}(v), \sum_{fA}^{(2)}(v), \sum_{fA}^{(3)}(v), \sum_{fA}^{(4)}(v)$

as functions of $v \geq 0, \leq v_0.^2$

(3) $v_0.$

(4) $\varphi_A(\nu), \varphi_T(\nu), \varphi_S(\nu)$

as functions of $\nu \geq 0, \leq 1.^2$

(5) $-10 \log \lambda$

as function of $\lambda \geq 0, \leq 1.^2$

Card C :

C_1	i
C_2	r
C_3	s
C_4	v
C_5	l

Random Variables :

R_1	λ
R_2	μ
R_3	ν
R_4	ρ'
R_5	ρ''
R_6	ρ'''
R_7	ρ''''

---

[1] Tabulated. (Discrete domain.)
[2] Tabulated, to be interpolated, or approximated by polynomials.
(Continuous domain.)

TENTATIVE COMPUTING SHEET (Cont'd.)

Calculation :
Instructions :   *Explanations* :

		Instructions	Explanations
	1	r of $C_1 - 1$, see (1)	r_{i-1}
	2	r of C_1, see (1)	r_i
	3	$(C_3)^2$	s^2
	4	$(C_2)^2$	r^2
	5	$3 - 4$	$s^2 - r^2$
	6	$C_3 \begin{cases} \geq 0 \therefore A \\ < 0 \therefore B \end{cases}$	$s \begin{cases} \geq 0 \therefore A \\ < 0 \therefore B \end{cases}$
Only for B:	7	$(1)^2$	r_{i-1}^2
Only for B:	8	$5 + 7$	$r_{i-1}^2 + s^2 - r^2$
Only for B:	9	$8 \begin{cases} \leq 0 \therefore B' \\ > 0 \therefore B'' \end{cases}$	$r_{i-1}^2 + s^2 - r^2 \begin{cases} \leq 0 \therefore B' \\ > 0 \therefore B'' \end{cases}$
	10	$\begin{array}{l} A \text{ or } B' \therefore 2 \\ B'' \therefore 1 \end{array}$	$\left. \begin{array}{l} A \text{ or } B' \therefore r_i = \\ B'' \therefore r_{i-1} = \end{array} \right\} r^*$
	11	$\begin{array}{l} A \text{ or } B' \therefore +1 \\ B'' \therefore -1 \end{array}$	$\left. \begin{array}{l} A \text{ or } B' \therefore +1 = \\ B'' \therefore -1 = \end{array} \right\} \epsilon$
	12	$(10)^2$	r^{*2}
	13	$5 + 12$	$r^{*2} + s^2 - r^2$
	14	$11(\text{sign}) \times \sqrt{13}$	s^*
	15	$14 - C_3$	d
	16	x of C_1, see (1)	x_i
	17	y of C_1, see (1)	y_i
	18	z of C_1, see (1)	z_i
	19	$\sum_{aA}$ of C_4, see (2)	$\sum_{aA}(v)$
	20	16×19	$\sum_{aA}(v)x_i$
	21	$\sum_{aT}$ of C_4, see (2)	$\sum_{aT}(v)$
	22	17×21	$\sum_{aA}(v)y_i$
	23	$20 + 22$	$\sum_{aA}(v)x_i + \sum_{aT}(v)y_i$
	24	$\sum_{aS}$ of C_4, see (2)	$\sum_{aS}(v)$
	25	18×24	$\sum_{aS}(v)z_i$

TENTATIVE COMPUTING SHEET (Cont'd.)

Calculation :
Instructions :   *Explanations :*

26	$23 + 25$	$f_1 = \sum_{aA}(v)x_i +$ $\sum_{aT}(v)y_i +$ $\sum_{aS}(v)z_i$
27	$\sum_{aA}$ *of* C_4, see (2)	$\sum_{aA}(v)$
28	16×27	$f_2 - f_1 = \sum_{sA}(v)x_i$
29	$26 + 28$	f_2
30	$\sum_{sT}$ *of* C_4, see (2)	$\sum_{sT}(v)$
31	17×30	$f_3 - f_2 = \sum_{sT}(v)y_i$
32	$29 + 31$	f_3
33	$\sum_{sS}$ *of* C_4, see (2)	$\sum_{sS}(v)$
34	18×33	$f_4 - f_3 = \sum_{sS}(v)z_i$
35	$32 + 34$	f_4
36	$\sum_{fA}^{(2)}$ *of* C_4, see (2)	$\sum_{fA}^{(2)}(v)$
37	16×36	$f_5 - f_4 = \sum_{fA}^{(2)}(v)x_i$
38	$35 + 37$	f_5
39	$\sum_{fA}^{(3)}$ *of* C_4, see (2)	$\sum_{fA}^{(3)}(v)$
40	16×39	$f_6 - f_5 = \sum_{fA}^{(3)}(v)x_i$
41	$38 + 40$	f_6
42	$\sum_{fA}^{(4)}$ *of* C_4, see (2)	$\sum_{fA}^{(4)}(v)$
43	16×42	$f - f_6 = \sum_{fA}^{(4)}(v)x_i$
44	$41 + 43$	f
45	$^{-10}\log$ of R_1, see (5)	$^{-10}\log \lambda$
46	$45 : 44$	d^1
47	$46 \begin{cases} \geq 15 \therefore P \\ < 15 \therefore Q \end{cases}$	$d_1 \begin{cases} \geq d \therefore P \\ < d \therefore Q \end{cases}$
48	$\left.\begin{array}{l} P \therefore 15 : \\ Q \therefore 46 : \end{array}\right\} C_4$	$\left.\begin{array}{l} P \therefore d : \\ Q \therefore d^1 : \end{array}\right\} v = \tau$
49	$C_5 + 48$	$t^* = t + \tau$

TENTATIVE COMPUTING SHEET (Cont'd.)

Calculation :

		Instructions :	Explanations :
Only for P:	50	$C_1 + 11$	$i^* = i + \epsilon$
Only for P:	51	$C_1' : 50$	$C_1' : i^*$
		$C_2' : 10$	$C_2' : r^*$
		$C_3' : 14$	$C_3' : s^*$
		$C_4' : C_4$	$C_4' : v$
		$C_5' : 49$	$C_5' : t^*$

From here on only Q:

52	$R_2 \times 44$	$\bar{\mu} = \mu f$
53	$52 < 26 \therefore Q_1$	$\bar{\mu} < f_1 \therefore Q_1$

$$\left.\begin{array}{l} \geq 26 \\ < 29 \end{array}\right\} \therefore Q_2 \qquad\qquad \left.\begin{array}{l} \geq f_1 \\ < f_2 \end{array}\right\} \therefore Q_2$$

$$\left.\begin{array}{l} \geq 29 \\ < 32 \end{array}\right\} \therefore Q_3 \qquad\qquad \left.\begin{array}{l} \geq f_2 \\ < f_3 \end{array}\right\} \therefore Q_3$$

$$\left.\begin{array}{l} \geq 32 \\ < 35 \end{array}\right\} \therefore Q_4 \qquad\qquad \left.\begin{array}{l} \geq f_3 \\ < f_4 \end{array}\right\} \therefore Q_4$$

$$\left.\begin{array}{l} \geq 35 \\ < 38 \end{array}\right\} \therefore Q_5 \qquad\qquad \left.\begin{array}{l} \geq f_4 \\ < f_5 \end{array}\right\} \therefore Q_5$$

$$\left.\begin{array}{l} \geq 38 \\ < 41 \end{array}\right\} \therefore Q_6 \qquad\qquad \left.\begin{array}{l} \geq f_5 \\ < f_6 \end{array}\right\} \therefore Q_6$$

$$\geq 41 \therefore Q_7 \qquad\qquad \geq f_6 \therefore Q_7$$

54	$C_3 \times 46$	sd^1
55	2×54	$2sd^1$
56	$(46)^2$	$(d^1)^2$
57	$55 + 56$	$2sd^1 + (d^1)^2$
58	$4 + 57$	$r^2 + 2sd^1 + (d^1)^2$
59	$\sqrt{58}$	r^*

TENTATIVE COMPUTING SHEET (Cont'd.)

Calculation :

	Instructions :	*Explanations* :

Only for Q_1:   60 $C_1' : C_1$ $C_1' : i$
                        $C_2' : 59$ $C_2' : r^*$
                        $C_3' : \cdots$ $C_3' : \cdots$
                        $C_4' : 0$ $C_4' : 0$
                        $C_5' : 49$ $C_5' : t^*$

From here on only $Q_2, \ldots, Q_7$:

Only for Q_2:	61	φ_A of R_3, see (4)	$\varphi_A(v) = \varphi$
Only for Q_3:	62	φ_T of R_3, see (4)	$\varphi_T(v) = \varphi$
Only for Q_4:	63	φ_S of R_3, see (4)	$\varphi_S(v) = \varphi$

Only for $\left.\begin{array}{c} \\ Q_2, Q_3, Q_4 \end{array}\right\}$:   64 $C_4 \times$ (61 or $v\varphi$
                                                              62 or 63)

65    $\begin{array}{l} Q_2, Q_3, Q_4 \therefore 64 \\ Q_5, Q_6, Q_7 \therefore (3) \end{array}$ $\left.\begin{array}{l} Q_2, Q_3, Q_4 \therefore v\varphi = \\ Q_5, Q_6, Q_7 \therefore v_0 = \end{array}\right\} v'$

66    $2 \times R_4$ $2\rho'$
67    $66 - 1$ $2\rho' - 1$
68   59×67 s'
69    $C_1' : C_1$ $C_1' : i$
       $C_2' : 59$ $C_2' : r^*$
       $C_3' : 68$ $C_3' : s'$
       $C_4' : 65$ $C_4' : v'$
       $C_5' : 49$ $C_5' : t^*$

From here on only Q_5, Q_6, Q_7:

70    $2 \times R_5$ $2\rho''$
71    $70 - 1$ $2\rho'' - 1$
72   59×71 s''
73    $C_1'' : C_1$ $C_1'' : i$
       $C_2'' : 59$ $C_2'' : r^*$
       $C_3'' : 72$ $C_3'' : s''$
       $C_4'' : 65$ $C_4'' : v'$
       $C_5'' : 49$ $C_5'' : t^*$

31

TENTATIVE COMPUTING SHEET (Cont'd.)

Calculation :

Instructions :	Explanations :

From here on only Q_6, Q_7:

74	$2 \times R_6$	$2\rho'''$
75	$74 - 1$	$2\rho''' - 1$
76	59×75	s'''
77	$C_1''' : C_1$	$C_1''' : i$
	$C_2''' : 59$	$C_2''' : r^*$
	$C_3''' : 76$	$C_3''' : s'''$
	$C_4''' : 65$	$C_4''' : v'$
	$C_5''' : 49$	$C_5''' : t^*$

From here on only Q_7:

78	$2 \times R_7$	$2\rho''''$
79	$78 - 1$	$2\rho'''' - 1$
80	59×79	s''''
81	$C_1'''' : C_1$	$C_1'''' : i$
	$C_2'''' : 59$	$C_2'''' : r^*$
	$C_3'''' : 80$	$C_3'''' : s''''$
	$C_4'''' : 65$	$C_4'''' : v'$
	$C_5'''' : 49$	$C_5'''' : t^*$

At right, sample page of John von Neumann's handwritten "tentative computing sheet" as it was used in this 1947 report.

	Instructions:	Explanations:

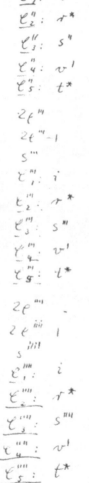

73   c_1'' : c_1 c_1'' : i

     c_2'' : 59 c_2'' : $r*$

     c_3'' : 72 c_3'' : s''

     c_4'' : 65 c_4'' : v'

     c_5'' : 49 c_5'' : $t*$

From here on only c_6, c_7 :

74   2 x R_6 $2c'''$

75   74 − 1 $2c''' - 1$

76   59 x 75 s'''

77   c_1''' : c_1 c_1''' : i

     c_2''' : 59 c_2''' : $r*$

     c_3''' : 76 c_3''' : s'''

     c_4''' : 65 c_4''' : v'

     c_5''' : 49 c_5''' : $t*$

From here on only c_7 :

78   2 x R_7 $2c''''$

79   78 − 1 $2c'''' - 1$

80   59 x 79 s''''

81   c_1'''' : c_1 c_1'''' : i

     c_2'''' : 59 c_2'''' : $r*$

     c_3'''' : 80 c_3'''' : s''''

     c_4'''' : 65 c_4'''' : v'

     c_5'''' : 49 c_5'''' : $t*$

April 2, 1947

Professor John von Neumann,
The Institute for Advanced Study,
School of Mathematics
Princeton, New Jersey

Dear Johnny:

As Stan told you, your letter has aroused a great deal of interest here. We have had a number of discussions of your method and Bengt Carlson has even set to work to test it out by hand calculation in a simple case.

It has occurred to us that there are a number of modifications which one might wish to introduce, at least for calculations for a certain type. This would be true, for example, if one wished to set up the problem for a metal system containing a 49 core in a tuballoy tamper. It seems to us that it might at present be easier to define problems of this sort than, for example, problems for hydride gadgets. It is not so much our intention to suggest that the method you are working on now should be modified as to suggest that perhaps alternative procedures should be worked out also. Perhaps one of us could do this with a little assistance from you; for example, during a visit to Princeton.

The specific points at which it seems to us modifications might be desired are as follows:

1. Of the three components A, T, S that you consider, only one is fissionable, whereas in systems of interest to us, there will be an appreciable number of fissions in the tuballoy of the tamper, as well as in the core material.

2. On the other hand, we are not likely for some time to have data enabling one to distinguish between the velocity dependence of the three functions

$$\sum_{fa}^{(2)}(v), \qquad \sum_{fa}^{(3)}(v), \qquad \sum_{fa}^{(4)}(v)$$

that you introduce so that for any particular isotope these might as well be combined into a single function of velocity with a random procedure used merely for determining the number of neutrons emerging. If there is a single such function of velocity for each of the three isotopes 25, 28, 49, the total number of function tables required would be the same as in your letter.

3. It is suggested that in the case of 25 or 28, one might wish to allow also for the possibility of one neutron emerging from fission. The dispersion of the number of neutrons per fission is not too well known but we think we could provide some guesses.

4. Because of the sensitive dependence of tamper fissions on the neutron energy spectrum, it might be advisable to feed in the measured fission spectrum at the appropriate point. This would, of course, require introduction of one or two additional random variables and would raise the nasty question of possible velocity correlation between neutrons emerging from a given fission.

5. Material S could, of course, be omitted for systems of this sort. On the other hand, when moderation really occurs, it seems to us there would have to be a correlation between velocity and direction of the scattered neutron.

6. For metal systems of the type considered, it would probably be adequate to assume just one elastically scattering component and just one inelastically scattering component. These could be mixed with the fissionable components in suitable proportions to mock up most materials of interest.

In addition, we have one general comment as follows: Suppose that the initial deck of cards represents a group of neutrons all having $t = 0$ as their time of origin. Then after a certain number of cycles of operations, say 100, one will have a deck of cards representing a group of neutrons having times of origin distributed from some earliest t_1, to a latest, t_2. Thus all of the multiplicative chains will have been followed until time t_1 and some of them will have been followed to various later times. Then if one wishes, for example, to find the spatial distribution of fissions, it would be natural to examine all fissions occurring in some interval Δt and find their spatial distribution. But unless the interval Δt is chosen within the interval $(0, t_1)$ one cannot be sure that he knows about *all* the fissions taking place on Δt, and the fissions that are left out of account may well have a systematically different spatial distribution than those that are taken into account. Therefore, if, as seems likely, $t_1 \ll t_2$, it would seem to be necessary to discard most of the data obtained by the calculation. The obvious remedy for this difficulty would seem to be to follow the chains for a definite time rather than for a definite number of cycles of operation. After each cycle, all cards having t greater than some preassigned value would be discarded, and the next cycle of calculation performed with those remaining. This would be repeated until the number of cards in the deck diminishes to zero.

These suggestions are all very tentative. Please let us know what you think of them.

Sincerely,

R. D. Richtmyer.

cc: S. Ulam
    C. Mark
    B. Carlson

3

MULTIPLICATIVE SYSTEMS IN SEVERAL VARIABLES I, II, III

With C.J. Everett
(LA-683, June 7, 1948)
(LA-690, June 11, 1948)
(LA-707, October 28, 1948)

These three reports with C.J. Everett form the foundation of a very large amount of work concerning cascades of particles, epidemics, and a great variety of other processes of this kind. As can be seen in the book on branching processes by Harris, quoted in the note for Chapter 1, it is the basis of several theories in probability theory. Much of the material in these reports is still unexploited and will give rise to further applications in problems in astronomy, chemistry, biology, and other fields where chain reactions are studied. (Author's note).

I

ABSTRACT

This report is the first of a series developing the probability theory of systems of particles of many types, differing in nature, position, and velocity, which undergo transformations of type. The principal technique is that of the generating transformation G whose iterates yield the probability distributions for higher generations. The properties of G with respect to fixed points which determine criticality, convergence under iteration, and relation to first moment matrix M are obtained. Necessary and sufficient conditions for supercriticality of the system in terms of M are given. A ratio theorem is proved for supercritical systems stating that, with overwhelming probability, the distribution of progeny among types in high generations will be essentially in the ratios of the unique characteristic vector of the first moment matrix M of G.

Introduction

The following report, first of a series, deals with some of the purely mathematical theory of multiplicative systems in several variables. The scheme admits various physical interpretations. The t variables may be regarded as different kinds of elementary particles, for example the electrons, photons, mesons, etc., occurring in cosmic ray showers, or as particles of the same nature but belonging to different velocity groups.

The methods and results generalize those of a report by D. Hawkins and S. Ulam on multiplicative systems involving one type of particle. The principal results are: (1) probability distributions for higher generations are given by iteration of the generating transformation G, (2) relation of first and second moments of the k-th generation to those of the first via the Jacobian $M = J(G)$ and Hessians of G, (3) existence of a unique fixed point $x^0 = G(x^0)$ of the unit cube in the supercritical case, (4) convergence of this cube to x^0 under iteration of G, (5) necessary and sufficient conditions for supercriticality in terms of the first moment matrix M, (6) existence and uniqueness of a positive characteristic vector v of M, (7) convergence of the positive sector to the v-ray under iteration of M, (8) flow of probability toward the v-ray with succeeding generations.

The next report will deal with problems arising in subcritical systems.

I. The Generating Transformation

1. Suppose that a system of particles consists of t distinct types, and that a particle of type i upon transformation has probability $p_1(i; j_1, \ldots, j_t)$ of producing $j_1 + \ldots + j_t$ new particles, j_i of type i, $i = 1, \ldots, t$. For every set of non-negative integers $j_1, \ldots, j_t$, we have then $p_1(i; j_1, \ldots, j_t) \geq 0$, and for each i, $\sum_j p_1(i; j_1, \ldots, j_t) = 1$.

With each i, we associate a *generating function* $g_i(x_1, \ldots, x_t) = \sum_j p_1(i; j_1, \ldots, j_t) x_1^{j_1} \ldots x_t^{j_t}$, which defines the probabilities of progeny at the end of one generation from one particle of type i. Hence $x' = G(x)$, explicitly,

$$x_1' = g_1(x_1, \ldots, x_t)$$
$$\vdots$$
$$x_t' = g_t(x_1, \ldots, x_t)$$

with $0 \le x_i \le 1$, defines a *generating transformation* of the unit cube I_t of euclidean t-space into itself.* Moreover, abbreviating $(1, \ldots, 1)$ as 1, we see that $G(1) = 1$, so that the point 1 is a fixed point of transformation G.

2. If, in a given generation, the generating function is $f(x_1, \ldots, x_t)$ $= \sum_k q(k_1, \ldots, k_t) x_1^{k_1} \ldots x_t^{k_t}$, i.e., the coefficient $q(k)$ is the probability in this generation of the state: k_i particles of type i, $i = 1, \ldots, t$, then the generating function of the *next* generation is $f(g_1, \ldots, g_t) =$

$$\sum_k q(k) \cdot \left[\sum_j p_1(1; j) x_1^{j_1} \ldots x_t^{j_t} \right]^{k_1} \ldots \left[\sum_j p_1(t; j) x_1^{j_1} \ldots x_t^{j_t} \right]^{k_t}.$$

3. If we begin with one particle of type i, then the generating functions are:

for 1st generation, $g_i(x_1, \ldots, x_t)$,
for 2nd generation, $g_i(g_1, \ldots, g_t)$,
for 3rd generation, $g_i(g_1(g_1, \ldots, g_t), \ldots, g_t(g_1, \ldots, g_t))$, and so on.

Hence, adopting the notation $G^k(x)$ for the k-th iterate of the transformation $G(x)$:

$$x' = G^k(x) : \begin{cases} x_i' = g_i^{(k)}(x_1, \ldots, x_t) = \sum_j p_k(i; j) x_1^{j_1} \ldots x_t^{j_t} \\ i = 1, \ldots, t \end{cases}$$

we have that $g_i^{(k)}(x)$ is the generating function for the k-th generation of progeny from one particle of type i.

II. First moments. Jacobian

1. Let $f(x_1, \ldots, x_t) = \sum_j q(j) x_1^{j_1} \ldots x_t^{j_t}$ be the generating function for a particular generation. Then $\partial f / \partial x_1 = \sum_j q(j) j_1 x_1^{j_1 - 1} x_2^{j_2} \ldots x_t^{j_t}$ and

$$\left. \frac{\partial f}{\partial x_1} \right]_{x=1} = \sum_j q(j) j_1 = \sum_{j_1=0}^{\infty} \left(\sum_{j_2, \ldots, j_t} q(j_1, \ldots, j_t) \right) j_1 = \sum_{j_1=0}^{\infty} P(j_1) j_1$$

where $P(j_1)$ is the probability of j_1 particles of type 1 in this generation.

---

* Euclidean t-space is the set of all real t-tuples, with the distance function $d(x, x') = \sqrt{\sum (x_i - x_i')^2}$. The unit cube I_t consists of all points x for which $0 \le x_i \le 1$, $i = 1, \ldots, t$.

Hence we define $\partial f/\partial x_1]_{x=1}$ as the first moment for particles of type 1, similarly, $\partial f/\partial x_j]_{x=1}$ as the first moment of particles of type j.

2. We adopt notation $\partial g_i^{(k)}/\partial x_j\Big]_{x=1} = m_{ij}^{(k)}$ as the first moment of particles of type j in the k-th generation of progeny from one particle of type i.

3. Recall that, for a transformation $G(x) : g_i(x_1, \ldots, x_t)$, the Jacobian matrix is $J(G) = [\partial g_i/\partial x_j]$ (i-row, j-column) and if $H(x)$ is a second transformation: $h_i(x_1, \ldots, x_t)$, then the Jacobian of the composite transformation $H(g(x)) : h_i(g_1, \ldots, g_t)$ is $J(H(G))]_x = J(H)]_{G(x)} J(G)]_x$, since $\partial h_i(g_1, \ldots, g_t)/\partial x_k = \sum_j \partial h_i/\partial x_j]_G \cdot \partial g_j/\partial x_k]_x$.

4. It follows, since $G^k = G(G^{k-1})$ that $J(G^k)_x = J(G)_{G^{k-1}} J(G)_{G^{k-2}} \ldots J(G)_G \cdot J(G)_x$, and for a fixed point,*$\bar{x} = G(\bar{x})$, $J(G^k)_{\bar{x}} = (J(G)_{\bar{x}})^k$, i.e., the Jacobian of the k-th iterate of G at a fixed point is the k-th power of the Jacobian matrix of G. In particular, since $1 = G(1)$, we have $J(G^k)_1 = (J(G)_1)^k$.

5. Now $J(G^k)_1 = \left[\partial g_i^{(k)}/\partial x_j\right]_1 = \left[m_{ij}^{(k)}\right]$, so that the relation $\left[m_{ij}^{(k)}\right] = \left[m_{ij}^{(1)}\right]^k$ exists between the first moments of the k-th generation and those of the first.

6. Since $(S^{-1}AS)^k = S^{-1}A^k S$ for matrices, we have $\left[m_{ij}^{(k)}\right] = S(S^{-1}M^k S)S^{-1} = S(S^{-1}MS)^k S^{-1}$ where $S^{-1}MS$ is the canonical form of $M = \left[m_{ij}^{(1)}\right]$. This permits more rapid computation of $m_{ij}^{(k)}$.

III. Second moments. Hessians

1. Let $f(x_1, \ldots, x_t)$ again be the generating function of a particular generation. Then

$$\partial^2 f/\partial x_1^2 = \sum_{j_1 \geq 2} q(j_1, \ldots, j_t) j_1(j_1 - 1) x_1^{j_1-2} x_2^{j_2} \ldots x_t^{j_t}$$

so that

---

*A fixed point $\bar{x} = (\bar{x}_i, \ldots, \bar{x}_t)$ of G is one such that $\bar{x}_i = g_i(\bar{x}_1, \ldots, \bar{x}_t)$, $i = 1, \ldots, t$. Clearly, $\bar{x} = G(\bar{x})$ implies $\bar{x} = G(\bar{x}) = G^2(\bar{x}) = \ldots$ so that a fixed point of G is also a fixed point of all iterates G^k of G.

$$\partial^2 f/\partial x_1^2]_1 = \sum_{j_1 \geq 2} q(j)j_1(j_1 - 1) + \sum_{j_1 \geq 1} q(j)j_1(j_1 - 1) =$$

$$\sum_{j_1 \geq 1} q(j)j_1^2 = \sum_{j_1 \geq 1} q(j)j_1 \, ,$$

where, just as in II, the former sum is the second moment of particles of type 1 in this generation.

2. Letting $t_{ij}^{(k)}$ represent the second moment of particles of type j in the k-th generation of progeny from one particle of type i, we have

$$\partial^2 g_i^{(k)}/\partial x_j^2]_1 = t_{ij}^{(k)} - m_{ij}^{(k)}.$$

3. Recall that, for a *single* function $g(x_1, \ldots, x_t)$, the Hessian is defined as the matrix $H(g) = [\partial^2 g/\partial x_i \partial x_j]$, ($i$-row, j-column). If $S(x)$ is a *transformation*: $s_i(x_1, \ldots, x_t)$, then for the *function* $g(S) = g(s_1, \ldots, s_t)$, we have $\partial g(S)/\partial x_i = \sum_j \partial g/\partial x_j]_S \cdot \partial s_j/\partial x_i$ and hence $\partial^2 g(S)/\partial x_i \partial x_k =$

$$\sum_{j,r} (\partial^2 g/\partial x_j \partial x_r)_S \, \partial s_r/\partial x_k \cdot \partial s_j/\partial x_i + \sum_j (\partial g/\partial x_j)_S \cdot \partial^2 s_j/\partial x_i \partial x_k \, .$$

In matrix notation, this reads *

$$H(g(S)) = J^\tau(S) \cdot H(g)_S \cdot J(S) + \sum_j (\partial g/\partial x_j)_S H(s_j) \, .$$

4. In particular, if we set $g = g_n(x_1, \ldots, x_t)$ and $S = G^{k-1}$, we have

$$H(g_n(G^{k-1})) = H(g_n^{(k)}) = [\partial^2 g_n^{(k)}/\partial x_i \partial x_k] =$$

$$J^\tau (G^{k-1}) H(g_n)_{G^{k-1}} \cdot J(G^{k-1}) + \sum_{j_1} \partial g_n/\partial x_{j_1} \cdot H(g_{j_1}^{(k-1)}) \, .$$

If we write $H_n^{(k)} = H\left(g_n^{(k)}\right)_1$ and $J = J(G)_1 = [m_{ij}]$, we have

$$H_n^{(k)} = (J^{k-1})^\tau H_n^{(1)} J^{k-1} + \sum_{j_1} m_{nj_1} \cdot H_{ji}^{(k-1)} = \ldots =$$

$$(J^{k-1})^\tau H_n^{(1)} J^{k-1} + \sum_{j_1} m_{nj_1} (J^{k-2})^\tau H_{ji}^{(1)} J^{k-2} + \sum_{j_1 j_2} m_{nj_1} m_{j_1 j_2} (J^{k-3})^\tau$$

$$H_{j_2}^{(1)} J^{k-3} + \ldots + \sum_{j_1, \ldots, j_{k-1}} m_{nj_1} m_{j_1 j_2} \ldots m_{j_{k-2} j_{k-1}} \cdot H_{j_{k-1}}^{(1)} \, .$$

---

* If $A = [a_{ij}]$, A^τ denotes the transposed matrix $[a_{ji}]$.

If we define m_{ij} as the element in the i,j position of M we obtain

$$H_n^{(k)} = \left(J^{k-1}\right)^\tau H_n J^{k-1} + \sum_j J_{nj} \left(J^{k-2}\right)^\tau H_j J^{k-2} + \sum_j (J^2)_{nj} \left(J^{k-3}\right)^\tau$$

$$H_j J^{k-3} + \ldots + \sum_j (J^{k-1})_{nj} H_j \cdot {}^*$$

5. Since $H_n^{(k)} = \left[\partial^2 g_n^{(k)} / \partial x_i \partial x_k\right]_1 =$

$$\begin{bmatrix} t_{n1}^{(k)} - m_{n1}^{(k)} & & * \\ & \ddots & \\ * & & t_{nt}^{(k)} - m_{nt}^{(k)} \end{bmatrix}$$

the preceding result gives the second moments $t_{nj}^{(k)}$ for the k-th generation in terms of the first and second partial derivatives associated with the first generation.

IV. Fixed Points of the Transformation $x' = G(x)$

1. We study properties of the transformation $x' = G(x)$:
$x_1' = g_i(x_1, \ldots, x_t) = \sum_j p_1(i; j_1 \ldots, j_t) x_1^{j_1} x_2^{j_2} \ldots x_t^{j_t}$, on the unit cube I_t.
We write $0 = (0, \ldots, 0)$, and as before, $1 = (1, \ldots, 1)$, $x = (x_1, \ldots, x_t)$.

2. We write $x = (x_1, \ldots, x_t) \prec x' = (x_1, \ldots, x_t')$ in case $x_i \leq x_i'$, $i = 1, \ldots, t$. Then $x \prec x'$ implies $G(x) \prec G(x')$, since, for each i, $g_i(x_1, \ldots, x_t) \leq g_i(x_1', \ldots, x_t) \leq g_i(x_1', x_2', \ldots, x_t) \leq g_i(x_1', \ldots, x_t')$. The individual inequalities hold because g_i has all coefficients non-negative, and hence is monotone non-decreasing in each variable separately.

3. We recall that a fixed point $\bar{x} = G(\bar{x})$ of G is a fixed point of all iterates $G^k(\bar{x})$, and hence that $G^k(1) = 1 : 1 = g_i^{(k)}(1, 1, \ldots, 1)$ all $k \geq 0$, $i = 1, \ldots, t$.

4. Since $g_i(0, \ldots, 0) = p_1(i; 0, \ldots, 0)$, we have $0 \prec G(0)$, and so, from $0 \prec x \prec 1$ follows $0 \prec G(0) \prec G(x) \prec G(1) = 1$.
Hence $G(I_t) \subset I_t$, G and all its iterates being transformations of the unit cube into itself.

---

*Note that in case $t = 1$, this reads $g^{k''} = (g')^{2k-2} g'' + \ldots + (g')^{k-1} g''$ at $x = 1$, a relation obtained in Chapter I.[3]

5. If $x^0 = \lim G^k(\bar{x})$ then x^0 is a fixed point of G. For $x^0 = \lim G^k(\bar{x}) = \lim G(G^{k-1}(\bar{x})) = G(\lim G^{k-1}(\bar{x})) = G(x^0)$. Thus all limit points under iteration of G are fixed points.

6. The set of all fixed points of $G(x)$ is closed, i.e., a limit x^0 of a sequence of fixed points $x^\nu = G(x^\nu)$ is itself a fixed point. For we have $x^0 = \lim x^\nu = \lim G(x^\nu) = G(\lim x^\nu) = G(x^0)$.

7. Since $0 \prec 1$, we have $0 \prec G(0) \prec G^2(0) \prec \ldots \prec G(1) = 1$, and for each i, $g_i^{(k)}(0) = p_k(i;0)$ is monotone non-decreasing and bounded above by 1. Hence exists $\lim g_i^{(k)}(0) = \lim p_k(i;0) = x_i{}^0 \le 1$. It follows that $\lim G^k(0) = x^0 = (x_1^0 \ldots, x_t^0)$ exists and $x^0 \prec 1$. Moreover, from 6, $x^0 = G(x^0)$ is a fixed point.

But $g_i^{(k)}(0) = p_k(i;0)$ is the probability of death in the k-th generation of progeny from one i-particle, and x_i^0 is the limit approached from below by this probability. For this reason we speak of x^0 as the *death fixed point* of G, and say i-progeny form a supercritical system in case $x_i^0 < 1$, subcritical if $x_i^0 = 1$.

8. From $0 \prec x \prec x^0$, follows $G^k(0) \prec G^k(x) \prec G^k(x^0) = x^0$ and since $\lim G^k(0) = x^0$, also $\lim G^k(x) = x^0$ for all x in the indicated range.

9. If the death fixed point x^0 is 1, then $\lim G^k(x) = 1$ for all x in I_t. This follows from 8.

10. If $0 \prec x \prec 1$ and $\lim G^k(x)$ exists, then $x^0 \prec \lim G^k(x) \prec 1$. For $G^k(0) \prec G^k(x) \prec G^k(1) = 1$ and $x^0 = \lim G^k(0) \prec \lim G^k(x) \prec 1$.

11. If $\bar{x}$ is a fixed point of G in I_t, then $x^0 \prec \bar{x} \prec 1$. For $\bar{x} = G^k(\bar{x})$ for all k, so $\lim G^k(\bar{x})$ exists, and we use 10.

12. If $0 \prec x = (x_1, \ldots, x_t) \not\gtrsim x' = (x_1', \ldots, x_t') \prec 1$, i.e., $x_i \le x_i'$, $i = 1, \ldots, t$, and $x_j \not\gtrsim x_j'$ for at least one j, then $g_i(x) < g_i(x')$ for *all* $i = 1, \ldots, t$. For, in the inequalities of 2, we have $g_i(x_1', \ldots, x_{j-1}', x_j, x_{j+1}, \ldots, x_t) < g_i(x_1', \ldots, x_{j-1}', x_j', x_{j+1}, \ldots, x_t)$, since by the law of the mean their difference is $\partial g_i(x_1', \ldots x_{j-1}', \xi_j, x_{j+1}, \ldots, x_t)/\partial x_j \cdot (x_j' - x_j)$ where $0 \le x_j < \xi_j < x_j'.{}^*$

---

    * In this paragraph and in the remainder of the paper we assume $\partial g_i/\partial x_j > 0$ for all i, j and all $x \ne 0$ of I_t.

13. If $x = (x_1, \ldots, x_t) \not\supseteq \bar{x} = (\bar{x}_i, \ldots, \bar{x}_t) = G(\bar{x})$, the latter being a fixed point, then $g_i(x) < g_i(\bar{x}) = \bar{x}_i$, $i = 1, \ldots, t$, as a consequence of 12. In particular, if the death fixed point x^0 is not $1 = (1, \ldots, 1)$, that is, if a least one $x_j^0 < 1$, we have $x_i^0 = g_i(x^0) < g_i(1) = 1$ for *all* i. This means that $x_i^0 = \lim p_k(i; 0)$ is 1 if and only if it is for every i, so that we may speak unambiguously of a system as subcritical (all $x_i^0 = 1$) or as supercritical (all $x_i^0 < 1$), regardless of which type of particle is considered as ancestor of the process.

14. We have seen that $x^0 \prec 1$ and that any other fixed point $\bar{x}$ of I_t must satisfy $x^0 \prec \bar{x} \prec 1$. Hence, for G subcritical, $x^0 = 1$ and there is only one fixed point. Moreover, if G is supercritical ($x_i^0 < 1$) and $\bar{x}$ is an additional fixed point, $x^0 \not\supseteq \bar{x} \not\supseteq 1$, we have $x_i^0 < \bar{x}_i < 1$, for all i, from 13. We shall see that no such third fixed point can exist.

15. If $0 \not\supseteq x^0$ then $0 \leq g_i(0) < g_i(x^0) = x_i^0$. Thus, if the death fixed point is not 0, all of its components are *positive*.

16. Suppose $x, \bar{x}$ are chosen in I_t with $\bar{x}_i \neq x_i$, $i = 1, \ldots, t$. Define $F_i(t) = g_i(\bar{x} + (x - \bar{x})t)$ on $0 \leq t \leq 1$. Taylor's expansion gives $F_i(1) = F_i(0) + F'(0) + 1/2 F_i''(\theta_i)$, $0 < \theta_i < 1$. Now

$F_i(1) = g_i(x)$,

$F_i(0) = g_i(\bar{x})$,

$F_i'(t) = \sum_j (\partial g_i / \partial x_j)_{\bar{x}} + (x - \bar{x})t^{(x_j - \bar{x}_j)}$,

$F_i'(0) = \sum (\partial g_i / \partial x_j)_{\bar{x}} (x_j - \bar{x}_j)$,

$F_i''(t) = \sum (\partial^2 g_i / \partial x_j \partial x_k)_{\bar{x} + (x - \bar{x})t} (x_j - \bar{x})(x_k - \bar{x}_k)$, and

$F_i''(\theta) = \sum (\partial^2 g_i / \partial x_j \partial x_k)_{\bar{x} + (x - x)\theta_i} (x_j - \bar{x}_j)(x_k - \bar{x}_k)$.

Hence $g_i(x) = g_i(\bar{x}) + \sum_j (\partial g_i / \partial x_j)_{\bar{x}} (x_j - \bar{x}_j) + 1/2 \sum (\partial^2 g_i / \partial x_k \partial x_k)_{\xi^{(i)}}$ $(x_j - \bar{x}_j)(x_k - \bar{x}_k)$ where $\xi_j^{(i)} = \bar{x}_j + (x_j - \bar{x}_j)t$, $0 < t < 1$.

In case $\bar{x}_j < x_j$ for all j, we have $\bar{x}_j < \xi_j^{(i)} < x_j$, and if $x_i < \bar{x}_j$ for all j, then $x_j < \xi_j^{(i)} < \bar{x}_j$.

17. We show now* that, in the supercritical case ($x^0 \not\supseteq 1$) the transformation G has no fixed point in I_t other than x^0 and 1. We have

---

* In this section and in the remainder of the paper we assume that for each $g_i(x)$ at least one of the $\partial^2 g_i / \partial x_j \partial x_k$ does not vanish on the range $x^0 \prec x_i \prec 1$. This is the last restriction we place on G.

already seen that if such a point $\bar{x}$ exists we have for all of its components the relation $x_i^0 < \bar{x}_i < 1$. In 16, let $\bar{x}$ be the hypothetical third fixed point, and set $x = 1$.

Then $1 = g_i(1) = \bar{x}_i + \sum \left(\partial g_i/\partial x_j\right)_{\bar{x}}(1 - \bar{x}_j) + 1/2 \sum \left(\partial^2 g_i/\partial x_j \partial x_k\right)_{\xi^{(i)}}$
$(1 - \bar{x}_j)(1 - \bar{x}_k)$ where $x_j^0 < \bar{x}_j < \xi_j^{(i)} < 1$.

Hence

$$1 - \bar{x}_i > \sum \left(\partial g_i/\partial x_j\right)_{\bar{x}}(1 - \bar{x}_j)\ldots\ldots\ldots(A)\ .$$

Now, in 16, let $\bar{x}$ again be the third fixed point, but set $x = x^0$.
Then we have $x_i^0 = g_i(x^0) = \bar{x}_i + \sum \left(\partial g_i/\partial x_j\right)_{\bar{x}}(x_j^0 - \bar{x}_j) +$
$1/2 \sum \left(\partial^2 g_i/\partial x_j \partial x_k\right)_{\bar{\xi}^{(i)}}(x_j^0 - \bar{x}_j)(x_k^0 - \bar{x}_k)$ where $x_j^0 < \bar{\xi}_j^{(i)} < \bar{x}_j < 1$.

Hence

$$x_i^0 - \bar{x}_i > \sum \left(\partial g_i/\partial x_j\right)_{\bar{x}}(x_j^0 - \bar{x}_j)\ ,\ \text{or}$$

$$\bar{x}_i - x_i^0 < \sum \left(\partial g_i/\partial x_j\right)_{\bar{x}}(\bar{x}_j - x_j^0)\ldots\ldots\ldots(B)\ .$$

Now set $u_i = 1 - \bar{x}_i > 0$, $v_i = \bar{x}_i - x_i > 0$, $p_{ij} = \left(\partial g_i/\partial x_j\right)_{\bar{x}} > 0$.
We have then, for every i,

$(A')\ u_i > \sum p_{ij} u_j > 0$

$(B')\ 0 < v_i < \sum p_{ij} v_j$

$(B'')\ 1/v_i > 1/\sum p_{ij} v_j > 0$

$(C)\ u_i/v_i > \sum p_{ij} u_j / \sum p_{ij} v_j > 0$.

Now define $u_m/v_m = \min\{u_1/v_1, \ldots, u_t/v_t\}$ so that for every j,
$u_j/v_j \geq u_m/v_m$ and hence $u_j \geq (u_m/v_m)v_j$.
Hence in (C) with $i = m$,

$$u_m/v_m > \left(\sum p_{mj} u_j\right)/\left(\sum p_{mj} v_j\right) \geq$$
$$\left(\sum p_{mj} v_j (u_m/v_m)\right)/\left(\sum p_{mj} v_j\right) = u_m/v_m,\ \text{a contradiction.}$$

V. On the $\lim_k p_k\ (i; j)$

1. We recall that $G^k(x)$ is denoted by

$$x_i' = g_i^{(k)}(x_1, \ldots, x_t) = p_k(i; 0) + \sum_{j \neq 0} p_k(i; j_1, \ldots, j_t) x_1^{j_1} \ldots x_t^{j_t},$$

where $\lim p_k(i; 0) = \lim g_i^{(k)}(0) = x_i^0$.

2. In the subcritical case, $x^0 = 1$, we have

$$g_i^{(k)}(1) - p_k(i;0) = 1 - p_k(i;0) = \sum_{j \neq 0} p_k(i; j_1, \ldots, j_t),$$

and hence $\lim_k \sum_{j \neq 0} p_k(i; j) = 0$.

3. In case $0 \not\geq x^0 \not\geq 1$, we have seen that $0 < x_i^0 < 1$ for all i. Hence, let $\xi = \min \{x_1^0, \ldots, x_t^0\}$ where $0 < \xi < 1$. Then

$$g_i^{(k)}(x^0) - p_k(i;0) = x_i^0 - p_k(i;0) = \sum_{j \neq 0} p_k(i; j)(x_1^0)^{j_1} \ldots (x_t^0)^{j_t},$$

and hence $\lim_k \sum_{j \neq 0} p_k(i; j) (x_1^0)^{j_1} \ldots (x_t^0)^{j_t} = 0$. For every integer $\sigma \geq 1$, we have*

$$\sum_{j \neq 0} p_k(i;j)(x_1^0)^{j_1} \ldots (x_t^0)^{j_t} \geq \sum_{1 \leq \sigma(j) \leq \sigma} p_k(i;j)(x_1^0)^{j_1} \ldots (x_t^0)^{j_t} \geq$$

$$\sum_{1 \leq \sigma(j) \leq \sigma} p_k(i;j)\xi^{\sigma(j)} \geq \sum_{1 \leq \sigma(j) \leq \sigma} p_k(i;j)\xi^{\sigma} \text{ since } 0 < \xi < 1 \text{ and } \sigma(j) \leq \sigma$$

on the range of summation.

It follows that $\lim_k \sum_{1 \leq \sigma(j) \leq \sigma} p_k(i;j)\xi^{\sigma} = 0$ and hence

$$\lim_k \sum_{1 \leq \sigma(j) \leq \sigma} p_k(i;j) = 0 \text{ since } \xi \neq 0.$$

4. In the only remaining case $x^0 = 0$, we have $G^k(0) = 0$ and so $g_i^{(k)}(0) = p_k(i;0) = 0$ and thus $g_i^{(k)}(x_1, \ldots, x_t) = \sum_{j \neq 0} p_k(i;j)x_1^{j_1} \ldots x_t^{j_t}$. We will show first the existence of a point x such that $G(x) \prec x \not\geq 1$. Since $G(x)$ has for its component functions the g_i indicated above, we have $\partial g_i/\partial x_1 = \sum p_1(i;j)j_1 x_1^{j_1-1} x_2^{j_2} \ldots x_t^{j_t}$ and so $(\partial g_i/\partial x_1)_0 = p_1(i; \frac{1}{j}, 0, \ldots, 0)$. Thus generally, $(\partial g_i/\partial x_j)_0 = p_1(i; 0, \ldots, \frac{1}{j}, \ldots, 0)$.

Letting $x = 1$ in $G(x)$, $1 = \sum_{j \neq 0} p_1(i;j) \geq \sum_{j=1}^t p_1(i; 0, \ldots, \frac{1}{j}, \ldots, 0) = \sum (\partial g_i/\partial x_j)_0$.

Moreover, if $1 = \sum (\partial g_i/\partial x_j)_0$, then all other $p_1(i;j) = 0$ whence $g_i(x) = \sum_{j=1}^t p_1(i; 0, \ldots, \frac{1}{j}, \ldots, 0)x_j$. But then $(\partial^2 g_i/\partial x_j \partial x_k) \equiv 0$ for all x of I_t and all j, k, thus violating our assumption on second partials. Hence $\sum_j (\partial g_i/\partial x_j)_0 < 1$ for each i.

Define $S_i(x_1, \ldots, x_t) = \sum (\partial g_i/\partial x_j)_x$ on I_t. Then $S_i(0) < 1$, hence for each i there exists a ξ_i, $0 < \xi_i < 1$, such that, for all x satisfying $0 \leq x_j \leq \xi_i$ we have $S_i(x) < 1$ (continuity of S).

---

* We use the notation $\sigma(j) = j_1 + \ldots + j_t$, for $j = (j_1, \ldots, j_t)$.

Let $\xi = \min\{\xi_1, \ldots, \xi_t\}$, $0 < \xi < 1$.

Then for all x such that $0 \le x_j \le \xi < 1$, we have $S_i(x) < 1$ for all i. Now the Taylor form for $g_i(x)$ about $x = 0$ may be written $g_i(x) = g_i(0) + \sum_j (\partial g_i/\partial x_j)_{\bar{x}} x_j = 0 + \sum_j (\partial g_i/\partial x_j)_{\bar{x}^{(i)}} x_j$, with $0 < \bar{x}_j^{(i)} < x_j$. Let $x = (\xi, \ldots, \xi)$ where ξ is that above. Then $g_i(\xi, \ldots, \xi) = \sum (\partial g_i/\partial x_j)_{\bar{x}^{(i)}} \xi < \xi$ since $0 < \bar{x}_j^{(i)} < \xi$.

Hence for $x = (\xi, \ldots, \xi)$ we have $G(x) \prec x \not\succeq 1$. It follows that $0 \prec \ldots \prec G^k(x) \prec G^{k-1}(x) \prec \ldots \prec G(x) \prec x \not\succeq 1$ so $\lim G^k(x) \not\succeq 1$ exists, and is a fixed point of G. Since there is only one fixed point $(x^0 = 0)$ other than 1, we have $\lim G^k(x) = 0$.

For $x = (\xi, \ldots, \xi)$ then, $\lim g_i^{(k)}(x) = \lim \sum_{j \ne 0} p_k(i; j) \xi^{\sigma(i)} = 0$. Again fix σ as any positive integer, and we have

$$\sum_{j \ne 0} p_k(i; j) \xi^{\sigma(j)} \ge \sum_{1 \le \sigma(j) \le \sigma} p_k(i; j) \xi^{\sigma(j)} \ge \sum_{1 \le \sigma(j) \le \sigma} p_k(i; j) \xi^{\sigma}.$$

So $\lim_k \sum p_k(i; j) \xi^{\sigma} = 0$ and thus $\lim \sum p_k(i; j) = 0$, where $1 \le \sigma_{(j)} \le \sigma$.

5. In summary, we have shown in this section that in all cases, for every σ and every i

$$\lim_k \sum_{1 \le \sigma(j) \le \sigma} p_k(i; j) = 0$$

and hence trivially, $\lim_k p_k(i; j_1, \ldots, j_t) = 0$ for all i and all $j \ne 0$.

VI. On $\lim G^k(x) = x^0$

1. In case $x^0 = 1$, we have already seen that, for all x in I_t, $\lim G^k(x) = x^0$. We prove in this section that, in the remaining case, $x^0 \not\succeq 1$, $\lim G^k(x) = x^0$ for all $x \ne 1$ of I_t. Indeed, we already know this when $0 \prec x \prec x^0$.

2. We shall use the following

Lemma. Let $0 \le x_i < 1$ for all i, and $\mu = \max\{x_1, \ldots, x_t\}$, $0 \le \mu < 1$. Then for every $\epsilon > 0$ there exists a $\sigma > 0$ such that

$$\sum_{\sigma(j) > \sigma} x_1^{j_1} \ldots x_t^{j_t} < \epsilon .$$

Proof. For arbitrary $\sigma > 0$ we have (Margenau, Murphy p. 417)

$$\sum_{\sigma(j)>\sigma} x_1^{j_1} \ldots x_t^{j_t} \leq \sum_{\sigma(j)>\sigma} \mu^{\sigma(j)} = \sum_{m=\sigma+1}^{\infty} \sum_{\sigma(j)=m} \mu^{\sigma(j)} = \sum_{m=\sigma+1}^{\infty} C_m^{t+m-1} \mu^m .$$

The series $\sum_{m=0}^{\infty} C_m^{t+m-1} \mu^m$ converges by the ratio test:

$$C_{m+1}^{t+m} \mu^{m+1} / C_m^{t+m-1} \mu^m = (t+m/m+1) \cdot \mu \to \mu < 1 .$$

3. Let $x^0 \nleq 1$ and fix x in I_t, with $0 \leq x_i < 1$ for all i. Let $\mu = \max\{x_1, \ldots, x_t\}$ and write

$$g_i^{(k)}(x_1, \ldots, x_t) - p_k(i; 0) = \sum_{j \neq 0} p_k(i; j) x_1^{j_1} \ldots x_t^{j_t}. \text{ Fix } \epsilon > 0 .$$

Then by 2, there exists a σ such that

$$\sum_{\sigma(j)>\sigma} x_1^{j_1} \ldots x_t^{j_t} < \epsilon/2 .$$

For this σ, there exists a K such that, for all $k \geq K$,

$$\sum_{1 \leq \sigma(j) \leq \sigma} p_k(i; j) < \epsilon/2 .$$

Hence, for all $k \geq K$,

$$g_i^{(k)}(x_1, \ldots, x_t) - p_k(i; 0) = \sum_{1 \leq \sigma(j) \leq \sigma} p_k(i; j) x_1^{j_1} \ldots x_t^{j_t} + \sum_{\sigma(j)>\sigma} p_k(i; j) x_1^{j_1} \ldots x_t^{j_t}$$

$$\leq \sum_{1 \leq \sigma(j) \leq \sigma} p_k(i; j) + \sum_{\sigma(j)>\sigma} x_1^{j_1} \ldots x_t^{j_t} < \epsilon/2 + \epsilon/2 = \epsilon .$$

Thus for every x such that $0 \leq x_i < 1$, all i, we have

$$0 = \lim_k \left(g_i^{(k)}(x) - p_k(i; 0) \right) \text{ since } \lim p_k(i; 0) = x_i^0 .$$

4. Now suppose $0 \prec x \nleq 1$. Then $g_i(x) < g_i(1) = 1$, for all i, so $G(x)$ is a point of the type considered in 3. Hence $x^0 = \lim G^k(G(x)) = \lim G^{k+1}(x) = \lim G^k(x)$.

This completes the proof that, in all cases, every point $x \neq 1$ of I_t has the $\lim G^k(x) = x^0$.

VII. Supercriticality Conditions

We give now necessary and sufficient conditions for supercriticality of the transformation $G(x)$, namely that $\lim G^k(0) = x^0 \not\equiv 1$. By definition, $m_{ij} = m_{ij}^{(1)} = (\partial g_i / \partial x_j)_1$. Since we have assumed all $(\partial g_i / \partial x_j)$ non-vanishing for $x \neq 0$ on I, clearly all $m_{ij} > 0$.

Theorem. The following conditions on $G(x)$ are equivalent:

(a) $\lim G^k(0) = x^0 \not\equiv 1$.

(b) there exists an $\bar{x}$ such that $0 \prec G(\bar{x}) \prec \bar{x} \not\equiv 1$, where $g_i(\bar{x}) < \bar{x}_i$, $0 < \bar{x}_i < 1$ for all i.

(c) there exists an $\bar{x}$ such that $0 \prec G(\bar{x}) \prec \bar{x} \not\equiv 1$, where $0 \leq \bar{x}_i < 1$ for all i.

(d) there exists an $\bar{x}$ such that $0 \prec G(\bar{x}) \prec \bar{x} \not\equiv 1$.

(e) there exists an $\bar{x}$ such that $\sum m_{ij}(1 - \bar{x}_j) > 1 - \bar{x}_i$, where $0 \leq \bar{x}_i < 1$, for all i.

(f) there exists a $\delta = (\delta_1, \ldots, \delta_t)$ such that $\sum m_{ij}\delta_j > \delta_i$, where $\delta_i > 0$ for all i.

(g) the matrix $[m_{ij} - \delta_{ij}]_1^t = J(G(1)) - I$ contains at least one upper principal minor $A_\nu = [m_{ij} - \delta_{ij}]_1^\nu$ ($1 \leq \nu \leq t$) such that $(-1)^{\nu+1} |A_\nu| \geq 0$ where $(-1)^{\nu+1} \cdot |A_\nu| > 0$ in case $\nu = t$.*

(h) there exists a $v = (v_1, \ldots, v_t)$ such that $\sum v_i m_{ij} > v_j$, where $v_j > 0$ for all j.

(i) there exists a characteristic root $r > 1$ of $M = [m_{ij}]$ and a corresponding characteristic vector $v = (v_1, \ldots, v_t)$ with all $v_i > 0$, that is, $\sum v_i m_{ij} = r v_j$ for such r and v.

In matrix notation

$$(v_1, \ldots, v_t) \begin{bmatrix} m_{11} & \cdots & m_{1t} \\ \vdots & & \vdots \\ m_{t1} & \cdots & m_{tt} \end{bmatrix} = (rv_1, \ldots, rv_t)$$

or briefly, $vM = rv$.

---

* $[a_{ij}]_m^n$ indicates a matrix whose indices i, j range from m to n. δ_{ij} is the Kronecker delta, 1 or 0 as i is or is not equal to j. I is the identity matrix $[\delta_{ij}]$. $|A|$ indicates determinant of A.

1. The proof consists of the following implications:

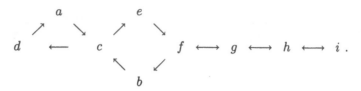

The theorem has been amplified to permit proof in easy stages, but the essential content is the reduction to matrix theory in $J(G)_1$ of the question of criticality of the (non-linear) transformation $G(x)$.

2. $(a \longrightarrow c)$ If $x^0 \not\prec 1$, then $0 \prec G(x^0) = x^0 \not\prec 1$, $0 \le x_i^0 < 1$, all i, so that x^0 serves as $\bar{x}$ in (c).

3. $(c \longrightarrow d)$ is of course trivial.

4. $(d \longrightarrow a)$ If $0 \prec G(\bar{x}) \prec \bar{x} \not\prec 1$, then $G^k(0) \prec G^k(\bar{x}) \prec G^{k-1}(x) \prec \ldots \prec G(\bar{x}) \prec \bar{x} \not\prec 1$, hence $x^0 = \lim G^k(0) \prec \bar{x} \not\prec 1$.

5. $(c \longrightarrow e)$ If $0 \prec G(\bar{x}) \prec \bar{x} \not\prec 1$, $0 \le \bar{x}_i < 1$, all i, we have
$\bar{x}_i \ge g_i(\bar{x}) = g_i(1) + \sum m_{ij}(\bar{x}_j - 1) + 1/2 \sum (\partial g_i / \partial x_j \partial x_k)_{\xi^{(i)}} (\bar{x}_j - 1)(\bar{x}_k - 1)$
where $\bar{x}_j < \xi^{(i)} < 1$ for all j. Hence $\bar{x}_i > 1 + \sum m_{ij}(\bar{x}_j - 1)$ and $1 - \bar{x}_i \le \sum m_{ij}(1 - \bar{x}_j)$ with $0 \le \bar{x}_j < 1$, all j.

6. $(e \longrightarrow f)$ is trivial, since we may define $\delta_i = 1 - \bar{x}_i > 0$ in terms of the $\bar{x}$ given in e.

7. $(f \longrightarrow b)$ Suppose $\sum_{j=1}^{t} m_{ij}\delta_j > \delta_i > 0$, $i = 1, \ldots, t$.
Since $m_{ij} = (\partial g_i / \partial x_j)_1$ and $\partial g_i / \partial x_j$ is continuous, there exists a ξ, $0 < \xi < 1$, such that $\sum (\partial g_i / \partial x_j)_{\bar{x}} \delta_j > \delta_i$, all i, whenever $\xi \le \bar{x}_j \le 1$, all j. Define $\rho = \min\{(1 - \xi)/\delta_i; \ i = 1, \ldots, t\} > 0$.
Then $\sum (\partial g_i / \partial x_j)_{\bar{x}} \cdot \rho\delta_j > \rho\delta_i$ all i, wherever $\xi \le \bar{x}_j \le 1$, all j.
Now define $\bar{x}_i = 1 - \delta_i\rho$; $i = 1, \ldots, t$. Then $\sum (\partial g_i / \partial x_j)_{\bar{x}}(1 - \bar{x}_j) > (1 - \bar{x}_i)$ all i, whenever $\xi \le \bar{x}_j \le 1$, all j. Since $0 < 1 - \bar{x}_i = \delta_i\rho \le 1 - \xi$ (because of definition of ρ), we have $0 < \xi \le \bar{x}_i < 1$, all i.
Now define $\bar{x} = (\bar{x}_1, \ldots, \bar{x}_t)$ where, as we have seen, $0 < \bar{x}_i < 1$. Then $1 = g_i(1) = g_i(\bar{x}) + \sum (\partial g_i / \partial x_j)_{\bar{x}}^{(i)}(1 - \bar{x}_j)/\bar{x}^{(i)}$ where $\bar{x}_j < \bar{x}_j^{(i)} < 1$, all i, j.
Since $\xi \le \bar{x}_j < x_j^{(i)} < 1$, we have $\sum (\partial g_i / \partial x_j)_{\bar{x}=^{(i)}}(1 - \bar{x}_j) > 1 - \bar{x}_i$ and $1 > g_i(\bar{x}) + (1 - \bar{x}_i)$. Hence $g_i(x) < \bar{x}_i$, all i, and $0 \prec G(\bar{x}) \prec \bar{x} \not\prec 1$ for an x of the type required by (b).

8. $(b \longrightarrow c)$ is trivial.

9. For the equivalence of (f) and (g) we need the following:

Lemma. Let $A = [a_{ij}]_1^t$ be a matrix in which $a_{ij} > 0$ for all $i \neq j$. Then there exists a point $\delta = (\delta_1, \ldots, \delta_t)$, with all $\delta_i > 0$, such that $\sum a_{ij}\delta_j > 0$, all i, if and only if, for some ν, $(1 \leq \nu \leq t)$ the upper principal minor $A_\nu = [a_{ij}]_1^\nu$ satisfies the relation $(-1)^{\nu+1}|A_\nu| \geq 0$, with $(-1)^{\nu+1}$. $|A_\nu| > 0$ in case $\nu = t$.

Proof. We first prove $\sum a_{ij}\delta_j > 0$ implies the condition on principal minors. Proof is by induction on the order t. If $t = 1$, we are given that $a_{11}\delta_1 > 0$ for $\delta_1 > 0$ so that $a_{11} > 0$, and $(-1)^2|A_1| > 0$ as required.

Assume the statement of the theorem for order $t - 1$, and suppose given $\sum_{j=1}^t a_{ij}\delta_j > 0$, $i = 1, \ldots, t$, $t \geq 2$, all $\delta_i > 0$, $a_{ij} > 0$ for $i \neq j$.

If $a_{11} \geq 0$, in the given system, we have at once $(-1)^2|A_1| \geq 0$ as required (since $t \geq 2$).

Now suppose $a_{11} < 0$. Then $\sum_2^t a_{1j}\delta_j > -a_{11}\delta_1$ and $a_{i1}\delta_1 + \sum_2^t a_{ij}\delta_j > 0$, $i = 2, \ldots, t$.

Since $-a_{11} > 0$ and $a_{i1} > 0$ $(i = 2, \ldots, t)$ we multiply to obtain $\sum_2^t a_{i1}a_{1j}\delta_j > a_{i1}(-a_{11})\delta_1$ and $(-a_{11})a_{i1}\delta_1 + \sum_2^t a_{ij}(-a_{11})\delta_j > 0$, $i = 2$, $\ldots, t$. Hence $\sum_2^t (-a_{11})a_{ij}\delta_j + \sum_2^t a_{i1}a_{1j}\delta_j = \sum_2^t (a_{11}a_{ij} + a_{i1}a_{1j})\delta_j > 0$; $i = 2, \ldots, t$.

Since, for $i \neq j$, $a_{ij} > 0$, and for $i, j = 2, \ldots, t$, $a_{i1} > 0$ and $a_{ij} > 0$ we have also $(-a_{11}a_{ij} + a_{i1}a_{1j}) > 0$ for $i \neq j$. By the induction assumption on systems of order $t - 1$ there exists a minor of order ν:

$$\mathcal{A}_\nu = [-a_{11}a_{ij} + a_{i1}a_{1j}]_2^{\nu+1}$$

such that $(-1)^{\nu+1}|\mathcal{A}_\nu| \geq 0$ with $(-1)^{\nu+1}|\mathcal{A}_\nu| > 0$ if $\nu = t - 1$.

But for the original matrix we have

$$|A_{\nu+1}| = |a_{ij}|^{\nu+1} = \begin{vmatrix} \cdots a_{1j} \cdots \\ \cdots a_{1j} \cdots \end{vmatrix} = 1/(-a_{11})^\nu \cdot \begin{vmatrix} \cdots a_{1j} \cdots \\ \cdots - a_{11}a_{ij} \cdots \end{vmatrix} =$$

$$1/(-a_{11})^\nu \cdot \begin{vmatrix} \cdots a_{1j} & \cdots \\ \cdots (-a_{11}a_{ij} + a_{i1}a_{ij} \cdots) \end{vmatrix} =$$

$$1/(-a_{11})^\nu \cdot \begin{vmatrix} a_{11}a_{12} \cdots a_{1t} \\ 0 \qquad \mathcal{A}_\nu \end{vmatrix} = a_{11}/(-a_{11})^\nu \cdot |\mathcal{A}_\nu|.$$

Hence $(-1)^{\nu+2} \cdot |A_{\nu+1}| = (-1)^{\nu+2} a_{11} / (-a_{11})^{\nu} |A_{\nu}| = (-1)^{\nu+1} / (-a_{11})^{\nu-1} |A_{\nu}| \geq 0$ and > 0 in case $\nu + 1 = t$ (i.e., $\nu = t - 1$).

We now prove the converse, namely, the condition on minors implies the existence of a δ such that $\sum a_{ij} \delta_j > 0$. Proof is again by induction on order t.

In case $t = 1$, we are given that $(-1)^2 |A_1| = a_{11} > 0$, so for (say) $\delta_1 = 1 > 0$ we have $a_{11} \delta_1 > 0$.

Assume the theorem true for order $t - 1$, and suppose given $A = [a_{ij}]_1^t$, with $t \geq 2$, having for a given $\nu (1 \leq \nu \leq t) (-1)^{\nu+1} \cdot |A_{\nu}| \geq 0 (> 0$ if $\nu + t)$.

If $a_{11} \geq 0$, let $\delta_2 = \ldots = \delta_t = 1$ and choose $\delta_1 > 0$ so that $\delta_1 > \max \{ -(1/a_{11}) \sum_2^t a_{ij}; i = 2, \ldots, t \}$. Then $\sum a_{ij} \delta_j = a_{11} \delta_1 + \sum_2^t a_{1j} \geq 0 + \sum_2^t a_{ij} > 0$ since $a_{1j} > 0$ for $j \neq 1$. Also, for $i = 2, \ldots, t, \sum_1^t a_{i1} \delta_j = a_{i1} \delta_1 = \sum_2^t a_{ij} > -\sum_2^t a_{ij} + \sum_2^t a_{ij} = 0$, by definition of δ_1. Thus the conclusion follows in case $a_{11} = (-1)^2 \cdot |A_1| \geq 0$.

Suppose then that $a_{11} < 0$. Then the *given* A_{ν} must have $2 \leq \nu \leq t$. Now

$$0 \leq (-1)^{\nu+1} \cdot |A_{\nu}| = (-1)^{\nu+1} |a_{ij}|_1^{\nu} = (-1)^{\nu+1} \cdot \begin{vmatrix} \cdots & a_{1j} & \cdots \\ \cdots & a_{ij} & \cdots \end{vmatrix}$$

$$(-1)^{\nu+1}/(-a_{11})^{\nu-1} \cdot \begin{vmatrix} \cdots & a_{1j} & \cdots \\ \cdots & (-a_{11} \ a_{ij}) & \cdots \end{vmatrix} = (-1)^{\nu+1}/(-a_{11})^{\nu-1}$$

$$\begin{vmatrix} \cdots & a_{1j} & \\ \cdots & (-a_{11} a_{ij} + a_{i1} a_{1j}) & \end{vmatrix} = (-1)^{\nu+1}/(-a_{11})^{\nu-1} \begin{vmatrix} a_{11} & a_{12} \cdots & a_{1t} \\ 0 & A_{\nu-1} & \end{vmatrix} =$$

$$(-1)^{\nu}/(-a_{11})^{\nu-2} \cdot |A_{\nu}|.$$

Hence $(-1)^{\nu} |A_{\nu-1}| \geq 0$ for $1 \leq \nu - 1 \leq t - 1$ (> 0 if $\nu - 1 = t - 1$) for the system of order $t - 1$:

$$A = [-a_{11} a_{ij} + a_{i1} a_{1j}]_2^t$$

in which $-a_{11} a_{ij} + a_{i1} a_{1j} > 0$ for $i \neq j$ just as before. By the induction hypothesis, $\delta_i > 0, i = 2, \ldots, t$ exist such that $\sum (-a_{11} a_{ij} + a_{i1} a_{1j}) \delta_j > 0$, $i = 2, \ldots, t$. Hence $a_{i1} (\sum_2^t a_{1j} \delta_j / -a_{11}) + \sum_2^t a_{ij} \delta_j > 0$, $i = 2, \ldots, t$. Let ϵ_1 represent the parenthesis of the last inequality. Then $\epsilon_1 > 0$, since $a_{1j} > 0$ for $j \neq 1$, and substituting, $a_{i1} \epsilon_1 + \sum_2^t a_{ij} \delta_j > 0$, $i = 2, \ldots, t$.

Choose T so that $0 < T < \epsilon_1$, and $T < (1/a_{i1})(a_{i1}\epsilon_1 + \sum_2^t a_{ij}\delta_j)$ for all $i = 2, \ldots, t$. Finally, define $\delta_1 = \epsilon_1 - T > 0$.
Then $\sum_1^t a_{1j}\delta_j = a_{11}\delta_1 + \sum_2^t a_{1j}\delta_j = a_{11}\epsilon_1 + \sum_2^t a_{1j}\delta_j + (-a_{11})T = (-a_{11})T > 0$, and for $i = 2, \ldots, t$, $\sum_1^t a_{ij}\delta_j = a_{i1}\delta_1 + \sum_2^t a_{ij}\delta_j = a_{i1}\epsilon_1 + \sum_2^t a_{ij}\delta_j - a_{i1}T > 0$ because of the second inequality defining T.
Hence the lemma is proved.

10. $(f \to g)$ If $\sum m_{ij}\delta_j > \delta_j > 0$, all i, then $\sum(m_{ij} - \delta_{ij})\delta_j > 0$, all i, and the matrix $[m_{ij} - \delta_{ij}]_1^t$ satisfied the condition of (g) because of the lemma.

11. $(g \to f)$ If the matrix $[m_{ij} - \delta_{ij}]_1^t$ satisfies the condition of (g), there exists by the lemma numbers $\delta_i > 0$ such that $\sum(m_{ij} - \delta_{ij})\delta_j > 0$ and hence $\sum m_{ij}\delta_j > \delta_i$, all i.

12. $(g \to h)$ If $[m_{ij} - \delta_{ij}]$ has the property $(-1)^{\nu+1}|A_\nu| \geq 0$ (> 0 if $\nu = t$) for some ν, then the matrix $[m_{ij} - \delta_{ij}]^\tau = [m_{ij} - \delta_{ij}]$ has* $(-1)^{\nu+1} \cdot |A_\nu^\tau| = (-1)^{\nu+1} \cdot |A_\nu| \geq 0$ (> 0 if $\nu = t$).
Write $\bar{m}_{ij} = m_{ji}$; then $[\bar{m}_{ij} - \delta_{ij}]$ has $(-1)^{\nu+1} \cdot |\bar{A}_\nu| \geq 0$ (> 0 if $\nu = t$).
But we have seen that $(g \to f)$ so that there exist $v_i > 0$ such that $\sum \bar{m}_{ij}v_j > v_i$. Hence $\sum m_{ji}v_j > v_i$, or changing notation, $\sum v_i m_{ij} > v_j$, all j.

13. $(h \to g)$ If $\sum v_i m_{ij} > v_j > 0$, all j, then $\sum m_{ji}v_j > v_i > 0$, and, defining $\bar{m}_{ij} = m_{ji}$, $\sum \bar{m}_{ij}v_j > v_i > 0$. But we know that $(f \to g)$ so the matrix $[\bar{m}_{ij} - \delta_{ij}]$ has the property $(-1)^{\nu+1} \cdot |\bar{m}_{ij} - \delta_{ij}|_1^\nu \geq 0$ (> 0 if $\nu = t$). But then $(-1)^{\nu+1} \cdot |\bar{m}_{ij} - \delta_{ij}|_1^\nu = (-1)^{\nu+1} \cdot |m_{ji} - \delta_{ji}|_1^\nu = (-1)^{\nu+1} \cdot |m_{ij} - \delta_{ij}|_1^\nu \geq 0$ (> 0 if $\nu = t$).

14. $(i \to h)$ If $\sum v_i m_{ij} = rv_j$ for $v_j > 0$, $r > 1$, we have $\sum v_i m_{ij} > v_j$, hence (h).

15. $(h \to i)$ Assume $\sum_{i=1}^t v_i m_{ij} > v_j > 0$. Since $(\sum v_i m_{ij})/v_j > 1$, $j = 1, \ldots, t$, we define r_0 so that $(\sum v_i m_{ij})/v_j > r_0 > 1$, $j = 1, \ldots, t$. Then $\sum v_i m_{ij} > r_0 v_j$, all j.
Write $\sum v_i m_{ij} = r_0 v_j + \epsilon_j'$.

---

* Recall that $|A^\tau| = |A|$.

Define $\|v\| = \sqrt{\sum v_j^2} > 0$. Then $\sum (v_i/2 \cdot \|v\|) m_{ij} = r_0(v_j/2 \cdot \|v\|) + (\epsilon_j'/2\|v\|)$.

Define $v_j^0 = (v_j/2\|v\|)$ and $\epsilon_j = (\epsilon_j'/2\|v\|)$. We have now $\sum v_i^0 m_{ij} = r_0 v_j^0 + \epsilon_j$, where $\epsilon_j > 0$, $v_j^0 > 0$, all j, and $\|v^0\| = 1/2$.

Define $\mu = \min \left\{ 1/2v_j^0; m_{ij} / \sqrt{\sum m_{ij}^2}; \text{ all } i,j = 1,\ldots,t \right\} > 0$.

Let $\Delta = \{v; v_j \geq \mu, \|v\| \leq 1, \sum v_i m_{ij} \geq r_0 v_j, \text{ all } j\}$.

(A) The set Δ is non-void for $v^0 \in \Delta$ and is a subset of euclidean t-space E_t. For $v_j^0 \geq 2\mu > \mu$ by definition of μ, $\|v^0\| = 1/2 < 1$, and $\sum v_i^0 m_{ij} > r_0 v_j^0$.

(B) The set Δ is bounded, since $\|v\| \leq 1$ for all $v \in \Delta$.

(C) The set Δ is closed. For, let $v = \lim v^a$ where $v^a \in \Delta$. We have to show $v \in \Delta$. Since $v_j^a \geq \Delta$, all a,j, we have $v_j = \lim_a v_j^a \geq \mu$. Also, $\|v\| = \|\lim_a v^a\| = \|\lim v^a\| \leq 1$, since $\| \ \|$ is continuous and all $\|v^a\| \leq 1$. Finally, $\sum v_i m_{ij} = \sum (\lim_a v_j^a) m_{ij} = \lim_a \sum v_i^a m_{ij} \geq \lim_a r_0 v_j^a = r_0 \lim_a v_j^a = r_0 v_j$. Hence $v \in \Delta$.

(D) The set Δ is convex. That is, for every $v, v' \in \Delta$, and every $a, a' \geq 0$ with $a + a' = 1$, we have $av + a'v' \in \Delta$. For $av_j + a'v_j' \geq a\mu + a'\mu = (a + a')\mu = \mu$. Also $\|av + a'v'\| \leq a \cdot \|v\| + a' \cdot \|v'\| \leq a + a' = 1$. And $\sum (av_i + a'v_i') m_{ij} = a \sum v_i m_{ij} + a' \sum v_i' m_{ij} \geq a r_0 v_j + a' r_0 v_j' = r_0(av_j + a'v_j')$.

(E) The set Δ has an inner point, e.g., v^0. For $v^0 \in \Delta$ there exists a neighborhood of v^0 contained in Δ. Let

$$\epsilon = \min \left\{ v_j^0 - \mu; \epsilon_j / \left(r_0 + \sum_i m_{ij} \right); \ j = 1, \ldots, t. \right\} .$$

Since $v_j^0 \geq 2\mu > \mu$, $v_j^0 - \mu > 0$ and $\epsilon > 0$. Now let v be any point of the ϵ- neighborhood of v^0, that is, $\|v - v^0\| < \epsilon$. We show that $v \in \Delta$. For $|v_j - v_j^0| \leq \sqrt{\sum (v_j - v_j^0)^2} = \|v - v^0\| < \epsilon$, so $v_j > v_j^0 - \epsilon \geq \mu$, by choice of ϵ, and thus v satisfies the first requirement for membership in Δ.

Next, $| \|v\| - \|v^0\| | \leq \|v - v^0\| < \epsilon$, so $\|v\| < \epsilon + \|v^0\| = 1/2 + \epsilon \leq 1$. The latter inequality obtains since $\epsilon \leq v_j^0 - \mu < v_j^0 < \sqrt{\sum (v_j^0)^2} = \|v^0\| = 1/2$. Hence v satisfies the second requirement.

Finally, $\sum v_i m_{ij} \geq \sum (v_i^0 - \epsilon) m_{ij} = r_0 v_j^0 + \epsilon_j - \epsilon \sum_i m_{ij} \geq r_0 v_j$. The latter inequality is true since it requires $r_0 (v_j^0 - v_j) \geq \epsilon \sum_i m_{ij} - \epsilon_j$, i.e.,

$v_j^0 - v_j \geq \left(\epsilon \sum_i m_{ij} - \epsilon_j\right)/r$. But we *know* $v_j^0 - v_j > -\epsilon$ and it suffices to prove $-\epsilon \geq \left(\epsilon \sum_i m_{ij} - \epsilon_j\right)/r_0$ or $\epsilon_j \geq \epsilon\left(\sum_i m_{ij} + r_0\right)$. But ϵ satisfies this requirement by definition.

The properties A–E show that Δ is a convex body in E_t. It is well known, therefore, that any continuous transformation T of Δ into Δ has a fixed point $T(v) = v$, $v \in \Delta$. (Alexandroff-Hopf).

For every $v \in \Delta$, define $T(v) = vM/\|vM\|$, that is, $T(v) = v'$ where $v_j' = \sum_i v_i m_{ij} \big/ \sqrt{\sum_j \left(\sum_i v_i m_{ij}\right)^2}$. Since for $v \in \Delta$, $v_i \geq \mu > 0$, and $m_{ij} > 0$, clearly $\|vM\| > 0$ on Δ.

First we show that if $v \in \Delta$, then $T(v) \in \Delta$. For $\left(v_j'\right)^2 = \left(\sum v_i m_{ij}\right)^2 \big/ \sum_j \left(\sum_i v_i m_{ij}\right)^2 \geq \sum v_i^2 m_{ij}^2 \big/ \sum_j \left(\sum v_i^2\right)\left(\sum_i m_{ij}^2\right) \geq 1/\|v\|^2 \cdot \sum_i \left\{m_{ij}^2 v_i^2 / \sum_{ij} m_{ij}^2\right\} \geq (1/\|v\|^2)\mu^2 \cdot \sum_i v_i^2 = \mu^2$ so $v_j' \geq \mu$ as required for $T(v) = v' \in \Delta$.

Moreover, $\|T(v)\| = \left(\|vM/\|vM\|\|\right) = 1$. Finally, $\sum_i v_i' m_{jk} = (1/\|vM\|) \sum_j m_{jk} \sum_i v_i m_{ij} \geq (1/\|vM\|) \sum m_{jk} r_0 v_j = r_0 v_k'$, so $T(v) \in \Delta$, and $T(\Delta) \subset \Delta$. Clearly, $T(v)$ is continuous on Δ. Hence, for some $\bar{v} \subset \Delta$, $T(\bar{v}) = \bar{v} = \bar{v}M/\|vM\|$, and thus $\bar{v}_j \cdot \|vM\| = \sum \bar{v}_i m_{ij} \geq r_0 \bar{v}_j$, so $\|vM\| \geq r_0 > 1$. Set $r = \|vM\|$ and we have $\sum \bar{v}_i m_{ij} = r \bar{v}_j$ with $\bar{v}_j > 0$ and $r > 1$. Thus the theorem is complete.

16. Some general remarks on positive matrices are now appropriate. If $[m_{ij}]$ is a matrix with all $m_{ij} > 0$, it can have at most one characteristic root $r > 0$ with a corresponding positive characteristic vector v, all $v_i > 0$, that is, the equations

$$\sum v_i m_{ij} = r v_j \text{ and } \sum v_j' m_{ij} = r' v_j'; \ v_j, v_j', r, r' > 0$$

imply $r = r'$.

For, choose λ, $\mu > 0$ so that $\lambda \leq \left(v_i'/v_i\right) \leq \mu$, $i = 1, \ldots, t$. Then $\lambda v_i \leq v_i' \leq \mu v_i$, and since $vM = rv$ implies $vM^k = r^k v$ and $v'M = r'v'$ implies $v'M^k = r'^k v'$, we have $\sum \lambda v_i M_{ij}^k \leq \sum v_i' m_{ij}^{(k)} \leq \sum \mu v_i m_{ij}^{(k)}$, and hence $\lambda r^k v_j \leq r'^k v_j' \leq \mu r^k v_j$, where $M^k = [m_{ij}^{(k)}]$ and clearly $m_{ij}^{(k)} > 0$. Then $\lambda v_j \leq \left(r'/r\right)^k v_j' \leq \mu v_j$, an impossibility for high k, if the ratio r'/r is either less than or greater than one.

17. We shall make use of the following

Lemma. If $B = [b_{ij}]_1^t$ is a matrix with $b_{ij} > 0$ for all $i \neq j$, and if $\sum v_i b_{ij} = 0 = \sum v_i' b_{ij}$ for $v_i, v' > 0$, then there exists an $a > 0$ such that $v_i' = av_i$, $i = 1, \ldots, t$.

Proof. Induct on order t. Suppose $t = 1$. Then we have $v_1 b_{11} = 0 = v_1' b_{11}$, $v_1, v_1' > 0$, so $v_i' = av_1$ where $a = v_1'/v_1$. Assume the lemma for systems of order $t - 1$, and suppose given $\sum_1^t v_i b_{ij} = 0 = \sum_1^t v_i' b_{ij}$ with $t \geq 2$. Since $\sum v_i b_{i1} = -b_{11} \cdot v_i$ and $b_{i1} > 0$ (for $i \neq 1$) and $v_i > 0$, all i, we must have $-b_{11} > 0$. Note also $v_i b_{ij} + \sum_2^t v_i b_{ij} = 0$, $i = 2, \ldots, t$.

Hence $\sum_2^t v_i b_{i1} b_{1j} = -b_{11} b_{ij} v_1$ and $v_1 b_{1j}(-b_{11}) + \sum_2^t v_i(-b_{11}) b_{ij} = 0$, $j = 2, \ldots, t$. Substituting, $\sum_2^t v_i(-b_{11} b_{ij} + b_{i1} b_{1j}) = 0$, $j = 2, \ldots, t$. For $i \neq j$, $b_{ij} > 0$ and $-b_{11} > 0$, $b_{i1} > 0$, $b_{1j} > 0$ for $i, j \geq 2$. Thus the matrix $[-b_{11} b_{ij} + b_{i1} b_{1j}]_2^t$ has the property of the lemma. In identical fashion we have also

$\sum_2^t v_i'(-b_{11} b_{ij} + b_{i1} b_{1j}) = 0$. Thus, by the induction hypothesis, $v_j' = av_j$, $j = 2, \ldots, t$, for some $a > 0$. But $-b_{11} v_i' = \sum v_i' b_{i1} = a \sum v_i b_{i1} = a(-b_{11} v_1)$ and thence also $v_i' = av_1$.

18. It follows that, if $m_{ij} > 0$, all i, j, and $\sum v_i m_{ij} = rv_j$, $\sum v_i' m_{ij} = rv_j'$ for $v_j, v_j' > 0$, then $v' = av$. That is, a positive matrix can have at most one positive characteristic vector (up to the scalar multiples) corresponding to any one of its characteristic roots.

For, the above equations may be written $\sum v_i(m_{ij} - \delta_{ij} r) = 0 = \sum v_i'(m_{ij} - \delta_{ij} r)$, where the non-diagonal elements of $[m_{ij} - \delta_{ij} r]$ are the $m_{ij} > 0$.

19. If M is the moment matrix of any G, there is one and only one solution for $r > 0$, $v_i > 0$ of the relation $vM = rv$. G is supercritical if and only if the solution r is greater than one. (See 20 for existence.)

20. The proof of property (i) for supercritical matrices was complicated by the fact that we wanted to prove $r > 1$. It is possible to show more easily the weaker result that a solution of the relation in 19 exists for arbitrary matrix $M = [m_{ij}]$ with all $m_{ij} > 0$.

Define $\mu = \min \{1/(4\sqrt{t}); m_{ij}/\sum_{ij} m_{ij}^2\} > 0$, $\Delta = \{v; v_i \geq \mu, \|v\| \leq 1\}$ and $T(v) = vM/\|vM\|$ for v in Δ. Then just as in the proof of (i) one verifies Δ is a convex body in E_t, and $T(v)$ is a continuous transformation of Δ into itself. As such, there exists a $\bar{v}$ in Δ such that $T(\bar{v}) = \bar{v}M/\|\bar{v}M\| = \bar{v}$, and for $r = \|\bar{v}M\| > 0$, $\bar{v}M = r\bar{v}$. Since $\bar{v} \in \Delta$, $\bar{v}_i \geq \mu > 0$.

21. Note that, if $\sigma(v) = \sum v_i$, σ is a linear functional:

$$\sigma(av + a'v') = \sum(av_i + a'v_i') = a\sum v_i + a'\sum v_i' a\sigma(v) + a'\sigma(v') .$$

If we define $T(v) = vM/\sigma(vM)$ on the sector $v_i \geq 0$ $(v \neq 0)$, where M has all $m_{ij} > 0$, we see that $\sigma(T(v))\sigma(vM)/\sigma(vM) = 1$ and $T(v)$ may be regarded as the central projection of vM onto the hyperplane $v_1+\ldots+v_t = 1$. Our operator T has some interest beyond that of the classical Markoff operators (see G. Birkhoff) since it need not be a contraction in the positive sector. For example, if $M = \begin{bmatrix} 1/4 & 1 \\ 19 & 1 \end{bmatrix}$ one finds that it is supercritical, has positive characteristic root $r = 5$, corresponding vector $= (4/5, 1/5) = \bar{v}$ where $\sigma(\bar{v}) = 1$, and $\bar{v}M = r\bar{v}$. Hence $T(\bar{v}) = \bar{v}$. Let $\delta - (1,0)$. Then $\|T(\delta) \quad T(v)\| - \sqrt{377/20} > \|\delta - v\| - \sqrt{2/5}$. Hence $T(\delta)$ is not closer to v than δ was. Nevertheless we prove $\lim T^k(v) = \bar{v}$ for all vectors v in the sector $v_i \geq 0$ $(v \neq 0)$. Our proof of this fact is lengthy but of a general topological character.

22. Let $\{y^i\}$ denote the simplex with vertices $y^1, \ldots, y^n$ in E, that is, the set of all points $y = \sum_1^n a_i y^i$, $a_i \geq 0$, $\sum a_i = 1$.

If A is an arbitrary matrix, we have $(\sum a_i y^i)A\sum a_i(y^i A)$ so that the map under A of the simplex $\{y^i\}$ is the simplex with vertices $y^i A$:

$$\{y^i\}A = \{y^i A\} .$$

23. For arbitrary x with $\sigma(x) \neq 0$ define the transformation $xS = x/\sigma(x)$. If the y^i are all $\neq 0$ and have all components non-negative, clearly the same properties are possessed by all points of $\{y^i\}$. Hence we may operate on $\{y^i\}$ with S and $\{y^i\}S = \{y^i S\}$. This says that the projection onto the hyperplane $\sum v_i = 1$ of the simplex $\{y^i\}$ is the simplex with vertices y_i/σ_i where $\sigma_i \equiv \sigma(y^i)$.

For, $\{y^i\}S$ is the set of all $\sum a_i y^i / \sum a_i \sigma_i$ and $\{y^i S\}$ is that of all $\sum b_i (y^i / \sigma_i)$ where $a_i, b_i \geq 0$, and $\sum a_i = 1 = \sum b_i$. We have to show that, given either the a_i or the b_i, the other set may be determined so that

$$\sum a_i y^i \Big/ \sum a_i \sigma_i = \sum (b_i / \sigma_i) y^i . \tag{1}$$

Given the a_i, definition of $b_i = a_i \sigma_i / \sum a_i \sigma_i$ implies $b_i \geq 0$, $\sum b_i = 1$ and validity of (1) so that $\{y^i\}S \subset \{y^i S\}$.

Given the b_i, it suffices to find a_i so that

$$a_i \sigma_i = \sum_j b_i a_j \sigma_j \quad \text{or} \quad \sum_j (b_i \sigma_j - \delta_{ij} \sigma_j) a_j = 0, \ i = 1, \ldots, n . \tag{2}$$

The determinant of the latter system is zero since addition of each row but the first successively *to* the first yields in the $(1, j)$ position $\sigma_j \sum b_i - \sigma_j = \sigma_j - \sigma_j = 0$.

Hence a solution $(a_1, \ldots, a_n) \neq (0, \ldots, 0)$ exists, and we see from $a_i / \sum a_i \sigma_i = b_i / \sigma_i > 0$ that all a_i have the same sign. Hence we can choose a solution $a_i \geq 0$ for all i, with $\sum a_i > 0$, and then $a_i / \sum a_i$ fulfills all requirements.

24. If $T(v) = vM / \sigma(vM)$ then $T^k(v)$, the k-th iterate of T, is given by $T^k(v) = vM^k / \sigma(vM^k)$. Proof is trivial by induction on k. Hence

$$T^k(v) = (vM^k)S \quad \text{and} \quad T^k\{y^i\} = (\{y^i\}M^k)S = \{y^i M^k\}S =$$

$$\{y^i M^k / \sigma(y^i M^k)\} = \{T^k(y^i)\} .$$

25. Define $\delta^j = (0, \ldots, \underset{j}{1}, \ldots, 0)$, the j-th unit vector of t-space, and $\Delta = \{\delta^i\}_1^t$, $\Delta_k = T^k(\Delta) = \{T^k(\delta^i)\}$. Since $m_{ij} > 0$, clearly $T(\Delta) \subset \Delta$ and hence $\Delta \supset T(\Delta) \supset T^2(\Delta) \supset \ldots$ is a "nest" of simplices. Define $D = \bigcap_k \Delta_k$, the intersection of all Δ_k.

We prove that $T(D) = D$. Let $d_e \in D$, i.e., $d_e = T(d_1) = T^2(d_2) = \ldots$. Then $T(d_0) = T^2(d_1) = T^3(d_2) = \ldots$ is in D and hence $T(D) \subset D$.

To obtain the other inclusion, we make use of the lemma. If $\{y^i\}_1^t$ is a simplex contained in Δ such that $T(y) = T(y')$ for $y \neq y'$ in $\{y^i\}$, then the linear dimension of the subspace spanned by the $T(y^i)$, $i = 1, \ldots, t$ is less than that of the space spanned by the y^i, $i = 1, \ldots, t$.

Proof. Let (say) $y^1, \ldots, y^d$ be linearly independent vertices of the simplex $\{y^i\}$ and the remaining y^{d+i} dependent upon them.

Write $y = \sum_1^d c_i y^i \neq y' = \sum_i^d c_i' y^i$ in $\{y^i\} \subset \Delta$, hence $\sigma(y) = \sigma(y') = 1$. Suppose that $T(y) = \sum c_i y^i M / \sum c_i \bar{\sigma}_i = T(y') = \sum c_i' y^i M / \sum c_i' \bar{\sigma}_i$ where $\bar{\sigma}_i = \sigma(y^i M)$. If we let $C = \sum c_i \bar{\sigma}_i$, $C' = \sum c_i' \bar{\sigma}_i$ then $\sum (c_i / C - c_i'/C') y^i M = 0$. Now not all coefficients here are zero, for if so, we should have $c_i' = (C'/C) c_i = R c_i$ and hence $y' = \sum c_i' y^i = R \sum c_i y^i = Ry$. But then $\sigma(y') = R\sigma(y)$, $R = 1$, and thus $c_i' = c_i$, $y = y'$, a contradiction. Hence the $y^i M$ are dependent $(i = 1, \ldots, d)$ and since the y^{d+i} are dependent on the y^i, $i = 1, \ldots, d$ so are the $y^{d+i} M$ dependent upon the $y^i M$, $i = 1, \ldots, d$. It follows that the space spanned by the $y^i M$, $i = 1, \ldots, t$ is of lower linear dimension than d. But this space is identical with that spanned by the $T(y^i) = y^i M / \bar{\sigma}_i$.

Since the dimension t of the space spanned by the δ^i can suffer only a finite number of reductions, it follows that, for all $k \geq K$, for some K, T is one to one on Δ_k to Δ_{k+1}.

Now we prove $D \subset T(D)$. Let $d \in D$ so that we have for all k, $d = T^k(d_k)$ for some $d_k \in \Delta$. Now for all $i \geq 0$,

$$T^{k+i+1}(d_{k+i+1}) = T^{k+i+2}(d_{k+i+2}) \text{ or } T\left(T^{k+i}(d_{k+i+1})\right)$$

$$= T\left(T^{k+i+1}(d_{k+i+2})\right)$$

where the operands of T are in $\Delta_{k+1} \subset \Delta_k$. Since T is one to one on Δ_k, we have $e_k = T^{k+i}(d_{k+i+1}) = T^{k+i+1}(d_{k+i+2})$ for all i. Hence $d = T(e_k)$ where $e_k \in \bigcap_{k \geq K+1} \Delta_k \subset \bigcap_k \Delta_k = D$, and thus $D \subset T(D)$.

26. We now show D is itself a simplex. To this end, consider the sequences $T^k(\delta^i)$, $i+1, \ldots, t$, of vertices of the Δ_k. There exists a common subsequence k_ν such that $\lim T^k(\delta^i) = \lambda^i$ all exist $i = 1, \ldots, t$. We claim that the simplex $\{\lambda^i\}_1^t$ is indeed the set D.

First, it is clear that $D = \bigcap_k T^k(\Delta) = \bigcap_\nu T^{k\nu}(\Delta)$ since the Δ_k are nested. Hence it suffices to prove that $\{\lambda^i\} = \bigcap_\nu T^{k\nu}(\Delta)$.

Since the $\lambda^i = \lim T^{k\nu}(\delta^i)$, λ^i is a limit of a sequence of points which are in $T^{k\nu}(\Delta) = \Delta_{k\nu}$ for all $\nu \geq N$. Since $T^k(\Delta)$ is closed, $\lambda^i \in T^{kN}(\Delta)$. Hence $\lambda^i \in \bigcap_\nu T^{k\nu}(\Delta)$. Now all $T^{k\nu}(\Delta)$ are convex, hence so is their intersection D. Since all λ^i are in the convex set $\bigcap_\nu T^{k\nu}(\Delta) = D$, and $\{\lambda^i\}$ is the convex hull of the λ^i, we have $\{\lambda^i\} \subset \bigcap_\nu T^{k\nu}(\Delta) = D$ (Alexandroff-Hopf).

We prove now $\bigcap_\nu T^{k\nu}(\Delta) \subset \{\lambda^i\}$. Suppose d in the intersection but not in $\{\lambda^i\}$. Then, since $\{\lambda^i\}$ is closed, there exists an ϵ such that the ϵ-neighborhood of d excludes $\{\lambda^i\}$. However, since $\lim T^{k\nu}(\delta^i) = \lambda^i$, we can find N so that, for $i+1, \ldots, t$, $\|T^{kN}(\delta^i) - \lambda^i\| < \epsilon/2$. Now $d \in T^{kN}(\Delta)$ hence write $d = \sum a_i T^{kN}(\delta^i)$. Then $\|d - \sum a_i \lambda^i\| = \|\sum a_i(T^{kN}(\delta^i) - \lambda^i)\| < \sum a_i \epsilon/2 = \epsilon/2$ and thus $\sum a_i \lambda^i$ of $\{\lambda^i\}$ is in the ϵ-neighborhood of d, a contradiction.

27. Since $\{\lambda^i\} = \cap T^{k\nu}(\Delta) = \cap T^k(\Delta) = D$, we have that T is one to one on the simplex $\{\lambda^i\}$ to all of itself. Now $\{\lambda^i\}$ is the convex hull of the finite set of points $\lambda^i, \ldots, \lambda^t$. As such it is a convex polyhedron, and the set of all its geometric vertices* is a subset of the original, say $\lambda^1, \ldots, \lambda^n$, and $\{\lambda^i\}_1^t = \{\lambda^i\}_1^n$. We know that $\{\lambda^i\}_1^n T = \{\lambda^i\}_1^n = \{T(\lambda^i)\}_1^n$. It follows that $T(\lambda^1), T(\lambda^2), \ldots, T(\lambda^n)$ is a permutation of the geometric vertices $\lambda^1, \ldots, \lambda^n$, and hence for some N, $T^N(\lambda^i) = \lambda^i = \lambda^i M^N / \sigma(\lambda^i M^N)$, $i = 1, \ldots, n$ or $\lambda^i M^N = \sigma(\lambda^i M^N) \cdot \lambda^i$. But M^N is a matrix with all elements positive. By the uniqueness results obtained previously, it follows that $n = 1$, and $D = \cap T^k(\Delta) = \{\lambda^1\}$, where $\lambda^1 M^N = \sigma(\lambda^1 M^N) \cdot \lambda^1$. But for M itself we have $\bar{v}M = r\bar{v}$ so $\bar{v}M^N = r^N \bar{v}$ and again by uniqueness, $\lambda^1 = \bar{v}$.

Hence $\lim T^k(v) = \lim vM^k / \sigma(vM^k) = \bar{v}$ for all v with $v_i \geq 0, (v \neq 0)$.

---

* A geometric vertex is the point of the polyhedron not an inner point of any segment $\{x, y\}$ of the polyhedron, P, y, x in P.

VIII. A Theorem on Ratios

1. Let $G_1, \ldots, G_t$ be arbitrary generating functions, each in the variables $x_1, \ldots, x_t$, and define $G = G_1 \ldots G_t$. Define

$$M_a = (\partial G/\partial x_a)_1, \quad T_a = (\partial^2 G/\partial^2 x_a)_1 + M_a$$

$$M_{ia} = (\partial G_i/\partial x_a)_1, \quad T_{ia} = (\partial^2 G_i/\partial x_a^2)_1 + M_{ia} \ .$$

Then $\partial G/\partial x_a = (\partial G_1/\partial x_a)G_2 \ldots G_t$

$$+ G_1(\partial G_2/\partial x_a)G_3 \ldots G_t$$

$$+ \ldots$$

$$+ G_1 G_2 \ldots G_{t-1}(\partial G_t/\partial x_a)$$

and $\partial^2 G/\partial x_a^2 = (\partial^2 G_1/\partial x_a^2)G_2 \ldots G_t$

$$+ (\partial G_1/\partial x_a)(\partial G_2/\partial x_a) \ldots G_t + \ldots$$

$$+ (\partial G_1/\partial x_a) \ldots (\partial G_t/\partial x_a) + \ldots$$

$$+ G_1(\partial G_2/\partial x_a) \ldots (\partial G_t/\partial x_a) + \ldots$$

$$+ G_1 \ldots (\partial^2 G_t/\partial x_a^2).$$

Hence $M_a = M_{1a} + M_{2a} + \ldots + M_{ta}$ and $T_a - M_a = (T_{1a} - M_{1a}) + M_{1a}M_{2a} + \ldots + M_{1a}M_{ta} + M_{1a}M_{2a} + (T_{2a} - M_{2a}) + \ldots + M_{2a}M_{ta} + \ldots + M_{1a}M_{ta} + \ldots M_{2a}M_{ta} + \ldots + (T_{ta} - M_{ta})$.

Define $D_a = T_a - M_a^2$ and $D_{ia} = T_{ia} - M_{ia}^2$ as the various dispersions for particles of type a. Then $D_a = \sum_{i=1}^t M_{ia} + \sum_{i=1}^t (T_{ia} - M_{ia}) + \sum_{i=j} M_{ia}M_{ja} = (\sum_i M_{ia})^2 = \sum_i T_{ia} - \sum_i M_{ia}^2 = \sum_i D_{ia}$.

2. Write $G(x) = \sum P(j_1, \ldots, j_t)x_1^{j_1} \ldots x_t^{j_t}$. Then

$$\sum P(j_1, \ldots, j_t)\left[(j_1 - M_1)^2 = + \ldots + (j_t - M_t)^2\right] =$$

$$(T_1 - M_1^2) + \ldots + (T_t - M_t^2) = \sum D_a = \sum_{i,a} D_{ia} \geq P(j \in S)K^2 \text{ *}$$

---

* $P(j \in S)$ means the probability that $j \in S$, i.e.,
$\sum P(j_1, \ldots, j_t)$ over all $j = (j_1, \ldots, j_t) \in S$.

where S denotes the set of all $j = (j_1, \ldots, J_t)$ such that

$$\sqrt{(j_1 - M_1)^2 + \ldots + (j_t - M_t)^2} \geq K .$$

Hence $P(j \in S) \leq \sum_{i,a} D_{ia}/K^2$.

3. We remark that, if $H(x) = h^n(x)$, where $h(x)$ and hence $H(x)$ is a generating function of $x_1, \ldots, x_t$, and as usual

$$M_a = (\partial H/\partial x_a)_1, \ T_a = (\partial^2 H/\partial x_a^2)_1 + M_a,$$

$$m_a = (\partial h/\partial x_a)_1, \ t_a = (\partial^2 h/\partial x_a^2)_1 + m_a, \text{ then}$$

$$\partial H/\partial x_a^2 = nh^{n-1}(\partial h/\partial x_a) \text{ and}$$

$$\partial^2 H/\partial x_a^2 = n\left[(n-1)h^{n-2}(\partial h/\partial x_a)^2 + h^{n-1}(\partial^2 h/\partial x_a^2)\right] .$$

Thus $M_a = nm_a$ and $T_a = M_a + n(m-1)m_a^2 + n(t_a - m_a)$. Hence $D_a = T_a - M_a^2 = n(t_a - m_a^2) = nd_a$.

4. Now let $G_i = (g_i^{(k)})$, $i = 1, \ldots, t$ for a fixed k, and $G = (g_1^{(k)})^{n_1} \ldots (g_t^{(k)})^{n_t}$ for a fixed $n = (n_1, \ldots, n_t) \neq 0$. Then with reference to this G, $P(j \in S) \leq \sum D_{ia}/K^2$ where

$$S = \left\{ j; \ \sqrt{(j_1 - M_1)^2 + \ldots + (j_t - M_t)^2} \geq K \right\} .$$

But now $M_a = n_1 m_{1a}^{(k)} + n_2 m_{2a}^{(k)} + \ldots + n_t m_{ta}^{(k)}$ and $D_{ia} = n_i d_{ia}^{(k)}$ where $d_{ia}^{(k)} = t_{ia}^{(k)} - (m_{ia}^{(k)})^2$. Thus $P(j \in S) \leq \sum_{ia} n_i d_{ia}^{(k)}/K^2$, where

$$S = \{j; \|j - nM^k\| \geq k\} .$$

We summarize: for every $n = (n_1, \ldots, n_t)$, k, and $K > 0$, in the probability distribution having generating function

$$g_1^{(k)n_1} \ldots g_t^{(k)n_t} = \sum P_{n_1 k}(j_i, \ldots, j_t)x_1^{j_1} \ldots x_t^{jt},$$

it is true that $P_{n,k}(\|j - nM^k\| \geq K) \leq d^{(k)} \sigma(n)/K^2$ where $d^{(k)} = \max_i \{\sum_a d_{ia}^{(k)}\}$.

But then $P_{n,k}\left(\|j/\sigma(nM^k) - T^k(n)\| \geq \epsilon\right) \leq d^{(k)}\sigma(n)/\epsilon^2(\sigma(nM^k))^2$.

Now $\sigma(nM^k) = \sum_{ij} n_i m_{ij}^{(k)} \geq \sum n_i m^{(k)} = m^{(k)}\sigma(n)$ where
$m^{(k)} = \min_i \{\sum_j m_{ij}^{(k)}\}$. Hence $P_{m,k}\left(\|j/\sigma(nM^k) - T^k(n)\| \geq \epsilon\right) \leq$
$d^{(k)}/\epsilon^2 (m^{(k)})^2 \sigma(n) = e^{(k)}/\epsilon^2 \sigma(n)$, where $e^{(k)} = d^{(k)}/(m^{(k)})^2$.

Then for every $n \neq 0, k, \epsilon > 0$, we have for
$$g_1^{(k)n_1} \dots g_t^{(k)n_k} = \sum P_{n,k}(j_1, \dots, j_t) x_1^{j_1} \dots x_t^{j_t}, \text{ that}$$
$$P_{n,k}\left(\|j/\sigma(nM^k) - T^k(n)\| \geq 1/2\epsilon\right) \leq e^k/1/4\epsilon^2 \sigma(n).$$

Now for arbitrary $\epsilon > 0$, $\eta > 0$, there exists a k (hereafter fixed) such that $\|T^k(n) - \bar{v}\| < 1/2\epsilon$ for all $n \neq 0$. For this k, and the original $\epsilon > 0$, $\eta > 0$, we can now fix σ so that for all n with $\sigma(n) > \sigma$, we have $e^{(k)}/1/4\epsilon^2 \cdot \sigma(n) < 1/2\eta$, and hence
$$P_{n,k}\left(\|j/\sigma(nM^k) - T^k(n)\| \geq 1/2\epsilon\right) \leq 1/2\eta.$$

But we have the set inclusion, for each n with $\sigma(n) > \sigma$:
$$\left\{j; \|j/\sigma(nM^k) - T^k(n)\| \geq 1/2\epsilon\right\} \supset \left\{j; \|j/\sigma(nM^k) - \bar{v}\| \geq \epsilon\right\}.$$
Hence $\sigma(n) > n$ implies $P_{n,k}\left(\| j/\sigma(nM^k) - \bar{v}\| \geq \epsilon\right) \leq 1/2\eta$.

For the fixed σ, and the original η, there exists an R so that for all $r > R$, $i = 1, \dots, t$, $\sum_{0 < \sigma(n) \leq \sigma} P_r(i; n_1, \dots, n_t) < 1/2\eta$.
Now for $r > R$, we have $G^{(k+r)} = G^r(G^k)$ so that $g_i^{(k+r)} =$
$g_i^{(r)}(g_1^{(k)}, \dots, g_t^{(k)}) =$
$$p_r(i; 0) + \sum_{1 \leq \sigma(n) \leq \sigma} p_r(i; n) g_1^{(k)n_1} \dots g_t^{(k)n_t} + \sum_{\sigma(n) > \sigma} p_r(i; n) g_1^{(k)n_1} \dots g_t^{(k)n_t},$$
and hence
$$1 \leq p_r(i; 0) + 1/2\eta + 1/2\eta + \sum_{\sigma(n) > \sigma} p_r(i; n) P_{n,k}\left(\|j/\sigma(nM^k) - \bar{v}\| < \epsilon\right).$$

It only remains to relate the last term to the probabilities $p_{k+r}(i; j)$ of the generation $k + r$. Consider the set $J(\bar{v}, \epsilon)$ of all $j = (j_1, \dots, j_t)$ such that $\|j/r - \bar{v}\| < \epsilon$ for some $r > 0$. Since $\|j/r - \bar{v}\| < \epsilon$ if and only if $\|j - r\bar{v}\| < r\epsilon$, this means the set of all j within an angle about $\bar{v}$ of opening arc-cos $\sqrt{1 - \epsilon^2}/\|\bar{v}\|^2$. Since every term $p_r(i; n_1, \dots, n_t) P_{n,k}(j_1, \dots, j_t)$ involved in the above sum represents part of the total probability $p_{k+r}(i; j_1, \dots, j_t)$ for a $j = (j_1, \dots, j_t) \in J(\bar{v}, \epsilon)$, we have

$$\sum_{j \in \bar{J}(\bar{v},\epsilon)} p_{k+r}\left(i; j_1, \ldots, j_t\right) \geq 1 - \eta.$$ Hence we have the

Theorem. For every $\epsilon > 0$, $\eta > 0$, and $i = 1, \ldots, t$, there exists a K such that for all $k \geq K$,

$$\sum p_k\left(i; j_1, \ldots, j_t\right) > 1 - \eta,$$

where the summation is over all vectors j such that $\|j - rv\| < r\epsilon$ for some $r \geq 0$.

This means that, with overwhelming probability the distribution of progeny in high generations among the t types will be essentially in the ratios of the unique characteristic vector of the first moment matrix M of the original generating function G.

II

ABSTRACT

We continue in this report the generalization of the methods and results of Hawkins and Ulam[2] which we began in I, being concerned principally with systems which are below critical. After deriving necessary and sufficient conditions for this state in terms of first moments, we study the direction of flow induced in the "unit cube" by the corresponding generating transformation. The latter results are used to show that the distribution of the generation in which death first occurs possesses moments of all orders.

Limits of expectations are obtained for the problem of subcritical system with source, and for that of total progeny (corpses) in subcritical systems.

Finally, it is shown how fictitious particles of new types may be introduced in such a way that certain more complicated problems may be reduced to the case of simple iteration of generating transformations. In particular we have shown how this may be accomplished for the system with source, and for the problem of total progeny.

The next section, III of this series, will deal with measure theorems on the space of all genealogies possible in multiplicative systems.

I. Some Properties of the Jacobian

1. We consider a multiplicative system involving t types of particles, in which a particle of type i has a fixed probability $p_1(i; j_1, \ldots, j_t)$ of producing a total of $j_1 + \ldots + j_t$ particles, j_ν of type ν, upon transformation. The corresponding generating transformation $G(x)$ of the unit cube I_t:

$$g_i(x) = \sum p_1(i; j_1, \ldots, j_t) x_1^{j_1} \ldots x_t^{j_t}$$

has the property that, upon iteration k times, the resulting transformation $G^k(x)$:

$$g_i^{(k)}(x) = \sum p_k\left(i; j_1, \ldots, j_t\right) x_1^{j_1} \ldots x_t^{j_t}$$

has as its coefficient $p_k(i; j)$ the probability that the state $(j_1, \ldots, j_t)$ shall exist in the k-th generation of progeny from one particle of type i.

2. The transformation $G(x)$ has for its Jacobian at $x = 1$ the first moment matrix

$$J(G(1)) = \left[\partial g_i / \partial x_j\right]_1 = \left[m_{ij}^{(1)}\right] = \left[m_{ij}\right],$$

and, more generally,

$$J\left(G^k(1)\right) = \left[\partial g_i^{(k)} / \partial x_j\right]_1 = \left[m_{ij}^{(k)}\right],$$

where $m_{ij}^{(k)}$ is the expectation of particles of type j in the k-th generation of progeny from one particle of type i. Under our assumptions on G (see I), all these moments are positive. Moreover, we have seen that $\left[m_{ij}^{(k)}\right] = [m_{ij}]^k$.

3. The importance of matrices with positive elements required study of their properties. We found that for such a matrix M, there is one and only one solution, $r, \bar{v}$ of the relations

$$\bar{v}M = r\bar{v}, \; r > 0, \; \bar{v} > 0 \; (i.e., \; \text{all } \bar{v}_i > 0) .$$

In similar fashion one shows the existence and uniqueness of r', v' such that

$$Mv' = r'v', \; r' > 0, \; v' > 0 .$$

That $r = r'$ in these equations is evident from the following

Lemma. If $\bar{v}M = r\bar{v}$, $r > 0$, $\bar{v} > 0$, and R is an arbitrary positive characteristic root of M, then $r \geq R$. Similarly, $Mv' = r'v'$, $r' > 0$, $v' > 0$, $|M - RI| = 0$, $R > 0$ implies $r' \geq R$.

Proof. First statement: Let $wM = Rw$, $w \neq 0$, and define $b = \min\{w_i/v_i\}$, $B = \max\{w_i/v_i\}$. Since all $v_i > 0$, $bv_i \leq w_i \leq Bv_i$ and hence, $b\sum v_i m_{ij}^{(k)} \leq \sum v_i m_{ij}^{(k)} \leq B\sum v_i m_{ij}^{(k)}$ and $br^k v_j \leq R^k w_j \leq Br^k v_j$, all j. Since $r > 0$, also $bv_j \leq (R/r)^k w_j \leq Bv_j$. Suppose $R/r > 1$. If at least one $w_j > 0$ the right member yields a contradiction for large k. If at least one $w_j < 0$ then the left member does. Hence all $w_j = 0$, contradicting choice of w. Thus $R/r \leq 1$.

The second statement of the lemma is proved similarly.

We include for later use the trivial remark: If M is a matrix with positive elements, and $Mv' = rv'$, $r > 0$, $v' > 0$, we have $M^n v' = r^n v'$ for all positive integers n, and thus $r^n v_i' = \sum_j m_{ij}^{(n)} v_j' \geq \sum_j m_{ij}^{(n)} \min(v_j')$. Hence $\sum_j m_{ij}^{(n)} \leq r^n v_i'/\min(v_j') \equiv r^n V_i \leq r^n \max(v_i) \equiv Vr^n$, where V is a positive constant.

Similarly, if $\bar{v}M = r\bar{v}$, $r > 0$, $\bar{v} > 0$, we have $\sum_i m_{ij}^{(n)} \leq Wr^n$, where W is a positive constant.

4. We shall also need the following

Theorem. If M is a matrix of positive elements with $Mv' = rv'$, $r > 0$, $v' > 0$, and $T^k(v)$ is the transformation $M^k v/s(M^k v)$, then $\lim T^k(v) - v'$ uniformly for all $v \neq 0$, $v_i \geq 0$.

Here $s(w)$ indicates the sum $\sum w_i$ of the components of the vector w. The proof is entirely analogous to that used in I to prove the same result for row vectors.

5. We proved in I that $\lim G^k(0) = x^0$ exists:

$$\lim g_i^{(k)}(0) = \lim p_k(i; 0) = x_i^0,$$

and defined G as supercritical in case all $x_i^0 < 1$. Under our conditions on G, the alternative case is that all $x_i^0 = 1$, hence $x^0 = 1$, and this case we called subcritical. Most of the present report will be devoted to systems of the latter type, in which the probability of death in generation k rises to limit one.

6. We have seen in I that a system is subcritical if and only if the maximal root r of the first moment matrix $M = J(G(1))$ is less than or equal to one. Equivalently G is subcritical if and only if the determinants $|A_n|$ of the upper principal minors $A_n = [m_{ij} - \partial_{ij}]_1^n$ of the matrix $A = [m_{ij} - \partial_{ij}]_1^t = M - I$ satisfy the relations:

$$(-1)^{n+1}|A_n| < 0, \ n = 1, \ldots, t - 1 \ ,$$

$$(-1)^{t+1}|M - I| \leq 0 \ .$$

7. We say a system is just-critical in case it is subcritical and the maximum positive characteristic root r of M is equal to one. A subcritical system with $r < 1$ is said to be below-critical. The just-critical case, while of theoretical interest is refractory, and we have limited ourselves for the most part to systems which are below-critical.

Theorem. A subcritical system is just-critical if and only if $|M - I| = 0$.

If $r = 1$, then of course $0 = |M - rI| = |M - I|$, r being a characteristic root of M. Conversely, if $|M - I| = 0$, $r \geq 1$ by maximality of r, and $r \leq 1$ by assumption of subcriticality; hence $r = 1$.

8. We include for future reference the trivial remark: If M is a matrix of positive elements with $Mv' = rv'$, $1 > r > 0$, $v' > 0$, then there exists an $e > 0$ such that all vectors w in the e-neighborhood of v' are positive and satisfy the inequalities

$$\sum m_{ij} \, w_j \leq \frac{1}{2}(1 + r)w_i < w_i \ .$$

It suffices to note that the functions $R_i(w) \equiv \sum m_{ij} \, w_j / w_i$ are continuous at $w = v'$ and there have value $R_i(v') = r < 1$. Hence there is an $e > 0$ so that whenever $\|w - v'\| \leq e$, w will have positive components and $R_i(w) \leq 1/2(1 + r)$.

II. Direction of Flow of $G^k(x)$ in Subcritical Systems

1. In this section we study properties of the vector $1 - G^K(\bar{x})$ with i-th component $1 - g_i^{(k)}(\bar{x})$, for $\bar{x} \neq 1$ in I_t, for a subcritical system, and of the vector $G^{k+1}(\bar{x}) = G^k(\bar{x})$ in a system below-critical. These results are of preliminary character, and are exploited in III.

Note that, for $\bar{x} \neq 1$ in I_t, we have $\bar{x} \not\geq 1$ and thence $g_i^{(k)}(\bar{x}) < g_i^{(k)}(1) = 1$ for all k so that the vector $1 - G^k(x)$ is not zero; in fact all its components are positive.

2. Recall that the e-cone of the vector v' consists of all vectors w such that $\|w/a - v'\| < e$ for some real positive a. We prove the

Theorem. If G is subcritical and $\bar{x} \neq 1$ is in I_t, then for every $e > 0$ there is a K so that for all $k \geq K$ the vector $1 - G^k(\bar{x})$ is in the e-cone of v'.

Here v' denotes the characteristic vector of the relation $Mv' = rv'$, $r > 0$, $v' > 0$, where M is the first moment matrix of G. The theorem asserts, in geometric terms, that the direction from $G^k(\bar{x})$ to 1 approaches the direction of v' with increasing k.

Proof. Fix $\bar{x} \neq 1$ in I_t, and $c > 0$. There exists a k (hereafter fixed) such that $\|T^k(v) - v'\| < e/2\sqrt{t}$ for all $v \neq 0$, $v_i \geq 0$. (The transformation T^k is that defined in I 4.) Since $\lim G^n(\bar{x}) = 1$ and the first partials of G^k are continuous at 1, there exists an N such that for, all $n \geq N$,

$$m_{ij}^{(k)} = \left(\partial g_i^{(k)} / \partial x_j \right) < e M_k / 2\sqrt{t} \quad (*)$$

where $M_k \equiv \min \left\{ \sum_i m_{ij}^{(k)}; j \right\}$, and the partial is evaluated at the point $G^n(\bar{x})$.

Fix $n \geq N$ and define $a_n^k = \sum m_{ij}^{(k)} \left(1 - g_j^{(n)}(\bar{x}) \right) = s \left(M^k \left(1 - G^n(\bar{x}) \right) \right)$. Then $\left| v_i' - \left(1 - g_i^{(k+n)}(\bar{x}) \right) / a_n^k \right| = \left| v_i' - \sum \left(\partial g_i^{(k)} / \partial x_j \right)_P \left(1 - g_j^{(n)}(\bar{x}) \right) / a_n^k \right|$ from Taylor's form of $g_i^{(k)}(x)$ expanded about 1, and evaluated at $x = G^n(\bar{x})$. Thus $G^n(\bar{x}) \prec P \prec 1$.

The latter absolute value does not exceed

$$\left| v_i' - \sum m_{ij}^{(k)} \left(1 - g_j^{(n)} (\bar{x}) \right) / a_n^k \right| +$$

$$\left| \sum m_{ij}^{(k)} - \left(\partial g_i^{(k)} / \partial x_j \right)_P \left(1 - g_j^{(n)} (\bar{x}) \right) / a_n^k \right| \leq$$

$$\left\| v' - T^{(k)} \left(1 - G^{(n)}(\bar{x}) \right) \right\| + \sum \left(m_{ij}^{(k)} - \left(\partial g_i^{(k)} / \partial x_j \right)_{G^{(n)}(\bar{x})} \right)$$

$$\left(1 - g_j^{(n)}(\bar{x}) \right) / a_n^k < e/2\sqrt{t} + eM_k/2\sqrt{t} \cdot s \left(1 - C^{(n)}(\bar{x}) \right) / a_n^k.$$

But $a_n^k \geq M_k s \left(1 - C^{(n)}(\bar{x}) \right)$ so finally the original absolute value is seen to be less than $e/\sqrt{t}$. But then $\left\| v' - \left(1 - G^{k+n}(\bar{x}) \right) / a_n^k \right\| < e$.

3. In the below-critical case in one variable ($t = 1$) the graph of the generating function $g(x)$ is monotone increasing and concave up on the interval $(0, 1)$ with $1 = g(1)$. Moreover, the Jacobian $J(g(1)) = g'(1)$ and hence the characteristic root r is $g'(1)$, which is therefore less than one when g is below-critical. It is obvious geometrically therefore that for every x satisfying $0 \leq x < 1$, the sequence of iterates $g^k(x)$ is monotone increasing: $x < g(x) < g^2(x) < \cdots$.

This simple situation need not obtain in case $t \geq 2$. For purposes of illustration we regard the following example.

Consider the transformation $G(x)$ of the unit square I_2 defined by

$$g_1(x) = \frac{1}{2} + \frac{1}{4}x_1 + \frac{1}{4}x_1 x_2$$

$$g_2(x) = \frac{1}{2} + \frac{1}{4}x_2 + \frac{1}{4}x_1 x_2 .$$

Computation of the four first partials at 1 shows that the first moment matrix is

$$M = \begin{bmatrix} \frac{1}{2} & \frac{1}{4} \\ \frac{1}{4} & \frac{1}{2} \end{bmatrix} \qquad \text{and thus } M - I = \begin{bmatrix} -\frac{1}{2} & \frac{1}{4} \\ \frac{1}{4} & -\frac{1}{2} \end{bmatrix}.$$

The upper minor determinants of the latter satisfy

$$(-1)^2 |A_1| = -\frac{1}{2} < 0, \ (-1)^3 |A_2| = -\frac{3}{16} < 0 .$$

Hence G is below-critical. Indeed the characteristic equation of M is $x^2 - x + 3/16 = 0$ with roots $1/4$ and $3/4$. Thus $r = 3/4 < 1$.

(The right hand characteristic vector v' is found to be $\begin{bmatrix} \frac{1}{2} \\ \frac{1}{2} \end{bmatrix}$ upon setting $r = 3/4$ in the matrix-vector equation $(M - rI)v' = 0$ and solving the resulting two homogeneous equations in v_1', v_2'.)

Note that (see I), M being subcritical, we must have $\sum m_{ij}\, v_j \le v_i$ for at least one $i = 1, 2$, whenever $v \ne 0$, $v_i \ge 0$, but not necessarily for both indices. In our case, for example,

$$\begin{bmatrix} \frac{1}{2} & \frac{1}{4} \\ \frac{1}{4} & \frac{1}{2} \end{bmatrix} \begin{bmatrix} 2 \\ 8 \end{bmatrix} = \begin{bmatrix} 3 \\ \frac{9}{2} \end{bmatrix} \qquad \text{where } \tfrac{9}{2} < 8 \text{ but } 3 > 2.$$

It is essentially this fact that causes the simplicity of the one variable case to break down. Specifically, let $x = (4/5, 1/5)$. Then our $G(x)$ at this point is $(74/100, 59/100)$ and $x \prec G(x)$ is false for this x.

We can however prove the following

Lemma. If G is below-critical, and $\bar{x} \ne 1$ is in I_t, the k-sequences $g_i^{(k)}(\bar{x})$ are *eventually* monotone increasing.

Proof. By I 8, there is an $e > 0$ such that $\|w - v'\| < e$ implies w positive and $\sum m_{ij}\, w_j < w_i$, all i.

By the preceding theorem, this e determines a K so that, for all $k \ge K$, $\| (1 - G^k(\bar{x}))/a_k - v'\| < e$. For all such k therefore,

$$\sum m_{ij} \bigl(1 - g_j^{(k)}(\bar{x})\bigr)/a_k < \bigl(1 - g_i^{(k)}(\bar{x})\bigr)/a_k$$

and the positive a_k may be deleted from this inequality. Under our assumptions on G, $g_i(x) > 1 + \sum m_{ij}(x_j - 1)$ for every x with all components less than one, and thus $1 - g_i(x) < \sum m_{ij}(1 - x_j)$. In this inequality we may set $x = G^k(\bar{x})$, since we have already shown that the latter enjoys this property of components (cf. II 1).

Thus $1 - g_i^{(k+1)}(\bar{x}) < \sum m_{ij}\bigl(1 - g_j^{(k)}(x)\bigr)$, and combining with the previous inequality, $1 - g_i^{(k+1)}(\bar{x}) < 1 - g_i^{(k)}(\bar{x})$, whence the result desired.

4. We have seen that the direction of the vector $1 - G^k(x)$ approaches that of v'. We intend to prove the same result for the "vector of flow" $G^{k+1}(\bar{x}) - G^k(\bar{x})$, with $\bar{x} \ne 1$ in I_t.

It is trivial that the latter vector is never zero for any k, and hence defines a direction. For otherwise, G would have a fixed point $G^k(\bar{x}) \not\geq 1$. Moreover, we know from the preceding result that for all $k \geq K$, all components of this vector are positive.

Theorem. If G is below-critical and $\bar{x} \neq 1$ is in I_t, the direction from $G^k(\bar{x})$ to $G^{k+1}(\bar{x})$ approaches that of v'.

Proof is entirely analogous to that for $1 - G^k(\bar{x})$. Fix $e > 0$ and $\bar{x} \neq 1$ in I_t. Then we may fix k so $\|T^k(v) - v'\|e/2\sqrt{t}$ for all $v \neq 0$, $v_1 \geq 0$, and next determine N so $n \geq N$ implies the inequality (*) of II 2, *and* $g_i^{(n+1)}(\bar{x}) > g_i^{(n)}(\bar{x})$.

Now define $A_n^k = \sum_{ij} m_{ij}^{(k)} \left(g_j^{(k+n+1)}(\bar{x}) - g_j^{(k+n)}(\bar{x}) \right) \geq$

$M_k s \left(G^{k+n+1}(\bar{x}) - G^{k+n}(\bar{x}) \right)$. Then $\left| v_i' - (g_i^{(k+n+1)}(\bar{x}) - g_i^{(k+n)}(\bar{x}))/A_n^k \right| =$

$\left| v_i' - \sum (\partial g_i^{(k)}/\partial x_j)_P (g_j^{(n+1)}(\bar{x}) - g_j^{(n)}(\bar{x}))/A_n^k \right|$

by Taylor's theorem where $G^n(\bar{x}) \prec P \prec G^{n+1}(\bar{x})$.

The remainder of the proof now proceeds just as in II 2. The essential point is that the T^k operates now on the vector $G^{k+n+1}(\bar{x}) - G^{k+n}(\bar{x})$ which we know by II 3 to be positive and hence subject to the inequality

$$\left\| v' - T^k \left(G^{n+1}(\bar{x}) - G^n(\bar{x}) \right) \right\| < e/2\sqrt{t}.$$

Thus the final result is

$$\left\| v' - \left(G^{k+1+n}(\bar{x}) - G^{k+n}(\bar{x}) \right)/A_n^k \right\| < e \text{ for all } n \geq N.$$

5. We now have immediately the main result which we want in III.

Theorem. If G is below-critical, and its first moment matrix satisfies the relation $Mv' = rv'$, $1 > r > 0$, $v' > 0$, and further, if $\bar{x} \neq 1$ is in I_t, then there is a K such that, for all $k \geq K$, the ratio of successive terms of the K-sequence $g_i^{(k+1)}(\bar{x}) - g_i^{(k)}(\bar{x})$ is less than $1/2(1 + r)$.

From I 8, there is an $e > 0$ such that $\|w - v'\| < e$ implies w positive and $\sum m_{ij} w_j < 1/2(1 + r)w_i$. This e determines K by the preceding theorem so that $k \geq K$ implies

$$\left\| \frac{G^{k+1}(\bar{x}) - G^k(\bar{x})}{A_k} - v' \right\| < e .$$

Hence

$$\sum m_{ij}\left(g_j^{(k+1)}(\bar{x}) - g_j^{(k)}(\bar{x})\right) < \tfrac{1}{2}(1+r)\left(g_i^{(k+1)}(\bar{x}) - g_i^{(k)}(\bar{x})\right)$$

since the positive A_k may be deleted. But

$$g_i^{(k+2)}(\bar{x}) - g_i^{(k+1)}(\bar{x}) = g_i\left(G^{k+1}(\bar{x})\right) - g_i\left(G^k(\bar{x})\right) =$$

$$\sum (\partial g_i/\partial x_j)_P\left(g_j^{(k+1)}(\bar{x}) - g_j^{(k)}(\bar{x})\right) \leq \sum m_{ij}\left(g_j^{(k+1)}(\bar{x}) - g_j^{(k)}(\bar{x})\right) .$$

Hence, combining, we have the desired result.

III. On the Distribution of Death in Subcritical Systems

1. Let G be subcritical, and define $q_k(i)$ as the probability that complete death of the system of progeny from one particle of type i should first occur in the k-th generation.

Clearly $p_k(i; 0)$, the probability of death in the k-th generation, may be expressed as

$$p_k(i; 0) = \sum_{j-1}^{k} q_j(i) .$$

Hence we have the relations:

$$q_k(i) = p_k(i; 0) - p_{k-1}(i; 0) = g_i^{(k)}(0) - g_i^{(k-1)}(0), \ k \geq 2 ,$$

$$q_1(i) = p_1(i; 0) = g_i(0) .$$

2. We recall that $x \ngeq x'$ implies $g_i(x) < g_i(x')$ under our assumptions on G. Now $0 \ngeq G(0)$, otherwise $G(0) = 0$ and G would have a fixed point in I_t besides 1 and would be supercritical. Hence, inductively

$$g_i(0) < g_i^{(2)}(0) < g_i^{(3)}(0) < \cdots .$$

It follows that the $q_k(i)$ are positive for $k \geq 2$, and $q_1(i) = g_i(0)$ is positive for at least one i. Also, $\sum_1^\infty q_j(i) = \lim_k \sum_1^k q_j(i) = \lim_k p_k(i;0) = 1$ for every i, and thus the sequence $q_1(i)$, $q_2(i)$, $q_3(i),\ldots$ is a probability density function on the positive integers.

3. Theorem. If G is below-critical, its first moment matrix M having maximal positive root $r < 1$, there is a K so that for all $k \geq K$, $q_{k+1}(i)/q_k(i) < 1/2(1+r)$. Consequently the density sequence $q_k(i)$ is eventually monotone decreasing, and all its moments

$$m_s(i) \equiv \sum_k k^s\, q_k\,(i), \quad s \geq 0$$

exist.

The first statement is an immediate consequence of II 5, with $\bar{x} = 0$. The finiteness of the moments follows from the ratio test:

$$(k+1)^s\, q_{k+1}/k^s q_k < (1+1/k)^s \tfrac{1}{2}(1+r) \longrightarrow \tfrac{1}{2}(1+r) < 1 \ .$$

4. In the case of a system below-critical in one variable ($t = 1$), it is geometrically obvious that $q_{k+1}/q_k = g^{k+1}(0) - g^k(0)/g^k(0) - g^{k-1}(0) = g(x') - g(x)/x' - x < g'(1) = r < 1$ for *all* k, so $q_k < q_1 r^{k-1}$ and
$m_1 = \sum k q_k < q_1\left(1 + 2r + 3r^2 + \ldots\right) = q_1\left(1 + x + x^2 + \ldots\right)'_{x=r} = q_1\left((1-x)^{-1}\right)'_{x=r} = q_1/(1+r)^2 \ .$
Hence in this case $m_1 \leq p(0)/(1+r)^2 = g(0)/(1+g'(1))^2$.

5. Examples show that, even in the one variable case if the system is just-critical, even m_1 may be infinite. We hope to study the one variable case more completely in a separate report.

IV. Subcritical System with Source

1. Consider t types of particles whose probabilities of transmutation are given by the generating transformation $G(x)$:

$$g_i(x) = \sum p_1(i; j_1, \ldots, j_t) x_1^{j_1} \ldots x_t^{j_t}$$

as before. Suppose further that we have a source which emits independently into the system $n_1 + \ldots + n_t$ particles, n_i of type i, with probability $s(n_1, \ldots, n_t) \geq 0$. We associate with the source the generating *function*

$$S(x) = \sum s\,(n_1, \ldots, n_t)\, x_1^{n_1} \ldots x_t^{n_t}, \quad S(1) = 1 \ .$$

Consider a process consisting of the following steps:

1. The source produces an initial set of $n_1 + \ldots + n_t$ particles, n_i of type i, with probability $s(n_1, \ldots, n_t)$. These particles transmute according to the G law to form a system which we regard as the first population.

2. The source again contributes new particles, and these together with the first population transform according to the G law to form the second population, and so on.

At the k-th step, the population $(m_1, \ldots, m_t)$, m_i of type i, will occur with some probability $h_k(m_1, \ldots, m_t)$, and we define the corresponding generating *function* $H_k(x) = \sum h_k(m_1, \ldots, m_t) x_1^{m_1} \ldots x_t^{m_t}$.

Now, from the elementary laws of probability, as we have pointed out in I, transmutation of any population with generating function $N(x)$ according to the G law gives a population with generating function $N(G(x)) = N(g_1(x), \ldots, g_t(x))$. Hence for the problem considered above, we see that

$$H_1(x) = S(G(x)),$$

$$H_2(x) = S(G(x)) \cdot S(G^2(x)), \ldots,$$

and, generally

$$H_k(x) = S(G(x)) \cdot H_{k-1}(G(x))$$

$$= S(G(x)) \cdot S(G^2(x)) \ldots S(G^k(x)).$$

For, if $H_{k-1}(x)$ is the generating function for the $(k-1)$st population, then the generating function for the intermediate population resulting from the contribution of the source to the $(k-1)$st is

$$T_k(x) = S(x) \cdot H_{k-1}(x),$$

and, upon transformation of this result by the G law, we must have for the generating function for the k-th population:

$$H_k(x) = T_k(G(x)) = S(G(x)) \cdot H_{k-1}(G(x)).$$

2. From $H_k(x) = \prod_1^k S\left(G^i(x)\right)$ follows

$$H_k(x) = H_{k-1}(x) \cdot S\left(G^k(x)\right) .$$

Since, for x in I_t, the latter factor is less than or equal to one, the k-sequence $H_k(x)$ is monotone non-increasing at x and $H(x) \equiv \lim H_k(x)$ exists in I_t. Clearly $0 \le H(x) \le 1$ and $H(1) = 1$.

If S satisfies the conditions

(S^*) at least one $\partial S/\partial x_j > 0$ for all $x \ne 0$ on I_t, then for every x' with all components less than one, we have $S(x') < 1$. For $S(x') = 1 + \sum (\partial S/\partial x_j)_P (x'_j - 1)$, where $0 \ngeq P$.

But for every $\bar{x} \ne 1$ in I_t, we have seen $x'_i \equiv g_i^{(k)}(\bar{x}) < 1$ for all i,k. Hence, for such $\bar{x}$, $S(x') = S\left(G^k(\bar{x})\right) < 1$ and $1 > H_1(\bar{x}) > H_2(\bar{x}) > \ldots$ so $H(\bar{x}) < 1$. Thus we have

Theorem. The function $H(x) \equiv \lim H_k(x)$ exists for all x in I_t, and $0 \le H(x) \le 1$, $H(1) = 1$. Moreover, $H(x)$ satisfies the functional equation

$$H(x) = S\left(G(x)\right) \cdot H\left(G(x)\right).$$

If the source function S satisfies condition (S^*), then $H(x)$ is not identically one on I_t, indeed, for every $\bar{x} \ne 1$ in I_t, $H(\bar{x}) < 1$.

3. If G is supercritical: $x^0 \ngeq 1$, then

$$H_k\left(x^0\right) = \prod S\left(G^i(x^0)\right) = \left(S(x^0)\right)^k \to 0, \text{ so } H\left(x^0\right) = 0.$$

Moreover, it is easy to see that $H(\bar{x}) \equiv 0$ for all $\bar{x} \ne 1$ in I_t. For $\lim G^k(\bar{x}) = x^0(\bar{x} \ne 1)$ and hence $S\left(G^k(\bar{x})\right)$ is bounded from 1. Thus only the subcritical case is of interest.

Theorem. If G is below-critical, and S is a polynomial of degree s, then $H(x)$ is continuous.

Proof. We have $H_{n-1}(x) - H_n(x) =$

$$S\left(G(x)\right)\ldots S\left(G^{n-1}(x)\right) \cdot \left(1 - S\left(G^n(x)\right)\right) \le 1 - S\left(G^n(x)\right) \le 1 - S\left(G^n(0)\right) .$$

Now, by Taylor's form, $g_i^n(0) \geq 1 + \sum_j m_{ij}^{(n)} (0-1) = 1 - \sum_j m_{ij}^{(n)} \geq 1 - Vr^n$. Since $r < 1$, the latter is positive for all n sufficiently large. Now $1 - S(G^n(0)) \leq 1 - \sum s(j_1, \ldots, j_t)(1 - Vr^n)^{s(j)} \leq 1 - \sum s(j_1, \ldots, j_t)$ $(1 - Vr^n)^s = 1 - (1 - Vr^n)^s \leq 1 - (1 - sVr^n) \leq sVr^n$. (See Appendix B). It follows that the sequence $H_n(x)$ is uniformly convergent to $H(x)$ on I_t, and hence the latter is continuous on this range (see Appendix A).

4. Since $H_n(x) = S(G) \ldots S(G^n) = H_{n-1}S(G^n)$, we have $\partial H_n/\partial x_j = \partial H_{n-1}/\partial x_j \cdot S(G^n) + H_{n-1}\partial(S(G^n))/\partial x_j$ and $|\partial H_{n-1}/\partial x_j - \partial H_n/\partial x_j|$ $\leq \partial H_{n-1}/\partial x_j \cdot (1 - S(G^n)) + H_{n-1}\partial(S(G^n))/\partial x_j$.

Now $\partial H_{n-1}/\partial x_j \leq \partial S(G)/\partial x_j + \ldots + \partial S(G^{n-1})/\partial x_j$ and $\partial S(G^n)/\partial x_j$ $= \sum(\partial S/\partial x_i)_{G^n} (\partial g_i^{(n)}/\partial x_j) \leq \sum(\partial S/\partial x_i)_1 m_{ij}^{(n)} \leq T\sum_i m_{ij}^{(n)} \leq TWr^n$.

Thus the above absolute value does not exceed $TW(r + \ldots + r^{n-1})sVr^n$ $+ TWr^n \leq Kr^n(1 + r + \ldots + r^{n-1}) < Kr^n/(1-r)$. Hence by Appendix A, the n-sequences $\partial H_n/\partial x_j$ are uniformly convergent on I_t, the partials $\partial H/\partial x_j$ exist, and $\lim \partial H_n/\partial x_j = \partial H/\partial x_j$ in I_t.

But $(\partial H_n/\partial x_j)_1$ is the expectation of particles of type j in generation n, and the limit approached by this expectation is $(\partial H/\partial x_j)_1$.

Since we know H satisfies the functional equation $H = S(G)H(G)$, we have $\partial H/\partial x_k = \sum \partial S/\partial x_j \cdot \partial g_j/\partial x_k H(G) + S(G)\sum \partial H/\partial x_j \partial g_j/\partial x_k$. Setting $x = 1$, we obtain

$$\sum_j (m_{jk} - \delta_{jk})(\partial H/\partial x_j)_1 = -\sum(\partial S/\partial x_j)_1 \cdot m_{jk}, \quad k = 1, \ldots, t.$$

Since G is below-critical, the determinant of the system is not zero and the expectation limits are uniquely determined. Thus follows the

Theorem. If G is below critical and the source function S is a polynomial, the limit function H possesses first partials on I_t, and the limit of the expectation of particles of type j in population n is the value at $x = 1$ of $\partial H/\partial x_j$. Moreover the latter limits are uniquely determined by the linear system

$$\sum (m_{jk} - \delta_{jk})(\partial H/\partial x_j)_1 = -\sum m_{jk}(\partial S/\partial x_j)_1$$

with non-vanishing determinant $|M - I|$.

V. Total Progeny for Systems Without Source

1. Returning to the simple problem without source, let $P_k\,(i; j_1, \ldots, j_t)$ be the probability that in the *total* progeny in all generations 1 through k produced by one particle of type i (generation 0), there should be j_1 particles of type $1, \ldots, j_t$ particles of type t. Define $c_i^{(k)}(x) = \sum P_k\,(i; j_1, \ldots, j_t)\,x_1^{j_1} \ldots x_t^{j_t}$ and $C^{(k)}(x)$ as the corresponding transformation of I_t. Here the upper k does *not* indicate iteration.

Clearly $P_1(i; j) = p_1(i; j)$, hence $c_i^{(1)}(x) = g_i(x)$ and $C^{(1)} = G(x)$.

Now let k be greater than 1. The production of the total state $J_1, \ldots, J_t$ at the end of the k-th generation from one particle of type i arises from the mutually exclusive states

$$j_1, \ldots, j_t; \ 0 \le j_h \le J_h$$

in the first generation. If this state is $0, \ldots, 0$, then and only then will the total state $J_1, \ldots, J_t$ be $0, \ldots, 0$, so $P_k(i; 0) = p_1(i; 0)$.

Suppose then that state J is not 0, and hence state $j \ne 0$. Each of the j_h particles of type h in the first generations acts independently of the others, and of those of other types to produce in the $k - 1$ next generations a total state of some $a_1, \ldots, a_t$ particles with probability $P_{k-1}\,(h; a_1, \ldots, a_t)$. We want the total state from the $j_1, \ldots, j_t$ particles of the first generation to be $J_1 - j_1, \ldots, J_t - j_t$ after the next $k - 1$ generations. It follows from the elementary laws of probability that, for $J \ne 0$,

$$P_k(i; J) = \sum_{\substack{0 \le j_k \le J_k \\ j \ne 0}} p_1(i; j) \sum_{\sum a_i = J_i - j_i} \prod^{j_i} P_{k-1}\,(1; a_1 \ldots a_t) \ldots \prod^{j_t} P_{k-1}\,(t; a_1, \ldots, a_t) \, .$$

But this is the coefficient of $x_1^{J_1} \ldots x_t^{J_t}$ in

$$g_i\big(x_1 c_1^{(k-1)}(x), \ldots, x_t c_t^{(k-1)}(x)\big) =$$

$$\sum p_1(i; j) x_1^{j_1} \ldots x_t^{j_t} \cdot \Big[\sum P_{k-1}\,(1; a) x^a\Big]^{j_1} \ldots \Big[\sum P_{k-1}(t; a) x^a\Big]^{j_t} \, ,$$

and $P_{k-1}(i; 0) = p_1(i; 0)$, which is the constant term of the above function. Hence the

Theorem. The generating transformation $C^{(k)}(x)$ for the total progeny in generations 1 through k satisfies the recursive relations

$$c_i^{(1)}(x) = g_i(x)$$

$$c_i^{(k)}(x) = g_i\left(x_1 c_1^{(k-1)}(x), \ldots, x_t c_t^{(k-1)}(x)\right).$$

2. If y, z are arbitrary points of I_t, we have from Taylor's form, $g_i(y) = g_i(z) + \sum (\partial g_i / \partial x_j)_P (y_j - z_j)$ and thus $|g_i(y) - g_i(z)| \leq \sum m_{ij} |y_j - z_j|$. It can be said therefore that the number $d_i^k \equiv |c_i^k(x) - c_i^{k-1}(x)| \leq \sum m_{ij} |x_j c_j^{k-1} - x_j c_j^{k-2}| \leq \sum m_{ij} d_j^{k-1}$, since x is in I_t. Iteration of this inequality yields $d_i^k \leq \sum m_{ij}^{(k-2)} d_j^{(2)}$, and eventually $d_i^k \leq \sum m_{ij}^{(k-2)} d_j^{(2)} = \sum m_{ij}^{k-2} |g_j (xG(x) - g_j(x))| \leq \sum m_{ij}^{k-2} \sum m_{jn}$ $|x_n g_n(x) - x_n| = \sum m_{in}^{(k-1)} x_n |g_n(x) - 1| \leq \sum_n m_{in}^{(k-1)} \leq V r^{k-1}.$
Thus we have the

Theorem. If G is below critical, the generating functions $c_i^k(x)$ are uniformly convergent to continuous limit functions $c_i(x)$ on I_t. The latter satisfy the functional equations $c_i(x) = g_i(x_1 c_1(x), \ldots, x_t c_t(x))$.

3. We seek now a dominating sequence for $D_{ij}^k \equiv |\partial c_i^k / \partial x_j - \partial c_i^{k-1} / \partial x_j|$.

First, note that $P_{ij}^k \equiv \partial c_i^k / \partial x_j = \sum (\partial g_i / \partial x_n)_p [\delta_{nj} c_n^{k-1} + x_n \partial c_n^{k-1} / \partial x_j] \leq m_{ij} + \sum m_{in} P_{nj}^{k-1}$, where $P = xC^{k-1}$. Iteration leads to $P_{ij}^k \leq m_{ij} + m_{ij}^{(2)} + \sum m_{in}^{(2)} P_{nj}^{k-2}$ and eventually $P_{ij}^k \leq m_{ij} + m_{ij}^{(2)} + \ldots + m_{ij}^{(k-1)} + \sum m_{in}^{(k-1)} (\partial g_n / \partial x_j) \leq m_{ij} + m_{ij}^{(2)} + \ldots + m_{ij}^{(k)}.$

For brevity, let A_n^k and B_n^k denote temporarily the round and square brackets involved in the P_{ij}^k sum above. Then

$$D_{ij}^k = |P_{ij}^k - P_{ij}^{k-1}| = \left| \sum A_n^k B_n^k - \sum A_n^{k-1} B_n^{k-1} \right| \leq$$

$$\sum A_n^k |B_n^k - B_n^{k-1}| + \sum B_n^{k-1} |A_n^k - A_n^{k-1}| \leq$$

$$\sum m_{in} |B_n^k - B_n^{k-1}| + \sum B_n^{k-1} |A_n^k - A_n^{k-1}|.$$

We obtain upper bounds for the three A, B expressions:

(1) $\left|B_n^k - B_n^{k-1}\right| = \left|\delta_{nj}\left(c_n^{k-1} - c_n^{k-2}\right) + x_n\left(\partial c_n^{k-1}/\partial x_j - \partial c_n^{k-2}/\partial x_j\right)\right| \leq$
$\delta_{nj}d_n^{k-1} + D_{nj}^{k-1} \leq \delta_{nj}Vr^{k-2} + D_{nj}^{k-1}$.

(2) $B_n^{k-1} \leq \delta_{nj} + \partial c_n^{k-2}/\partial x_j \leq \delta_{nj} + \left(m_{nj} + \ldots + m_{nj}^{(k-2)}\right)$.

(3) $\left|A_n^k - A_n^{k-1}\right| = \left|(\partial g_i/\partial x_n)_{xc^{k-1}} - (\partial g_i/\partial x_n)_{xc^{k-2}}\right| \leq$
$\sum \partial g_i/\partial x_n \partial x_p \left|x_p c_p^{k-1} - x_p c_p^{k-2}\right| \leq B\sum d_p^{k-1} \leq BtVr^{k-2}$.

Hence, combining, $D_{ij}^k \leq \sum \min\left(\delta_{nj}Vr^{k-2} + D_{nj}^{k-1}\right) +$
$\sum BtVr^{k-2}\left(\delta_{nj} + m_{nj} + \ldots + m_{nj}^{(k-2)}\right) \leq$
$m_{ij}Vr^{k-2} + \sum \min D_{nj}^{k-1} + BtVr^{k-2} + BtVr^{k-2}\sum_n\left(m_{nj} + \ldots + m_{nj}^{(k-2)}\right) \leq$
$m_{ij}Vr^{k-2} + BtVr^{k-2} + BtVr^{k-2}\left(Wr + \ldots + Wr^{k-2}\right) + \sum \min D_{nj}^{k-1} \leq$
$m_{ij}Vr^{k-2} + BtVr^{k-2} + BtVr^{k-2}Wr/(1-r) + \sum \min D_{nj}^{k-1}$.
Thus we have $D_{ij}^k \leq Kr^{k-2} + \sum \min D_{nj}^{k-1}$.

Iteration leads to $D_{ij}^k \leq$
$Kr^{k-2} + Kr^{k-3}\sum \min + Kr^{k-4}\sum m_{in}^{(2)} + \ldots + Kr\sum m_{in}^{(k-3)} + \sum m_{in}^{(k-2)}D_{nj}^2$.

We obtain an upper bound for $D_{nj}^2 = \left|\partial c_n^2/\partial x_j - \partial c_n^1/\partial x_j\right| =$
$\left|(\partial g_n/\partial x_j)_{xG}\cdot g_j + \sum(\partial g_n/\partial x_p)_{xG}\cdot x_p \partial g_p/\partial x_j - \partial g_n/\partial x_j\right| =$
$(\partial g_n/\partial x_j)_{xG}|g_j - 1| + |(\partial g_n/\partial x_j)_{xG} - (\partial g_n/\partial x_j)| +$
$\left|\sum(\partial g_n/\partial x_p)_{xG}x_p \partial g_p/\partial x_j\right| \leq$
$m_{nj} + B\sum|x_p g_p - x_p| + \sum m_{np}m_{pj} \leq m_{nj} + m_{nj}^{(2)} + Bt$.

Therefore, substituting gives $D_{ij}^k \leq$

$\bar{K} r^{k-2}(k-2) + \sum m_{in}^{(k-2)} m_{nj} + \sum m_{in}^{(k-2)} m_{nj}^{(2)} + Bt \sum_n m_{in}^{(k-2)} \leq$

$\bar{K} r^{k-2}(k-2) + V r^{k-1} + V r^k + Bt r^{k-2}$. Since each of these terms defines a convergent series so does their sum and, we have

Theorem. If G is below-critical, the sequences $\partial c_i^k / \partial x_j$ are uniformly convergent on I_t. Hence the partials $\partial c_i / \partial x_j$ exist on I_t, and are the limit functions of the corresponding sequences. Since $\lim \left(\partial c_i^k / \partial x_j \right)_1 = \left(\partial c_i / \partial x_j \right)_1$ the latter is the limit approached by the expectation of particles of type j in the total progeny at the end of k generations from one particle of type i. From the functional equation satisfied by $c_i(x)$ follows

$$\sum_n \left(m_{in} - \delta_{in} \right) \left(\partial c_n / \partial x_j \right)_1 = -m_{in}, \quad i = 1, \ldots, t \text{ where } |M - I| \neq 0,$$

and the expectation limits are uniquely determined.

VI. Total Progeny in Subcritical System with Source

1. Consider again the process described in IV. We found that the probability distribution in the n-th generation had generating function

$$H_n(x) = S(G) S \left(G^2 \right) \ldots S \left(G^n \right).$$

Here $S \left(G^n \right)$ gives the distribution for the isolated system produced by the initial action of the source, $S \left(G^{n-1} \right)$ is that for the component due to the second action of the source, and so on. Regard the isolated component produced by the $(n - k + 1)$st action of the source, with generating function at the n-th level (of the whole system) $S \left(G^k \right)$. By an argument exactly analogous to that of V 1, one sees that the generating function for total progeny at the n-th level for this isolated component of the system is given by $S \left(x_1 c_1^k(x), \ldots, x_t c_t^k(x) \right)$.

It follows from the elementary laws of probability that the generating function $U_n(x) \equiv \sum u_n \left(j_1, \ldots, j_t \right) x_1^{j_1} \ldots x_t^{j_t}$ is $U_n(x) = \Pi S \left(x_1 c_1^k(x), \ldots, x_t c_t^k(x) \right)$. Here $u_n \left(j_1, \ldots, j_t \right)$ denotes the probability that, at the n-th level of the entire system, there should be a total of j_1 particles of type $1, \ldots, j_t$ particles of type t produced altogether, counting particles contributed by the source as well as all progeny of such particles.

2. Since $U_n(x) = U_{n-1}(x)S\left(x_i c_i^n(x)\right)$, and for arbitrary x in I_t with all components less than one, the latter factor is less than or equal to $S(x)$ which in turn is less than one (supposing condition (S^*)), it is evident that $\lim U_n(x) = 0$ for all such x. Since however $U_n(1) = 1$ for all n, $U_n(x)$ converges to a discontinuous limit function on I_t. Moreover, it is manifest that the expectations approach infinity so we cannot expect a simple theory.

Appendix A. We collect here some standard results from classical analysis.

A sequence of functions $F_n(x)$ on I_t is said to be uniformly convergent on I_t in case (1) $\lim F_n(x) = F(x)$ exists for each x of I_t, and (2) for every $e > 0$ exists N so that $n > N$ implies $\left|F_n(x) - F(x)\right| < e$, all x in I_t.

Theorem. If all $F_n(x)$ are continuous on I_t, and the sequence is uniformly convergent there, then the limit function $F(x)$ is continuous on I_t.

Theorem. If $K_n(x)$ is a uniformly convergent sequence of continuous functions on I_t, with limit function $K(x)$, then the sequence of functions

$$I_n(x) = \int_0^{x_1} K_n\left(z_1, x_2, \ldots, x_t\right) dz_1$$

is uniformly convergent to

$$\int_0^{x_1} K\left(z_1, x_2, \ldots, x_t\right) dz_1 ,$$

and similarly for the other variables.

Theorem. If (1) $F_n(x)$ converges pointwise on I_t to the limit function $F(x)$, (2) the partials $\partial F_n/\partial x_1$ exist and are continuous on I_t, (3) the sequence $\partial F_n/\partial x_1$ converges uniformly on I_t, then $\partial F/\partial x_1$ exists on I_t and is equal to the limit of the sequence $\partial F_n/\partial x_1$. Similar statements hold for the other variables.

Theorem. $F_n(x)$ is uniformly convergent on I_t if and only if $e > 0$ implies existence of N such that for all $n \geq N$ and all positive integers p, $\left|F_{n+p}(x) - F_n(x)\right| < e$ on I_t.

Theorem. If $F_n(x)$ is a sequence of functions defined on I_t, $\sum M_n$ is a convergent series of non-negative numbers, and for all sufficiently large n,

$$\left|F_n(x) - F_{n+1}(x)\right| \leq M_n \text{ on } I_t$$

then $F_n(x)$ is uniformly convergent on I_t.

Appendix B

Theorem. If $1 + h \geq 0$ and n is a positive integer, then $(1 + h)^n \geq 1 + nh$.

Proof trivial by induction on n.

VII. The "Time" Particle

1. It is of interest to note that, in the system with source, it is possible to regard the n-th population as essentially the n-th generation of progeny of a simple system of $t + 1$ types of particles produced from one particle of new type $t + 1$. Suppose that we associate $x_1, \ldots, x_t$ as before with the t original types of actual particles, and introduce a new type of particle with variable x_{t+1}. Consider then the transformation $V(x)$ of I_{t+1}, defined by the component functions

$$
\begin{aligned}
v_1(\bar{x}) &= g_1(x) \\
&\vdots \\
v_t(\bar{x}) &= g_t(x)
\end{aligned}
$$

$$v_{t+1}(\bar{x}) = x_{t+1}S(G(x)), \qquad \begin{aligned} \bar{x} &\equiv (x_1, \ldots, x_t, x_{t+1}) \\ x &\equiv (x_1, \ldots, x_t,) \end{aligned}$$

One verifies easily that the $(t + 1)$st component of the k-th iterate $V^k(x)$ of $V(x)$ is

$$V_{t+1}^k = x_{t+1}S(G(x)) \ldots S\left(G^k(x)\right).$$

Hence v_{t+1}^k, the result of a simple iteration satisfies the relation

$$v_{t+1}^k(x_1, \ldots, x_t, x_{t+1}) = x_{t+1}H_k(x_1, \ldots, x_t).$$

The transformation $V(x)$ fails to satisfy the restrictions we have imposed throughout on our generating transformations, for example $\partial v_1 / \partial x_{t+1}$ vanishes identically on I_{t+1}, and we have treated the problem independently for this reason. Nevertheless we shall be able to make use of the indicated simplification to the iterative case in a future report on the space of histories of a multiplicative system.

VIII. Total Progeny as an Iterative Problem

1. In a similar way, the transformation $C^k(x)$ of V may be produced by an iterative process. Let $x_1, \ldots, x_t$, be the usual variables associated with the t types of actual particles, and $z_1, \ldots, z_t$, be variables for a set of t types of dummy particles. Suppose probabilities of transmutation among the $2t$ types are defined by the following generating transformation $L(x, z)$ of I_{2t},

$$
\begin{aligned}
L_1(x, z) &= z_1 g_1\left(x, \ldots, x_t\right) \\
&\vdots \\
L_t(x, z) &= z_t g_t\left(x, \ldots, x_t\right) \\
L_{t+1}(x, z) &= z_1 \\
L_{2t}(x, z) &= z_t \ . \qquad\qquad x = (x_1, \ldots, x_t) \\
&\qquad\qquad\qquad\quad\ z = (z_1, \ldots, z_t)
\end{aligned}
$$

where the $g_i(x)$ are the components of the usual $G(x)$. One verifies easily that the i-th component (for $i = 1, \ldots, t$) $L_i^k(x, z)$ of the k-th iterate of the transformation $L(x, z)$ satisfies the relation

$$
L_i^k(x, x) = x_i c_i^k(x) \ .
$$

If one examines the nature of the process induced by $L(x, z)$ one finds that each time an actual particle of type x_i is produced in a generation k, it is forced to produce in generation $k+1$ a dummy particle of type z_i, as well as its actual progeny. Moreover, every dummy particle of type z_i is forced to just reproduce itself in one for one fashion. Thus the total actual progeny is tallied by means of the one for one reproduction of the dummies through all generations. Thus if one sets $z_i = x_i$ in generation k one totals the entire progeny including the actual particles produced from actual particles in the $(k-1)$st generation and the dummies which total the whole previous progeny of actual particles. The extra x_i factor is due to counting the zero-generation particle in the L process.

III

ABSTRACT

The set Γ_i of all possible genealogies or graphs z of a multiplicative system produced from one particle of type i is here introduced as a fundamental concept in the theory of such systems (see I, II). This set possesses a natural intrinsic distance function $d(z, z')$ under which it is a complete zero-dimensional metric space satisfying the second axiom of countability.

Simple axioms on (A) intervals and (B) measure of intervals are given for an abstract set from which the classical theory of completely additive measure is derived.

Intervals in the set Γ_i are defined intrinsically and shown to satisfy the axioms A. If now a particular multiplicative system with given generating transformation $G(x)$ is given, the transition probabilities $p_1(i;j_1,\ldots,j_t)$ serve to define a measure for the intervals of Γ_i satisfying the axioms B. Proof of the latter is non-trivial due to non-local-compactness of the space Γ_i.

With this mathematical structure at hand it becomes possible to state in a simple way some of the striking properties of multiplicative systems.

If $x^0 = G(x^0)$ is the death-fix-point of $G(x)$, then the set of graphs of Γ_i which terminate in death has measure x_i^0.

If $v = (v_1,\ldots,v_t)$ is the characteristic vector corresponding to the maximal positive characteristic root $r > 1$ of supercritical system, then the set of all graphs of Γ_i whose k-th generation population approaches the ratios $v_1:v_2:\ldots:v_i$ has measure $1 - x_i^0$. Thus, almost all graphs (genealogies) either terminate in death or approach the mode v as limit. These results are trivial for subcritical systems, in which by definition, $x_i^0 = 1$.

I. A Remark on Measure Theory

1. It is convenient for our purposes to have a simple set of axioms on which measure theory may be shown to rest. To this end, let $\Gamma = \{z\}$ be a set of points z, and $I = \{i\}$ a class of special subsets i of Γ called intervals. We demand that the entire set Γ and the empty set ϕ be intervals. Denote by J the class of all subsets $S = \sum i$ of Γ which are set sums of a finite or countable number of intervals. (All set sums hereafter are understood to operate on finitely or countably many summands. The notation $\dot{\sum}_a$ is used to indicate that the summands are mutually disjoint.)

We suppose that intervals satisfy the following axioms:

I 1. Every set $S = \sum i$ of J can be represented as a sum $\dot{\sum} j$ of disjoint intervals.

I 2. The complement $i' = \Gamma - i$ of an interval i in J.

I 3. The set product ij of two intervals i and j is an interval k.

Moreover, we assume

m 1. To every interval i is assigned a non-negative real number $m(i) \geq 0$ called its measure.

m 2. $M(\Gamma) = 1$, $m(\emptyset) = 0$.

m 3. If $i = \dot{\sum} j$, where the i and j are intervals, then $m(i) = \sum m(j)$.

An additive class C of subsets of Γ is one such that

C 1. All intervals belong to C.

C 2. $\sum A$ is in C whenever all sets A are in C.

C 3. If A is in C, so is A'.

The Borel sets are those common to all additive classes, hence, themselves form an additive class.

We shall define a property of subsets U of Γ called measurability, and prove that the class of measurable sets is additive. Simultaneously we define a measure $m(M)$ for every measurable set M and prove

M 1. $0 \leq m(M) \leq 1$ for M measurable.

M 2. $m(\sum M) = \sum m(M)$ for M measurable.

M 3. If M is an interval, the measure assigned to M as a measurable set coincides with its intitially given measure.

2. In stating the above axioms, we have attempted to focus attention on the fundamental assumptions. It remains to show that the essential theorems follow from our axioms. This we do in the remainder of this chapter, for the sake of completeness, following the classical arguments for the real line[1].

We note first the following properties of intervals:

I 4. $\sum S$ is in J whenever all the summands S are in J.

For $\sum S$ is an at most countable sum of at most countable sums of intervals.

I 5. $\prod_1^N i$, then S' is in J whenever all N factors are in J.

It suffices to see this for $N = 2$. Hence let $S_1 = \sum i_m$, $S_2 = \sum j_n$. Then $S_1 S_2 = \sum i_m \sum j_n = \sum (i_m j_n)$. By I 3, $i_m j_n$ is an interval, so $S_1 S_2$ is in J, the range (m, n) being at most countable.

I 6. If $S = \sum_1^N i$, then S' is in J.

For $S' = \prod_1^N i'$, each i' is in J by I 2, and thus S' is in J by I 5.

3. We now assign a measure to every set S of J. Note first that if $i \supset \sum_1^N i_n$, then $i = \dot{\sum} i_n \dot{+} i(\dot{\sum} i_n)'$. By I 6 and I 5, the latter set is in J, so by I 1, we may write $i = \dot{\sum}_1^N i_n \dot{+} \dot{\sum} j_m$, and by m 3, and m 1,
$$m(i) = \sum m(i_n) + \sum m(j_m) \geq \sum m(i_n).$$

Hence, if $\dot{\sum} i_n$ is any sum of disjoint intervals, we have for every N, $\Gamma \supset \dot{\sum}_1^N i_n$ and hence $1 \geq \sum_1^N m(i_n)$. Thus $\sum m(i_n)$ exists, less than or equal to 1, for every such disjoint sum.

Now suppose $\dot{\sum} i_m = \dot{\sum} j_n$. Then $i_m = i_m (\dot{\sum} j_n) = \dot{\sum}(i_m j_n) = \dot{\sum}_n k_{mn}$, k_{mn} being the *interval* i_m, j_n. (See I 3.) Hence by m 3, $m(i_m) = \sum_n m(k_{mn})$ and $\sum m(i_m) = \sum_m \sum_n m(k_{mn})$. Similarly $\sum m(j_n) = \sum_n \sum_m m(k_{mn})$. Now the double sum $\sum_{mn} m(k_{mn})$ converges to a limit ≤ 1 by the preceding paragraph since the k_{mn} are disjoint. Hence the iterated sums are both equal to this limit and hence to each other. Thus $\sum m(i_m) = \sum m(j_n)$.

It is therefore clear that, if S is any set of J we can write $S = \dot{\sum} i$ and define $m(S) = \sum m(i)$ unambiguously. Clearly $0 \leq m(S)$, and if S is an interval i, $m(S)$ coincides with the initially given measure $m(i)$.

4. We establish in this section some properties of the measure $m(S)$ for sets S of J. First,

m 4. $m(\overset{\cdot}{\sum} S) = \sum m(S)$.

For, write $S_m = \overset{\cdot}{\sum} i_{mn}$. Then $m(\overset{\cdot}{\sum} S) \equiv m(\sum_{mn} i_{mn}) \equiv \sum_{mn} m(i_{mn}) = \sum_m \sum_n m(i_{mn}) = \sum_m (S_m)$. Next

m 5. If S and T are in J and $S \subset T$, then $m(S) \le m(T)$.

For, let $S = \overset{\cdot}{\sum} i_m$ and define $S_n = \overset{\cdot}{\sum_1^n} i_m$. Then $T = S_n \overset{\cdot}{+} T S_n'$, the summands being disjoint and in J. Hence, by m 4, $m(T) = m(S_n) + m(TS_n') \ge m(S_n) = \sum_1^n m(i_m)$. Hence $m(S) = \sum m(i) \le m(T)$. Thus follows

m 6. For all S in J, $0 \le m(S) \le 1$.

Next we prove

m 7. For arbitrary sets S of J, disjoint or not, $m(\sum S) \le \sum m(S)$, where the latter sum may of course be infinite.

Write $S_m = \overset{\cdot}{\sum} i_{mn}$. Then $\sum S_m = \sum_{mn} i_{mn} = \sum i_p = i_1 \overset{\cdot}{+} i_1' i_2 \overset{\cdot}{+} i_1' i_2' i_3 \overset{\cdot}{+} \ldots$, the latter summands being disjoint and in J. Hence, by m 4, m 5, $m(\sum S_m) = m(i_1) + m(i_1' i_2) + \ldots \le m(i_1) + m(i_2) + \ldots = \sum m(i_p) = \sum_{mn} m(i_{mn}) = \sum_m \sum_n m(i_{mn}) = \sum_m m(S_m)$, the latter possibly being infinite.

Finally we have

m 8. If S and T are in J, then $m(S + T) + m(ST) = m(S) + m(T)$.

First suppose S and T are each finite sums of intervals. Write

$$S + T = S \overset{\cdot}{+} (T - ST)$$
$$ST \overset{\cdot}{+} (T - ST) = T$$

where S and ST are in J, and, ST being a finite sum of intervals, also $T - ST = T(ST)'$ is in J. Thus by m 4,

$$m(S + T) = m(S) + m(T - ST)$$
$$m(ST) + m(T - ST) = m(T),$$

and addition yields the equation of the theorem.

Now let $S = \sum i_m$, $T = \sum j_n$, where we make either sum terminate in $\emptyset$ in case it happens to be finite. Define S_n and T_n as the corresponding partial sums through the first n terms. Then by the first part of the proof we have for every n, $m(S_n + T_n) + m(S_n T_n) = m(S_n) + m(T_n)$. But $\lim m(S_n) = \lim \sum_1^n i_m = m(S)$, and $\lim m(T_n) = m(T)$. (Recall that $m(\emptyset) = 0$.) Also, $S + T = \sum i_m + \sum j_n = \sum (i_m + j_m) \equiv \sum \bar{S}_m = \bar{S}_1 \dotplus \bar{S}_1' \bar{S}_2 \dotplus \dots$. Thus $m(S + T) = m(\bar{S}_1) + m(\bar{S}_1' \bar{S}_2) + \dots =$ $\lim_n \left(m(\bar{S}_1) m(\bar{S}_1' \bar{S}_2) + \dots + m(\bar{S}_1' \dots \bar{S}_{n-1}' \bar{S}_n) \right) = \lim_n m(\bar{S}_1 \dotplus \bar{S}_1' \bar{S}_2 \dotplus \dots \dotplus \bar{S}_1' \dots \bar{S}_{n-1}' \bar{S}_n) = \lim_n m(\bar{S}_1 + \dots + \bar{S}_n) = \lim_n m(S_n + T_n)$.

Finally $ST = \sum i_m \sum j_n = \sum_{mn} i_m j_n = \sum_{mn} k_{mn} = \sum k_p = k_1 + k_1' k_2 + \dots$ where the intervals $k_{mn} = k_p$ are listed in sequential order by upper squares: $(1,1)$, $(1,2)$, $(2,1)$, $(2,2)$; $\dots$. Then $m(ST) =$ $m(k_1) + m(k_1' k_2) + \dots = \lim_{n^2} \left(m(k_1) + \dots + m(k_1' \dots k_{n^2-1}' k_{n^2}) \right) =$ $\lim_{n^2} m(k_1 + \dots + k_1' \dots k_{n^2-1}' k_{n^2}) = \lim_{n^2} m(k_1 + \dots + k_{n^2}) \lim_n m(S_n T_n)$.

Hence we obtain m 8 in general by taking limits of both sides of the finite relation.

5. Let U be an arbitrary subset of Γ and define the outer measure $O(U) = \mathrm{glb}(m(S); U \subset S \in J)$, and the inner measure $I(U) = 1 - O(U')$. From m 6 we have for all subsets U,

m 9. $0 \le O(U) \le 1$, and $0 \le I(U) \le 1$.

Moreover, if $U_1 \subset U_2$, every S of J which contains U_2 also contains U_1, so the numbers $m(S)$ defining $O(U_2)$ form a subset of those defining $O(U_1)$, and hence

m 10. If $U_1 \subset U_2$, then $O(U_1) \le O(U_2)$ and $I(U_1) \le I(U_2)$.

For every $S_1 \supset U$, $S_2 \supset U'$, S_1, S_2, in J, we have $S_1 + S_2 \supset U + U' = \Gamma$ and $M(\Gamma) = 1 = m(S_1 + S_2) \le m(S_1) + m(S_2)$. Fix $S_1 \supset U$. Then $1 - m(S_1) \le m(S_2)$ for all $S_2 \supset U'$, so $1 - m(S_1) \le O(U'), 1 - O(U') \le m(S_1)$ for all $S_1 \supset U$, and $1 - O(U') \le O(U)$. Thus follows

m 11. For every subset U, $I(U) \le O(U)$.

Now let $\{U_n\}$ be a sequence of arbitrary subsets U_n of Γ. Fix $e > 0$. For every n there exists an S_n in J such that $S_n \supset U_n$ and $m(S_n) < O(U_n) + e/2^n$. Hence $\sum S_n \supset \sum U_n$, $\sum S_n$ is in J, and $m(\sum S_n) \le \sum m(S_n) < \sum O(U_n) + e$. Thus $O(\sum U_n) \le m(\sum S_n) < \sum O(U_n) + e$ for every $e > 0$ and we have m 12. For arbitrary subsets U, $O(\sum U) \le \sum O(U)$.

Suppose U_1, U_2 are disjoint. For every $S_1 \supset U_1'$, $S_2 \supset U_2'$, S_1, S_2 in J, we have $S_1 + S_2 \supset U_1' + U_2' = (U_1 U_2)' = \emptyset = \Gamma$ so $m(S_1 + S_2) = 1$. Also, $(U_1 + U_2)' = U_1' U_2' \subset S_1 S_2$ is in J, hence $O(U_1 + U_2)' \le$ glb $\left(m(S_1 S_2)\right)$. Then $I(U_1 + U_2) - I(U_1) - I(U_2) = 1 - O(U_1 + U_2)' - 1 - O(U_1') - 1 - O(U_2') = O(U_1') + O(U_2') - O(U_1 + U_2)' - 1 \ge$ glb $m(S_1) +$ glb $m(S_2) -$ glb $m(S_1 S_2) - 1 \ge$ glb $\left(m(S_1) + m(S_2) - m(S_1 S_2)\right) - 1 =$ glb $m(S_1 + S_2) - 1 = 1 - 1 = 0$. We therefore obtain

m 13. $U_1 U_2 = \emptyset$ implies $I(U_1) + I(U_2) \le I(U_1 + U_2)$.

Generalizing this, we have $I(\dot{\sum} U) \ge I(U_1) + I(\sum_2 U) \ge I(U_1) + I(U_2) + (I \sum_3 U) \ge \ldots \ge I(U_1) + \ldots + I(U_N) + I(\sum_{N+1} U) \ge \sum_1^N I(U)$ for all N, and

m 14. $\sum I(U) \le I(\dot{\sum} U)$, for disjoint summands.

6. We say a set M is measurable in case $I(M) = O(M)$ and denote by K the class of all measurable sets M. For a measurable set M we define a measure $m(M) = I(M) = O(M)$.

From m 9 follows

m 15. For a measurable set M, $0 \le m(M) \le 1$.

Let i be an interval. Since i is in J and contains itself, we have $O(i) \le m(i)$ since $O(i)$ is a lower bound. Now suppose $i \subset S \in J$. Then $m(i) \le m(S)$ and $m(i) \le O(i)$ since $O(i)$ is a greatest lower bound. Hence $m(i) = O(i)$.

Since i' is in J, we may write $\Gamma = i \dotplus \sum j_n$, so $m(\Gamma) = 1 = m(i) + m(i')$. Now $I(i) = 1 - O(i') = 1 - O(\sum j_n) \ge 1 - \sum (O(j_n)) = 1 - \sum m(j_n) = 1 - m(i') = m(i) = O(i) \ge I(i)$, so we have established

m 16. Every interval is measurable with $O(i) = I(i) = m(i)$, its initially given measure.

Also, we see

m 17. If M_n are disjoint measurable sets, then $\sum M_n$ is measurable and $m(\dot{\sum} M_n) = \sum m(M_n)$.

For, $\sum m(M_n) = \sum I(M_n) \le I(\dot{\sum} M_n) \le O(\dot{\sum} M_n) \le \sum O(M_n) = \sum m(M_n)$, and $I(\sum M_n) = O(\sum M_n) = \sum m(M_n)$.

As a corollary we get

m 18. Every set $S = \dot{\sum} i$ of J is measurable and its measure as such coincides with that previously defined.

If M is measurable, $I(M') = 1 - O(M) = 1 - I(M) = 1 - (1 - O(M')) = O(M')$, and so

m 19. The complement of a measurable set is measurable and $m(M) + m(M') = 1$.

Finally , we have to prove

m 20. A sum of measurable sets, disjoint or not, is measurable.

Let M_1, M_2 be measurable, and hence also M_1', M_2'. Fix $e > 0$. Then there exist sets S_1, S_2, T_1, T_2 in J such that $S_1 \supset M_1$, $S_2 \supset M_2$, $T_1 \supset M_1'$, $T_2 \supset M_2'$ and

$$m(S_1) < O(M_1) + e \; ; \; m(T_1) < O(M_1') + e$$
$$m(S_2) < O(M_2) + e \; ; \; m(T_2) < O(M_2') + e \; .$$

Now $S_1 + T_1 \supset M_1 + M_1'$ and $S_2 + T_2 \supset M_2 + M_2'$ so $S_1 + T_1 = \Gamma = S_2 + T_2$. But $1 + m(S_1 T_1) = m(S_1 + T_1) + m(S_1 T_1) = m(S_1) + m(T_1) \le O(M_1) + e + O(M_1') + e = m(M_1) + m(M_1') + 2e = 1 + 2e$. Thus $m(S_1 T_1) < 2e$ and similarly $m(S_2 T_2) < 2e$.

But $M_1 + M_2 \subset S_1 + S_2$ and $(M_1 + M_2)' = M_1' M_2' \subset T_1 T_2$, so $\Gamma = (M_1 + M_2) + (M_1 + M_2)' \subset S_1 + S_2 + T_1 T_2$. Moreover, from the first two inclusions, $O(M_1 + M_2) \le m(S_1 + S_2)$ and $I(M_1 + M_2) = 1 - O(M_1 + M_2)' \ge 1 - m(T_1 T_2)$. So, $O(M_1 + M_2) - I(M_1 + M_2) \le m(S_1 + S_2) + m(T_1 T_2) - 1 = m(S_1 + S_2 + T_1 T_2) + m((S_1 + S_2) T_1 T_2) - 1 = m(\Gamma) + m((S_1 + S_2) T_1 T_2) - 1 = m(S_1 T_1 T_2 + S_2 T_1 T_2) \le m(S_1 T_1 + S_2 T_2) \le m(S_1 T_1) + m(S_2 T_2) < 4e$. Thus $O(M_1 + M_2) \le I(M_1 + M_2)$ and $M_1 + M_2$ is measurable. Accordingly, so is every sum of a finite number of measurable sets.

But we see now that a sum of a countable number of measurable sets $\sum M_n = M_1 \dot{+} M_1' M_2 \dot{+} (M_1 + M_2)' M_3 \dot{+} \ldots$ is measurable by m 17 and the fact that each of the latter summands $(M_1 + \ldots + M_{n-1})' M_n = (M_1 + \ldots + M_{n-1} + M_n')'$ is measurable by m 19 and the preliminary results for finite sums.

II. The Set of Graphs

1. Consider t types of particles, such that a particle of type i may produce, upon transformation, an arbitrary number $j_1 + \ldots + j_t \geq 0$ of such particles, of which j_s are of type s, $s = 1, \ldots, t$. We suppose that transformation times are the same for each type, and hence that generations may be counted unambiguously. We agree to consider zero-generation as consisting of one particle of a fixed type i, and then consider the set Γ_i of all conceivable genealogies or histories proceeding from it, that is, the infinite record of the transformations of this particle and of all its progeny through all generations $k = 0, 1, 2, \ldots$

2. We may represent such a genealogy in the plane by a graph or lattice if we agree on the following conventions:

(a) A particle of type i in the k-th generation is represented by a number i in the k-th row.

(b) If a particle of type i in generation k is transformed into no particle, that is, if it dies in generation $k + 1$, this is so indicated by a sequence of zeros proceeding from it to the $(k + 1)$st and thence successively to all lower rows, thus:

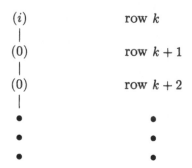

(c) If a particle of type i in generation k is transformed into $j_1 + \ldots + j_t > 0$ particles, j_s of type s, in the $(k + 1)$st generation, this is indicated by a branching from the corresponding number i in the k-th row into a group of $j_1 + \ldots + j_t$ numbers in row $k + 1, j_s$ being the quantity of numbers s, identical numbers being grouped consecutively, and different numbers ranging from left to right in increasing order.

Thus

indicates that a particle of type 1 was transformed into two particles of type 1 and one particle of type 2. Note that the two 1's represent different particles, so that the events

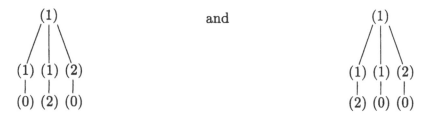

are counted as different.

Consideration shows that the set Γ_i of all genealogies is uniquely represented by the set of all such graphs z in the plane, and we speak hereafter of the set Γ_i of all graphs z. We will not change the type of the zero-generation ancestor during the subsequent discussion.

3. The set Γ_i has at least the power of the continuum. For to every sequence of 0's and 1's $\{a_n\}$, we may make correspond a particular graph (there are many) which contains a total of one particle in generation n if a_n is 0 and a total of two particles in generation n if a_n is 1. This correspondence is one-one on the set of all sequences $\{a_n\}$ which has the power of the continuum to a subset of the set Γ_i. That the set Γ_i has power less than or equal to that of the continuum may be seen directly or from a topological result which we wish to establish anyway in the next section. It will thus appear that Γ_i has power of the continuum.

4. If z is a graph, z_n denotes its upper segment from generation 0 through n, both inclusive. Thus, if $z = z'$, then $z_n = z'_n$ for all n, whereas, if $z \neq z'$, there exists a first integer $k = k(z, z')$ such that $z_k \neq z'_k$. Since $z_0 = (i)$ for all graphs z of Γ_i, the integer $k(z, z') \geq 1$.

5. Suppose z, z', z'' are all different, where $m = k(z, z') \geq n = k(z', z'')$. Then $z_{n-1} = z'_{n-1} = z''_{n-1}$ and $k(z, z'') \geq n$. Hence we have the

Lemma. If z, z', z'' are three different graphs, then $k(z, z'') \geq \min\left(k(z, z'), \; k(z, z'')\right)$.

6. A graph z is said to terminate in case it contains no particles (only 0's) in some generation. We can write the set T of all terminating graphs in the form of a disjoint sum

$$T = T_0 \dot{+} T_1 \dot{+} T_2 + \ldots$$

where T_n is the set of all graphs y^n which contain at least one particle in all generations through the n-th and die completely in generation $n + 1$. The set of T_0 consists of the single graph

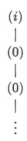

We prove the

Lemma. The sets $T_1, T_2, \ldots$ are all countable and thus so is T.

Proof. Induct on n of T_n. First, T_1 is countable, since it is in one-one correspondence with the countable set of all integer-component vectors $(j_1, \ldots, j_t)$, $j_s \geq 0$, $\sum j_s > 0$. Suppose T_n is countable. Every graph of T_{n+1} is the continuation to one more "live" generation of the section y_n^n of some y^n of T_n. This serves to split T_{n+1} into a countable (since T_n is) number of subsets S, all y^{n+1} in a fixed S being continuations of the same y_n^n of T_n to one more live generation $n+1$, followed by death. Now the n-th generation of y_n^n contains only a finite number of particles each of which can branch into the $(n+1)$st generation in only a countable number of ways. Hence each subset S is countable and so is T_{n+1}.

7. We developed in I an abstract theory of measure, which we are eventually to apply to the set Γ_i of graphs z. To this end we define intervals in Γ_i as follows. By an interval of order $n(n = 0, 1, \ldots)$ is meant the set $i(y^n)$ of all graphs z such that $z_n = y^n_n$, where y^n is some terminating graph of T_n. Note that the interval of order 0 is Γ_i itself. By an interval we mean an interval of order n, or a single graph z, terminating or not, or the null set $\emptyset$. Define J as the class of all sets $S = \sum i$ which are the sums of an at most countable number of intervals, just as in I. We now have to verify the axioms I 1, 2, 3.

8. We have in the present case the simple and unusual situation that, if i and j are intervals then they are either disjoint or one is contained in the other, hence $ij = \emptyset$ or i or j and I 3 follows. For suppose $ij \neq \emptyset$ so that some z is in both. If i or j is a single graph it must be z itself and hence lies in the other interval.

Suppose then $i = i(y^m)$, $j = i(\bar{y}^n)$, with $m \leq n$. Then $y^m_m = z_m = \bar{y}^n_m$. Let z' be in $i(\bar{y}^n)$. Then $z'_m = \bar{y}^n_m = y^m_m$ and z' is in $i(y^m)$. Thus $i(y^m) \supset i(\bar{y}^n)$. Moreover, we see that if $m = n$, $i(y^m) = i(\bar{y}^n)$.

9. Suppose $S = \sum i(y^n) + \sum z + \sum \emptyset$ is any sum of intervals where the $i(y^n)$ are summed according to increasing n. Without altering the sum S we may successively

(a) delete the $\emptyset$ summands,

(b) delete duplicate z's,

(c) delete duplicate $i(y^n)$'s,

(d) delete z's contained in the remaining $i(y^n)$'s,

(e) delete every $i(y^n)$ which intersects a preceding $i(y^m)$ with $m < n$, since then the former is contained in the latter. The resulting summands are now disjoint with sum S, and we have I 1. Moreover, and unusually, the final summands are a subclass of those originally given.

10. We have to show now (I 2) that the complement of an interval is in J.

(a) If $i = \emptyset$, $i' = \Gamma_i = i(y^0)$.

(b) If $i = i(y^0) = \Gamma_i$, $i' = \emptyset$, an interval by definition.

(c) If $i = i(\bar{y}^n)$, where $\bar{y}^n$ is in T_n and $n \geq 1$, we may write the disjoint countable sum

$$\Gamma_i = \sum_{k=0}^{n-1} \sum_{y^k \in T_k} y^k + \sum_{y^n \in T_n} i(y^n)$$

since every graph z either terminates at some generation less than or equal to n, or lives through generation n and is thus in some $i(y^n)$ with y_n in T_n. The complement of $i(\bar{y}^n)$ is manifestly in J.

(d) If $i = y^{n-1}$, where $n - 1 \geq 0$, the above decomposition of Γ_i shows that i' is in J.

(e) If i is a non-terminating graph z, define y^n in T_n so $y_n^n = z_n$, $n = 0, 1, \ldots$. We claim that $z = \prod_0^\infty i(y^n)$. Clearly z is in every $i(y^n)$. Moreover if z' is in $i(y^n)$ for all n, then $z'_n = y_n^n = z_n$ and $z' = z$. It follows that $z' = \sum_0^\infty (i(y^n))'$, the summands being in J by b, c. Hence z' is in J.

11. We saw in I that if $S = i_1 + \ldots + i_n$ then S' is in J. This may be false in our case for countable sums. For example, if $T = \sum y^n$ is the set of all terminating graphs, then T' is the set of all non-terminating graphs. The latter cannot be a sum of a countable or finite number of intervals, for if $T' = \sum_i i(y^n) + \sum z$, we have T' with power of the continuum, the z summands are at most countable, hence at least one $i(y^n)$ must occur in the T' sum. But then y^n is in T', a contracdiction.

12. For use in IV we include the

Lemma. An interval of finite order cannot be expressed as a finite sum of two or more disjoint non-null intervals.

For suppose $i(\bar{y}^n) = \dot{\sum} i(y^m) \dot{+} \dot{\sum} z$ is such a finite sum. Let $\bar{T}$ be the set of all terminating $\bar{y}^{n+1}$ such that $\bar{y}_n^{n+1} = \bar{y}_n^n$. Of these there are countably many, all in $i(\bar{y}^n)$. Since the z's above are finite in number, an infinite subset $\bar{\bar{T}}$ of $\bar{T}$ must be in the $i(y^m)$; if such a $\bar{y}^{n+1}$ is in $i(y^m)$ we must have $n + 1 \geq m$ since $\bar{y}_m^{n+1} = y_m^m$ and y^m is alive in generation m. But since y^m is in $i(\bar{y}^n)$, we have also $m \geq n$. Moreover if $m = n$, $i(\bar{y}^n) = i(y^m)$ and we should have only one summand above. Hence $m = n + 1$ and $\bar{y}^{n+1} = y^m$. So to each $\bar{y}^{n+1}$ of the infinite set $\bar{\bar{T}}$

corresponds a summand $i(y^m)$. The correspondence being one-one a contradiction arises.

Corollary. If $\dot{\sum}_1^N i_n \supset j$, where i, j are intervals then j is in some i_n.

If j is $\emptyset$ or a single graph, this is trivial. Let j be an interval of finite order. Then $j = j(\dot{\sum}_1^N i_n) = \sum_1^N (ji_n)$, where ji_n is an interval. By the preceding result, all $ji_n = \emptyset$ except one, so (say) $j = ji_1 \subset i_1$.

13. Let $i(\bar{y}^n)$ be written as the disjoint sum $i(\bar{y}^n) = \bar{y}^n \dot{+} \sum i(\bar{y}^{n+1})$ where $\bar{y}^{n+1}$ ranges over the set $\bar{T}$ of the preceding paragraph. Then if j is an interval properly contained in $i(\bar{y}^n)$, it must be contained in one of the summands. We need only consider the case $j = i(y^m)$. Since y^m is in $i(\bar{y}^n)$, y^m must be a continuation of $\bar{y}^n$, and since $y^m \neq \bar{y}^n$, y^m must be a continuation of some $\bar{y}^{n+1}$ of $\bar{T}$. Thus y^m is in some $i(\bar{y}^{n+1})$ and so is j.

14. If we are given an arbitrary finite class N of disjoint intervals $i(y^m)$ and y^p, there exists a disjoint decomposition of Γ_i into such intervals among which appear all those of the given class N.

First write

$$\Gamma_i - y^0 \dot{+} \sum_{y' \in T_1} i(y') \tag{1}$$

We leave unaltered all $i(y')$ which occur in the class N. All others we decompose further thus:

$$i(y') = y' \dot{+} \sum i(y^2)$$

where y^2 ranges over all graphs of T_2 which are continuations of y'_1, and substitute the results into (1). Again we retain all $i(y^2)$ occurring in N and decompose all other $i(y^2)$ one more step as indicated. This procedure eventually obtains a disjoint sum of the desired sort including all $i(y^m)$ of N as summands. The y^p of N, being disjoint from the given $i(y^m)$ must occur in one of the other intervals of the sum before us. Each such interval may be decomposed until the contained y^p's are expressed as summands.

15. If $\overset{.}{\Sigma}\,i\,(y^n) \supset i(\bar{y}^m)$, then the latter is contained in one of the summands. For $\bar{y}^m$ is some $i(y^n)$ and hence, so is $i(\bar{y}^m)$.

III. The Space of Graphs

1. On Γ_i we define a metric or distance function as follows: If $z = z'$, $d(z, z') = 0$, whereas, if $z \neq z'$, $d(z, z') = 1/k(z, z')$ where $k(z, z')$ is the first integer $k(\geq 1)$ for which $z_k \neq z'_k$. It is clear that d satisfies the axioms for a metric:

(A) $d(z, z') \geq 0$, for if $z = z'$, $d = 0$, while if $z \neq z'$, $d > 0$. Hence also $d(z, z') = 0$ if and only if $z = z'$.

(B) $d(z, z') = d(z', z)$ since $k(z, z')$ is symmetrically defined.

(C) $d(z, z'') \leq \max\,(d(z, z'),\ d(z', z''))$. If any two of the three graphs involved are equal, this inequality is trivial. If all three graphs are different, C follows from the lemma of II 5. This is indeed stronger than the customary triangle inequality: $d(z, z'') \leq d(z, z') + d(z', z'')$.

2. If z and z' are graphs, the following are equivalent

(a) $d(z, z') < e$.

(b) $z = z'$ or $z \neq z'$ and $k(z, z') > 1/e$.

(c) $z = z'$ or $z \neq z'$ and $k(z, z') \geq [1/e] + 1$,

where the notation $[1/e]$ indicates the greatest integer in $1/e$.

(d) $z_n = z'_n$ where $n = [1/e]$.

To each graph z and real positive number $e > 0$ we assign the e-neighborhood $N_e(z)$ of z, namely the set of all z' such that $d(z, z') < 0$. Thus $N_e(z)$ consists of all graphs z' such that $z'_n = z_n$ where $n = [1/e]$. It is clear that if z' is in $N_e(z)$ then $N_e(z') = N_e(z)$.

3. A sequence of graphs $\{z^{(n)}\}$ converges to limit z in case $d(z^{(n)}, z)$ $\to 0$, that is, for every $e > 0$ there exists an N such that for all $n \geq N$, $z^{(n)}$ coincides with z through generation $[1/e]$.

4. The space Γ_i is complete in the sense that every sequence $(z^{(n)})$ such that $d(z^{(n)}, z^{(m)}) \to 0$ has a sequential limit z. For the given condition implies that for every $n = 1, 2, \ldots$ there is an $N(n)$ increasing with n such that $z_n^{N(n)} = z_n^m$, $m \geq N(n)$. From this one sees that the subsequence $z^{N(n)}$, $n = 1, 2, \ldots$ has the property $z_n^{N(n)} = z_n^{N(n+p)}$ and hence defines a graph z such that $z_n \equiv z_n^{N(n)}$ which is the limit of the subsequence and hence of the original sequence.

5. Every $N_e(z)$ is a closed set. For let $z' = \lim z^{(n)}$ where $z^{(n)} \in N_e(z)$. Then for n sufficiently large, $z^{(n)}$ coincides with z' through generation $[1/e]$. But $z^{(n)}$ is in $N_e(z)$, hence coincides also with z through generation $[1/e]$. Thus z' coincides with z through $[1/e]$ and is in $N_e(z)$. Hence the space Γ_i is zero-dimensional, every neighborhood being both open and closed.[4]

6. The space Γ_i is not compact. For example, the sequence consisting of an arbitary sequential ordering of the countable set of graphs y' or T_1 can have no convergent subsequence, since no two y' have the same section y_1'. For the same reason, no neighborhood $N_e(z)$ of a graph z which is alive in generation $[1/e]$ is compact, and therefore the space Γ_i is not even locally compact.

7. A space is said to satisfy the second axiom of countability in case there exists an at most countable set C of neighborhoods of the original system such that for every z and $N_e(z)$ there is a neighborhood N of the system C such that $z \in N \subset N_e(z)$. Our space Γ_i satisfies this axiom in the trivial sense that the whole set of original neighborhoods is itself countable. Consider an arbitrary $N_e(z)$. If $[1/e] = 0$, then $N_e(z) = \Gamma_i$, and there is only one such neighborhood. If $[1/e] = n \geq 1$, clearly $N_e(z) = N_{\frac{1}{n}}(z)$. In this case define a terminating graph x so that $x_n = z_n$ and x has only 0's in generation $n + 1$. Then $N_e(z) = N_{\frac{1}{n}}(x)$, where x is in T. Since T is countable, so is the entire neighborhood system.

Note that it is not implied in the above that x is in T_n. If z is dead in generation n, $x = z$ will be in some T_m with $m < n$, and $N_e(z)$ will consist of z alone.

8. It is well known[5] that a space satisfying the second axiom of countability has at most the power of the continuum. The argument rests on assigning to every z the class of all neighborhoods in C which contain z. This is one-one (by the separation axiom) on Γ_i to some subsets of a countable set. The class of all subsets of a countable set has power of the continuum so the power of Γ_i cannot exceed this. Combining with the opposite inequality obtained earlier, we see that Γ_i has the power of the continuum.

9. A word may be said about the relation of intervals to neighborhoods. Every neighborhood is either an interval of finite order or a terminating graph, hence an interval. Every interval of order n is either $\Gamma_i = N_2(y^0)$(when $n = 0$) or a $N_{\frac{1}{n}}(y^n)$ (when $n \geq 1$) hence a neighborhood. Every terminating graph is a neighborhood, e.g., $y^n = N_{\frac{1}{n+1}}(y^n)$, but a non-terminating graph cannot be realized as a neighborhood, and of course, neither can the null-set.

IV. Measure in the Space of Graphs

1. Whereas the notions of the set Γ_i, its intervals, and its topology are intrinsic in character, depending only on the number t of types of particles considered, we may establish a measure for intervals, and hence for the class of measurable sets in various ways. We are concerned at present only with the following procedure. As in I 3, we consider a fixed generating transformation $G(x)$ of the unit cube I_t of t-space:

$$g_i(x_1,\ldots,x_t) = \sum p_1(i; j_1,\ldots,j_t)x_1^{j_1}\ldots x_t^{j_t}$$
$$i = 1,\ldots,t,$$

defining the probabilities of transmutation $p_1(i; j)$ of a particle of type i into $j_1 + \ldots + j_t \geq 0$ particles, j_s of type s, $s = 1,\ldots,$,t. Then every finite upper section z_n of a graph z of Γ_i has an associated probability $p(z_n)$ that the event defined by z_n should occur.

For example, if $G(x)$ is

$$g_1(x_1, x_2) = \frac{1}{2} + \frac{1}{4}x_1 + \frac{1}{4}x_1 x_2$$
$$g_2(x_1, x_2) = \frac{1}{2} + \frac{1}{4}x_2 + \frac{1}{4}x_1 x_2 ,$$

and $z_0 = (1)$; $z_1 = (1)$; $z_2 = (1)$

$$
\begin{array}{ccc}
| & / & \backslash \\
(2) & (1) & (2) \\
 & | & | \\
 & (0) & (2)
\end{array}
$$

we have $p(z_0) = 1$, $p(z_1) = 0$, and $p(z_2) = 1/4(1/2 \cdot 1/4) = 1/32$.

2. We assign a G-measure to intervals of Γ_i thus:

(a) $m(\emptyset) = 0$,

(b) $m(z) = \lim p(z_n)$,

(c) $m(i(y^n)) = p(y_n^n)$.

Note that $m(z)$ may be non-zero, for example for a terminating y^n, $m(y^n) = p(y_{n+1}^n)$, and $m(i(y^n))$ may be zero. Clearly m1, m2 of I are satisfied and it remains to prove m 3. This is non-classical because of the non-local compactness of the space Γ_i and we divide the proof into a number of steps. Since $m(\emptyset) = 0$, we have only to consider the case $i(\bar{y}^n) = \dot{\sum} i(y^n) + \dot{\sum} z$ of a countable number of summands (see II 12).

3. First suppose $i \supset j_1 \dot{+} \ldots \dot{+} j_s$, the j's being disjoint and non-null. Then $m(i) \geq \sum m(j)$. Proof is by induction on s. If $s = 1$, we have the case $i \supset j$. If j has finite order, so does i and we write $i(y^m) \supset i(y^n)$ where necessarily $n \geq m$. Then $m(i(y^n)) = p(y_n^n) \leq p(y_m^n) = p(y_m^m) = m(i(y^m))$. If j is a graph z, the only case to consider is $i = i(y^m) \supset z$. Then $m(z) = \lim p(z_n) \leq p(z_m) = p(y_m^m) = m(i(y^m))$. Now suppose the theorem true for $\leq s - 1$ summands and consider $i(y^n) \supset j_1 + \ldots + j_s$ with $s \geq 2$. Write $i(y^n) = y^n + \sum i(y^{n+1})$, where $m(i(y^n)) = m(y^n) + \sum m(i(y^{n+1}))$. Then since $s \geq 2$, each j is properly contained in $i(y^n)$ and is therefore in one of its summands. If not all j are in the same summand, we invoke the induction assumption and obtain the result. If all j are in the same summand $i(y^{n+1})$ we decompose it and repeat the argument. Eventually the j's must be split between two summands, since the intersection of a nest $i(y^n) \supset i(y^{n+1}) \supset i(y^{n+2}) \supset \ldots$ is a single point. Hence eventually we obtain the result from the induction hypothesis and the fact that $m(i(y^n)) \geq m(i(y^{n+p})) = m(y^{n+p}) + \sum m(i(y^{n+p+1}))$.

4. Hence we see that $i = \dot{\sum} j$ implies $m(i) \geq \sum m(j)$. The opposite inequality remains to be proved. Because of non-local compactness we need the lemma of the next section.

5. Lemma. For every $e > 0$ there exists a set $\mathcal{Y}$ of disjoint intervals $i(y^n)$ such that

(a) $\sum m(i(y^n)) < e$,

(b) $C_e \equiv \left(\sum i(y^n)\right)'$ is closed and compact.

Denote by s_n an arbitrary finite graph from generation 0 through n. Since $1 = \sum p(s_1)$ where s_1 ranges over all finite graphs of order 1, we have $\sum p(s_1) < e/2$ for all but a finite number N_1 of s_1. The intervals or order 1 corresponding to the former s_1 we throw into $\mathcal{Y}$.

Consider for each of the N_1 remaining s_1 all of its continuations s_{12}. Then $p(s_1) = \sum p(s_{12})$ and we have $\sum p(s_{12}) < e/2^2 N_1$ for all but some finite number of s_{12}. Collecting the former s_{12} for all s_1 of the finite set, we have $\sum \sum p(s_{12}) < e/2^2$. The corresponding intervals of order 2 we throw into $\mathcal{Y}$, and retain the remaining finite number of s_{12}. We proceed in this way to obtain the set $\mathcal{Y}$ with $\sum m(i(y^n)) < e/2 + e/2^2 + \ldots = e$, at each stage retaining a finite set of finite graphs $s_{12\ldots n}$. We see now the C_e is compact, since an infinite set of graphs z in C_e may be thrown into a finite number of disjoint classes according to the agreement of their segments z_1 with the retained s_1. At least one class must be infinite. The latter class is further subdivided into a finite number of subclasses according to z_2, and so on. The process defines a graph z with segments s_1, s_{12}, s_{123}, which is clearly a limit point of the given infinite set of z's. Thus C_e is compact. It is manifestly closed since it is a complement of an open set. (Recall that every interval of finite order is a neighborhood, hence an open set, and the sum of open sets is open).

6. We saw in 4 that if $i = \dot{\sum} j$ then $m(i) \geq \sum m(j)$. We have now to prove the opposite inequality. We do this first for the case where $i = \Gamma_i$, that is, we suppose $\Gamma_i = \sum j$, explicitly

$$\Gamma_i = \sum i(y^m) \dotplus \sum y^p \dotplus \sum z^{(\nu)} \equiv \sum j$$

the z^ν, $\nu = 1, 2, \ldots$ being non-terminating. Fix $e > 0$. Since $m(z^{(\nu)}) = \lim_n p(z_n^{(\nu)})$ and the latter sequence is monotone non-increasing in n, we can fix $N(\nu)$ such that $p(z_{N(\nu)}^{(\nu)}) < m(z^{(\nu)}) + e/2^\nu$. Define $y^{N(\nu)}$ so that it is in $T_{N(\nu)}$ and $y_{N(\nu)}^{N(\nu)} = z_{N(\nu)}^{(\nu)}$. Then $m(i(y^{N(\nu)})) = p(y_{N(\nu)}^{N(\nu)}) = p(z_{N(\nu)}^{(\nu)}) < m(z^{(\nu)}) + e/2^\nu$ and $z^{(\nu)}$ is in $i(y^{N(\nu)})$.
For the given $e > 0$ we define the set C_e as in 5, and we have

$$C_e \subset \Gamma_i \subset \sum i(y^m) + \sum y^p + \sum i(y^{N(\nu)}).$$

Now we saw in III 9, that every interval $i(y^m)$ of order m is a neighborhood of our space namely $N_{1/m}(y^m)$, and every terminating graph y^p is a neighborhood, namely $N_{1/(p+1)}(y^p)$. Hence we have the closed compact set C_e covered by a countable class of open sets. A well known topological theorem tells us that C_e is covered by a finite sub-class of the original open sets, *i.e.*,

$$C_e \subset \sum{}' i(y^m) + \sum{}' y^p + \sum{}' i(y^{N(\nu)})$$

the primed sigmas indicating summation over a sub-class of the original summands. Moreover we know that every sum of intervals can be expressed as a disjoint sum of a subclass of the original intervals so that we have

$$C_e \subset \sum{}'' i(y^m) \dotplus \sum{}'' y^p \dotplus \sum{}'' i(y^{N(\nu)}). \text{ The latter sum we call } S.$$

Since there are only a finite number of intervals in S, consisting exclusively of either intervals of finite order or terminating graphs, we may decompose Γ_i into a disjoint sum of such intervals, among which will be those of S (see II 14.):

$$\Gamma_i = S \dotplus \sum \bar{y}^p \dotplus \sum i(\bar{y}^m)$$

and such that

$$1 = \sum{}'' m(i(y^m)) + \sum{}'' m(y^p) + \sum{}'' m(i(y^{N(\nu)})) + \sum m(\bar{y}^p) + \sum m(i(\bar{y}^m)).$$

Taking complements in the preceding inclusion we obtain

$$C'_e \supset S'$$

that is,

$$\sum_y \dot{i}(y^n) \supset \sum \dot{\bar{y}}^p \dotplus \sum \dot{i}(\bar{y}^m) \ .$$

Now each summand on the right is contained in one on the left (II 15) and thus using the results of (5) and (4)

$$e > \sum_y m(i(y^n)) \geq \sum m(\bar{y}^p) + \sum m(i(\bar{y}^m))$$

$$= 1 - \left(\sum{}'' m(i(y^m)) + \sum{}'' m(y^p) + \sum{}'' m(i(y^{N(\nu)})) \right)$$

$$\geq 1 - \left(\sum m(i(y^m)) + \sum m(y^p) + \sum m(i(y^{N(\nu)})) \right)$$

$$\geq 1 - \left(\sum m(i(y^m)) + \sum m(y^p) + \sum m(z^\nu) + e \right)$$

$$= 1 - \sum m(j) - e.$$

So, $\sum m(j) \geq 1 - 2e$ for every $e > 0$ and $m(j) \geq 1$, as was to be proved. From (4) we therefore see that $\Gamma_i = \dot{\sum} j$ implies $1 = \sum m(j)$.

7. Finally, suppose $i(y^n) = \dot{\sum} j$. Then we may decompose Γ_i into a sum of disjoint intervals (see II 14) in two ways:

$$\Gamma_i = i(y^n) \dotplus \sum \dot{k} = \sum \dot{j} \dotplus \sum \dot{k}.$$

By (6) we know that $1 = m(i(y^n)) + \sum m(k)$ and also $1 = \sum m(j) + \sum m(k)$, so that $m(i(y^n)) = \sum m(j)$. This completes the proof of property m 3 of I.

8. Hence we see that every generating transformation G defines a G-measure for intervals of Γ_i, and thus a measure for the additive class of measurable sets as indicated in I.

V. $m(T) = x_i^0$

1. Let T be the set of all terminating graphs y. It is trivial that T is measurable since it may be regarded as the sum of its countably many points each of which is an interval. Now $p_k(i; 0)$ is the probability of death in the k-th generation and as such represents the sum of the measures of all graphs in the set $T_0 + \ldots + T_{k-1}$. It follows that $m(T) = \sum_{k=0}^{\infty} \sum_{y \in T_k} m(y) = \lim_k \sum_{y \in T_k} m(y) = \lim_k p_k(i; 0) = x_i^0$. Thus we have

Theorem. The G-measure of the set T of all terminating graphs of Γ_i is x_i^0, the i-th component of the death-fixed-point x^0 of G. Hence, in a subcritical system ($x^0 = 1$), almost all graphs terminate.

VI. A Strong Ratio Theorem for Supercritical Systems

1. Let G be the generating transformation: $g_i(x) = \sum p_1(i; j) x^j$ for a supercritical system with first moment matrix $M = [m_{ij}]$. We recall that $m_{ij} = (\partial g_i / \partial x_j)_1$ and the maximal positive characteristic root r of M is greater than unity, possessing a unique left characteristic vector v with positive components v_i, and having norm $\|v\| = 1$.

By the e-cone C_e of the system ($e > 0$) is meant the vector $j = 0$ together with all vectors j such that $\|j/a - v\| < e$ for some positive a.

We proved in I 3, a weak ratio theorem to the effect that, for every $e > 0$, $f > 0$, there exists a K such that, for every $k \geq K$,

$$\sum_{j \notin C_e} p_k(i; j) < f.$$

2. We prove in this section a much stronger result. Consider the set of all graphs z of Γ_i. Let T be the set of all terminating graphs y of Γ_i. For an arbitrary graph z denote by $j(z_k)$ the vector $j = (j_1, \ldots, j_t)$ whose component j_s equals the number of particles of type s in the k-th generation of z. Let L be the set of all non-terminating graphs z such that $\lim_k$ ray $j(z_k) = $ ray v. Precisely, L consists of all non-terminating z such that, for every $e > 0$, there exists K such that, for all $k \geq K$, $j(z_k)$ is in C_e. Define $N = \Gamma_i - (T + L)$. We thus have

$$\Gamma_i = T \dot{+} L \dot{+} N$$

where N consists of all non-terminating z for which $e > 0$ exists so that for all K, there exists a $k \geq K$ for which $j(z_k)$ is outside C_e. We may now state the

Strong Ratio Theorem (SRT). For a super-critical system, the set N of non-terminating graphs of Γ_i which do not approach the ray v is measurable and $m(N) = 0$. Hence L is measurable and $m(L) = 1 - x_i^0$.

The proof is difficult and we may proceed by means of lemmas 1–20.

Lemma 1. The set N may be decomposed into subsets: $N = N_1 + N_2 + \ldots$ where N_n is the set of all non-terminating graphs such that for every K, there is a $k \geq K$ such that $j(z_k)$ is outside $C_{1/n}$.

Clearly every N_n is a subset of N. Moreover if z is in N, where, for every K there exists a $k \geq K$ for which $j(z_k)$ is outside C_{e_z}, then, defining n so that $1/n < e_z$, and noting that $C_{1/n}$ is in C_{e_z}, we see that z is in N_n.

Lemma 2. To prove the SRT, it suffices to show that if N_e is the set of all non-terminating graphs such that for every K, a $k \geq K$ exists for which $j(z_k)$ is outside C_e, then outer measure $O(N_e) = 0$.

For then $O(N_n) = 0$ for every N_n of the decomposition of Lemma 1, and so $O(N) \leq \sum O(N_n) = 0$, whence N is measurable and $m(N) = 0$.

We must now adopt some notations: Let $g_i^K(x) = \sum p_K(i;j)x_1^{j_1} \ldots x_t^{j_t}$. Then $g_i^{K+1}(x) = g_i^K(g_1,\ldots,g_t) = \sum p_K(i;j)g_1^{j_1} \ldots g_t^{j_t} \equiv \sum p_K(i;j) \sum P_j(j^1) x_1^{j_1} \ldots x_t^{j_t}$ and so on. Generally,

$$g_i^{K+s}(x) = \sum_j p_K(i;j) \sum_{j^1} P_j(j^1) \ldots \sum_{j^s} P_{j^{s-1}}(j^s) x_1^{j_1^s} \ldots x_t^{j_t^s}.$$

Lemma 3. Let N_e be given as in Lemma 2. To prove $O(N_e) = 0$, it suffices to show that, for every $f > 0$ there exists a K such that

$$\sum_{j \notin C_e} p_K(i;j) + \sum_{\substack{j \neq 0 \\ j \in C_e}} p_k(i;j) \sum_{j^1 \notin C_e} P_j(j^1) +$$

$$\sum_{\substack{j \neq 0 \\ j \in C_e}} p_K(i;j) \sum_{\substack{j^1 \neq 0 \\ j^1 \in C_e}} P_j(j^1) \sum_{j^2 \notin C_e} P_{j^1}(j^2) + \ldots \leq f.$$

For, let N_e be fixed, and fix $f > 0$. Then, assuming such a K exists, we have for *this* K that every z in N_e has $z(j_k)$ not in C_e for some *first* $k \geq K$. Define y^k in T_k so $y_k^k = z_k$. Then $z \in i(y^k)$. But the above sum is the sum of all $m\big(i(y^k)\big)$ for all y^k so defined. Hence $O(N_e) \leq f$ for every $f > 0$, and so $O(N_e) = 0$.

It is the condition of Lemma 3 which makes clear the relation of the SRT with the weaker form.

Lemma 4. To prove $m(N_e) = 0$, it suffices to show that, for every f there is a K such that for all $s \geq 1$,

$$
S_{K,s} \equiv \sum_{\substack{j \notin C_e}} p_K(i;j) + \sum_{\substack{j \neq 0 \\ j \in C_e}} p_K(i;j) \sum_{\substack{j^1 \notin C_e}} P_j(j^1) + \ldots +
$$

$$
\sum_{\substack{j \neq 0 \\ j \in C_e}} P_K(i;j) \sum_{\substack{j^1 \neq 0 \\ j^1 \in C_e}} P_j(j^1) \ldots \sum_{\substack{j^s \notin C_e}} P_{j^{s-1}}(j^s) \leq f \ .
$$

Now, setting $x = 1$ in the form obtained for $g_i^{K+s}(x)$, we see that

$$
1 = p_K(i;0) + \sum_{\substack{j \notin C_e}} p_K(i;j) + \sum_{\substack{j \neq 0 \\ j \in C_e}} p_K(i;j) \geq
$$

$$
p_K(i;0) + \sum_{\substack{j \notin C_e}} p_K(i;j) + \sum_{\substack{j \neq 0 \\ j \in C_e}} p_K(i;j) \sum_{\substack{j^1 \notin C_e}} P_j(j^1) +
$$

$$
\sum_{\substack{j \neq 0 \\ j \in C_e}} p_K(i;j) \sum_{\substack{j^1 \neq 0 \\ j^1 \in C_e}} P_j(j^1) \geq \ldots \geq p_K(i,0) + S_{K,s} +
$$

$$
\sum_{\substack{j \neq 0 \\ j \in C_e}} p_K(i;j) \sum_{\substack{j^1 \neq 0 \\ j^1 \in C_e}} P_j(j^1) \ldots \sum_{\substack{j^s \neq 0 \\ j^s \in C_e}} P_{j^{s-1}}(j^s) \ .
$$

The latter expression we denote by $I_{K,s}$.

Lemma 5. To prove $O(N_e) = 0$, it suffices to show that, for every $f > 0$ there exists a K such that, for all $s \geq 1$,

$I_{K,s} \geq 1 - p_K(i;0) - f$.

For then $S_{k,s} \leq 1 - p_K(i;0) - I_{K,s} \leq 1 - p_K(i;0) - \big(1 - p_K(i;0) - f\big) = f$.

We must now consider more closely than we have done hitherto the transformations $T(n) = nM/s(nM)$ and $S(n) = nM/\|nM\|$, each of which we regard as operating on the set of non-null vectors with non-negative components. One sees easily that the k-th iterates are given by $T^k(n) = nM^k/s(nM^k)$ and $S^k(n) = nM^k/\|nM^k\|$. We have already denoted by v the characteristic vector of norm 1, for which $vM = rv$, and for our present purposes we let $\tilde{v} = v/s(v)$ so that $s(\tilde{v}) = 1$, and $\tilde{v}M = r\tilde{v}$. Clearly $T^k(\tilde{v}) = \tilde{v}$ and $S^k(v) = v$. The transformation $T(n)$ is a mapping onto the plane $s(x) = 1$, $S(n)$ is onto the sphere $\|x\| = 1$. Clearly $S^k(n)/s(S^k(n)) = T^k(n)$ while $T^k(n)/\|T^k(n)\| = S^k(n)$. We first note some trivial relations.

Lemma 6. If j has non-negative components, then

$$\|j\| \leq s(j) \ .$$

For $\|j\|^2 = \sum j_m^2 \leq \left(\sum j_m\right)^2 = s^2(j) \ .$

Lemma 7. If j is an arbitrary vector, then $\|j\| \geq |s(j)|/t \geq s(j)t$.

For $\|j\| = \left(\sum j_m^2\right)^{\frac{1}{2}} \geq |j_m|$, $m = 1,\ldots,t$, and $t\|j\| \geq \sum |j_m| \geq |\sum j_m| = |s(j)| \geq s(j)$.

Lemma 8. If x and y are arbitrary vectors, $\left(\sum x_i y_i\right)^2 \leq \|x\|^2 \|y\|^2$.

This is the well-known Schwarz inequality.[2]

Lemma 9. If x is an arbitrary vector, then $\|xM\| \leq \|x\|\left(\sum_{ij} m_{ij}^2\right)^{\frac{1}{2}}$

$\equiv \|x\|\sigma$.

For, $\|xM\|^2 = \sum_j \left(\sum_i x_i m_{ij}\right)^2 \leq \sum_j \sum_i x_i^2 \sum_i m_{ij}^2 = \|x\|^2 \sigma^2$. We have used the Schwarz inequality here.

Lemma 10. If n is a vector with non-negative components, then $\|nM\| \geq \|n\|\left(\min_i \sum_j m_{ij}^2\right)^{\frac{1}{2}} \equiv \|n\|\sigma_1$.

For $\|nM\|^2 = \sum_j \left(\sum_i n_i m_{ij}\right)^2 \geq \sum_j \sum_i n_i^2 m_{ij}^2 = \sum_i n_i^2 \sum_j m_{ij}^2 \geq \sum_i n_i^2 \min_i \sum_j m_{ij}^2 = \|n\|^2 \cdot \sigma_1^2$.

We have now the means of establishing the existence of a Lipschitz constant for the transformation $S(n)$.

Lemma 11. There exists a constant $L > 1$ such that $\|S(n) - S(n')\|$ $\leq L\|n-n'\|$ for all vectors n, n' which are non-null and have non-negative components, $\|n'\|$ being unity.

For $\|S(n) - S(n')\| = \|nM/\|nM\| - n'M/\|n'M\|\| \leq$ $\|nM/\|nM\| - nM/\|n'M\|\| + \|nM/\|n'M\| - n'M/\|n'M\|\| =$ $|1/\|nM\| - 1/\|n'M\|| \cdot \|nM\| + \|nM - n'M\|/\|n'M\| =$ $|\|nM\| - \|n'M\||/\|n'M\| + \|nM - n'M\|/\|n'M\| \leq 2\|(n - n')M\|/\|n'M\| \leq$ $2\sigma\|n - n'\|/\|n'\|\sigma_1$, by Lemmas 9, 10. Now $\|n'\| = 1$. Hence $L \equiv 2\sigma/\sigma_1$ serves.

Lemma 12. If n and n' are non-null, non-negative component vectors and $\|n'\| = 1$, then $\|S^k(n) - S^k(n')\| \leq L^k\|n - n'\|$, $k = 1, 2, \ldots$.

For $\|S^2(n) - S^2(n')\| \leq L\|S(n) - S(n')\| \leq L^2\|n - n'\|$, and so on. Note that $\|S^k(n')\| = 1$.

The transformation $T(n)$ has topological and algebraic advantages since its denominator is a linear functional (see I 3). However $S(n)$ has simplicity from the point of view of norm inequalities. We work with whichever seems more adaptable to the current purpose. The following two lemmas give inequalities connecting the two operators.

Lemma 13. Let j and k be vectors with non-negative components and having $s(j) = 1 = s(k)$. Define $j' = j/\|j\|$ and $k' = k/\|k\|$ as their projections on the unit sphere. Then $\|j' - k'\| \leq 2t\|j - k\|$.

For $\|j' - k'\| = \|j/\|j\| - k/\|j\|\| \leq \|j/\|j\| - j/\|k\|\| + \|j/\|k\| - k/\|k\|\|$ $= |1/\|j\| - 1/\|k\|| \cdot \|j\| + \|j - k\|/\|k\| = |\|j\| - \|k\||/\|k\| + \|j - k\|/\|k\| \leq$ $2\|j - k\|/\|k\| \leq 2t\|j - k\|/s(k) = 2t\|j - k\|$. Here we have used Lemma 7.

Lemma 14. Let j' and k' be vectors with non-negative components and $\|j'\| = 1 = \|k'\|$. Define $j = j'/s(j')$, $k = k'/s(k')$, their projections on the unit plane $s(x) = 1$. Then $\|j - k\| \leq (t + 1)\|j' - k'\|$.

For $\|j - k\| = \|j'/s(j') - k'/s(k')\| \leq \|j'/s(j') - j'/s(k')\| +$ $\|j'/s(k') - k'/s(k')\| = |1/s(j') - 1/s(k')| \cdot \|j'\| + \|j' - k'\|/s(k') =$ $|s(j') - s(k')|/s(j')s(k') + \|j' - k'\|/s(k') = |s(j' - k')|/s(j')s(k') +$ $\|j' - k'\|/s(k')$.

But $s(k') \geq \|k'\| = 1, s(j') \geq \|j'\| = 1$ by Lemma 6, and $|s(j' - k')| \leq$ $t\|j' - k'\|$, by Lemma 7.

Lemma 15. For every $e > 0$ there exists a K such that for all $k \geq K$ and all non-null, non-negative component vectors n,

$$\|S^k(n) - v\| < e .$$

We saw in I 3, that for $e/2t > 0$ there exists a K such that for all $k \geq K$ and all non-null non-negative component vectors n, $\|T^k(n) - \tilde{v}\| < e/2t$. Referring to Lemma 13, we see that $T^k(n)$ and $\tilde{v}$ are on the plane $s(x) = 1$, and are vectors with non-negative components. Moreover $S^k(n) = T^k(n)/\|T^k(n)\|$ and $v = \tilde{v}/\|\tilde{v}\|$. Hence $\|S^k(n) - v\| \leq 2t\|T^k(n) - \tilde{v}\| < e$.

Thus the sequence of the transformations $S^k(n)$ is uniformly convergent to v.

We saw in Lemma 11 that $\|S(n) - S(n')\| \leq L\|n - n'\|$ where $L > 1$. Examples show that $S(n)$ need not be a contraction in the sense that the above relation holds for some $L < 1$. We can however show that there exists an iterate $S^q(n)$ which is a contraction toward v. It seems simpler first to establish the analogous result for $T(n)$.

We recall that the simplex $\{\delta^1, \ldots, \delta^t\}$ means the set of all vectors $n = \sum n_i \delta^i$, $n_i \geq 0$, $\sum n_i = 1$, that is, all vectors $n = (n_1, \ldots, n_t)$ with non-negative components and $s(n) = 1$. By the boundary of the simplex we mean all of its points having at least one component zero. Since $\tilde{v}M = r\tilde{v}$, $s(\tilde{v}) = 1$, and M has all elements positive, v can have no component zero and is therefore not a boundary point. One sees therefore that there is a minimum distance $D > 0$ from $\tilde{v}$ to the boundary of the simplex $\{\delta^1, \ldots, \delta^t\}$.

Lemma 16. For every $f > 0$ there exists a Q such that for all $q \geq Q$, $\|T^q(n) - \tilde{v}\| \leq f\|n - \tilde{v}\|$ for all vectors n with non-negative components and having $s(n) = 1$.

Proof. Let $V = \max \tilde{v}_i / \min \tilde{v}_i \geq 1$. Let $D = \min \{\|b - v\|\} > 0$ where b ranges over the boundary of the simplex $\{\delta^1, \ldots, \delta^t\}$. If f be given, we know that for the positive constant fD/Vt there exists a Q such that for all $q \geq Q$, $\|T^q(n) - \tilde{v}\| < fD/Vt$ for all non-zero n with non-negative components.

Now let $n \neq \tilde{v}$ be an arbitrary such vector with $s(n) = 1$. The half-ray $\tilde{v} + \alpha(n - \tilde{v})$, $\alpha \geq 0$, cuts the boundary of the simplex $\{\delta^1, \ldots, \delta^t\}$ in a unique point b, and n may thus be written

$$n = h\tilde{v} + kb, \quad h \geq 0, \quad k > 0, \quad h + k = 1.$$

Thus for $q \geq Q$, $\|T^q(n) - \tilde{v}\| = \|nM^q/s(nM^q) - \tilde{v}\| =$

$$\|h\tilde{v}M^q + kbM^q/hs(\tilde{v}M^q) + ks(bM^q) - \tilde{v}\| =$$

$$\|hr^q\tilde{v} + kbM^q/hr^q + ks(bM^q) - \tilde{v}\| =$$

$$k\|bM^q - s(bM^q)\tilde{v}\|/hr^q + ks(bM^q) =$$

$$ks(bM^q)\|T^q(b) - \tilde{v}\|/hr^q + ks(bM^q) \leq$$

$$ks(bM^q)f\|b - \tilde{v}\|/Vt(hr^q + ks(bM^q)) =$$

$$f\|kb - k\tilde{v}\|/Vt(hr^q/s(bM^q) + k).$$

Now $\|kb - k\tilde{v}\| = \|kb - (1-h)\tilde{v}\| = \|h\tilde{v} + kb - \tilde{v}\| = \|n - \tilde{v}\|$. Also, note that $\min \tilde{v} \sum_i m_{ij}^q \leq \sum_i \tilde{v}_i m_{ij}^q = r^q \tilde{v}_j \leq r^q \max \tilde{v}$, and hence $\sum_i (m_{ij}^q)^2 \leq$ $(\sum_i m_{ij}^q)^2 \leq r^{2q}V^2$ and $(\sum_i (m_{ij}^q)^2)^{\frac{1}{2}} \leq r^q V$. Thus $s(bM^q) = \sum_j \sum_i b_i m_{ij}^q$ $\leq \sum_j \|b\|(\sum_i (m_{ij}^q)^2)^{\frac{1}{2}} \leq s(b) \sum_j (\sum_i (m_{ij}^q)^2)^{\frac{1}{2}} \leq tr^q V$.

Hence $hr^q/s(bM^q) + k \geq h/tV + k = 1 - k + ktV/tV = 1 + k(tV-1)/tV \geq 1/tV$. Thus we see that $\|T^q(n) - \tilde{v}\| \leq f\|n - \tilde{v}\|$.

Lemma 17. For every $e > 0$, there exists a Q such that for all $q \geq Q$, $\|S^q(n') - v\| \leq e\|n' - v\|$, for all n having non-negative components and $\|n'\| = 1$.

Fix $e > 0$. Define $f = e/2t(t+1)$. Then by Lemma 16, Q exists so that for all $q \geq Q$, $\|T^q(n) - \tilde{v}\| \leq e\|n - \tilde{v}\|/2t(t+1)$ for all n with non-negative components and $s(n) = 1$.

For arbitrary n' with $\|n'\| = 1$ and non-negative components, define $n = n'/s(n')$. Then we have $\|S^q(n') - v\| \leq 2t\|T^q(n') - \tilde{v}\| \leq 2t\|T^q(n) - \tilde{v}\| \leq 2te\|n - \tilde{v}\|/2t(t+1) \leq e(t+1)\|n' - v\|/t+1 = e\|n' - v\|$. We have used Lemmas 13, 14, and the fact that $T^q(n') = n'M^q/s(n'M^q) = nM^q/s(nM^q) = T^q(n)$.

Just as the possible failure of $S(n)$ to contract all vectors toward v forced us to prove Lemma 17, so does the possible failure of the transformation nM to increase the norm of n cause difficulties for which we must provide in the next two lemmas.

Lemma 18. There exists a constant $e > 0$ such that for all vectors n satisfying $\|n-v\| < e$, one has $\|nM\| \geq \|n\|(1+r)/2 \equiv m\|n\|$, $1 < m < r$.

For the function $R(x) \equiv \|xM\|/\|x\|$ is continuous at $x = v$ and $R(v) = r > 1$. Thus for $m = (1+r)/2 < r$ there exists an $e > 0$ such that $\|n - v\| < e$ implies $R(x) \geq m$.

Lemma 19. There exists a K such that for all $k \geq K$, $\|nM^k\| \geq \|n\|$ for all n, with non-negative components.

Note that if w is the right characteristic root of M, with $s(w) = 1$, $M^k w = r^k w$, we have max $w \sum_j m_{ij}^k \geq \sum_j m_{ij}^k w_j = r^k w_i \geq r^k$ min w so $\sum_j m_{ij}^k \geq r^k W$ where $W \equiv \min w / \max\ w$. Now $\|nM^k\| \geq s(nM^k)/t = \sum_{ij} n_i m_{ij}^k/t = \sum_i n_i \sum_j m_{ij}^k/t \geq r^k W s(n)/t \geq r^k W \|n\| t$. Take K so $r^K W/t > 1$.

By virtue of Lemma 5, it now remains to prove the final lemma for which we have prepared all the essential tools.

Lemma 20. Let $e > 0$, $f > 0$ be fixed. Then there is a K such that, for all $s \geq 1$,

$$\sum_{\substack{j \neq 0 \\ j \in C_e}} p_K(i;j) \sum_{\substack{j^1 \neq 0 \\ j^1 \in C_e}} P_j(j^1) \cdots \sum_{\substack{j^s \neq 0 \\ j^s \in C_e}} P_{j^{s-1}}(j^s) \geq 1 - p_K(i;0) - f.$$

Our proof is based upon a complicated construction of K which was, of course, obtained by looking from the other end. In spite of its glaring artificiality we give the direct route:

1. By Lemma 18, we fix $\bar{e}$ so that $0 < \bar{e} < e$ and $\|n - v\| < \bar{e}$ implies $\|nM\| \geq m\|n\|$ where $m = (1+r)/2 > 1$.

2. By Lemma 17, there exists a q such that $\|S^q(n') - v\| \leq \|n' - v\|/4$ for all non-negative n' with $\|n'\| = 1$. We determine an $\bar{\bar{e}}$ satisfying the following conditions:

    a. $0 < \bar{\bar{e}} < \bar{e} < e$, and $\bar{\bar{e}} < 1$.

    b. $\bar{\bar{e}} L^{q-1} < \bar{e}/2$, that is $\bar{\bar{e}} < \bar{e}/2L^{q-1}$ where L is the Lipschitz constant of Lemma 11, and q is the constant just defined.

3. Fix $E > 0$ so that

    a. $E(1 + L + L^2 + \ldots + L^{q-1}) < \bar{\bar{e}}/2$,

    b. $m(1 - E) > 1$, that is $E < (m-1)/m < 1$.

Note that we now have automatically, for all $s \leq q$, $E(1 + L + L^2 + \ldots + L^{s-2}) + \bar{\bar{e}}\, L^{s-1} \leq E(1 + L + L^2 + \ldots + L^{q-2}) + \bar{\bar{e}}\, L^{q-1} < \bar{\bar{e}}/2 + \bar{e}/2 < \bar{e}/2 + \bar{e}/2 = \bar{e}$.

4. Fix k so that

    a. $\|S^k(n) - v\| < \bar{\bar{e}}/2$ for all non-negative $n \neq 0$ (see Lemma 15).

    b. $\|nM^k\| \geq \|n\|$, for all non-negative n. (see Lemma 19).

5. We define $d_k = \sum_j \left(\sum_i (d_{ij}^k)^2 \right)^{\frac{1}{2}}$, d_{ij}^k being the dispersion of g_i^k with respect to x_j, and define $m_k = \left(\min_i \sum_j (m_{ij}^k)^2 \right)^{\frac{1}{2}}$.

We now fix $A > 0$ so

    a. $4d_k / \bar{\bar{e}}^2 m_k^2 A < f/4$.

    b. $d_1 / E^2 m_1^2 (1 - \bar{\bar{e}}) A < 1$.

    c. $d_1 m(1 - E) / E^2 m_1^2 (1 - \bar{\bar{e}}) A (m(1 - E) - 1) < f/2$.

6. Fix r so that
$$\sum_{0 < \|n\| \leq A} p_r(i; n) < f/4, \text{ see I 3.}$$

7. Fix $K = r + k$.

We contend that this K satisfies the condition of the lemma.
First note that $g_i^K(x) - \sum p_K(i;j)x_1^{j_1} \ldots x_t^{j_t} = g_i^r(G^k(x)) = \sum p_r(i;n) \sum P_{k,n}(j)x_1^{j_1} \ldots x_t^{j_t}$, where $(g_1^k)^{n_1} \ldots (g_t^k)^{n_t} = \sum P_{k,n}(j)x_1^{j_1} \ldots x_t^{j_t}$ by definition. Setting $x = 1$, we may write

$$1 = p_r(i;0) + \sum_{0 < \|n\| \leq A} p_r(i;n) \sum_j P_{k,n}(j) +$$

$$\sum_{\|n\| > A} p_r(i;n) \sum_{\|j/\|nM^k\| - v\| \geq \bar{\bar{e}}} P_{k,n}(j) +$$

$$\sum_{\|n\| > A} p_r(i;n) \sum_{\|j/\|nM^k\| - v\| < \bar{\bar{e}}} P_{k,n}(j)$$

Now just as in I 3, we see that, for arbitrary $R > 0$,

$$\sum_{\|j - NM^k\| \geq R} P_{k,n}(j) \cdot R^2 \leq \sum P_{k,n}(j) \|j - nM^k\|^2 = \sum_{ij} n_i d_{ij}^k \leq$$

$$\|n\| \sum_j \left(\sum_i (d_{ij}^k)^2 \right)^{\frac{1}{2}} = \|n\| d_k .$$

Thus, $\qquad \displaystyle\sum_{\|j - nM^k\| \geq R} P_{k,n}(j) \leq \|n\| d_k / R^2 .$

Setting $R = \|nM^k\| \, \bar{\bar{e}} \, / 2$, we have

$$\sum_{\|j/\|nM^k\| - S^k(n)\| \geq \bar{\bar{e}} \, /2} P_{k,n}(j) \leq \;\; 4\|n\| d_k / \, \bar{\bar{e}}^{\,2} \|nM^k\|^2 \leq 4\|n\| d_k / \, \bar{\bar{e}}^{\,2} \|n\|^2 m2_k =$$

$4 d_k / \, \bar{\bar{e}}^{\,2} m_k^2 \|n\|$, where $m_k = \left(\min_i \sum_j (m_{ij}^k)^2 \right)^{\frac{1}{2}}$ (see Lemma 10).

However, if $\|j/\|nM^k\| - v\| \geq \bar{\bar{e}}$ then j is on the preceding range, for if not, $\|j/\|nM^k\| - S^k(n)\| < \bar{\bar{e}} \, /2$ and $\|S^k(n) - v\| < \bar{\bar{e}} \, /2$ by choice of k, whence $\|j/\|nM^k\| - v\| < \bar{\bar{e}}$, a contraction. Hence

$$\sum_{\|j/\|nM^k\| - v\| \geq \bar{\bar{e}}} P_{k,n}(j) \leq \; 4 d_k / \, \bar{\bar{e}}^{\,2} m_k^2 \|n\|.$$

Thus $\displaystyle\sum_{\|n\| > A} p_r(i;n) \displaystyle\sum_{\|j/\|nM^k\| - v\| \geq e} P_{k,n}(j) \; \leq \; \displaystyle\sum_{\|n\| > A} p_r(i;n) \cdot 4dk / \, \bar{\bar{e}}^{\,2} m_k^2 A \leq$

$(f/4) \displaystyle\sum_{\|n\| > A} p_r(i;n) \leq f/4 .$

But by choice of r

$$\sum_{0 < \|n\| \leq A} p_r(i;n) \sum_j P_{k,n}(j) = \sum_{0 < \|n\| \leq A} p_r(i;n) < f/4 .$$

And, as always, $p_r(i;0) \leq p_{r+k}(i;0)$. Thus we obtain from the original equation that

$$\sum_{\|n\| > A} p_r(i;n) \sum_{\|j/\|nM^k\| - v\| < \bar{\bar{e}}} P_{k,n}(j) \; \geq 1 - p_{r+k}(i;0) - \frac{1}{2} f .$$

Every vector j involved on the range above has the property that $\|j/a - v\| < \bar{\bar{e}}$ for some $a > A$, for $\|nM^k\| \geq \|n\| > A$ by choice of k. Thus

$$\sum_{\substack{\{\|j/a-v\|<\bar{\bar{e}} \\ a>A}} p_K(i;j) \geq 1 - p_K(i;0) - \frac{1}{2}f \ .$$

Since $\|v\| = 1 > \bar{\bar{e}}$ by choice of $\bar{\bar{e}}$, $j = 0$ is not on this range.

Note the two trivial remarks:

A. If $\|\alpha j' - v\| < e$, $\alpha > 0$, $\|j'\| = 1$, then $\|j' - v\| < 2e$.

For $e > \|\alpha j' - v\| \geq \left| \|\alpha j'\| - \|v\| \right| = |\alpha - 1|$, and $\|j' - v\| = \|j' - \alpha j'\| + \|\alpha j' - v\| = |1 - \alpha| \cdot \|j'\| + e < 2e$.

B. Corollary. If $\|j/a - v\| < e$ and $j' = j/\|j\|$ then $\|j' - v\| < 2e$.

Consider now the product

$$\sum_{\substack{\{\|j/a-v\|<\bar{\bar{e}} \\ a>A}} p_K(i;j) \qquad \sum_{\|j^1/\|jM\|-S(j)\|<E} P_j(j^1) \qquad \sum_{\|j^1/\|j^1M\|-S(j^1)\|<E} P_{j^1}(j^2) \quad \cdots$$

$$\sum_{\|j^{q-1}/\|j^{q-2}M\|-S(j^{q-2})\|<E} P_{j^{q-2}}(j^{q-1}) \qquad \sum_{\|j^q/\|j^{q-1}M\|-S(j^{q-1})\|<E} P_{j^{q-1}}(j^q) \quad \cdots \qquad (*)$$

Since $\bar{\bar{e}} < 1$, $E < 1$, and $\|v\| = 1 = \|S(j^s)\|$, we see that no j or j^s is ever zero on these ranges. Applying Lemma 12 we find that

$$\|j^q/\|j^{q-1}M\| - S(j^{q-1})\| < E,$$

$$\|S(j^{q-1}) - S^2(j^{q-2})\| < LE,$$

$$\|S^2(j^{q-2}) - S^3(j^{q-3})\| < L^2 E \ldots$$

$$\|S^{q-2}(j^2) - S^{q-1}(j^1)\| < L^{q-2}E$$

$$\|S^{q-1}(j^1) - S^q(j)\| < L^{q-1}E, \text{ and hence}$$

$$\|j^q/\|j^{q-1}M\| - S^q(j)\| < E(1 + L + \ldots + L^{q-1}) \ .$$

But since $\|S^q(j/\|j\|) - v\| < \|j/\|j\| - v\|/4 < 2\bar{\bar{e}}/4 < \bar{\bar{e}}/2$ (see remark B and definition of q) we have $\|j^q/\|j^{q-1}M\| - v\| < E(1 + L + \ldots + L^{q-1}) + \bar{\bar{e}}/2 < \bar{\bar{e}}/2 + \bar{\bar{e}}/2 = \bar{\bar{e}}$. (see definition of E).

Similarly we shall find, working from $2q$ to q, that $\|j^{2q}/\|j^{2q-1}M\| - v\| < \bar{\bar{e}}$ and so on for all multiples of q.

Note also, for $s \leq q$,

$$\|j^{s-1}/\|j^{s-2}M\| - S(j^{s-2})\| < E$$

$$\|S(j^{s-2}) - S^2(j^{s-3})\| < EL \ldots$$

$$\|S^{s-2}(j^1) - S^{s-1}(j)\| < EL^{s-2}$$

$$\|S^{s-1}(j) - v\| < \bar{\bar{e}}\,L^{s-1}, \text{ so}$$

$$\|j^{s-1}/\|j^{s-2}M\| - v\| < E(1 + L + \ldots + L^{s-2}) + \bar{\bar{e}}\,L^{s-1} < \bar{e}.$$

(see definition of E). Similarly all j^{nq+r} on the non-q-multiple postions are in $C_{\bar{e}}$. It is clear now that the ranges for $j, j^s, \ldots$ are all in $C_{\bar{e}} \subset C_e$ and exclude zero. Hence the product of Lemma 20 is greater than or equal to that of (*) carried to the corresponding place.

Now note the elementary result that if $\|j/a - k\| < e_0$ where $\|k\| = 1$, we have $a - \|j\| = \|ak\| - \|j\| \leq |\|j\| - \|ak\|| \leq \|j - ak\| < ae_0$ and hence $\|j\| > a(1 - e_0)$.

We have seen that, on the ranges involved in the (*) product, all j and j^s are in $C_{\bar{e}}$, indeed $\|j/a - v\| < \bar{\bar{e}} < \bar{e}$, and $\|j^s/\|j^{s-1}M\| - v\| < \bar{e}$, all s. By choice of $\bar{e}$, $\|(j/a)M\| \geq m\|j/a\|$ and $\|(j^s/\|j^{s-1}M\|)M\| \geq m\|(j_s/\|j^{s-1}M\|)\|$. But then $\|jM\| \geq m\|j\|$ and $\|j^sM\| \geq m\|j^s\|$. Thus we see that

$\|j/a - v\| < \bar{\bar{e}}$ implies $\|j\| \geq a(1 - \bar{\bar{e}}) > A)(1 - \bar{\bar{e}})$.

$\|j^1/\|jM\| - S(j)\| < E$ implies $\|j^1\| \geq \|jM\|(1 - E) \geq$

$m(1 - E)\|j\| > m(1 - E)A(1 - \bar{\bar{e}})$,

$\|j^2/\|j^1M\| - S(j^1)\| < E$ implies $\|j^2\| \geq \|j^1M\|(1 - E) \geq$

$m(1 - E)\|j^1\| > m^2(1 - E)^2 A(1 - \bar{\bar{e}}), \ldots,$

$\|j^s/\|j^{s-1}M\| - S(j^{s-1})\| < E$ implies $\|j^s\| > m^s(1 - E)^s A(1 - \bar{\bar{e}})$.

Letting $C = A(1 - \bar{\bar{e}})$ and $R = m(1 - E) > 1$ by choice of E, we have on the (*) ranges, $\|j\| > C$, $\|j^1\| > CR$, $\|j^2\| > CR^2, \ldots \|j^{s-1}\| > CR^{s-1}$.

Now regard the distribution defined by $g_1^{j_1^{s-1}} \dots g_t^{j_t^{s-1}} = \sum P_{j^{s-1}}(j^s) x_1^{j_1^s} \dots x_t^{j_t^s}$. We see as before, for arbitrary $n > 0$,

$$\sum_{\|j^s - j^{s-1}M\| \geq N} P_{j^{s-1}}(j^s) \cdot N^2 \leq \sum P_{j^{s-1}}(j^s) \|j^s - j^{s-1}M\|^2 = \sum_{\mu\nu} j_\mu^{s-1} d_{\mu\nu}^1 \leq$$

$\|j^{s-1}\| \cdot d_1$, and hence

$$\sum_{\|j^s - j^{s-1}M\| < N} P_{j^{s-1}}(j^s) \geq 1 - \|j^{s-1}\| d_1 / N^2 .$$

Setting $N = \|j^{s-1}M\| \cdot E$ for all $s \geq 1$

$$\sum_{\|j^s / \|j^{s-1}M\| - S(j^{s-1})\| < E} P_{j^{s-1}}(j^s) \geq 1 - \|j^{s-1}\| d_1 / E^2 \|j^{s-1}M\|^2 \geq$$

$1 - \|j^{s-1}\| d_1 / E^2 \|j^{s-1}\|^2.$ $m_1^2 = 1 - d_1 / E^2 m_1^2 \|j^{s-1}\|.$

Beginning at the last (s)-th term, we have

$$\sum_{\|j^s / \|j^{s-1}M\| - S(j^{s-1})\| < E} P_{j^{s-1}}(j^s) \geq 1 - d_1 / E^2 m_1^2 C R^{s-1} > 0$$

$$\sum_{\|j^{s-1} / \|j^{s-2}M\|\ S(j^{s-2})\| < E} P_{j^{s-2}}(j^{s-1}) \geq 1 - d_1 / E^2 m_1^2 C R^{s-2} > 0$$

. . . .

$$\sum_{\|j^1 / \|jM\| - S(j)\| < E} P_j(j^1) \geq 1 - d_1 / E^2 m_1^2 C > 0 .$$

Note that $d_1 / E^2 m_1^2 C R^i \leq d_1 / E^2 m_1^2 C = d_1 / E^2 m_1^2 (1 - \bar{e}) A < 1$ by choice of A. Hence the (*) product through s factors is greater than or equal to

$$(1 - p_K(i;0) - f/2)(1 - d_1 / E^2 m_1^2 C)(1 - d_1 / E^2 m_1^2 C R) \dots (1 - d_1 / E^2 m_1^2 C R^{s-1}).$$

It is trivially proved by induction that if all p_i and $1 - p_i$ are positive, then $(1 - p_1)(1 - p_2) \dots (1 - p_n) \geq 1 - p_1 - p_2 - \dots - p_n$. Hence the preceding product is greater than or equal to

$$1 - p_K(i;0) - f/2 - \left(d_1/E^2 m_1^2 C\right)\left(1 + 1/R + 1/R^2 + \ldots + 1/R^{s-1}\right) \geq$$

$$1 - p_K(i;0) - f/2 - d_1/E^2 m_1^2 C(1 - 1/R) =$$

$$1 - p_K(i;0) - f/2 - d_1 R/E^2 m_1^2 C(R - 1) =$$

$$1 - p_K(i;0) - f/2 - d_1 m(1 - E)/E^2 m_1^2 (1 - \bar{\bar{e}}) A\left(m(1 - E) - 1\right) \geq$$

$$1 - p_K(i;0) - f/2 - f/2 = 1 - p_k(i;0) - f, \text{ by choice of } A. \textbf{ Q.E.D.}$$

VII. Remarks on Systems Below Critical

1. For a subcritical system $(x^0 = 1)$, we have seen in V that $m(T) = x_i^0 = 1$ so that $m(T') = 0$, and since $N \subset T'$, $O(N) \leq O(T') = m(T') = 0$. Thus, in such a system, it is trivial that $m(N) = 0$. Similarly $L \subset T'$ and $m(L) = 0 = 1 - x_i^0$. Therefore the SRT is self-evident for subcritical systems.

References

I

1. P. Alexandroff, H. Hopf, *Topologie*, Berlin, 1935
2. G. Birkhoff, *Lattice Theory*, Am. Math. Soc. Colloq. Publ. XXV, New York, 1940.
3. D. Hawkins, S. Ulam, Theory of Multiplicative Processes, see Chapter 1.
4. H. Margenau, G. M. Murphy, *Mathematics of Physics and Chemistry*, Van Nostrand, 1943.

II

1. C. J. Everett, S. Ulam, Multiplicative Systems in Several Variables, I, (LA 693) see Part I.
2. D. Hawkins, S. Ulam, Theory of Multiplicative Processes, see Chapter 1.

III

1. H. Cramer, *Mathematical Methods of Statistics*, Princeton University Press, 1946.
2. C. J. Everett and J. R. Ryser, The Gram Matrix and Hadamard Theorem, Am. Math. Monthly, LIII, 1946.
3. C. J. Everett, S. Ulam, Multiplicative Systems in Several Variables, I and II.
4. W. Hurewicz and H. Wallman, *Dimension Theory*, Princeton University Press, 1941.
5. W. Sierpinski, *Introduction to General Topology*, University of Toronto Press, 1934.

4

HEURISTIC STUDIES IN PROBLEMS OF MATHEMATICAL PHYSICS ON HIGH SPEED COMPUTING MACHINES

With John Pasta
(LA-1557, 1953)

This report introduces some elementary but basic methods of a specific version of "brute force" approaches in problems of hydrodynamics which do not yield to analytical methods.

The ideas in this report were further developed, generalized, and modified by Frank Harlow, and continue to play an important role in modern computer calculations in continuum mechanics. (Author's note.)

This paper, the first of a series, is intended primarily to illustrate some possible uses of electronic computers as a means of performing "mental experiments" on mathematical theories and methods of calculation, for a variety of physical phenomena.

There is no unifying principle in the few problems selected for this first report; on the contrary, the aim is to illustrate a certain variety in the problems which one can consider by performing model calculations. Mathematically, these problems are mainly attempts to solve special cases of various partial differential equations; in some cases one prefers to have direct recourse to the physical formulation and perform, as it were, the model experiments on paper rather than solving, by standard conversion of differential to difference expressions, the equations themselves. One general tendency might be noted—we stress an attempt to "observe" a few functionals of the unknown functions, rather than to put credence in the solutions themselves "at each individual point."

The calculations were performed on the Los Alamos electronic computer. It is a pleasure to express our thanks to N. Metropolis for making it available and for his generous help in general. We are indebted to Miss Mary Tsingou who performed much of the work of "coding" the problems, running the machine and checking the procedure and results.

One purpose of this work was to gain a feeling for: first, the time necessary to formulate and prepare the problems for such machine study; second, the time of computing as a function of the size and complexity of the formulation. In general, we restrict ourselves in this first paper to problems where meaningful results could be obtained in a few hours of actual computing.

All the problems discussed in this paper were run on our machine. We append a small selection of the results which we intend to discuss at greater length in our second paper. We want to thank Mrs. Connie Snowden for preparation of the graphs.

1. Hydrodynamical Problems; Heuristic Considerations

In general for problems involving two or more spatial variables, there is at present little hope for obtaining solutions analytically—in closed form. In the several sections that follow, heuristic considerations are set forth with exploratory calculations on the high speed computing machines performed in several cases. The purpose of these was primarily to establish the feasibility of such calculations, to estimate the sizes of problems which can be handled in a reasonable time, and in general to gain experience in the new methods and new fields which, in our opinion, are now open for investigation.

Our approach to the problem of dynamics of continua can be called perhaps "kinetic"—the continuum is treated, in an approximation, as a collection of a finite number of elements of "points;" these "points" can represent actual points of the fluid, or centers of mass of zones, i.e., globules of the fluid, or, more abstractly, *coefficients* of functions, representing the fluid, developed into a series, e.g., Fourier or Rademacher series. This corresponds to the use of general Lagrangian coordinates in classical mechanics; their use can be rigorously justified in problems where entropy is constant—i.e., holonomic systems.

They are always functions of time, which proceeds by discrete intervals.

We thus replace partial differential equations or integer-differential equations by *systems* of total differential equations. The number of elements which we can at present handle is always less than 1,000, the limitation being primarily in the "memory" of the machines, the number of time intervals ("cycles") of the order of 100.

The problems which we study are characterized by lack of symmetry. The positions of our points become, as a rule, more and more irregular as time goes on. This has the consequence that the meaningful results of the calculations are not so much the precise positions of our elements themselves as the behavior, in time, of a _few functionals_ of the motion of the continuum.

Thus in the problem relating to the _mixing_ of two fluids, it is not the exact position of each globule that is of interest but quantities such as the _degree_ of mixing (suitably defined); in problems of turbulence, not the shapes of each portion of the fluid, but the _overall_ rate at which energy goes from simple _modes_ of motion to higher frequencies; in problems involving dynamics of a star cluster, not the individual positions, but quantities like angular momenta of _subsystems_ of the whole system, of size smaller but still comparable to the whole system, etc.

Needless to say, our investigations are in a most preliminary and rudimentary state; we have not made rigorous estimates of the disturbing or smoothing action of the roundoff errors which accumulate, nor the effects of finiteness of the time interval (i.e., the errors due to replacement of differential by differences expressions). In the individual cases that follow the reader will be able to judge for himself how far elementary common sense permits to estimate these effects.

We hope in the future to multiply considerably the number of calculations performed for _each_ of our problems to assay the influence of changes in initial conditions on our conclusions. We repeat that so far the value, if any, of this work is only heuristic.

The first problem considered concerns the behavior of a gas confined in a vessel, expanding into vacuum under its own pressure and weight. The surface is not plane but has an irregularity of a finite size. In other words we consider the problem of instability in the compressible case.

2. Instability and Mixing

The conditions at time $t = 0$ are the following: a vessel (two-dimensional) contains a gas filling it partly, with a boundary against a vacuum. This boundary is not flat but has an irregularity in it in form of a triangular prominence jutting out with dimensions comparable to the diameter of the vessel (about one-fourth of its width).

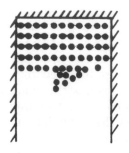

We want to follow the behavior of the expansion of this gas, under its own pressure, in the vacuum below it. Two problems were considered:

1) the gas was assumed weightless, i.e., there was no external force acting on it;

2) the gas, in addition to its own pressure, was acted upon by a constant gravitational field.

The hydrodynamical setup used was the following: the gas was represented by 256 material points. These represent centers of mass of regions in the gas. The treatment is Lagrangian; that is, the calculation follows, in time, the position of these masses. The pressure gradients are represented by forces which our points exert on each other. These are repulsive forces, depending only on the distance between points (and thus having a potential "pressure") and are of the form $F_{ij} = a/r_{ij}^{\propto}$; where the exponent $\propto$ depends on the adiabatic equation of the state of the gas considered.

We chose $\propto = 1$ for our first problem. Given a point $p_i = (x_i, y_i)$, one considers all forces exerted on it by other points (x_j, y_j) and computes their resultant vector.

Actually, we limit the p_j in computing the forces to the "neighbors" of p_i; these are defined as the nearest eight points to p_i. This is done for two reasons: the economy of the computation—we have to calculate only eight instead of 256 of the total number of forces for each point under consideration; the second, more fundamental reason is, of course, that in the gradient of pressure only the local configuration matters. The $1/r$ force law gives divergence at infinity. We might mention here parenthetically that in the search or scanning of points for the nearest eight to the given one, the following was adapted for purely practical, economy, reasons. The points really scanned were 50 "candidates" for the nearest neighbor. They were the 50 *originally* nearest to p_i; the problem was not run long enough so that we had to relocate the original candidates but it is, of course, possible to recheck this periodically. In addition, in order to avoid the use of multiplications and operate only with the much shorter *addition* times, we used, in the search for the nearest eight points, the non-Euclidean metric $\rho(\nu_i, \nu_j) = |\ (x_i - x_j)\ | + |\ (y_i - y_j)\ |$. Once the eight points were found, however, the true Euclidean distances were computed in order to find correctly the resulting force.

We shall not describe here the special treatment which has to be accorded to points which adjoin the walls of the vessel and the points on the boundary of the gas (with the vacuum).

Among the quantities that were printed as the results of the calculation we shall mention here only these: an interesting functional of the motion is the kinetic energy of the gas divided into two parts: the energy of the motion in the x direction and in the y direction. We study $E_1 = 1/2\ \Sigma m_i \dot{x}_i^2$ and $E_2 = 1/2\ \Sigma m_i \dot{y}_i^2$.

The ratio of the two is a function of time and can serve, in a way, as a measure of mixing or irregularity of the motion. One expects due to the initial irregularity of the boundary this ratio to be positive. From the beginning, sidewise motions ensue. Later on one would expect, of course, the motion to be predominantly downward; as the irregularity increases, the ratio of the two quantities should, in Problem I, increase again and approach a constant less than unity.

It is perhaps remarkable that the time behavior found for this ratio was extremely regular; a graph (Fig. 1) for the first 36 cycles is appended.

One word to explain the need to resort to the rather unorthodox procedures outlined above:

It was found impractical to use a "classical" method of calculation for this hydrodynamical problem, involving two independent spatial variables in an essential way (since the gas interface had an irregularity assumed from the beginning). This "classical" procedure, correct for infinitesimal steps in time and space, breaks down for any reasonable (i.e., practical) finite length of step in time. The reason is, of course, that the computation of Jacobians which define the compression assumes that "neighboring" points, determining a "small" area, stay as neighbors for a considerable number of cycles. It is clear that in problems which involve mixing specifically this is not true. Calculations were made just in order to observe the rapidity of change in the "neighbor" relations on the classical pattern and have shown just what was expected to happen: the proximity relations change radically, for points near the boundary, just when the mixing to be studied is starting and the neighboring relation of our points has to be redefined; i.e., the classical way of computing by referring to initial (at time $t = 0$) ordering of points becomes meaningless.

The problem can be treated, of course, using the Eulerian variables (of a set-up due to von Neumann to be discussed in our next report)* where this difficulty does not occur. This Eulerian treatment

---

* There was no subsequent report. (Eds.)

is not suitable, however, for the study of the shape of the boundary—a fictitious, i.e., purely calculational, diffusion and mixing obscures the very phenomenon one wants to study. We shall return to this question in our second report.

In the "classical" set-up involving calculation of Jacobians for determining pressure gradients the cycle was approximately two minutes.

In the problem involving the calculation of forces from the eight "neighbors," the cycle time was three minutes. Meaningful results are obtained in about 100 cycles. Appended are graphs showing the results of a few dozen cycles.

3. Billowing Transformations

The aim of the model calculation here is to exhibit transformations of space which, when applied iteratively to a sphere (or, actually, so far, to a circular region in a plane) will show a sequence of regions, imitating the familiar phenomenon of a ball of smoke billowing outwards. The parameter of iteration is, of course, the time again increasing by discrete intervals.

We assume that the billowing is due to initial irregularities or deviations from a spherical form.

The gas is contained in a region R, and is under internal pressures. We assume first, for simplicity, that it expands into vacuum and we want to study the motion in a highly stylized form—keeping, we hope, the quantitative phenomena essentially correct. The situation then is the following: R is the region occupied by the gas under pressure in the initial position. The following assumptions are made and made plausible qualitatively:

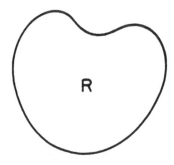

1. The accelerations of points on the boundary are in the direction of the outer normal to the boundary.

2. The motion is computed only for the points on the boundary above; the accelerations at each point depend on the shape of the boundary which is approximately correct in the following approximations:

The density distribution and the pressure well inside the ball is essentially uniform—if the time elapsed since the origin of our ball is long enough—i.e., the instantaneous changes in the position of the particles on the boundary are small compared to the dimensions of our object. "The pressures inside have had time to equalize."

3. On the boundary itself—the accelerations along the normal to each point depend on the curvature at each point in the following way:

There is a positive term outward in n with an additive term whose sign is that of the *curvature* c of the boundary at the point which is considered. This is due to the convergence (or divergence) effects of the streamlines.

E.g., in case 2 the local density at P is probably greater than in the nearby points because of converging motions of points in a band near the surface; in case 1 there is correspondingly a local rarefaction. So we set

$$n_{tt} = p + ac \qquad (1)$$

where c is the curvature of the boundary. The terms p and u are not constant but vary, say, with the volume enclosed by the boundary—the pressure diminishing, e.g., inversely with the volume.

A qualitatively similar formulation, (in a discussion with John von Neumann) *in the small*, would be this: consider the part of the curve given by $y = f(x)$

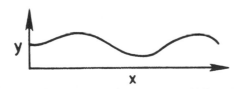

f is positive and the direction of expansion is upward.

We write

$$y_{ttt} = y_{xx} \qquad (2)$$

with the initial conditions at $t = 0$, $y_{tt} = 1$, $y_t = 0$, say, and $y = f(x)$. The tendency is again the same in convex points. The accelerations upwards are decreased, while in concave ones (from which one expects "jetting") they are increased.

Setting $y = \phi(t)\,\chi(x)$ and say $\chi(x) = cos(nx - \beta)$, one gets $\phi'''(t) = -n^2\,\phi$. Setting $\phi = e^{\alpha(t)}$, α is the cubic root of $-n^2$. It follows, since the dominant part will be due to the root with positive real and imaginary parts that the motion is highly unstable. A kind of billowing or swirling will take place—the concave parts will puff out, giving rise later to at least two concavities on its sides. This will repeat the same phenomenon later, etc. It would seem then that a multiplication of irregularities takes place—their number will increase rapidly—accompanied with a continuous increase in size.

In three dimensions the problem is much harder to treat. The additional term in the acceleration must be set up in terms of the two principal radii of curvature at each point.

Our computation program started in two dimensions as follows. Initially the boundary is taken to be composed of sixty-four equal segments defined by points (x_i, y_i) on the surface. The magnitude and sense of the curvature at (x_i, y_i) is derived from the cross product of the two vectors $(x_i - x_{i-1}, y_i - y_{i-1})$ and $(x_{i+1} - x_i, y_{i+1} - y_i)$. From the coordinates (x, y) at two consecutive time levels the positions at the next time level are found by integrating $n_{tt} = p + ac$.

Because of the difficulty of plotting many points every time and in order to observe the motion, it was decided to display the points on one of the cathode ray tubes of the computer memory section. Each such tube has a 32×32 array of points, each representing a binary digit in one of the machine's 1024 forty bigit numbers. There are forty such tubes. The most convenient tube to use is tube number one, the "sign bigit" tube. For the proper display it is convenient that all constants of the problem be taken as positive, which may be done without loss of generality. In the computer used here, machine orders are independent of the bigit in the sign position. Thus, the abscissa and ordinate of a point are transformed into an instruction for changing the sign of the appropriate memory position, which, in turn, lights that spot on the tube face. In this way a picture of the boundary is "painted" on the tube face. The picture is displayed for a fixed time, is erased by dropping the sign of all memory positions, and a picture of the new position is then plotted and displayed. A scaling routine keeps the surface within the confines of the tube matrix.

This procedure should be useful in many problems of motion of gases or liquids.

It is certainly of value in obtaining a quick overall check on the correctness of the code; reasonableness of the time intervals, etc.

The cycle time on this problem is of the order of ten seconds. No printing is involved. Meaningful results are obtained in a matter of minutes.

4. Problems on Rotational Motions in Gravitating Systems

An interesting set of questions in statics concerns the properties of moments of forces exerted on each other by randomly distributed points forming a system Σ. In dynamics the questions concern angular momenta of subsystems σ contained in Σ; as a function of the time. Let us imagine the following situation: Σ consists of a number of mass points m_1, m_2, ... m_n located at $t = 0$ at positions $\bar{r}_1, \ldots \bar{r}_n$ given at random, say, in a unit sphere (let us assume, for example, a uniform probability distribution for the position of each point). We assume further Newtonian attractive forces F_{ij} acting between any two points m_i, m_j. Denote by $\bar{G}_i$ the sum over all j of forces acting upon the point m_i and let us imagine the vector $\bar{G}_i$ applied at the point $\bar{r}_i$. It is clear that the sum over all i of $\bar{G}_i$ is equivalent to zero. What we propose to study is, at first, the statistical behavior of the forces G_i if summed over subsystems σ of the whole set with the following questions to be investigated: let ρ be any number and let us consider subsystems located in a circle with radius ρ and an arbitrary center $\bar{r}_o$. Let us form the sum of all $\bar{G}_i$ located in such and we obtain, in general, a single force Φ and a couple Ψ referred to $\bar{r}_o$ with magnitude which we shall call Θ. Both Φ and Θ thus computed are functions of ρ and $\bar{r}_o$ but we can integrate these quantities over all initial positions $\bar{r}_o$ and will obtain Φ_ρ and Θ_ρ, a single force and moment, which will be now functions of ρ alone. It is our aim to obtain these functions for a random dynamical system, that is to say, the expected values in a random distribution. This can be obtained in practice by computing, on the machine, these quantities for a large number of systems each chosen by a random process. Our statistical computations will be confined at first to plane cases, the three dimensional systems requiring too great a memory at present.

The next, more interesting, thing to study is the following. In a situation as described above at time $t = 0$ let t increase. Motions of our points will ensue and we intend to investigate the angular momenta of subsystems σ as functions of the size of σ and of Φ in time.

Such a situation is perhaps exemplified by star clusters. What we want to study are dynamical systems with many particles, but not gases; that is to say, by mean free path for "collision" we mean an appreciable change in the velocities due to gravitational forces acting between just two of our mass points. It is known that clusters or galaxies possess a rotational motion as a whole. These could, perhaps, originate as follows: the original distribution of matter now separated in galaxies was more or less uniform and random-like. Our system Σ can be imagined infinite. Finite subsystems have angular momenta as a result of fluctuations in the distribution and then, if fluctuations of density occurred also, some subsystems would isolate themselves, stay together due to gravitation (cf. the work of Jeans) and if the whole space expanded with time these condensations would have kept all or most of their angular momenta due to the original fluctuations in the system of vector forces and as the condensations receded from each other, their non-zero angular momenta would have stayed constant in time. Another way to look upon the problem is to study the distribution of vorticity of finite subsystem σ in a very large or infinite system Σ whose points exert forces on each other.

Our proposed calculations consist then of producing a sizeable number of randomly chosen systems Σ and following the behavior of subsystems in time, i.e., using a discrete series of time intervals or "cycles" on the machine and computing the following set of averages: $L(\bar{r}_0, \rho)$, the angular momentum as a function of the radius of the subsystem ρ and position $\bar{r}_0$. For a system of 100 points four positions along a radius with four values of ρ at each point are adequate. The running time per cycle for 100 mass points is less than five minutes with this machine. The running time increases as the square of the number of mass points, but statistics can be improved by running many problems with few mass points, in preference to the less economical method of increasing the number of mass points.

The total kinetic and potential energy is calculated. The system should stabilize at some given size. In this steady state the number of double and triple "stars" is of interest. The total energy serves as a check on the problem.

The moment of inertia of the complete system is calculated, and finally the number of particles in each of the sixteen subsystems mentioned above. The latter numbers permit an approximate plot of the density distribution of the system. The largest value of ρ is chosen to be of the order of the dimension of the system.

Several hundred cycles will be necessary for results on this problem.

Appended are graphs (Figs. 2 to 5) showing the results of 100 cycles run so far.

5. Magnetic Lines of Force

It appears that our computing machines are especially well adapted to the study of properties *in the large* of the system of lines of magnetic force, due to given currents in space. The renewed interest in the qualitative, ergodic, or even just topological behavior of such families of lines is due to studies in magneto-hydrodynamics, applications in astronomy, questions of origin of the cosmic rays (Fermi), not to mention the importance of such knowledge for applications in the construction of high energy accelerating machines (cyclotrons, synchrotons, etc).

In order to study this subject systematically, it is best to consider first steady currents following through *given* lines (wires) in space. If there is only one current through a straight line extending infinitely, the system of lines of magnetic force is, of course, very well known. They form circles linking the line; the same is, or course, true for a current in a single closed circle.

However, the topology of the system of lines of force seems to be very complicated and the ergodic behavior of single lines of force unknown for the case where the single closed curve, through which the current flows, forms a knotted loop, say in the simplest case a clover leaf knot. Some single lines of force will probably be dense on two dimensional surfaces, probably some singularities in the field of lines exist, independently of the *metric* appearance of the knot, but are present necessarily in every topological knot of this sort. There seems to be little hope of obtaining analytically closed expressions describing the system of lines of force.

The situation is complicated in case of two given currents. It is easily seen, in the case given below, that *almost all* lines of force will be bounded, not closed and each dense on a surface! Let the two currents flow as follows: current 1 on a straight line, say the z axis, current 2 on a circle $x^2 + y^2 = 1$. In general, except at points where the ratio of the two current strengths is rational, a line of force will exhibit an "ergodic" behavior on a surface of a torus.

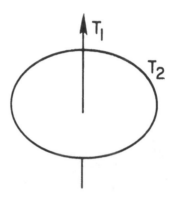

We propose to investigate, on the computing machines, the properties of lines of force due to two currents—each on a straight line extending to infinity. The two lines are skew. T_1 flows on the y axis, T_2 on the line $z = d$, $y = 0$.

One is interested, among other things, in the following questions: Do there exist lines which, although *not closed*, cross a surface of a fixed sphere infinitely many times? Are there lines going arbitrarily far from a fixed point and returning to its neighborhood? Do there exist lines braiding or linking both given wires any number of times?

The computations of such lines of force do not involve much of the "memory" of the machine. The procedure is this: starting at a point $(x_0,\ y_0,\ z_0)$ we compute the direction of the magnetic fields, simply adding the two field strengths, given elementarily from each wire; we perform "short" step $(\Delta x,\ \Delta y,\ \Delta z)$ in the direction of the (constant in time!) force. The computation of this step is done as follows: we calculate a provisional set of increments $(\Delta x)'$, $(\Delta y)'$, $(\Delta z)'$ of the variables, in the new position we calculate the new set $(\Delta x)''$, $(\Delta y)''$, $(\Delta z)''$; we then take $\Delta x = ((\Delta x)' + (\Delta x)'')/2$, $\Delta y, = ((\Delta y)' + (\Delta y)'')/2$, $\Delta z = ((\Delta z' + (\Delta z)'')/2$ and proceed anew. In general this way of solving a system of equations $dx/X = dy/Y = dz/Z$ works well. In our case it is seen that each step is computed in the order of 50 milliseconds; to perform, say, 1,000 steps will take of the order of a minute. The idea is now that with the order of 10^4 steps we shall be able to get some qualitative information about a single line of force as follows: it seems practical to take each step long enough so that in, say, 50 to 100 steps one complete "loop" can be described around a wire (in positions where it is expected that the lines of force surround the current). One would expect then to obtain a number of "loops" of the order of a few hundred.

The quantities *printed* as a result of each such calculations could be for instance:

1. The number of "returns" of a line to a given sphere. One simply has to record on the machine the number of times our line crosses the surface of a given sphere.

2. The number of times and the *sense* in which a line loops the two wires separately and the number of loops surrounding both together. This can be done simply by computing the work done by moving on the line of force, calculating the loops around each wire, as if it alone had current flowing through it. The Gaussian looping coefficients for any two given curves in space can be quickly computed on the machine.

It is convenient to take the length of the "step" along the magnetic field vector to be inversely proportional to the magnetic field strength at that point. In this way the step is proportional to the distance from the wire since the magnetic field is inversely proportional to that distance. The number of steps per looping of the wire is then constant and the step is appropriately shorter at points where the curvature is greater.

It is worthwhile to point out that some integer valued topological invariants may be computed *exactly* even though we use difference expressions instead of differential ones, and have also round-off errors. This is due to "ϵ-invariance" theorems on simplicial approximations in topology[1]. Also, so to say, the field of "error vectors" is in general "curl-free."

Reference

1. K. Borsuk and S. Ulam, "Über Gewisse Invarianten der Abbildungen," Math. Ann. **108**, 311-318, 1933.

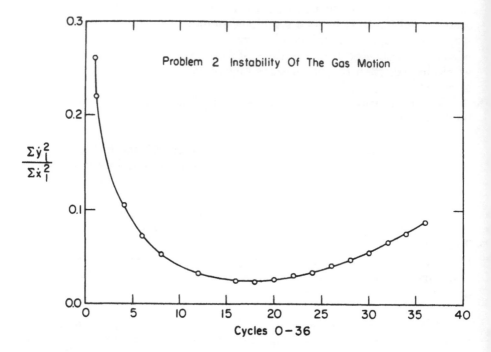

Fig. 1.

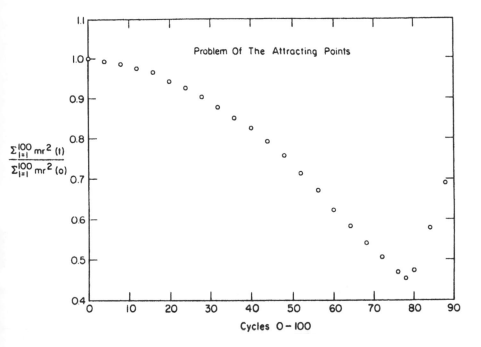

Fig. 2.

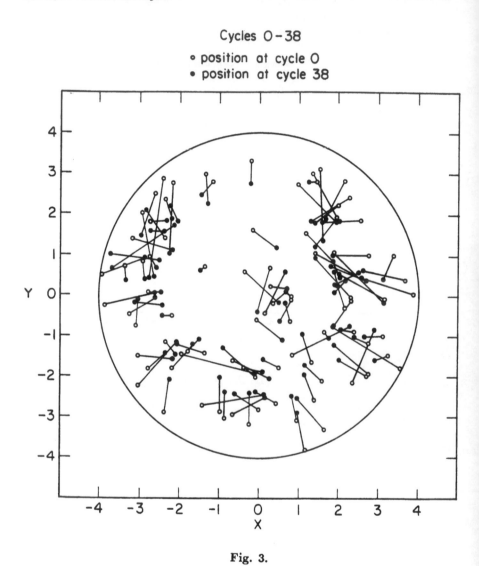

Cycles 0 – 38
∘ position at cycle 0
• position at cycle 38

Fig. 3.

Cycle #55

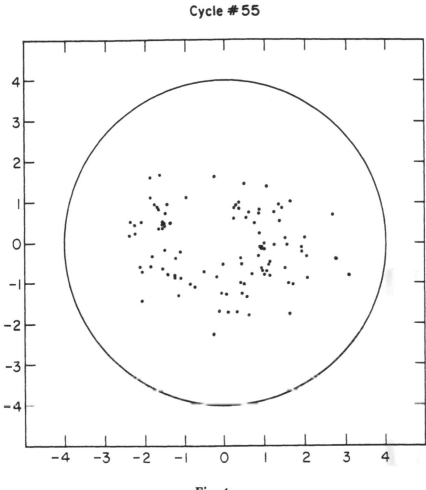

Fig. 4.

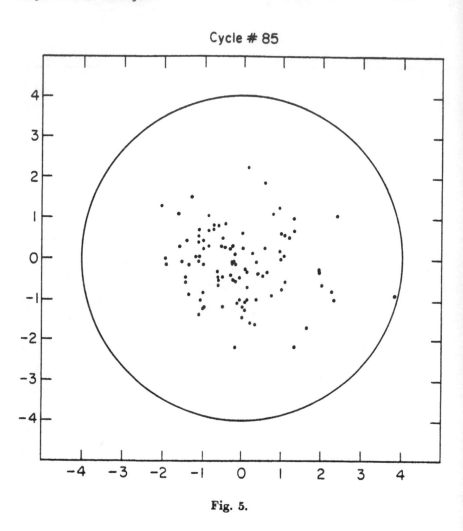

Fig. 5.

5

STUDIES OF NON LINEAR PROBLEMS

With E. Fermi and J. Pasta
(LA-1940, May 1955)

This report details the first attempt to study by computer experimentation the asymptotic behavior of nonlinear dynamical systems for which closed analytic solutions were not available. The results, both interesting and surprising, led to the examination of other examples of such physical systems (e.g., reports 6, 10 and 11.) The impact of these pioneering efforts is reflected in continuing research and publications. The evolution of the concept of solitons, as well as much of the current experimentation—aided by vastly improved computer graphics—on chaos and turbulence, can be traced to this source. (Eds.)

ABSTRACT

A one-dimensional dynamical system of 64 particles with forces between neighbors containing nonlinear terms has been studied on the Los Alamos computer MANIAC I. The nonlinear terms considered are quadratic, cubic, and broken linear types. The results are analyzed into Fourier components and plotted as a function of time.

The results show very little, if any, tendency toward equipartition of energy among the degrees of freedom.*

---

*After the untimely death of Professor Fermi in November 1954, the calculations were continued in Los Alamos. The last few examples were calculated in 1955.

This report is intended to be the first one of a series dealing with the behavior of certain physical systems where the nonlinearity is introduced as a perturbation to a primarily linear problem. The behavior of the systems is to be studied for times which are long compared to the characteristic periods of the corresponding linear problems.

The problems in question do not seem to admit of analytic solutions in closed form, and heuristic work was performed numerically on a fast electronic computing machine (MANIAC I at Los Alamos).* The ergodic behavior of such systems was studied with the primary aim of establishing, experimentally, the rate of approach to the equipartition of energy among the various degrees of freedom of the system. Several problems will be considered in order of increasing complexity. This paper is devoted to the first one only.

We imagine a one-dimensional continuum with the ends kept fixed and with forces acting on the elements of this string. In addition to the usual linear term expressing the dependence of the force on the displacement of the element, this force contains higher order terms. For the purpose of numerical work this continuum is replaced by a finite number of points (at most 64 in our actual computation) so that the partial differential equation defining the motion of this string is replaced by a finite number of total differential equations. We have, therefore, a dynamical system of 64 particles with forces acting between neighbors with fixed end points. If x_i denotes the displacement of the i-th point from its original position, and α denotes the cofficient of the quadratic term in the force between the neighboring mass points and β that of the cubic term, the equations were either

$$\ddot{x}_i = (x_{i+1} + x_{i-1} - 2x_i) + \alpha[(x_{i+1} - x_i)^2 - (x_i - x_{i-1})^2]$$
$$i = 1, 2, \ldots 64, \tag{1}$$

or

$$\ddot{x}_i = (x_{i+1} + x_{i-1} - 2x_i) + \beta\ [(x_{i+1} - x_i)^3 - (x_i - x_{i-1})^3]$$
$$i = 1, 2, \ldots 64, \tag{2}$$

α and β were chosen so that at the maximum displacement the nonlinear term was small, e.g., of the order of one-tenth of the linear term. The corresponding partial differential equation obtained by letting the number of particles become infinite is the usual wave equation plus nonlinear terms of a complicated nature.

---

* We thank Miss Mary Tsingou for efficient coding of the problems and for running the computations on the Los Alamos MANIAC machine.

Another case studied recently was

$$\ddot{x}_i = \delta_1(x_{i+1} - x_i) - \delta_2(x_i - x_{i-1}) + c \tag{3}$$

where the parameters δ_1, δ_2, c were not constant but assumed different values depending on whether or not the quantities in parentheses were less than or greater than a certain value fixed in advance. This prescription amounts to assuming the force as a broken linear function of the displacement. This broken linear function imitates to some extent a cubic dependence. We show the graphs representing the force as a function of displacement in three cases.

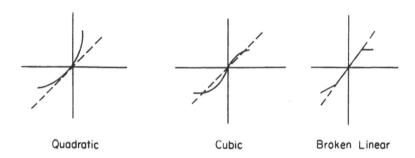

Quadratic   Cubic   Broken Linear

The solution to the corresponding linear problem is a periodic vibration of the string. If the initial position of the string is, say, a single sine wave, the string will oscillate in this mode indefinitely. Starting with the string in a simple configuration, for example in the first mode (or in other problems, starting with a combination of a few low modes), the purpose of our computations was to see how, due to nonlinear forces perturbing the periodic linear solution, the string would assume more and more complicated shapes, and, for t tending to infinity, would get into states where all the Fourier modes acquire increasing importance. In order to see this, the shape of the string, that is to say, x as a function of i and the kinetic energy as a function i were analyzed periodically in Fourier series. Since the problem can be considered one of dynamics, this analysis amounts to a Lagrangian change of variables: instead of the original $\dot{x}_i$ and x_i, $i = 1, 2, \ldots 64$, we may introduce a_k and $\dot{a}_k$, $k = 1, 2, \ldots 64$, where

$$a_k = \sum x_i \, sin \, \frac{ik\pi}{64} . \tag{4}$$

The sum of kinetic and potential energies in the problem with a quadratic force is

$$E_{x_i}^{kin} \; + \; E_{x_i}^{pot} = \frac{1}{2} \; \dot{x}_i^2 + \frac{(x_{i+1} - x_i)^2 + (x_i - x_{i-1})^2}{2} \tag{5a}$$

$$E_{a_k}^{kin} \; + \; E_{a_k}^{pot} \; = \frac{1}{2} \dot{a}_k^2 \; + \; 2a_k^2 \; sin^2 \; \frac{\pi k}{128} \tag{5b}$$

if we neglect the contributions to potential energy from the quadratic or higher terms in the force. This amounts in our case to at most a few per cent.

The calculation of the motion was performed in the x variables, and every few hundred cycles the quantities referring to the a variables were computed by the above formulas. It should be noted here that the calculation of the motion could be performed directly in a_k and $\dot{a}_k$. The formulas, however, become unwieldy and the computation, even on an electronic computer, would take a long time. The computation in the a_k variables could have been more instructive for the purpose of observing directly the interaction between the a_k's. It is proposed to do a few such calculations in the near future to observe more directly the properties of the equations for $\ddot{a}_k$.

Let us say here that the results of our computations show features which were, from the beginning, surprising to us. Instead of a gradual, continuous flow of energy from the first mode to the higher modes, all of the problems show an entirely different behavior. Starting in one problem with a quadratic force and a pure sine wave as the initial position of the string, we indeed observe initially a gradual increase of energy in the higher modes as predicted (e.g., by Rayleigh in an infinitesimal analysis). Mode 2 starts increasing first, followed by mode 3, and so on. Later on, however, this gradual sharing of energy among successive modes ceases. Instead, it is one or the other mode that predominates. For example, mode 2 decides, as it were, to increase rather rapidly at the cost of all other modes and becomes predominant. At one time, it has more energy than all the others put together! Then mode 3 undertakes this role. It is only the first few modes which exchange energy among themselves and they do this in a rather regular fashion. Finally, at a later time mode 1 comes back to within one per cent of its initial value so that the system seems to be almost periodic. All our problems have at least this one feature in common. Instead of gradual increase of all the higher modes, the energy is exchanged, essentially, among only a certain few. It is, therefore, very hard to observe the rate of "thermalization" or mixing in our problem, and this was the initial purpose of the calculation.

If one should look at the problem from the point of view of statistical mechanics, the situation could be described as follows: the phase space of a point representing our entire system has a great number of dimensions. Only a very small part of its volume is represented by the regions where only one or a few out of all possible Fourier modes have divided among themselves almost all the available energy. If our system with nonlinear forces acting between the neighboring points should serve as a good example of a transformation of the phase space which is ergodic or metrically transitive, then the trajectory of almost every point should be everywhere dense in the whole phase space. With overwhelming probability this should also be true of the point which at time $t = 0$ represents our initial configuration, and this point should spend most of its time in regions corresponding to the equipartition of energy among various degrees of freedom. As will be seen from the results this seems hardly the case. We have plotted (Figs. 1-9) the ergodic sojourn times in certain subsets of our phase space. These may show a tendency to approach limits as guaranteed by the ergodic theorem. These limits, however, do not seem to correspond to equipartition even in the time average. Certainly, there seems to be very little, if any, tendency towards equipartition of energy among all degrees of freedom at a given time. In other words, the systems certainly do not show mixing.*

The general features of our computation are these: in each problem, the system was started from rest to time $t = 0$. The derivatives in time, of course, were replaced for the purpose of numerical work by difference expressions. The length of time cycle used varied somewhat from problem to problem. What corresponded in the linear problem to a full period of the motion was divided into a large number of time cycles (up to 500) in the computation. Each problem ran through many "would-be-periods" of the linear problem, so the number of time cycles in each computation ran to many thousands. That is to say, the number of swings of the string was of the order of several hundred, if by a swing we understand the period of the initial configuration in the corresponding linear problem. The distribution of energy in the Fourier modes was noted after every few hundred of the computation cycles. The accuracy of the numerical work was checked by the constancy of the quantity representing the total energy. In some cases, for checking purposes, the corresponding linear problems were run and these behaved correctly within one per cent or so, even after 10,000 or more cycles.

It is not easy to summarize the results of the various special cases. One feature which they have in common is familiar from certain problems in mechanics of systems with a few degrees of freedom. In the

---

* One should distinguish between metric transitivity or ergodic behavior and the stronger property of mixing.

compound pendulum problem one has a transformation of energy from one degree of freedom to another and back again, and not a continually increasing sharing of energy between the two. What is perhaps surprising in our problem is that this kind of behavior still appears in systems with, say, 16 or more degrees of freedom.

What is suggested by these special results is that in certain problems which are approximately linear, the existence of quasi-states may be conjectured.

In a linear problem the tendency of the system to approach a fixed "state" amounts, mathematically, to convergence of iterates of a transformation in accordance with an algebraic theorem due to Frobenius and Perron. This theorem may be stated roughly in the following way. Let A be a matrix with positive elements. Consider the linear transformation of the n-dimensional space defined by this matrix. One can assert that if $\bar{x}$ is any vector with all of its components positive, and if A is applied repeatedly to this vector, the directions of the vectors $\bar{x}_i$, $A(\bar{x})$, ..., $A^i(\bar{x})$, ..., will approach that of a fixed vector $\bar{x}_0$ in such a way that $A(\bar{x}_0) = \lambda(\bar{x}_0)$. This eigenvector is unique among all vectors with all their components non-negative. If we consider a linear problem and apply this theorem, we shall expect the system to approach a steady state described by the invariant vector. Such behavior is in a sense diametrically opposite to an ergodic motion and is due to a very special character, linearity of the transformations of the phase space. The results of our calculation on the nonlinear vibrating string suggest that in the case of transformations which are approximately linear, differing from linear ones by terms which are very simple in the algebraic sense (quadratic or cubic in our case), something analogous to the convergence to eigenstates may obtain.

One could perhaps conjecture a corresponding theorem. Let Q be a transformation of a n-dimensional space which is nonlinear but is still rather simple algebraically (let us say, quadratic in all the coordinates). Consider any vector $\bar{x}$ and the iterates of the transformation Q acting on the vector $\bar{x}$. In general, there will be no question of convergence of these vectors $Q^n(\bar{x})$ to a fixed direction.

But a weaker statement is perhaps true. The directions of the vectors $Q^n(\bar{x})$ sweep out certain cones C_α or solid angles in space in such a fashion that the time averages, i.e., the time spent by $Q^n(\bar{x})$ in C_α, exist for n $\rightarrow \infty$. These time averages may depend on the initial $\bar{x}$ but are able to assume only a finite number of different values, given C_α. In other words, the space of all direction divides into a finite number of regions R_i, $i = 1, \ldots k$, such that for all vectors $\bar{x}$ taken from any one of these regions the percentage of time spent by images of $\bar{x}$ under the Q^n is the same in any C_α.

The graphs which follow show the behavior of the energy residing in various modes as a function of time; for example, in Fig. 1 the energy content of each of the first 5 modes is plotted. The abscissa is time measured in computational cycles, δt, although figure captions give δt^2 since this is the term involved directly in the computation of the acceleration of each point. In all problems the mass of each point is assumed to be unity; the amplitude of the displacement of each point is normalized to a maximum of 1. N denotes the number of points and therefore the number of modes present in the calculation, α denotes the coefficient of the quadratic term, and β that of the cubic term in the force between neighboring mass points.

We repeat that in all our problems we started the calculation from the string at rest at $t = 0$. The ends of the string are kept fixed.

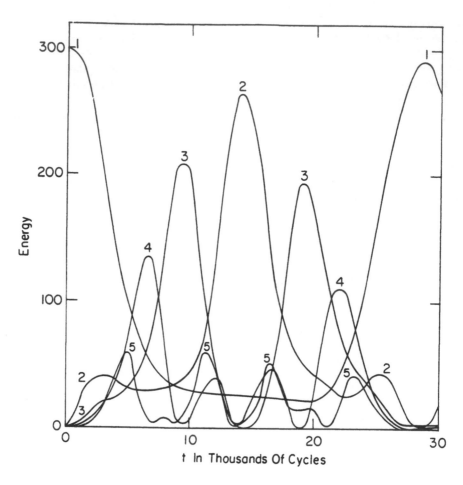

Fig. 1. The quantity plotted is the energy (kinetic plus potential in each of the first five modes). The units for energy are arbitrary. $N = 32$; $\alpha = 1/4$; $\delta t^2 = 1/8$. The initial form of the string was a single sine wave. The higher modes never exceeded in energy 20 of our units. About 30,000 computation cycles were calculated.

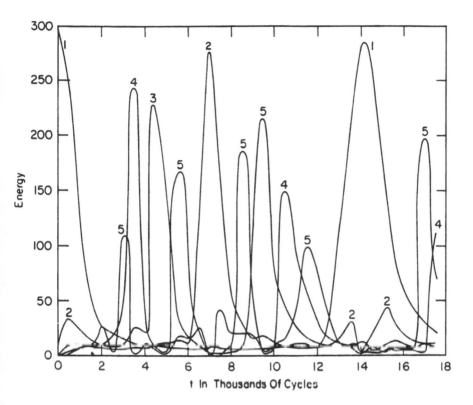

Fig. 2. Same conditions as Fig. 1, but the quadratic term in the force was stronger.
$\alpha = 1$. About 14,000 cycles were computed.

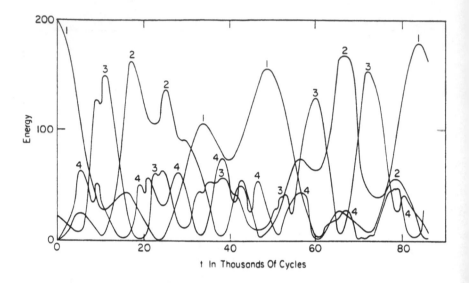

Fig. 3. Same conditions as in Fig. 1, but the initial configuration of the string was a "saw-tooth" triangular-shaped wave. Already at $t = 0$, therefore, energy was present in some modes other than 1. However, modes 5 and higher never exceeded 40 of our units.

148

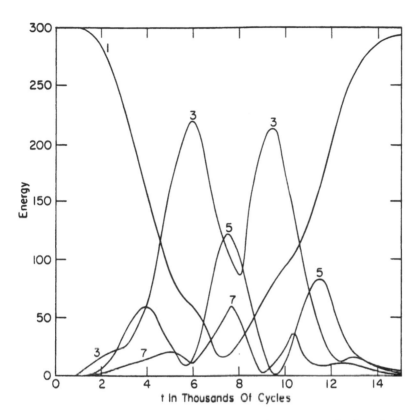

Fig. 4. The initial configuration assumed was a single sine wave; the force had a cubic term with $\beta = 8$ and $\delta t^2 = 1/8$. Since a cubic force acts symmetrically (in contrast to a quadratic force), the string will forever keep its symmetry and the effective number of particles for the computation $N = 16$. The even modes will have energy 0.

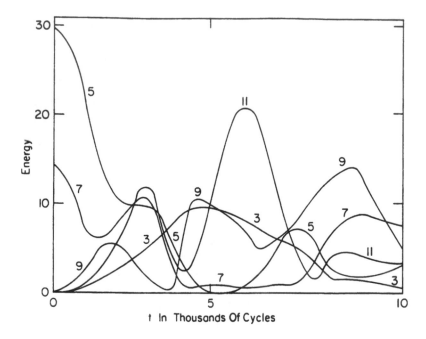

Fig. 5. $N = 32$; $\delta t^2 = 1/64$; $\beta = 1/16$. The initial configuration was a combination of 2 modes. The initial energy was chosen to be 2/3 in mode 5 and 1/3 in mode 7.

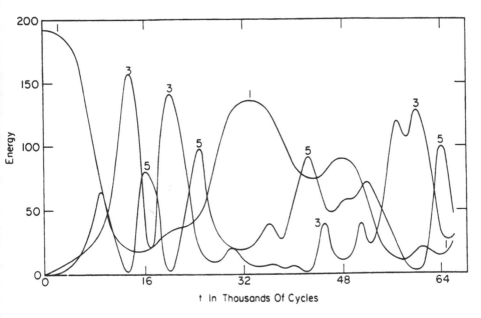

t In Thousands Of Cycles

Fig. 6. $\delta t^2 = 2^{-6}$. The force was taken as a broken linear function of displacement. The amplitude at which the slope changes was taken as $2^{-5} + 2^{-7}$ of the maximum amplitude. After this cut-off value, the force was assumed still linear but the slope increased by 25 per cent. The effective $N = 16$.

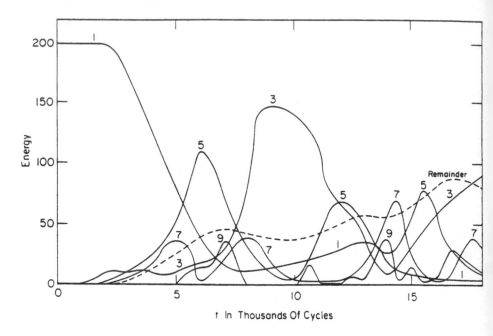

Fig. 7. $\delta t^2 = 2^{-6}$. Force is again broken linear function with the same cut-off, but the slope after that increased by 50 percent instead of the 25 percent charge as in problem 6. The effective $N = 16$.

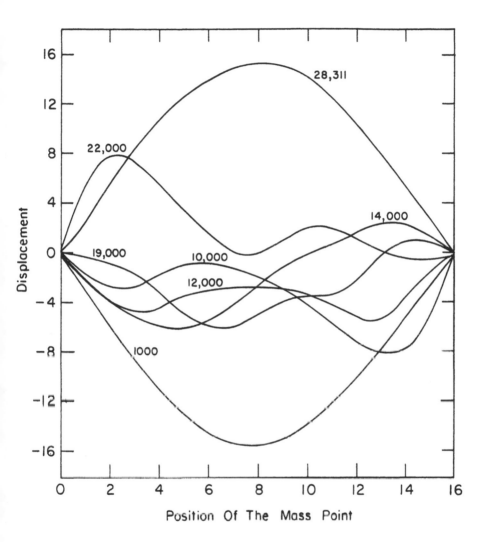

Fig. 8. This drawing shows not the energy but the actual *shapes*, i.e., the displacement of the string at various times (in cycles) indicated on each curve. The problem is that of Fig. 1.

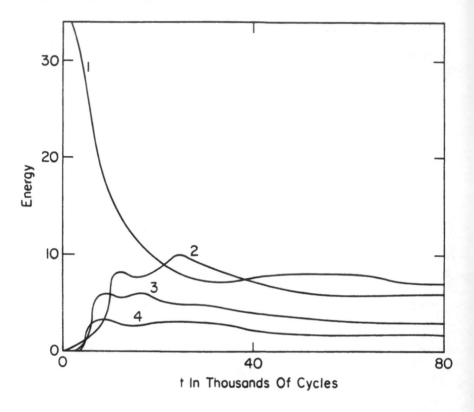

Fig. 9. This graph refers to the problem of Fig. 6. The curves, numbered 1, 2, 3, 4, show the time averages of the kinetic energy contained in the first 4 modes as a function of time. In other words, the quantity is

$$\frac{1}{\nu} \sum_{i=1}^{\nu} T_{a_k}^i \, .$$

ν is the cycle number, $\quad k = 1, 3, 5, 7$.

6

ON THE ERGODIC BEHAVIOR
OF DYNAMICAL SYSTEMS

(LA-2055, May 10, 1955)

This is a lecture which is included in a series of lectures on the physics of ionized gases given in 1955. It presents ideas and remarks on general properties of ionized gases connected with the "Sherwood" project. "Sherwood" was one of the early attempts to use fusion for the peaceful production of energy by the confinement of thermonuclear reactions. (Author's note.)

The purpose of this lecture is to review the present status of the so-called ergodic hypothesis, summarize the mathematical results of the last twenty years or so, and indicate briefly the nature of the difficulties that still remain in applying the general theorems to specific physical situations.

As has been pointed out in previous lectures the ergodic hypothesis can serve as a fundamental point on which to base the entire structure of statistical mechanics. (See, e.g., the derivation of the H-theorem from the ergodicity assumption by ter Haar.) My own feeling, to anticipate the conclusions of this lecture, is that the mathematical work of the last twenty years has brought complete rigor to only a small part of the theory of statistical mechanics, "the equivalent of the first twenty pages or so of a standard book on the subject." One could say here that, as is often the case, the mathematicians know a great deal about very little and the physicists very little about a great deal. The ergodic theorem itself and the subsequent proof of existence and more: the prevalence of ergodic transformations among all volume or measure preserving flows assert, roughly speaking, the legitimacy of assumptions one makes in physical theories about the limiting or equilibrium states of physical systems. The next question of importance equally essential for our understanding of statistical mechanics, concerns the rate of approach to the equilibrium, if this approach, indeed, does take place. This problem seems much more difficult and only very incomplete results exist. In the second half of this lecture, I shall mention some recent results obtained with Fermi and Pasta on the would-be approach to equilibrium in certain simple nonlinear systems. These were obtained on the calculating machine here in Los Alamos.

To start with, we have a space located in a Euclidean space E of $6n$-dimensions. This space is the phase of a dynamical system which we shall assume for most of the talk to be conservative. The Hamilton equations define a flow in the space E. This flow preserves the volume in the space E. This is the theorem of Liouville and it follows directly from the Hamilton equations. The measure or volume in the space is preserved exactly, not only infinitesimally to the first order, but, of course, for volumes of any finite size. I shall not discuss here the definition of the volume in the space E. It is obtained from the ordinary Euclidean volume in the $6n$-dimensional space by the most elementary geometrical considerations. The space E represents the entire available phase space of the dynamical systems. It is divided onto a one-parameter family of subspaces E_k, corresponding to fixed values k of the entire energy of the physical systems. Each of these subspaces, of one less dimension than E, undergoes a flow into itself. Each of these spaces has its own volume and this volume is also preserved under the dynamical flow. It is important to note that, *in special cases*, each of these spaces E_k may again be decomposed into a family of subspaces of still lower dimensions, each of which flows into itself. Indeed, if there are further integrals of the given dynamical problem in addition to the integral of energy, we shall obtain such further decompositions.

The ergodic theory deals with the asymptotic properties of a flow (volume-preserving in such spaces). One is interested in the behavior of trajectories of single points under the given flow. A single point represents one possible initial condition of the dynamical system. That is to say, the position of the n-particles at time $t = 0$ and all the velocities at this time. As time proceeds, this representative point will describe a line in the phase space. One is interested in how this line behaves in the given space E_k (or in case of existence of further integrals, in the "irreducible" subspace of it, E_k).

It is simpler if only for typographical reasons to consider, instead of a continuous flow in phase space, one mapping and a discrete sequence of time intervals. That is to say, instead of the family of transformations T^λ where λ is any real number, we consider T^n which means we look at the flow at intervals of one "second" each. Of course, we have in both cases either $T^{\lambda+\mu} = T^\lambda(T^\mu)$ or $T^{m+n} = T^m(T^n)$. These relations merely express that we have a one-parameter group flow or a sequence of powers of one transformation. All the mathematical results proved so far are valid in both formulations. We shall deal with the discrete one here.

The so-called ergodic theorem proved by G. D. Birkhoff asserts the following: the time averages of functions $f(T^n(p))$ exists for almost every point p if T is a volume preserving transformation of a space on which a measure or volume is defined and f any integrable function. In particular (which is equivalent to the more general statement) it is true that if T is a transformation of the above sort

$$\lim_{n\to\infty} \sum_{i=1}^{n} \frac{f_A T^i(p)}{n}$$

exists for almost every point p. f_A is the characteristic function of any set A in our phase space. That is to say, $f(p) = 1$ if p does, $f(p) = 0$ if p does not belong to A. The sum written above merely counts how many of the first n-iterates of the point fall into the set A. The theorem asserts the existence of a sojourn time for almost every point p in any volume A in phase space. This theorem is necessary to have in order to formulate rigorously the Boltzmann hypothesis according to which this sojourn time is equal to the relative volume of the region A in phase space. So, at least the existence of the sojourn time has been proved. (It should be pointed out here that a somewhat weaker form of the theorem which we have just stated, a so-called weak ergodic theorem, was proved several months before Birkhoff by John von Neumann. His theorem asserted the convergence of our sums in the mean and not for almost every point p, as did Birkhoff, which latter is a stronger statement. The difference, however, may be of less importance to physicists than to mathematicians.)

Birkhoff noticed also that the hypothesis of Boltzmann is equivalent to the following property of the transformation T. There is no subregion of subset $E' \subset E$ which has positive measure less than the measure of the whole space and which goes into itself under the transformation T. Such transformations T are called by him metrically transitive.

One might say that the result of Birkhoff, as far as applications of the ergodic theorem are concerned, shifted the emphasis from the existence of limits to the search for transformations which would be metrically transitive and so satisfy the Boltzmann hypothesis. Only very special such transformations were known and only on very special manifolds E. For example, if E should be the circumference of a circle

157

in one-dimension and T a rotation of it through an angle α, irrational with respect to the length of the circle, then the famous results of H. Weyl on equipartition of numbers n, α modulo 1 (circumference was length 1) establish the metric transitivity in this case and the ergodic theorem with it.

The generalization of this to n dimensions has the form of the well-known theorem of Kronecker about vectors of the form $(k\alpha_1, k\alpha_2, \ldots \alpha_n)$, $k = 1, 2, \ldots$, where the α's are rationally independent of each other. Another case known was that of a flow along geodetic lines on a surface of two dimensions and of constant negative curvature. This result asserted the transitivity of such a flow, i.e., its ergodic properties, and was established by Hopf, Hedlund, and others.

In 1941, a rather general result was established by Oxtoby and the speaker,[1] and can be described as follows: let E be a manifold in any number of dimensions with the volume defined for its subsets. Consider all possible continuous and volume-preserving transformations of such a manifold E into itself. Most of such transformations will be metrically transitive, that is to say, satisfy Boltzmann hypothesis. The expression "most" is defined rigorously and, in particular, implies that arbitrarily near to any transformation T, given in advance, one will be able to find transformations which are metrically transitive. (Two transformations S and T are said to be within an ϵ of each other, if for every point p, $S(p)$ and $T(p)$ are within the same ϵ.) I shall not go into the technical definition of the word "most," etc., but would like to stress that the above result shows the existence of ergodic transformations on any manifold (e.g., sphere, ellipsoid, etc.), in any number of dimensions and what is more, the *prevalence* of such transformations among all possible ones. It is important to stress here that the transformations which are given by actual dynamical flows defined by Hamilton's equation could, nevertheless, form exceptions. An analogy: most real numbers are transcendental, but, of course, there exists a "small minority" of algebraic numbers. The question is still open whether or not the actual dynamical transformations are, in general, metrically transitive, i.e., ergodic. Our result above merely makes it very probable or, if one may say so, the probability *a priori* is equal 1 that most of these transformations will be ergodic. It is certainly true that every flow may be perturbed arbitrarily little to become ergodic. The actual criteria, given a dynamical transformation, are, however, still lacking.

In what now follows, I shall discuss a number of examples of physical systems where the ergodic behavior seems *a priori* inevitable and which were considered during the last two years by E. Fermi, John Pasta, and the speaker. Numerical computations were performed on the MANIAC here in Los Alamos. The results were most unexpected to all of us mainly because, as we shall see, they show a certain hesitancy on the part of these particular physical systems to behave ergodically, or, more exactly, the times taken for mixing or equipartition seem to be unduly long, if not infinite.

Before we take up these examples, we make two more remarks about the generality of the ergodic behavior of dynamical flows in phase space:

1) The subject of this seminar is mainly the study of behavior of charged particles or plasma in the presence of magnetic fields. The discussion given above applies to the case of motion of charged particles in a fixed given magnetic field if one neglects the changes in the field due to the motions of the particles themselves. A system of such particles acted upon by the external field and their mutual electrostatic interactions "should" behave ergodically independently of given constraints in form of walls, etc. The ergodic behavior has to be understood relative to the phase space after it has been reduced by a number of integrals of the motion imposed by such constraints.

2) It may be worthwhile pointing out here that the proof of the prevalence of ergodic flows among all possible volume preserving continuous flows as given in the paper quoted above, establishes more than equality of the time and space averages for most transformations. The construction used in the proof exhibits the general transformation as having a "turbulent" character. That is to say, a roughly periodic motion which does not, however, exactly close the orbits of points in phase space, but feeds a large periodic motion into smaller "rotational" motions which in turn do not have the orbits quite closed but feed still smaller rotations in turn and this process continues indefinitely. Fourier analysis in n dimensions of such a transformation which is the general one would show the feeding of vortex-type flows into successively smaller vortices. This fact had not been noticed or exploited by the authors at the time the paper was written but may be useful in discussing a statistical mechanical type of treatment for motions of fluids; that is to say, systems with infinitely many degrees of freedom which, in general, it is believed tend to become turbulent. Parenthetically, we may add here that a statistical mechanical type of treatment of systems with infinitely many degrees of freedom is required if one wants to include the radiation effects due to the motion of charges. The magnetic field itself would have, of course, infinitely many degrees of freedom.

The problem which was studied numerically on the MANIAC is the following: we have a continuous string with ends kept fixed and with

159

forces acting between its elements which are nonlinear as functions of displacements. This continuous string is replaced, for the purpose of the computations, by a finite number of points, 64 in our actual work, and the equations describing its motion were:

$$\ddot{x}_i = \alpha(x_{i+1} + x_{i-1} - 2x_i) + \beta[(x_{i+1} - x_i)^2 - (x_i - x_{i-1})^2]$$
$$i = 1, 2, \ldots, 64 .$$

or

$$\ddot{x}_i = \alpha(x_{i+1} + x_{i-1} - 2x_i) + \gamma[(x_{i+1} - x_i)^3 - (x_i - x_{i-1})^3]$$
$$i = 1, 2, \ldots, 64 .$$

β and γ were chosen so that at the maximum displacement x_i, the nonlinear term, was small, e.g., of the order of one-tenth of the linear term. The corresponding partial differential equation obtained by letting the number of particles become infinite is the usual wave equation plus nonlinear terms of a complicated nature. There seems to be, of course, very little hope for obtaining explicit solutions and what was done was to run a great number of special problems.

These were studied as follows: the initial position of the string, that is to say, the distribution of the x_1 at time $t = 0$ was assumed to be in a form of a single sine wave or in some other simple forms, e.g., a sum of two sine waves of low frequencies or triangular. Most problems were run under the assumption of constant masses (equal to 1, say) for each i. The position of this string and the total energy of it, kinetic plus potential, were analyzed during the course of each problem in Fourier series. That is to say, we studied the 64 Fourier coefficients A_k, the total energy residing in each mode E_k, $k = 1, 2, \ldots$, 64, as functions of time. Since the problem is approximately linear, for times not very long the string vibrates periodically; the length of the numerical run of the problem was, in each case, equal to several hundred or more of what would be full vibration periods from the initial condition in the corresponding linear problem, if the nonlinear terms in the force had been neglected. Of course, each single vibration period in the calculation corresponded to a hundred or several hundred time cycles on the machine so the number of computation cycles in each problem was several tens of thousands. The initial motivation for the problem was to observe how, in the course of time, the energy of the system, initially contained in the first or the first few modes, in time flows to other modes and to observe the rate at which equipartition of energy among all modes becomes established. This problem was to serve as

the first one of a series in a systematic investigation of the question
of rates of approach to equilibrium, so important in many problems of
statistical mechanics.

The actual results were somewhat surprising to us since, instead
of what one expected—a continual and steady flow of energy, say from
the first mode to all higher modes and an asymptotically uniform approach to equipartition—something entirely different seems to happen.
For example, in the problem where the initial position was a pure sine
wave and all of the energy was in the first mode, the behavior of the
system was the following: initially, as predicted by Rayleigh's perturbation analysis, that is to say, the infinitesimal study for short t,
the other modes grow in energy one by one, the first mode feeds the
second, the first and second together the third, the second feeds the
fourth and so on. This, indeed, was observed but later on it was only
single modes, say, mode No. 3, that continued to grow in energy systematically and for many dozens of the periods of the vibration the
higher modes not growing at all. Then the energy in the third mode
was dropping steadily and mode No. 5 was increasing. The first few
modes, that is 1, 3, 5, 7, were exchanging energy among themselves
slowly with the higher modes not obtaining any sizeable contributions
and after 30,000 cycles or 300 "would be" full vibrations of the string,
the system came back within one percent of the total energy, to its
original shape, that is to say, a pure sine wave.

This behavior seems to be typical in other cases, too. It is the first
few modes that exchange energy among themselves in a somewhat er-
ratic fashion but it is always one or the other or very few of them
that seem to predominate and, far from a tendency towards equipartition among all modes proceeding steadily, one sees an almost periodic
exchange between the low modes. In the case of initially triangular
shape where modes 1, 3, 5, 7, are mainly involved in the beginning,
again these played among themselves and the string did not show, up
to times where the computation was stopped, any tendency to become
really turbulent or energy thermalized.

Another problem recently calculated was the following: instead of
assuming the nonlinear term to be quadratic or cubic in displacement,
we took this nonlinear part to be represented by a broken polygon
imitating the shape of a cubic (and also small compared to the linear
term). This was done because perhaps the quadratic or cubic form
of the forces would introduce some analytic peculiarities which could
possibly explain this almost periodic and non-mixing or only slowly
mixing behavior, and here a non-analytic form of the force would possibly remove this special character and the unknown analytic reason
for the unexpected behavior of the motion. The results were again very

similar. Starting with a single wave, only the first few Fourier modes exchanged energy among themselves. There seems to be one difference; the system does not come back to its original position, but again the first few modes only seem to exchange energy significantly. The behavior of this system thus may seem to serve as a warning against relying too much on the statistical arguments for an approach to equilibrium for systems of many degrees of freedom. These results cannot be discussed here in any detail, but a report on all the work done will be available shortly.* It was merely intended to point out the existence of cases where the estimate, *a priori*, based on the usual volume in phase space considerations and estimates of relaxation times seems quite inadequate. Problems which are nonlinear, but still are algebraic or "simple" in terms of the forces involved, may not be good examples of the general or random flows which Boltzmann or Gibbs had in mind but instead show an almost periodic behavior, a slow transfer of energy between the degrees of freedom. More generally, one might suspect, on certain mathematical grounds, the existence, instead of, as in the linear case, *states* of a system, the appearance of *quasi-states* between which the system oscillates. These quasi-states apparently need not form a continuum or be too dense, but may, approximately, consist of combinations of a few states in the corresponding linear problem.

Reference

1. J. C. Oxtoby and S. Ulam, "Measure-Preserving Homeomorphisms and Metrical Transitivity," Ann. Math., **40**, 2, 874-920, 1941.

---

* This refers to the preceding report which was distributed after this lecture was delivered, but before this collection of lectures was issued. (Eds.)

7

ON A METHOD OF PROPULSION OF PROJECTILES BY MEANS OF EXTERNAL NUCLEAR EXPLOSIONS

With C. J. Everett
(LAMS-1955, August 1955)

This report outlines the methods and proposals which led to "Project Orion" on which vast efforts were expended at General Dynamics, Westinghouse and other places, by scientists like Ted Taylor and Freeman Dyson. Everett and the author hold a patent on the idea. (Author's note).

ABSTRACT

Repeated nuclear explosions outside the body of a projectile are considered as providing means to accelerate such objects to velocities of the order of 10^6 cm/sec. A few schematic calculations are presented, showing the dependence of the mass ratios ("propellant" to the final mass), accelerations, etc, on the various free parameters entering in this scheme.

1. Introduction

It is the purpose of this report to summarize certain considerations and proposals, some of which originated as long as ten years ago, and to discuss additional ideas concerning the attempt to attain velocities in the range of the missiles considered for intercontinental warfare and even more perhaps, for escape from the earth's gravitational field, for unmanned vehicles.

The methods most frequently proposed for obtaining such vehicles involve expulsion of material at high velocity from rocket motors. This ejected material is heated in the rocket itself, either by a chemical reaction, or, in more recent schemes, by nuclear reactors.[1] In both cases there is a severe limitation on motor temperature and thus also on the velocity of material ejected. The well-known exponential rocket formula* then demands impractical mass ratios for the attainment of final velocities V_f in the desired ranges, and multi-stage vehicles become necessary. The advantage of the nuclear rocket of this kind over the chemical type lies paradoxically not so much in its potentially enormous power source, which is limited by chamber temperature T to much the same range as chemical motors, but in its ability to use hydrogen as propellant, with molecular weight μ lower than the average of chemical reaction products[2], thus permitting operation at higher specific impulse, which is a function of $\sqrt{T/\mu}$.

The scheme proposed in the present report involves the use of a series of expendable reactors (fission bombs) ejected and detonated at a considerable distance from the vehicle, which liberate the required energy in an external "motor" consisting essentially of empty space. The critical question about such a method concerns its ability to draw on the real reserves of nuclear power liberated at bomb temperatues without smashing or melting the vehicle.

General proposals of this sort were first made by S. Ulam in 1946, and some preliminary calculations were made by F. Reines and S. Ulam in a Los Alamos memorandum dated 1947. More recently, an additional idea was advanced, which consists in placing between each bomb and the rocket a "propellant" consisting of water or some plastic, which will be heated by the bomb, and which will propel the vehicle during its subsequent explosive expansion. Some of the advantages of this proposal will be mentioned in the final section.

In any such device, one of the principal difficulties is the heating of the rocket by the propellant. We seem to encounter a situation in which the base of the rocket will be, periodically, at one second intervals, in the proximity of a very hot gas for durations of about one millisecond each. Study of the effects of such a variable wall temperature on various materials will be made, and reported on subsequently.

The most recent idea is that the use of a sufficiently powerful magnetic field shielding the base of the rocket will have the effect of reflecting the (ionized) atoms of the hot propellant gas before they reach the rocket, thus avoiding heating of the base and incidentally gaining a factor on momentum transfer. It is hoped that this possibility

---

* M_0/M_f = mass-ratio = $\exp(V_f/I)$; I = specific impulse.

also may be investigated at least schematically and reported on in Part II.* However, there appear to be many difficulties in such a study, involving the reaction of a plasma to the magnetic field. Whether the field strength required is impractically large remains to be seen. There is, it seems, the possibility of the formation of a powerful plasma current at the base of the rocket and a pinch effect, which may mean that the magnetic field becomes compressed to a smaller volume and the magnetic pressure considerably increased.

2. Kinematics

In order to gain some quantitative insight into the elements of such a system, we propose to adopt a particular set of assumptions and to study numerically the effect of variation of parameters. The Eqs. (1–7) which follow are obviously highly tentative and subject to many questions here unresolved.

The vehicle is considered to be saucer-shaped, of diameter about 10 meters, sufficient at any rate to intercept all or most of the exploding propellant. Its final mass M_f is perhaps 12 tons, which must cover structure, payload, instruments, storage for propellant and bombs, and, if required, apparatus for maintaining the magnetic field. The initial mass M_0 of the vehicle exceeds this by the mass of bombs and propellant.

The bombs are ejected at something like one second intervals from the base of the rocket and are detonated at a distance of some 50 meters from the base. Synchronized with this, disk-shaped masses of propellant are ejected in such a way that the rocket-propellant distance is about 10 meters at the instant the exploding bomb hits it. The propellant is raised to high temperature, and, in expanding, transmits momentum to the vehicle. The final velocity V_f is attained after N ($\sim$50) such explosions.

We regard now the i-th stage of the process. From the rocket, traveling at velocity V^{i-1} with respect to the earth, are ejected first the i-th bomb (mass m_B) and then the i-th mass of propellant m_P^i at some small velocity v_0 relative to the rocket. It is supposed that, upon detonation, a certain fraction σ of the mass of the bomb collides inelastically with the ejected propellant mass. This fraction could be made, in our case, perhaps as much as 1/10, which is considerably more than the factor given by the solid angle. This could probably be achieved by a suitable distribution of the mass of the tamper surrounding the core

---

* Part II never appeared.(Eds.)

of the bomb. In this way, a larger fraction of the mass of the bomb would hit the propellant. (It is easy to make the distribution of the mass involved in the bomb explosion nonisotropic; the energy distribution is probably essentially isotropic.) If v_B^i is the average velocity of explosion of the bomb in the sector reaching the propellant, we have

$$\sigma m_B(V^{i-1} - v_0 + v_B^i) + m_P^i(V^{i-1} - v_0) = (\sigma m_B + m_P^i)V_P^i \,,$$

where V_P^i is the velocity relative to the earth of the center of mass of the combined system $(\sigma m_B, m_P^i)$. If we introduce a velocity v_P^i by means of the relation

$$V_P^i = V^{i-1} - v_0 + v_P^i$$

we obtain

$$\sigma m_B v_B^i = (\sigma m_B + m_P^i)v_P^i \,. \tag{1}$$

The excess kinetic energy in this transfer is supposed to appear initially as thermal energy H^i in the propellant

$$H^i = \frac{1}{2}\sigma m_B(v_B^i)^2 - \frac{1}{2}(\sigma m_B + m_P^i)(v_P^i)^2 \,. \tag{2}$$

It is assumed that about half of this heat H^i reappears in kinetic energy of expansion of the propellant, with an expansion velocity v_E relative to its own center

$$\frac{1}{2}H^i = \frac{1}{2}(\sigma m_B + m_P^i)(v_E^i)^2 \,. \tag{3}$$

We assume, arbitrarily, that in the expansion of the propellant, one half of its internal energy becomes converted to kinetic energy of expansion. This fraction depends obviously on the distance d and is, in our case, higher.

In our schematic computation we prefer to adopt this much too conservative value.

We may consider that the upper and lower halves of the exploding propellant travel with average velocities

$$V_P^i \pm v_E^i \,,$$

respectively. Now Eqs. (1), (2), (3) show that

$$(v_E^i)^2 = \frac{1}{2}\left(\frac{m_P^i}{\sigma m_B}\right)(v_P^i)^2$$

and since, in all cases we consider, $m_P^i > 2\sigma m_B$, we have $v_E^i > v_P^i$. Thus $V_P^i - v_E^i = V^{i-1} - v_0 + v_P^i - v_E^i < V^{i-1}$, and the lower half of the exploding propellant will not reach the rocket.

The momentum conservation equation for the rocket and upper half of the propellant should read

$$\frac{1}{2}(\sigma m_B + m_P^i)(V^{i-1} - v_0 + v_P^i + v_E^i) + M^i V^{i-1} =$$

$$\frac{1}{2}(\sigma m_B + m_P^i)(V^{i-1} - (-v_0 + v_P^i + v_E^i)) + m^i V^i \; ,$$

or, simplifying,

$$\frac{1}{2}(\sigma m_B + m_P^i) \cdot 2(-v_0 + v_P^i + v_E^i) = M^i \Delta_i V \; ,$$

where M^i is the present mass of the rocket, and $\Delta_i V$ is the i-th increment in its velocity relative to the earth. This assumes total reflection of the propellant. To allow for side effects and imperfect reflection, we use the equation

$$\frac{1}{2}(\sigma m_B + m_P^i)(v_P^i + v_E^i) = M^i \Delta_i V \; . \tag{4}$$

Finally, we assume the time $\Delta_i t$ for the i-th acceleration to be

$$\Delta_i t = \frac{2d}{(v_P^i + v_E^i)} \tag{5}$$

where d is the distance from propellant to rocket. The i-th acceleration is thus

$$\alpha_i = \frac{\Delta_i V}{\Delta_i t} \; . \tag{6}$$

There are two cases of mathematical simplicity which we outline, and for which we include some numerical examples. (Tables 1 and 2 for the cases 1 and 2, respectively.)

Case 1. Constant Acceleration

We take as independent parameters:

V_f the final velocity
M_f the final mass of the rocket
N the number of stages (bombs)
α the acceleration at each stage (assumed constant)
d distance from propellant to rocket
m_B mass of each bomb
σ fraction of m_B hitting propellant

and show how all other parameters may be expressed in terms of these.

Thus each change in velocity will be

$$\Delta_i V = \frac{V_f}{N} \tag{7}$$

over a time interval

$$\Delta_i t = \frac{\Delta_i V}{\alpha} = \frac{V_f}{\alpha N} . \tag{8}$$

The propelling velocity $v_P^i + v_E^i \equiv w^i$ is thus

$$\omega_i = \frac{2d}{\Delta_i t} = \frac{2\alpha N d}{V_f} . \tag{9}$$

We now consider Eq. (4), setting

$$C = (1 - \sigma)m_B \tag{10}$$

and $m_i = m_B + m_P^i$, the total ejected i-th mass. Thus (4) becomes

$$m_i - C = k \left\{ M_0 - \sum_{j=1}^{i} m_j \right\} \tag{4*}$$

where

$$k = \frac{2\Delta_i V}{\omega_i} = \frac{1}{\alpha d} \left(\frac{V_f}{N} \right)^2 \tag{11}$$

and M_0 is the initial mass of the rocket.

Writing the equation (4*) for $i + 1$ and subtracting shows that $m_{i+1} = m_i \rho$ where

$$\rho = \frac{1}{1 + k} . \tag{12}$$

Thus $m_i = m_1 \rho^{i-1}$, $i = 1, 2, \ldots, N$. We determine M_0 and m_1 as follows. Substituting $\sum_{j=1}^{i} m_j = m_i(1 - \rho^i)/(1 - \rho)$ into (4*) shows that

$$m_1(1 + k) = kM_0 + C ,$$

while, by definition,

$$M_0 - M_f = \sum_{j=1}^{N} m_j = m_1(1 - \rho^N)/(1 - \rho) = \frac{m_1}{k}(1 + k)(1 - \rho^N) .$$

Eliminating M_0 between these two relations yields

$$m_1 = (kM_f + C)(1 + k)^{N-1} \qquad (13)$$

and so

$$M_0 = [m_1(1 + k) - C]/k . \qquad (14)$$

Thus we have trivially the i-th mass:

$$m_i = m_1 \rho^i , \qquad (15)$$

the mass ratio:

$$M.R. = \frac{M_0}{M_f} , \qquad (16)$$

the total expelled mass:

$$T = M_0 - M_f , \qquad (17)$$

the total bomb mass:

$$M_B = Nm_B , \qquad (18)$$

the total propellant mass:

$$M_P = T - M_B , \qquad (19)$$

and the i-th mass of propellant:

$$m_P^i = m_i - m_B . \qquad (20)$$

Now, solving equations (1), (2), and (3) for v_P^i and v_E^i in terms of v_B^i, we get

$$v_P^i = \sigma m_B v_B^i / m_i - C \qquad (21)$$

169

and

$$v_E^i = \frac{1}{m_i - C} \sqrt{\frac{\sigma m_B m_P^i}{2}} \cdot v_B^i \ .$$

Substitution into

$$v_E^i + v_P^i = \omega \tag{22}$$

yields

$$v_B^i = \omega(m_i - C)/\left\{ \sigma m_B + \sqrt{\frac{\sigma m_B m_P^i}{2}} \right\} \tag{23}$$

whence the values of v_P^i and v_E^i may now be obtained, using (21) and (22), respectively.

Thus all parameters are determined in terms of the fundamental set $V_f, M_f, N, \alpha, d, m_B, \sigma$. It is interesting to note that the mass ratio

$$M.R. = \frac{m_1(k+1) - C}{m_1(k+1)\rho^N - C}$$

is (approximately in general and exactly when $C = 0$)

$$(1 + k)^N$$

where $k = (1/\alpha d)(V_f/N)^2$, which indicates the extreme sensitivity of the mass ratio to α, N, d, and especially to V_f, in the constant acceleration case.

A rough indication of the energy of the i-th bomb is given by the $k_B^i = (1/2)m_B(v_B^i)^2$ included in the tables. The actual yield of each bomb is several times greater since we assumed a special shaping of the tamper to concentrate as much as possible the mass, but not the energy of the exploding bomb, towards the propellant.

Table 1 is intended to show how the various factors in the problem depend on the initial parameters N, α, d and m_B. None of the twelve "problems" is intended as an optimum case. It may be noted that problems 1 and 2 with $V_f = .7 \times 10^6$ are included for the sake of comparison with various intercontinental ballistic missiles schemes. It should be noted that our mass ratios are considerably less than those contemplated in such cases, while the accelerations are very much more ($\approx 10,000$ g's), lasting for periods of about 1 millisecond each. One also notes that the bombs are rather "small" ($10^{19} - 10^{20}$ ergs).

TABLE 1 (c.g.s. units)

Problem#	1	2	3	4	5	6	7	8	9	10	11	12
$V_f \times 10^{-6}$	.7	.7	1.2	1.2	1.2	1.2	1.2	1.2	1.2	1.2	1.2	1.2
$M_f \times 10^{-6}$	12	12	12	12	12	12	12	12	12	12	12	12
N	40	80	30	30	30	40	40	40	60	100	100	100
$a \times 10^{-6}$	20	10	100	50	20	30	20	20	10	8	8	5
$d \times 10^{-2}$	6	10	6	10	6	10	6	10	10	10	10	10
$m_B \times 10^{-6}$	.5	.2	.5	.5	.5	.3	.5	.3	.3	.3	.2	.3
σ	.1	.1	.1	.1	.1	.1	.1	.1	.1	.1	.1	.1
$\Delta V \times 10^{-6}$	.0175	.00875	.04	.04	.04	.03	.03	.03	.02	.012	.012	.012
$\Delta t \times 10^3$	.875	.875	.4	.8	2.0	1.0	1.5	1.5	2.0	1.5	1.5	2.4
$\omega \times 10^{-6}$	1.37	2.29	3.00	≈.50	.60	2.00	.80	1.33	1.00	1.33	1.33	.833
k	.0255	.0077	.0267	.032	.133	.030	.075	.045	.04	.018	.018	.0288
$m_1 \times 10^{-6}$	2.02	.498	1.65	2.08	76.5	2.00	22.7	4.51	7.59	2.84	2.32	10.2
$m_N \times 10^{-6}$	.756	.272	.770	.334	2.05	.63	1.35	.810	.750	.486	.396	.616
$M_o \times 10^{-6}$	63.6	42.1	46.8	53.0	648.	59.5	319.	98.7	190.5	146.	121.	356.
$M.R.$	5.3	3.5	3.9	4.4	54.0	5.0	26.6	8.2	15.9	12.1	10.1	29.7
$T \times 10^{-6}$	51.6	30.1	34.8	41.0	636.	47.5	307.	86.7	178.5	134.	109.	344.
$M_B \times 10^{-6}$	20	16	15	15	15	12	20	12	18	30	20	30
$M_P \times 10^{-6}$	31.6	14.1	19.8	26.0	621	35.5	287.	74.7	160.5	104	89.	314
$m_P^1 \times 10^{-6}$	1.52	.298	1.15	1.58	76.0	1.70	22.2	4.21	7.29	2.54	2.12	9.94
$m_P^N \times 10^{-6}$	.256	.0719	.270	.334	1.55	.33	.85	.51	.45	.186	.196	.316
$v_B^1 \times 10^{-6}$	8.78	9.75	16.4	16.4	32.0	18.2	22.4	20.1	20.3	15.2	17.2	20.0
$v_B^N \times 10^{-6}$	3.23	4.49	7.27	6.79	3.88	7.17	3.68	6.13	4.28	3.48	4.48	2.92
$k_B^1 \times 10^{-18}$	19.3	9.51	67.5	67.1	255.	49.8	125.	60.5	61.8	34.8	29.6	59.8
$k_B^N \times 10^{-18}$	2.60	2.01	13.2	11.5	3.77	7.72	3.38	5.63	2.75	1.81	2.01	1.28
$v_P^1 \times 10^{-6}$	.280	.613	.682	.503	.0210	.317	.0504	.142	.0832	.178	.161	.0601
$v_P^N \times 10^{-6}$	.527	.977	1.14	.884	.122	.598	.204	.341	.268	.483	.415	.253
$v_E^1 \times 10^{-6}$	1.09	1.67	2.32	2.00	.579	1.68	.750	1.19	.917	1.16	1.17	.773
$v_E^N \times 10^{-6}$	.844	1.31	1.87	1.62	.478	1.40	.596	.993	.733	.850	.919	.580

Case 2. Constant Mass

In this case, which closely corresponds to the usual rocket assumption, we take as independent parameters $M_f, N, d, m_B, \sigma,$ and now m_P, v_B (assumed constant) instead of α and V_f.

Thus we have for the mass expelled at each stage:

$$m = m_B + m_P , \tag{24}$$

the total bomb mass:

$$M_B = N m_B , \tag{25}$$

and the total propellant mass:

$$M_P = N m_P , \tag{26}$$

the total mass expelled:

$$T = M_B + M_P , \tag{27}$$

the initial rocket mass:

$$M_0 = M_f + T , \tag{28}$$

and the mass ratio:

$$M.R. = M_0/M_f . \tag{29}$$

Since v_B is given, we find from Eq. (1) that

$$v_P = \sigma m_B v_B/(\sigma m_B + m_P) , \tag{30}$$

while Eqs. (2) and (3) show that

$$H = \frac{1}{2} \sigma m_B v_B^2 \left(\frac{m_P}{\sigma m_B + m_P} \right) , \tag{31}$$

and

$$v_E = \sqrt{H/(\sigma m_B + m_P)} . \tag{32}$$

Hence we again have a constant propelling velocity given by

$$\omega = v_P + v_E . \tag{33}$$

The "rocket equation" (4) now becomes

$$L = \frac{1}{2} (\sigma m_B + m_P)\omega = M^i \Delta_i V ,$$

the left side being a known constant, and $M^i = M^0 - im$ being a known function of $i = 1, \ldots, N$. Hence we can compute the i-th increment of velocity

$$\Delta_i V = L/M^i , \tag{34}$$

and the velocity after i stages:

$$V_i = \sum_{j=1}^{i} \Delta_j V . \tag{35}$$

In particular the final velocity is

$$V_f = V_N = \sum_{j=1}^{N} \Delta_j V . \tag{36}$$

The time $\Delta_i t$ is given by the constant

$$\Delta_i t = 2d/\omega , \tag{37}$$

and hence we have the i-th acceleration

$$\alpha_i = \Delta_i V/\Delta_i t . \tag{38}$$

In particular,

$$\alpha_{min} = \alpha_1 - \left(\frac{L\omega}{2d}\right)/(M_0 - m) , \tag{39}$$

and

$$\alpha_{max} = \alpha_N = \left(\frac{L\omega}{2d}\right)/M_f . \tag{40}$$

In analogy with the usual rocket equation, our Eq. (34) might be written

$$\left(\frac{L}{M}\right) m = M^i \Delta_i V ,$$

or, letting

$$\beta = L/M = \frac{1}{2}\left(\frac{\sigma m_B + m_P}{m}\right) \cdot \omega - \beta \, dM = M \, dV$$

whence

$$\frac{dM}{M} = -\beta^{-1} \, dV$$

and

$$ln \frac{M_0}{M} = \beta^{-1}V$$

or

$$\frac{M_0}{M_f} = e^{V_f/\beta} ,$$

which affords a rough estimate of V_f, namely

$$V_f \sim \frac{L}{m} \, ln \, (M.R.) .$$

   In Table 2, Problem #4′ is intended to be an analogue of Problem #4 of Table 1, while Problem #12′ is intended as a companion to Problem #12 of the former table. It may be noted that in order to duplicate the performance of a given rocket of constant acceleration α by the second method, one requires accelerations whose average is $\sim \alpha$ and which, therefore, individually greatly exceed α in the final stage. It may be that the method of Case 1, although unorthodox, has advantages in this sense which might justify the use of bombs of variable yield.

3. Remarks

   1. The mass of each fission bomb is assumed to be of the order of 500 kg, including tamper and explosive. Since these bombs are of small yield and many of them are required, they might be of hydride composition. Certainly a disadvantage of our scheme is its wastefulness of fissionable material.

   2. The figure of 12 tons for the final mass of the projectile was assumed arbitrarily in our computations. Actually increasing this number with a proportional increase in the mass of the propellant is very advantageous since the mass of the bombs need hardly be increased even though their yields can be made considerably greater. Thus with, say, 20 tons for the vehicle the mass ratio will be more favorable.

   3. Assuming ~ 1 second intervals between explosions, the total duration of the process will be less than 100 seconds, and the resulting loss of velocity due to the earth's gravitational pull will not exceed 10^5 cm sec^{-1}. Thus the velocity V_f of Section 2 should be taken as the actual desired final velocity plus 10^5. This explains our use of $V_f = 1.2 \times 10^6 = 1.1 \times 10^6 + .1 \times 10^6$.

TABLE 2   (c.g.s. units)

Problem#	4'	12'
$M_f \times 10^{-6}$	12	12
N	30	100
$d \times 10^{-2}$	10	10
$m_B \times 10^{-6}$	.5	.3
σ	.1	.1
$m_P \times 10^{-6}$	1	3
$v_B \times 10^{-6}$	10	10
$M_B \times 10^{-6}$	15	30
$M_P \times 10^{-6}$	30	300
$T \times 10^{-6}$	45	330
$M_o \times 10^{-6}$	57	342
M.R.	4.75	28.5
v_P	.476	.0990
$k_B \times 10^{-18}$	25	15
$v_E \times 10^{-6}$	1.51	.700
$\omega \times 10^{-6}$	1.98	.799
$L \times 10^{-12}$	1.041	1.211
$\Delta t \times 10^3$	1.0	2.5
$\Delta_1 V \times 10^{-6}$	.0188	.00357
$^*\Delta_N V \times 10^{-6}$	.0868	.1009
$\alpha_1 \times 10^{-6}$	18.8	1.43
$\alpha_N \times 10^{-6}$	86.8	40.4
$V_f \times 10^{-6}$	1.12	1.28
$(\frac{L}{m} ln M.R.) \times 10^{-6}$	1.08	1.23

*The complete $\Delta_1 v$ table is not included.

4. The accelerations of the order of 10,000 g are certainly large, and must be rather uniform over the entire structure or breakage is inevitable. The question of the necessary strength for our structure under such accelerations has not been studied. Shock heating in these accelerations is believed to be small.

5. The problem of predetonation of remaining bombs by neutron flux from previously exploded ones must be considered. Strong source bombs and suitable shielding should overcome this difficulty. One should also consider the heating of the vehicle by neutrons and γ-rays. Solid angle considerations insure that this effect will be small.

6. The propellant could be made of a solid material fabricated in N sheets which are placed at the bottom of the projectile. They are detached one by one and expelled to the desired distance. They could be separated by very thin ceramic layers. The placing of the propellant at the bottom of the structure has the advantage that the problem of heating of the permanent structure is attenuated. After each explosion only a small fraction of the next sheet of the propellant would be lost by evaporation and melting.

7. The problem of heating by the propellant and the possible avoidance of this difficulty by the use of magnetic fields have yet to be studied and will be reported in Part II as indicated previously.

8. The whole scheme presupposes elevation of the entire structure beyond the earth's atmosphere by a chemical booster rocket. On the other hand, for the first few explosions we could use air as the propellant with a resultant gain in our mass ratio and with smaller accelerations.

9. We have assumed that the expansion of the thin propellant layer will be essentially perpendicular to its disk surfaces. The losses due to sidewise expansion beyond the base of the rocket were treated summarily by halving the momentum imparted each time to the base of the projectile.

10. The problem of stability has not been seriously studied. The saucer must be so designed that the "center of push" is ahead of the center of mass. Since the immediate impact is at the base of the rocket, stability will probably be a major problem.

11. At little additional cost in mass a V-2 or Viking type of vehicle

could be carried as part of the payload and the saucer jettisoned after the escape velocity is attained. The standard rocket could then proceed under its own power with greater control over its trajectory.

12. The position of the propellant provides a given momentum with larger mass and smaller velocity than would be the case if the same mass of propellant surrounded the bomb, where solid angle losses are considerable. Moreover, it is presumably easier to eject the heavy propellant mass to the smaller distance.

One could even consider iterating this scheme by providing a propellant in two parts at distances of, say, 10 and 20 meters from the rocket, thus increasing the contribution of v_P and decreasing that of v_E to the velocity ω.

References

1. Lee Aamodt, The Feasibility of Nuclear Powered Long Range Ballistics Missiles, LAMS-1870, (Del.) 3/24/1955.
   B. B. McInteer, G. I. Bell, R. M. Potter and E. S. Robinson, A Pachydermal Rocket Motor, and Appendix on Porous Tube Criticality Calculations, LAMS-1887, 5/24/55.
2. LA-714 T-Division Progress Report, Feb.-Ma. 1948.

8

SOME SCHEMES FOR NUCLEAR PROPULSION, PART I

With C. Longmire
(LAMS-2186, March 1958)

*Part I of this report is a discussion of the proposals made in the preceding
report. (Author's note).**

Introduction

It is intended to present here a qualitative description of certain
schemes for nuclear propelled rockets. The ideas sketched in the sequel
stem from the schemata proposed by some of us in the past. Various
details and technical points were discussed in a Rocket Group which
meets weekly in our Laboratory.

The scheme discussed here might be considered as intermediate
between the one outlined in report 7^1 and the ones where the idea is to
propel a nuclear rocket by having a gaseous fission reactor operating
inside the vehicle.[2]

Part I. C. Longmire and S. Ulam—Internal Explosions

Briefly speaking, we imagine a great number N of very mild explosions taking place in succession. These explosions involve bomb-like
assemblies of either metal surrounded by a small amount of high explosive and essentially hydrogenous material or UD_k cores. Each of
these explosions is supposed to heat the total mass involved only to

---

* Only Part I is reproduced here, as Part II is by F. Reines. (Eds.)

very moderate temperatures. To fix the ideas we consider temperatures of the order of 3/4 ev, i.e., 9,000°C, although temperatures up
to a few ev may be useful. Each of these explosions will involve only
several kilograms of active material and several tens of kilograms of
hydrogenous material, and therefore the total yield of the order of a
few *hundreds* of kilograms (sic!) of TNT equivalent. These explosions
are, properly speaking, "fizzles" resembling burning rather than a true
nuclear detonation. One imagines a large chamber with steel walls of
roughly paraboloidal shape with the "explosions" taking place at its
focus. The chamber may be considered, for the purpose of this discussion, as being evacuated except for the material to be exploded. The
linear dimensions of the chamber are large compared to the assembly
which is exploding. For orientation, we may assume the diameter of
the chamber to be of the order of 4 meters, whereas the diameter of
the bomb, together with the enclosing hydrogen, is say of the order of
40 cm. Each of our bombs should be thought of as being in a liquid or
solid state before the explosion. This explosion will convert its whole
mass into gas which will expand and fill the chamber with high velocity particles impinging on the walls and ultimately escaping from the
chamber.

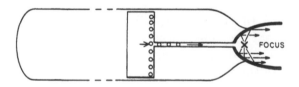

The "bombs" are brought in in rapid succession from a storage
chamber and brought to the "nozzle" chamber where they are exploded. Compared to the proposals made in the preceding report the
present scheme differs in the following respects: The explosions are of
smaller yield. Their number is greater by a factor of 10 or 20. They
will be of longer duration and lesser violence, and therefore, by order
of magnitude, the individual accelerations given to the body of the
rocket in each push will be smaller. Secondly, they are made internally, which allows a greater fraction of mass to be used in imparting
the momentum. This, of course, is more than counter-balanced by the
greater number of supercritical assemblies that one has to employ. Let

us say from the beginning that the total amount of fissionable material expended will be of the order of a few tons, at least for a first design. This makes it appear, offhand, that the primary use of such rocket motors would be to have large satellites and vehicles for interplanetary travel, rather than for stockpiling in large numbers.

We shall employ the following notations:

N = total number of exploding assemblies
$E_i \; i=1 \ldots N$ = the energy release in the i-th explosion
M_i = the total amount of material exploded
m_i = the mass of fissionable material in the i-th "bomb"
R = the diameter of the nozzle chamber
P_i = the pressure on the wall of the nozzle
d = the thickness of the wall
v_e = the velocity of propellant mass escaping the chamber
μ = mean molecular weight of the bomb material
T_i = the temperature to which the mass M_i is brought as a result of the nuclear reaction
W_w = weight of the walls of the paraboloid
W_p = weight of the propellant
W_a = weight of the structure of housing of the bombs, and injecting mechanisms, instruments and "payload"
W = $W_w + W_p + W_w$ total weight.

We now give a tentative set of values in c.g.s. units for our quantities.

N = 10^3
m_i = $m_1 = 5.10^4$ gms
R = 2.10^2 cm
T = $3/4$ ev
d = 3cm
μ = 3.

The effective volume V of our paraboloid with a length of 300 cm would be

V ~ $\pi R^2 \ell = 3.14 \times (2.10^2)^2 \cdot 3.10^2 \sim 4.10^7$ cc.
W_w ~ $2\pi R d \ell \rho \cong 6.3 \times 2.10^2 \cdot 3.10^2 \times 3 \times 8 \cong 10$ tons
W_p ~ $10^3 \times 5.10^4 \sim 50$ tons
W_a ~ 10 tons
W ~ 70 tons.

181

The exit velocity v_e of the propellant will be sensibly higher than the thermal velocity of our material at the temperature obtained in the nuclear explosion. This is so because of the effects of the recombination of the molecules and ions. If $T = 3/4$ ev, $\mu = 3$, the thermal velocity v is about 6 km/sec., and the final v_e about 10 km/sec. The energy E_i of each explosion is then given by $E \sim 1/2$ $mv_e^2 = 1/2 \times 5 \times 10^4 \cdot 10^{12} = 2.5 \times 10^{16}$ ergs, about 500 kgs of TNT equivalent. The pressure on the walls will be of the order of $P \sim (\gamma - 1) \, E/V = (.4 \times 2.5 \times 10^{16})/(4 \times 10^7) \sim 2.5 \times 10^8 \sim 250$ atmospheres ~ 4000 lbs/sq.in.

In the first discussion we shall assume that the quantities are independent of i, that is to say, each assembly and explosion have constant characteristics.

The numerical data above represent merely an order of magnitude orientation about the scheme and are, of course, in no way optimal. There are many degrees of freedom in this scheme. Obviously, most of the fissionable material is "wasted" and we could choose our yields E_i within a very wide range of values—also the composition of the hydrogenous material surrounding the bomb and its mass in proportion to the mass of U^{235} is at our disposal, in a large measure. The geometries of the chamber, etc., seem not to be limited from above by the numbers adopted here.

Speaking qualitatively, the possible advantages of our scheme are as follows:

1. If we admit that the temperature of the material heated by the nuclear explosion is of the order of $1/2 \sim 1$ ev, the expansion of this material in the vacuum of the nozzle chamber will convert most of the energy released and initially present in the form of thermal energy to kinetic energy of the particles with the corresponding cooling of the gas. The velocity of the escape of the propellant will be therefore of the order of 10 kilometers per second, that is to say, the velocity of a satellite. For the velocity of the final "payload" to be of this order, one needs only a ratio e between the mass of the propellant and the mass of the installation and instruments, etc.

2. We mentioned a ratio of about 10 between the linear dimensions of the nozzle chamber and those of the exploding assembly. The density of the gas which will fill the chamber before impinging on the walls will be therefore 1/1000 of the original density. This means that the pressure of the wall will be moderate. The tensile strength of the wall of a fixed thickness depends, inversely, linearly on the inner diameter. If we assume that the pressure on the wall is given by the

Bernoulli formula $P = 1/2 \rho (v^2)$ — since ρ depends inversely on the cube of the linear expansion there is obviously a gain by having the walls of given tensile strength far apart. This gain obtains as long as the total weight of the propellant, auxiliary equipment, and the "payload" exceeds sensibly the weight of the walls of the chamber where the explosions take place. Heating by neutrons and gammas becomes even less of a problem when, with the weight of the payload and equipment essentially constant, the chamber is large.

Considerable computational and experimental work seems necessary to provide a design of the above sort. First of all one should try to calculate individual explosions which are to heat the material to be reproducible and as precise as possible. This should be done with the greatest possible economy of the fissionable material. Probably experiments have to be made with actually exploding such assemblies in order to learn about their characteristics. The action of the expanding gas on steel or tungsten covered structures has to be studied in order to understand the erosion of the material by successive explosions of this sort. The velocity of the propellant leaving the chamber has to be calculated—possible benefits from shaping the exit of the nozzle should be studied. One should discuss the possibilities of cooling of the walls by "sweating" if that should be necessary. We have not discussed the problem of "pumping" individual assemblies at a sufficiently fast rate and the concomitant engineering difficulties. At any rate, the problem here involves a shoving in of masses of the order of 50 kilograms each in intervals of about 1/10 sec. The problem of neutron heating should be calculated in detail, also the problem of the residual gas remaining after the $(i - 1)$th explosion at the time when the i-th explosion is to take place, etc.

It seems likely that a shock absorber between the thrust chamber and the remainder of the missile is desirable, to spread the sharp impulses out over time as well as possible. The number of g's that the main structure has to stand can thus be reduced to a small number.

It appears that steel of average 3 cm thickness will, for our choice of the radius of the wall, contain 4000 lbs/sq.in.

The wall can be coated with tungsten to resist the temperature of the gas accumulating on its surface. The contact of the gas with the wall is of extremely short duration—in a case like the one illustrated above about 1 millisecond for each explosion, so that heating by conduction is seemingly negligible.

Materials other than steel could be considered for confining the exploding gas, with greater strength for weights than steel.

The main problem is the construction of "economic" bombs giving yields of $\sim$1 ton of TNT equivalent.

183

We had the benefit of conversation with George Bell on this problem and C. B. Mills is in the process of calculating critical masses and "alphas" for such UD_k assemblies.

References

1. On a Method of Propulsion of Projectiles by Means of External Nuclear Explosions, Part I, Everett and Ulam, September, 1955.
2. T-821, Nuclear Chinese Rockets, C. Longmire, May, 1956.

9

ON THE POSSIBILITY OF EXTRACTING ENERGY FROM GRAVITATIONAL SYSTEMS BY NAVIGATING SPACE VEHICLES

(LAMS-2219, April 1, 1958)

This report contains the outline of calculations connected with ideas for using suitable orbits so space vehicles can gain energy from near encounters with stars, planets, asteroids, and such.

Orbits like these have since been used repeatedly in some U.S. planetary missions. (Author's note.)

It is intended to outline in this brief report a number of problems of the following type: We assume an astronomical system composed of two or more stellar bodies and a space vehicle which, as an additional body of infinitely small mass compared to the celestial objects, forms part of a many-(e.g., 3) body system. We assume that the "rocket" not only describes the trajectory under the action of the gravitational forces, but also that it has still a reserve energy available for steering by suitably emitted impulses. This energy in the discussion below will be assumed to be roughly of the order of the kinetic energy which the rocket already possesses. The problem, broadly speaking, involves the possibility of using this reserve energy in such a way as to acquire, by suitable near collisions with one or the other of the celestial bodies, much more kinetic energy than it possesses—more by an order of magnitude than the available reserve energy would allow it to acquire by itself.

As examples of the situation we have in mind: Assume a rocket cruising between the sun and Jupiter, i.e., in an orbit approximately that of Mars, with an energy in reserve which would allow the kinetic energy of the vehicle to increase by a factor like 2. The question is whether, by planning suitable approaches to Jupiter and then closer approaches to the sun, it could acquire, say, 10 times more energy. Another example would be a space vehicle moving in a double star system "half-way in between." Then the question is whether, by using additional impulses of its own, it could acquire again a kinetic energy much greater than what it already possesses.

As a purely mathematical problem we could consider the case of two mass points each of mass 1 forming a Keplerian system, and a rocket of mass vanishingly small compared to 1 in an orbit which forms a curve between the two mass points. Suppose that the reserve power of the rocket is such that it could double its kinetic energy. Question: Can one, in this idealized condition, obtain a velocity arbitrarily large (i.e., close to light velocity)?

That this possibility exists seems extremely probable from the theorems of ergodic transformations.[1] It has been shown that arbitrarily near to any given transformation, like the one given by the Hamiltonian describing the n-body system above, there exist transformations which are metrically transitive, that is to say, in particular, Liouville flows such that the trajectory of the system will penetrate arbitrarily near any point on the phase space. The theorem has been proved for bounded phase spaces. This does not make our theorem inapplicable to the problem. We could put in cutoffs in the distance of approach and assume at a finite but very great distance from the gravitational bodies another cutoff. The theorem would imply that arbitrarily near the given dynamical motion there exists one which will make the rocket approach as close to the cutoff sphere surrounding any one of the given mass points as we please, which would in particular imply obtaining arbitrarily high velocities. The theorem asserts the existence of such motions arbitrarily near given ones. The question whether these can be obtained by changes effected through emitting additional impulses inside the rocket is not essentially answered, but in view of the prevalence of ergodic motions near a given one, this seems extremely likely. Such an ergodic trajectory would, of course, in particular provide arbitrarily high velocities. Nothing is said, however, about the times necessary for effecting this. They might be of super-astronomical lengths. It is clear, on the other hand, on general thermodynamical grounds, that "in general" the equipartition of energy may take place. This implies that the body with the small mass of the rocket will acquire very high velocities. This is well known even in systems of a moderate number of particles. The energy distribution is Maxwellian, again tending to provide the small masses with high velocity. The problem is whether, by steering the rocket, one can to some modest extent acquire the properties of a Maxwell demon, i.e., plan the changes in the trajectory in such a way as to shorten by many orders of magnitude the time necessary for acquisition of very high velocities.

As is well known, the perturbations of Jupiter on the motion of some comets provide them occasionally with velocities of escape from the solar system. It has been noticed also[2] that one can use the attraction of the moon to provide a rocket with additional kinetic energy, enabling it to escape from the earth's gravitational field even if it did not have enough energy to do that to begin with.

Our problem is whether one can do it repeatedly to obtain essentially arbitrary kinetic energies by repeated and suitably timed approaches to the two or more celestial bodies.

The question is that of finding general recipes for a 3-(or more) body problem to achieve that aim as quickly as possible. It is proposed to calculate some very schematic, simple, but perhaps instructive cases, for a "strategy" of steering the rocket.

1. The first case involves a problem in one dimension. Suppose two masses oscillate at the end of a segment with given amplitude, say, harmonically; a point of vanishingly small mass is rocketed in between and, possessing some initial kinetic energy, collides elastically with the two oscillating end points. If these should be in phase, the calculation will show the increase of kinetic energy of the small mass. If the phases of the two oscillators should be randomly independent, the question arises how to plan the emission of additional small impulses by the middle point, so as to make it increase its kinetic energy most efficiently. Obviously, one should plan to collide head on with the two oscillators as much as possible. In other words, through additional impulses, collisions that lead to a gain in energy for the "free" point should be maximized as far as possible. The ones that involve colliding by overtaking the receding end point should be diminished in their effect. Without a strategy of changes in the velocity of the "rocket", the gain of energy towards an eventual equipartition would be a very slow process—the rates at which this approach to equipartition takes place are unknown in statistical mechanics, but certainly the gains in a random process increase with the square root of time or slower. With an operating intelligence perhaps this approach to near-equilibrium could be made vastly more rapid.

2. This problem will involve two mass points describing quite elongated Keplerian ellipses around their center of mass. The rocket moves initially in a roughly circular orbit in between the two masses

around their common center. Of course, the actual trajectory in this 3-body problem is very complicated. The question is again to plan a strategy of changing, by small amounts, the energy of the small object so as to approach one or the other of the large bodies to gain kinetic energy. If this is to take place, the approach to either of the two bodies must be increasingly closer. This involves great elongation and an increase in apastron of the rocket. Again the plan is to make near collisions head on. Presumably the planned changes, that is to say, the emitted impulses from the rocket, will be most efficient when the body is at maximum distance from the center of gravity of the two celestial points. It is there that a small increase in velocity will enable one to make changes in the time of the next approach.

The above discussion is, of course, intended for a purely theoretical, mathematical question. Even so, during the next few decades large objects may be constructed with a cruising velocity of 20 kilometers a second, and there will be still some additional energy left for changes in this velocity. It is obvious that the process of increasing the kinetic energy of the rocket by such extraction of gravitational energy from celestial motions, is, at best, very slow. The computations required to plan changes in the trajectory might be of prohibitive length and complication. This little note is meant merely as an introduction to exploratory analyses and calculations undertaken with Kenneth W. Ford and C. J. Everett of LASL.

References

1. J. Oxtoby and S. Ulam, "Measure-Preserving Homeomorphisms and Metrical Transitivity," Annals of Mathematics, **42** 874 (1941).
2. Krafft A. Ehricke and George Gamow, "A Rocket around the Moon," Scientific American **196**, No. 6, 47 (1957).

10

QUADRATIC TRANSFORMATIONS
PART I

With P. R. Stein and M. T. Menzel
(LA-2305, March 1959)

This report is the original study of properties of iterations of non-linear transformations. It has given rise to a very large body of work, and by now, an extensive literature is still appearing on the subject at an ever increasing rate. (Author's note).

ABSTRACT

This report deals with the properties of a restricted class of homogeneous quadratic transformations, with interesting physical and biological analogues, which we have called Binary Reaction Systems. All possible transformations of this class in 3 variables have been studied numerically on a computing machine, and the limiting behavior of random initial vectors under iteration of each of these transformations is tabulated. Some examples of 4-variable Binary Reactions Systems are also studied, and a few generalizations of the notion of Binary Reaction System are investigated for particular cases. Some remarks and results concerning the behavior in the large are presented, and examples of the mode of approach to the limit are given. Several of the more interesting phenomena are illustrated graphically.

The appendix deals with a different class of homogeneous quadratic transformations (of arbitrary dimension) which arise naturally from the study of a

simple evolutionary mode. For this class of transformations, the limiting behavior of arbitrary vectors under iteration can be given explicitly.

Introduction

This report summarizes and discusses some recent studies of the
properties of quadratic transformations in several variables under iteration. This report is of an interim nature, and consists mainly in
a presentation of "experimental" (*i.e.,* numerical) results. A general
theory of quadratic transformations (in contrast to the linear case) is
essentially non-existent, and from the theoretical point of view, the
work summarized below does little to improve the situation. It can
only be hoped that as more facts become known, some outlines of a
theory—at least a classification or "descriptive theory"—will emerge.

The motivation for the considerations which follow lies in the combinatorial problems suggested by genetic or biological systems. One
has to deal with large populations of individuals (or particles) present
in a given generation. Those may combine in pairs and produce, in the
next generation, new particles. Suppose the original particles are of N
different types. Given a rule for the type i $(i = 1, \ldots N)$ produced by
mating of individuals of type j and k, the proportion or fraction x_i of
a given type in the next generation will be a quadratic function of the
two fractions x_j and x_k.

More generally, one could consider a system (gas) of physical particles with N possible characteristics which collide in pairs and produce
through the collision, say, a pair of particles with new characteristics.
There could be many different values of the momenta, but possibly
the "type" of the particle resulting from the collision could be different
from the original ones.

Our present considerations concern the averages or expected values
of the fraction x_i in the next generation. The role of fluctuations or
deviations from the fractions given by the quadratic formulae will be
studied in a subsequent report.

As suggested in the title, this report is the first of a series (one
might add, of unspecified length). By the time these remarks appear
in print, much of the work will probably have been generalized and
the results extended. It is hoped that at least some of the tentative
conclusions presented below will stand.

190

I. Homogeneous Quadratic Transformations

We begin by defining a homogeneous quadratic transformation as a set of N coupled non-linear first-order difference equations of the form:

$$x_i' = \sum_{k,\ell=1}^{N} \gamma_i^{k\ell} x_k x_\ell \quad i = 1, \ldots N \tag{1}$$

where the γ_i^k are some real numerical coefficients with the property:

$$\gamma_i^{k\ell} = \gamma_i^{\ell k} \tag{2}$$

In the present work we restrict ourselves to the case of *non-negative* coefficients:

$$\gamma_i^{k\ell} \geq 0, \ \text{all} \ k, \ell, i \tag{3}$$

Then if we choose the x_j as non-negative real numbers, the x_j' will also have this property. In the following we shall always restrict our x_j in this manner.

In order that the system (1) be of dimension N, we insist that for a given index i, not all $\gamma_i^{k\ell}$ can vanish, *i.e.*,

$$\sum_{k,\ell=1}^{N} \gamma_i^{k\ell} > 0 \tag{4}$$

Systems of the form (1) (not necessarily with the restrictions of reality or non-negativeness) can be considered from two points of view:

(a) As the transformation of all vectors $\begin{pmatrix} x_1 \\ \cdot \\ x_N \end{pmatrix}$ into vectors $\begin{pmatrix} x_1' \\ \cdot \\ x_N' \end{pmatrix}$ or, more generally, as a mapping of some specified region X into a region X'.

(b) As a set of difference equations which determine the value of a vector $\begin{pmatrix} x_1 \\ \cdot \\ x_N \end{pmatrix}$ at "time" n from the value of some initial vector $\begin{pmatrix} x^{(0)} \\ \cdot \\ x_N^{(0)} \end{pmatrix}$.

If we take the first point of view we are, in effect, studying a single iteration, the "mapping" in explicit form. The literature does contain some sporadic work on this problem for low values of N.[1] On the other hand, the second point of view does not seem to have been considered except for the case of one dimension.* The general "solution" to the problem posed by (b) is the explicit construction of the vector $\bar{x}^{(n)}$ in terms of the initial vector $\bar{x}^{(0)}$ and the time-variable n. Such a solution can only, in general, be presented as an iterative procedure. The most one can hope to do is to predict the limiting behavior of the system as $n \to \infty$. Luckily, for practical applications this is usually the only thing of interest. However, even this problem has hardly been touched upon.

In the limit $n \to \infty$ a variety of behaviors is possible; the vectors may, for example, converge under iteration to a limiting vector $\bar{x}$, they may oscillate between a finite set of limit vectors $\bar{x}_i$, or they may exhibit a more or less chaotic behavior, *i.e.*, do neither, but have an ergodic behavior, *i.e.*, the limit in time of average $1/N \sum_{i=1}^{N} x_i$ will exist. (The last is familiar from Kronecker-Weyl's theorem[4] of irrational rotations; see also Ref. 5.) The type of behavior, as well as the numerical value of the limit (if it exists) may or may not depend on the initial vector $\bar{x}^{(0)}$ All the examples (with one exception—see Section XI) studied in this report turn out to be of the first two types; *i.e.*, they either converge to a limit vector or converge to an oscillation between a finite set of vectors in a definite order, *i.e.*, with a definite period.

II. Normalization

Formally, the system (1) contains $N^2(N+1)/2$ parameters, the $\gamma_i^{k\ell}$. It proves convenient to reduce this number (somewhat arbitrarily) by postulating the condition:

$$\sum_{i=1}^{N} \gamma_i^{k\ell} = 1, \text{ all } k, \ell \tag{5}$$

---

* To be sure, there is an extensive literature on coupled non-linear differential equations of 1st order in 2 variables.[2] For N variables, a special class of differential equations is treated in V. Volterra.[3]

This has the great practical advantage that if we now normalize the initial vector by the conditions:

$$\sum_{i=1}^{N} x_i = S \tag{6}$$

then if we divide the right-hand side of each equation of (1) by the constant S, we have:

$$\sum_{i=1}^{N} x_i' = S \tag{7}$$

In biological terminology (see below) this means that we restrict ourselves to the case of a "constant population." Of course, there is no loss of generality in taking $S = 1$. In the sequel we therefore always impose condition (5), and furthermore, take:

$$0 \le x^{(0)} \le 1, \text{ for all } i \text{ with}$$
$$\sum_{i=1}^{N} x_i^{(0)} = 1 \tag{8}$$

This, of course, reduces the number of variables to $N - 1$. (For an example of a different procedure when (5) is not postulated, see the last section of the Appendix.) The x_i are now restricted to lie on the positive portion of the hyperplane:

$$\sum_{i=1}^{N} x_i = 1$$

Even with all these restrictions, it has so far proved impossible to give any general theory of the limiting behavior of the systems (1). However, one sub-class, defined by certain reasonable restrictions on the coefficient $\gamma_i^{k\ell}$, has been completely studied for all N. The results for this case are summarized in the Appendix. The rest of the report is concerned with a discussion of a different sub-class which seems to be of considerable interest, but for which no general theory as yet exists.

III. Binary Reaction Systems

1. Equations (1) have a natural interpretation in terms of biological or genetic language. Consider a large population which consists of N different "types" of individuals.* Let x_j represent the fraction of male individuals of type j, x_ℓ be the fraction of the population of type ℓ (we assume that there are equal numbers of males and females of every given type, hence we may represent them both by the same letter x). Then the system (1), defined by the coefficients $\gamma_i^{j\ell}$, determines the composition of the next generation, which results from a random pairing (once for each individual) of the population at the present generation. If we assume that the members of the old generation do not survive into the next, then for very large populations the expected value of the fractions of the individuals of type j will be given by the system (1). By virtue of our restrictions:

$$\sum_{i=1}^{N} \gamma_i^{k\ell} = 1, \text{ all } k, \ell \tag{5}$$

$$\sum_{i=1}^{N} x_i = 1, \ 0 \le x_i \le 1 \tag{8}$$

the size of the population is constant. The system (1) can therefore be looked upon simply as defining a "mating rule," *i.e.*, determining the characteristics of the offspring from the characteristics of parents. The problem, of course, is to determine the composition of the limiting population.

2. As mentioned in II, the set of equations (1) is, in general, very difficult to study. Consequently, we decided to restrict ourselves initially to a sub-class of systems defined by the following additional restrictions on the coefficients $\gamma_i^{k\ell}$:

> For each pair k, ℓ, there is exactly one value of i for which $\gamma_i^{k\ell} = 1$. For all other values of i the coefficient is zero; *i.e.*,
>
> $$\gamma_i^{k\ell} = \delta_{i i_1} \tag{9}$$
>
> where each pair (k, ℓ) determines one value of i_1.

---

* These "types" are not, of course, meant to correspond to the genotypes or phenotypes of Mendelian genetics!

This means that every term in the product $(x_1+x_2\ldots+x_N)^2$ will appear in exactly one row of the set (cross-terms appearing with the factor 2). For example,

$$x_1' = x_1^2 + x_2^2 + x_4^2 + 2x_1x_4 + 2x_2x_4 + 2x_3x_4$$
$$x_2' = 2x_1x_3 + 2x_2x_3$$
$$x_3' = 2x_1x_2$$
$$x_4' = x_3^2$$

Since there are $N(N+1)/2$ terms in the square of the sum, the number of different possible systems of this sort is clearly equal the number of ways of placing $N(N+1)/2$ different objects in N boxes, no box being empty. Setting $P \equiv N(N+1)/2$, this number is easily shown to be:

$$T_N = N^P - \binom{N}{1}(N-1)^P + \binom{N}{2}(N-2)^P - \ldots \tag{10}$$

For example,

$$T_3 = 540$$
$$T_4 = 818,520.$$

Naturally, many of these are equivalent to each other under permutation of the indices $1, 2, \ldots N$. A *lower limit* to the number S_N of different systems, inequivalent by permutation, is:

$$T_N^* = \frac{T_N}{N!} \tag{11}$$

In fact, the number S_N of inequivalent systems (in this sense) will be somewhat higher than this because of the fact that some of the systems are formally invariant under certain of the $N!$ permutations. Thus, for $N = 3$, $T_N^* = 90$, but by actual enumeration it is found that there are 97 inequivalent systems, i.e., $S_N = 97$. For $N = 4$, $T_N^* = 34,105$; at the present writing the actual number S_N of inequivalent systems has not been determined by us.

We have called systems restricted by condition (9) *"Binary Reaction Systems."* The reason for this name is that such a system associates with each pair (x_k, x_ℓ) a unique result, say x_j'. Symbolically:

$$k \oplus \ell \to j \tag{12}$$

This seems to us to be a natural and simple definition for binary reactions in which "particles" of types k and ℓ produce by "collision" particles of type j. (The "genetic" case is inherently more complicated; see, for instance, the Appendix.)

More general and natural, though less simple, would be a rule of the form:

$$k \oplus \ell \to (j, m) \tag{13}$$

i.e., a pair of particles produces a pair of not necessarily similar particles. We have studied a few generalizations of the simple scheme (12), though not yet in comparable detail (see below, Section VIII.2).

The reaction rule may be presented in tabular form. Consider, *e.g.*, the system:

$$\left. \begin{aligned} x_1' &= x_1^2 + x_2^2 + 2x_2 x_3 \\ x_2' &= 2x_1 x_3 + x_3^2 \\ x_3' &= 2x_1 x_2 \end{aligned} \right\} \tag{14}$$

The table for this would be:

	x_1	x_2	x_3
x_1	x_1	x_3	x_2
x_2	x_3	x_1	x_1
x_3	x_2	x_1	x_2

(15)

Considered as an algebraic system with a law of composition (multiplication) given by the table, this scheme is seen to be commutative $(x_i x_j = x_j x_i)$ but *non-associative*, *e.g.*,

$$(x_1 x_2)x_3 = x_3 x_3 = x_2$$
$$x_1(x_2 x_3) = x_1 x_1 = x_1$$

Binary reaction systems, as defined above, are always commutative (since each product occurs in only one of the set of equations) but are not in general associative. Indeed, for $N = 3$ there are just five associative schemes of this sort:

	x_1	x_2	x_3
x_1	x_1	x_2	x_1
x_2	x_2	x_1	x_2
x_3	x_1	x_2	x_3

	x_1	x_2	x_3
x_1	x_1	x_1	x_1
x_2	x_1	x_2	x_2
x_3	x_1	x_2	x_3

	x_1	x_2	x_3
x_1	x_1	x_1	x_1
x_2	x_1	x_2	x_3
x_3	x_1	x_3	x_1

	x_1	x_2	x_3
x_1	x_1	x_1	x_1
x_2	x_1	x_2	x_1
x_3	x_1	x_1	x_3

	x_1	x_2	x_3
x_1	x_2	x_3	x_1
x_2	x_3	x_1	x_2
x_3	x_1	x_2	x_3

This last table corresponds to the finite group in 3 variables. Although this classification is suggestive, it does not appear that these associative systems are distinguished from the non-associative ones as regards their convergence properties. This is at least the case for $N = 3$, and the presumption is that no particular significance will attach to the associative property for higher N either. However, if the "reaction table" possesses the properties of a group table, certain special properties are easy to establish; *e.g.*:

The fixed point of the transformation has coordinates $x_1 = x_2 \ldots x_N = 1/N$ and is attractive, *i.e.*, the iterates of any vector in its neighborhood converge to it.

IV. Procedure and Results

At this point it is necessary to describe our experimental procedure and results in some detail, since they will be referred to frequently in what follows.

As stated above, for the case of 3 variables, x_1, x_2, x_3, there are 97 binary reaction systems inequivalent to each other under permutation of the labels 1, 2, 3. Each one of these has been studied numerically (on an IBM 704) by having the machine select randomly three initial vectors with coordinates $x_1^{(0)}$, $x_2^{(0)}$, $x_3^{(0)}$ (satisfying $\sum x_i^{(0)} = 1$, $0 \le x_i^{(0)} \le 1$) and letting the computer iterate the transformation in question "as long as necessary," *i.e.*, until some definite limiting behavior was observed.* In all but two cases such limiting behavior became evident without further analysis (one "ambiguous" case is discussed in Section IX.1; the other is mentioned in Section XI). All other transformations eventually either:

(a) Reached a stable distribution or a "fixed point" of the transformation, or

(b) Oscillated between two or three fixed sets of values.

Only in (relatively) few cases did the behavior depend on the choice of the initial vector. However, the rate of convergence (used in the generalized sense to refer to both fixed points and "fixed" oscillations) often varied considerably with this choice.

The main results are contained in Table II. Here each system is written down in symbolic form, the results of iteration being given below. Each system has attached to it a conventional symbol, *e.g.*, I.5.p, II.1.d, *etc.* The Roman numerals I, II, III refer to a distribution of quadratic terms on the right-hand side corresponding respectively to the three partitions of 6 into exactly 3 parts, *viz.*: $(3, 2, 1)$, $(4, 1, 1)$, $(2, 2, 2)$. The other symbols refer to distributions within these main divisions, and are purely conventional. (They correspond to a particular order in which we have examined these cases on computing machines.)

A few examples will serve to illustrate the notation.

(a) Consider the system:

$$x_1' = 2x_1x_2 + 2x_1x_3 + 2x_2x_3$$
$$x_2' = x_1^2 + x_3^2 \qquad\qquad \text{Conventional name: I.6.b}$$
$$x_3' = x_2^2$$

In Table II this appears symbolically as:

$$\left.\begin{array}{l} 2(12) + 2(13) + 2(23) \\ (11) + (33) \\ (22) \end{array}\right\} \begin{array}{l}\text{b.d.p.}\\ \text{not degenerate}\end{array}$$

i.f.p. given by $(x = x_2)$:

$$2x^4 + 2x^3 - x^2 - 3x + 1 = 0$$

$$m \to \text{i.f.p.}$$

$$\left.\begin{array}{l} x_1 = .56311573 \\ x_2 = .32878482 \\ x_3 = .10809945 \end{array}\right\} \text{i.f.p.}$$

---

* Although the sum $x_1 + x_2 + x_3 = 1$ is formally conserved, it is necessary to normalize at each step to avoid loss of accuracy by round-off errors in the last digit.

The notation b.d.p. means that if any of the initial x's $= 1$, the limiting configuration will be an oscillation between the two states: $(x_1 = 0,\ x_2 = 1,\ x_3 = 0)$ and $(x_1 = 0,\ x_2 = 0,\ x_3 = 1)$. This we call a "boundary double point" (b.d.p.). Since it is evident from the structure of the system which variables will assume these values, it is not necessary to specify the b.d.p. more completely. In some systems the b.d.p. will only be reached if either of some two rather than any of all 3 variables is initially equal to 1. These cases are always immediately obvious from the structure of the system.

The words "not degenerate" mean that if initially we have some $x_i^{(0)} = 0$, it will not automatically remain so for all time, *i.e.*, that x_i is not a factor of the r.h.s. of the ith equation. The notation i.f.p. $(x = x_2)$ followed by an equation means that there exists an *interior fixed point* (i.f.p.), that is, a fixed point with no $x_i = 0$, and that the value of one of the variables (in this case x_2) is given by the relevant root of the equation. This equation is simply gotten by suppressing the primes on the l.h.s. and eliminating two of the variables. By relevant root we mean a real root between 0 and 1 which satisfies the set (sometimes extraneous roots are introduced in the elimination process; these do not satisfy the original set). To the right of this is given the resulting fixed point obtained from this equation. The notation $m \to$ i.f.p. means that the transformations as carried out on the machine actually converged to this value (to 8 decimal places) for three random initial vectors.

(b) As a second example, consider:

$$x_1' = x_1^2 + 2x_1x_2 + 2x_1x_3$$

$$x_2' = x_2^2 + x_3^2 \qquad\qquad \text{Conventional name: I.4.a}$$

$$x_3' = 2x_2x_3$$

In Table II this appears as:

$$
\left.
\begin{array}{l}
(11) + 2(12) + 2(13) \\
(22) + (33) \\
2(23)
\end{array}
\right\}
\quad
\begin{array}{l}
\text{2 n.f.p.'s} \\
\text{b.f.p.} \\
\text{doubly degenerate}
\end{array}
$$

no i.f.p.

$m \to$ n.f.p. $(x_1 = 1)$

n.f.p. means "nodal fixed point" and refers to the fact that $x_i = 1$, for at least one value of i, is a fixed point. b.f.p. or "boundary fixed point" means that there exists a fixed point for which one $x = 0$. In view of the explanation of "not degenerate," the term "doubly degenerate" is self-explanatory. "No i.f.p." means that there is no interior fixed point, and $m \rightarrow$ n.f.p. $(x_1 = 1)$ means that the system converged to $x_1 = 1$ for 3 randomly chosen initial vectors.

(c) Consider finally:

$$x_1' = x_1^2 + x_3^2$$

$$x_2' = 2x_1x_3 + 2x_2x_3 \qquad \text{Conventional name: III.1.f}$$

$$x_3' = 2x_1x_2 + x_2^2$$

In Table II (leaving out some explanatory material with regard to the b.d.p.):

$$\left.\begin{array}{l} (11) + (33) \\ 2(13) + 2(23) \\ 2(12) + (22) \end{array}\right\} \begin{array}{l} \text{n.f.p.} \\ \text{b.d.p.} \\ \text{not degenerate} \end{array}$$

i.f.p. given by $(x = x_3)$:
$$1 - 4(1 - x)^3 = 0$$
$m \rightarrow$ i.d.p.

$$\left.\begin{array}{l} x_1 = .16374000 \\ x_2 = .46622052 \\ x_3 = .37003948 \end{array}\right\} \text{i.f.p.}$$

$$\begin{array}{ll} x_1 = .17899745 & \left.\begin{array}{l} x_1 = .40604579 \\ x_2 = .20944248 & x_2 = .47510870 \\ x_3 = .61156007 & x_3 = .11884551 \end{array}\right. \end{array}$$

$$\left.\begin{array}{l} x_1 = .40604579 \\ x_2 = .47510870 \\ x_3 = .11884551 \end{array}\right\} \text{i.d.p.}$$

In this case the transformation did *not* converge to the i.f.p., but rather ended up oscillating between two sets of values, that is, it achieved an "interior double point" (i.d.p.).

We hope that with these examples in mind, it will be possible to interpret Table II. In a few instances, some remarks are appended, but these are self-explanatory.

V. Convergence Behavior

To date we know of no criterion which enables one to predict combinatorially, *i.e.*, from an inspection of the reaction table, whether or not the limiting behavior of a given system will be true convergence (fixed point attained), oscillation between a finite set of limit vectors (periodic point), or neither. For general N, such a criterion will certainly not be simple; it would, for instance, have to take into account the various boundary fixed points and boundary periodic points which are, of necessity, present in any binary reaction system. Of course, every binary reaction system has either boundary fixed points or boundary periodic points (or both). (from Brouwer's fixed point theorem it follows that at least one fixed point must exist.) In other words, the behavior of boundary points under iteration will always have to be treated specially.

An example of this type of complication is provided by system I.1.c. Transcribed from Table II, this reads:

$$x_1' = x_1^2 + x_2^2 + x_3^2$$

$$x_2' = 2x_1x_3 + 2x_2x_3$$

$$x_3' = 2x_1x_2$$

First of all, it is clear that $x_i^{(0)} = 1$ will lead to the n.f.p. $x_1 = 1$. Furthermore, it is clear that a b.d.p. exists, namely:

$x_1 = 1/2$	$x_1 = 1/2$
$x_2 = 0$	$x_2 = 1/2$
$x_3 = 1/2$	$x_3 = 0$

Experimentally, for three different randomly chosen *interior* vectors, the systems converged to the i.f.p. given in the table.

Other special cases have a behavior more difficult to discover; *e.g.*, the system III.1.f, in addition to the i.f.p., has the periodic solution:

$x_1 = .31944846$	$x_1 = .56519772$
$x_2 = 0$	$x_2 = .43480228$
$x_3 = .68055154$	$x_3 = 0$

which is attained if $x_2^{(0)}$ or $x_3^{(0)} = 0$.

For our randomly selected initial vectors, however, the system attains the i.d.p. given in the table.

One might think that things get simpler if we consider only *interior* initial points. Several examples, however, show that even here nothing universally true can be asserted. For example, the system I.2.m attained its i.f.p. with two different initial vectors, but went to the b.d.p. from a third initial vector.

The situation clearly gets more complicated in higher dimensions, where the classification of special boundary solutions in general depends on a complete knowledge of the behavior of lower-dimensional systems.

One may consider, in order to determine whether the fixed point of the transformation is "attractive" or "repellent," the value of the Jacobian of the transformation or the fixed point. If, for example, the absolute value of the Jacobian is > 1, then, in general, iterates of points in every neighborhood of the fixed point will diverge from it.

A summary of the convergence behavior in all 97 three-variable systems for random initial vectors is given in Table I. In 23 systems there is convergence to $x_i = 1$, for some i. Twelve converged to a b.f.p., 15 to a b.d.p., 4 to an i.d.p., 4 to a b.t.p. (boundary triple point), one to an interior triple point, while 6 showed varying behavior depending on the initial conditions. (This class may turn out to be larger with more initial points sampled.) One does not converge at all. All the rest converged to an i.f.p.

The two systems II.1.d and II.1.f showed a continuum of i.f.p.'s and i.d.p.'s, respectively, but this behavior is easy to understand, and not specially significant (see remarks in Table II).

Except in 2 cases, systems I.2.j and III.2.3.a, convergence was numerically evident (though occasionally extremely slow). I.2.j is particularly interesting, and is discussed in detail in Section IX.1. III.2.3.a is briefly discussed in Section XI.

VI. The Nature of the Interior Fixed Points

1. Leaving aside for the moment the question of convergence, it is of interest to inquire into the nature of the various i.f.p. Since for a given N, there are a finite number of different systems, there are only a finite number of i.f.p. These are, of course, defined by the set of algebraic equations obtained on suppressing the primes on the left-hand side of the systems in question. As mentioned above, from Brouwer's theorem it follows that there exists at least one fixed point, but it need not lie in the interior. Frequently these systems have no solutions such that $0 < x_i < 1$, all x_i, which means that no i.f.p. exists. Consider, for example, system I.1.a. The set of equations defining the fixed point is:

$$x_1 = x_1^2 + x_2^2 + x_3^2$$

$$x_2 = 2x_1x_2 + 2x_1x_3$$

$$x_3 = 2x_2x_3$$

Since, by definition of an i.f.p., $x_3 \neq 0$, we must have $x_2 = 1/2$; the second equation then implies:

$$1/2 = 2x_1(1 - x_1), \quad i.e., \ x_1 = 1/2, \text{ so that}$$

$$x_1 + x_2 = 1, \text{ implying } x_3 = 0. \text{ Thus no i.f.p. exists.}$$

In general, to find the i.f.p. we must eliminate two of the variables. The resulting equation is then of 4th order in the remaining variable, say x_i, although it may have factors corresponding to $x_i = 1$, $x_i = 0$, or perhaps to extraneous roots like $x_i = -1$. (In Table II the equation listed is always in "reduced" form, with these factors removed.) Occasionally the equation may have two real roots in the interval 0 to 1. For $N = 3$, in all such cases one of the roots proved to be spurious, *i.e.*, it did not satisfy the original system. In fact, excepting the case II.1.d, mentioned above, which had a continuum of i.f.p., no system had more than one i.f.p. Although it is doubtless possible to give a complete theory of these equations for $N = 3$, a similar treatment for general N seems beyond reach. Here the elimination process can yield an (unreduced) equation of order 2^{N-1}.

2. Bounds for the i.f.p.

Consider an i.f.p. satisfying:

$$1 > x_1 \geq x_2 \geq x_3 \geq \ldots \geq x_N > 0 \tag{16}$$

Clearly, we lose no generality by specifying this ordering, since we can always carry out a permutation on the system so that (16) holds. For a given N, the "largest" i.f.p. will be defined as that i.f.p. for which $x_1 < 1$ has the largest numerical value as we range over all possible systems. (Since the number of systems is finite for finite N, there will always exist a largest i.f.p.) The question then arises: Given N, which system has the largest i.f.p., and what is the corresponding value of

x_1? In view of the astronomical number of inequivalent systems (for even moderate values of N) it is of some interest that a partial answer can be given to this question.

For $N = 3$, our complete study reveals that the system possessing the largest i.f.p.—hereafter called the "maximal system"—is II.3.d, for which the defining equations are (we interchange x_2 and x_3 for convenience):

$$x_1 = x_3^2 + 2x_1 x_2 + 2x_1 x_3 + 2x_2 x_3$$
$$x_2 = x_1^2 \qquad (17)$$
$$x_3 = x_2^2$$

The (unreduced) equation is clearly $(x \equiv x_1)$:

$$x + x^2 + x^4 = 1 \qquad (18)$$

which yields:

$$x = .56984029 \qquad (19)$$

The natural generalization of this system to N dimensions is:

$$x_1 = x_N^2 + \ldots = \left(\sum_{i=1}^{N} x_i \right)^2 - \sum_{i=1}^{N-1} x_i^2$$
$$x_2 = x_1^2$$
$$x_3 = x_2^2 \qquad (20)$$
$$\cdot$$
$$\cdot$$
$$\cdot$$
$$x_N = x_{N-1}^2$$

The root $x = x_1$ is then given as a root of the equation:

$$f_N(x) \equiv \sum_{N=0}^{p=N-1} x^{2^p} = 1 \qquad (21)$$

This root converges very rapidly as $N \to \infty$; for example,

$$N = 4: \quad x = .566160865\ldots$$
$$N = 5: \quad x = .566123797\ldots \qquad (22)$$
$$N = \infty: \quad x = .566123792\ldots$$

It is tempting to consider this last number as an N-independent bound for all binary reaction systems. Unfortunately, this is false, as will be shown below.

Consider a system with $x_1 > 1/2$ and satisfying the ordering (16). Let us assume the "skeleton":

$$x_1 = x_1^2 + 2x_1 x_2 + \ldots$$

$$x_2 = 2x_1 x_3 + \ldots$$

$$x_3 = 2x_1 x_4 + \ldots$$

$$\cdot$$
$$\cdot \qquad\qquad\qquad\qquad\qquad\qquad (23)$$
$$\cdot$$

$$x_{N-1} = 2x_1 x_N + \ldots$$

$$x_N = \ldots$$

Clearly:

$$1 - x_1 = x_2 + x_3 + x_4 + \ldots + x_N \le x_2 \left[1 + \frac{1}{2x_1} + \frac{1}{4x_1^2} + \frac{1}{8x_1^3} \cdots \frac{1}{(2x_1)^{N-2}} \right] \quad (24)$$

But $\qquad x_2 \le \dfrac{1 - x_1}{2}$ $\qquad\qquad\qquad\qquad\qquad\qquad (25)$

or $\qquad \dfrac{1 - x_1}{2} \ge x_2 \ge \dfrac{1 - x_1}{\sum_{p-0}^{N-2} (2x_1)^{-p}}$ $\qquad\qquad (26)$

If we equate these bounds and set $y = 2x_1$, we obtain the equation:

$$y^{N-1} - 2y^{N-2} + 1 = 0 \qquad\qquad\qquad\qquad\qquad (27)$$

Calling the root of this equation y_N, it is evident from (26) that:

$$x_1 \le \frac{y_N}{2} \qquad\qquad\qquad\qquad\qquad\qquad (28)$$

Clearly, $y_N \to 2$ as $N \to \infty$, i.e., the bound is N-dependent. One might suspect at first that this bound is a very weak one, and that the actual maximal system has a much lower i.f.p. However, that the bound is the best possible is proven by exhibiting a system for which $y_N/2 = x_1$ is actually obtained. In fact, such a system is:

$$x_1 = x_1^2 + 2x_1x_2$$

$$x_2 = 2x_1x_3$$

$$x_3 = 2x_1x_4$$

$$\cdot$$
$$\cdot \qquad\qquad\qquad\qquad\qquad\qquad (29)$$
$$\cdot$$

$$xx_{N-1} = 2x_1x_N$$

$$xx_N = \sum_{i=2}^{N} x_i^2 + 2 \sum_{i<j=2}^{N} x_ix_j = \left(x_2 + \ldots x_N\right)^2 = \left(1 - x_1\right)^2$$

It is easily verified that this system has (27) as its i.f.p. equation.

$$\text{For } N = 4 \quad x_1 = \frac{1 + \sqrt{5}}{4} = .809016995$$
$$N = 5 \quad x_1 = .919643378 \qquad\qquad (30)$$

Experimentally ($N = 4, 5, 6$), this i.f.p. is *not* attained on iteration starting from a general point, but these converge to the n.f.p. $x_N = 1$. This is to be contrasted with the behavior of the system (20), which actually attained its i.f.p. ($N = 3, 4, 5$). Indeed, for system (29) it turns out that the absolute value of the Jacobian at the i.f.p. is $(y = 2x_1)$:

$$|J| = y^{N-2}(2 - y)\left(\frac{3y}{2} - 1\right) > 1$$

which makes it reasonable that this i.f.p. is not attractive.

On the other hand, it is clear that for a "skeleton" of the form:

$$x_1 = \ldots$$

$$x_2 = x_1^2 + \ldots$$

$$x_3 = x_2^2 + \ldots$$

$$\cdot$$
$$\cdot \qquad\qquad\qquad\qquad\qquad\qquad (31)$$
$$\cdot$$

$$x_N = x_{N-1}^2 + \ldots$$

we have:

206

$$x_2 \geq x_1^2$$
$$x_3 \geq x_2^2 \geq x_1^4$$
$$\cdot$$
$$\cdot$$
$$\cdot \qquad\qquad\qquad\qquad (32)$$
$$x_N \geq x_{N-1}^{\,2} \geq x_1^{2^{N-1}}$$

whence: $1 = x_1 + x_2 \ldots + x_N \geq \displaystyle\sum_{p=0}^{N-1} x_1^{2^p} = f_N(x_1)$

Therefore, for such a skeleton, the root of $f_N(x) = 1$ does indeed provide an upper bound (attained for the system (20)).

At this writing it has not yet been shown that the system (29) is actually maximal. However, a weak upper bound can be obtained for all systems such that x_1 does *not* contain the term x_1^2 (skeletons of the form (31) are a sub-class of these).

Namely, in this case

$$x_1 \leq 2x_1(1 - x_1) + (1 - x_1)^2 = 1 - x_1^2 \qquad\qquad (33)$$

Therefore, clearly

$$x_1^2 + x_1 \leq 1$$

or

$$x_1 \leq \frac{\sqrt{5} - 1}{2} = .61803399 \qquad\qquad (34)$$

However, we can do much better for this case. In fact, we can show that for $x_1 > 1/2$, we must have, under the ordering (16):

$$x_k > x_{k-1}^2 \qquad\qquad (35)$$

which then establishes the bound (21) by the previous argument.

VII. Periodic Limits

For a large number of 3-dimensional systems, randomly chosen initial vectors iterated to a periodic limit, *i.e.*, the limiting behavior was an oscillation of period 2 or 3 between fixed points. Twenty-four systems exhibited this behavior for three initial vectors, while six others achieved a similar limiting configuration for at least *one* choice of initial vector (with no coordinates lying on the boundary). In most of these cases the final state was of the form:

$$x_i^{(n)} = 1, \ x_j^{(n)} = x_k^{(n)} = 0; \ x_i^{(n+1)} = 0, \ x_j^{(n+1)} = 1, \ x_k^{(n+1)} = 0;$$

$$x_i^{(n+2)} = 1, \ x_j^{(n+2)} = x_k^{(n+2)} = 0;$$

i.e., a boundary double point. A few cases of a boundary triple point were also observed (*cf.* Table I). Such final states we call "trivial," for the reason that the algebraic structure of the transformation alone indicates that such a final state is at least possible. We may contrast this "trivial" type of oscillatory final state with those for which the oscillation takes place between two or three *interior* points. The latter we call an "interior double point" (i.d.p.) or "interior triple point" (i.t.p.).

For $N = 3$ we found just four examples of an i.d.p. and one of an i.t.p. There was also one case of a "non-trivial" b.d.p. (system I.2.e) for which the final state was oscillatory with period 2, but between two "non-trivial" boundary points, *viz.*:

$x_1 = .56519772$	$x_1 = .31944846$
$x_2 = 0$	$x_2 = .68055154$
$x_3 = .4348022$	$x_3 = 0$

The existence of an interior double (or triple) point means that the second (or third) power of the transformation possesses these limit values as fixed points. The algebraic difficulty of finding such points is in general prohibitive. For example, in the unique case of the interior triple point (system I.3.g), if we let:

$$x_1^{(n+3)} = x_1^{(n)} = x, \ x_1^{(n+1)} = y, \ x_1^{(n+2)} = z$$

then one coordinate of the triple point is determined by the set of equations:

$$x = 1 - z^2 - 2y^2(1 - y^2)$$
$$y = 1 - x^2 - 2z^2(1 - z^2) \tag{36}$$
$$z = 1 - y^2 - 2x^2(1 - x^2)$$

It may be verified that the successive values of x_1 given in Table II indeed satisfy this set of equations.

Although there are no oscillatory limiting configurations with periods greater than 3 for $N = 3$, one can, of course, find such by going to higher N. Indeed, we discovered, by chance, a particularly interesting case, *viz.*:

$$x_1' = x_2^2 + 2x_1x_2 + 2x_2x_3 + 2x_1x_4 + 2x_2x_4 + 2x_3x_4$$

$$x_2' = x_4^2 + 2x_1x_3$$

$$x_3' = x_1^2$$

$$x_4' = x_3^2$$

(I.5.j — ext — 1)

This can be considered as one particular generalization of the 3-variable system I.5.j. This generalization—which we applied to several of our original systems (Table III)—consists in setting $x_4' = x_3^2$, replacing x_3^2 in the original system by x_4^2, and putting the new cross-terms $2x_4(x_1 + x_2 + x_3)$ in the top line. When we generalized the triple periodic case, I.3.g, in this manner, the resulting limiting behavior (for 3 randomly chosen initial vectors) was still periodic with period 3, but the configuration was of the "trivial" sort, *i.e.*, the b.t.p. $(1,0,0,0)$, $(0,0,1,0)$, $(0,0,0,1)$. However, in the case of I.5.j, whose resulting generalization is given above, the limiting configuration was oscillatory with period 12. (The values of the coordinates are given in Table III.) A further generalization to 5 variables, following the same prescription, yields the system:

$$x_1' = x_2^2 + 2x_1x_2 + 2x_2x_3 + 2x_4(x_1 + x_2 + x_3) + 2x_5(x_1 + x_2 + x_3 + x_4)$$

$$x_2' = x_5^2 + 2x_1x_3$$

$$x_3' = x_1^2$$

$$x_4' = x_3^2$$

$$x_5' = x_4^2$$

(I.5.j — ext — 2)

In this case, 3 random initial vectors achieved an oscillatory limiting configuration of period 6. (See Table III for the numerical values.)

In our opinion, it is not likely that the behavior observed in these two cases could be predicted by means of any simple criteria.

VIII. Form Stability

1. In a few cases we investigated the effect of making slight changes in the *form* of the equations themselves. One way of doing this, which makes the change of form depend on a single parameter, is as follows: Multiply each term by the factor $1 - \epsilon$, and add $\epsilon/2$ times the term in each of the two other rows.

For example, the system I.5.0 is:

$$x_1' = x_3^2 + 2x_1x_2 + 2x_2x_3$$
$$x_2' = x_2^2 + 2x_1x_3 \tag{I.5.0}$$
$$x_3' = x_1^2$$

This was now modified to:

$$x_1' = (\epsilon/2)x_1^2 + (\epsilon/2)x_2^2 + (1 - \epsilon)x_3^2 + 2(1 - \epsilon)x_1x_2 + \epsilon x_1x_3 + 2(1 - \epsilon)x_2x_3$$
$$x_2' = (\epsilon/2)x_1^2 + (1 - \epsilon)x_2^2 + (\epsilon/2)x_3^2 + \epsilon x_1x_2 + 2(1 - \epsilon)x_1x_3 + \epsilon x_2x_3$$
$$x_3' = (1 - \epsilon)x_1^2 + (\epsilon/2)x_2^2 + (\epsilon/2)x_3^2 + \epsilon x_1x_2 + \epsilon x_1x_3 + \epsilon x_2x_3 \tag{37}$$

In this case it is easy to carry out the elimination of x_2, x_3 to obtain the equation for the i.f.p. as a function of ϵ. There results:

$$b_1x_1^4 + b_2x_1^3 + b_3x_1^2 + b_4x_1 + b_5 = 0 \tag{38}$$

where the b_i are given in terms of the parameter:

$$\alpha \equiv \frac{2 - 3\epsilon}{2} \tag{39}$$

as follows:

$$b_1 = \alpha^4$$
$$b_2 = 4\alpha^3$$
$$b_3 = 2\alpha^2 - \frac{2\alpha^3}{3}(2 + \alpha)$$
$$b_4 = \alpha(1 + 2\alpha) - \frac{4\alpha^2}{3}(2 + \alpha) \tag{40}$$
$$b_5 = -\frac{\alpha}{3}(2\alpha + 1) + \frac{\alpha^2}{9}(2 + \alpha)^2$$

For $\epsilon = 0$ ($\alpha = 1$) we get (dividing out x_1) the original i.f.p. equation:

$$x_1^3 + 4x_1^2 - 1 = 0 \tag{41}$$

At $\epsilon = 2/3$, $\alpha = 0$, all the coefficients b_i vanish. This corresponds to the set:

$$x_1' = x_2' = x_3' = \frac{1}{3}\left(x_1^2 + x_2^2 + x_3^2\right) + \frac{2}{3}\left(x_1 x_2 + x_1 x_3 + x_2 x_3\right) \tag{42}$$

which reaches the i.f.p. (1/3, 1, 3, 1/3) on a single iteration starting from any initial vector. In fact, as $\epsilon \to 2/3$, all systems to which this generalization is applied tend toward this simple case (since at $\epsilon = 2/3$, $1 - \epsilon = \epsilon/2$).

For the system I.5.0, generalized in this manner, we investigated the convergence for several values of ϵ, *viz.*: $\epsilon = .001, .01, .05, .1$. In each case, randomly chosen initial vectors iterated to the i.f.p. predicted by equation (38). In other words, there is a sort of continuity in the convergence behavior as the form of the equation is changed in this simple manner.

Slightly more interesting is the result of applying this one-parameter generalization to the system I.3.g. As mentioned above, random initial vectors iterated according to this transformation reached an oscillatory final state with period 3. Vectors under the generalized transformation behaved in the same fashion, but as ϵ was increased the final-state oscillations decreased in amplitude, until at $\epsilon \simeq .045$ general initial vectors appeared to converge to the i.f.p. predicted by the corresponding i.f.p. equation. Presumably, the final state is still oscillatory (with period 3), but the oscillations are too small to observe with 8-decimal-place accuracy.*

The conclusion to be drawn from these experiments is that binary reaction systems are "stable" under small perturbations of formal structure.

---

*For $\epsilon > 0$, some initial points converged to the i.f.p. In other words, the i.f.p. (which is a function of ϵ) appears to be attractive for $\epsilon > 0$. More detailed analysis would probably show that what is happening is that the area of the region of the triangle for which convergence (of a point in this region) to the i.f.p., is increasing with increasing ϵ. Correspondingly, one is less likely to pick an initial point outside this region, *i.e.*, a point which will iterate to the i.t.p. For further discussion of behavior in the large, see Section XI.

2. When more radical changes of form are made, we do get correspondingly greater changes in behavior. For instance, we took the above-mentioned system I.3.g and kept only the skeleton:

$$x_1' = \frac{1}{2}x_2^2 + \frac{1}{2}x_3^2 + 2x_1x_2 + \dots$$
$$x_2' = 2x_1x_3 + 2x_2x_3 + \dots \tag{43}$$
$$x_3' = \frac{1}{2}x_1^2 + \dots$$

To this we added the missing terms $(1/2)\,x_1^2$, $(1/2)\,x_2^2$, $(1/2)\,x_3^3$ in all possible ways (*i.e.*, 27 ways, of which one corresponds to the original system). For 25 of the resulting 26 new transformations, random initial vectors iterated to an i.f.p. (The results are summarized in Table IV.) One system, however, gave an oscillatory final state with period 3, a behavior analogous to that of the original system, *viz.*:

$$x_1' = x_2^2 + \frac{1}{2}, x_3^2 + 2x_1x_2$$
$$x_2' = \frac{1}{2}\,x_3^2 + 2x_1x_3 + 2x_2x_3 \tag{44}$$
$$x_3' = x_1^2$$

The configuration of the final state (starting from 3 randomly chosen initial vectors) was:

$$x_1^{(n+3)} = x_1^{(n)} = .87777286, \quad x_1^{(n+1)} = .00766150, \quad x_1^{(n+2)} = .34944206$$
$$x_2^{(n+3)} = x_2^{(n)} = .00011739, \quad x_2^{(n+1)} = .22185332, \quad x_2^{(n+2)} = .65049924$$
$$x_3^{(n+3)} = x_3^{(n)} = .12210975, \quad x_3^{(n+1)} = .77048518, \quad x_3^{(n+2)} = .00005870$$

It is clear that a "rule" which allows $x_3 \oplus x_3 \to x_1$ or x_2 (with equal probability) is a rather unnatural one. A somewhat more logical modification is to assume a skeleton in which the cross-terms appear with coefficient unity, and to add the missing terms x_1x_2, x_1x_3, x_2x_3, in all possible ways. When this was done for I.3.g, all the resulting transformations had an i.f.p., and in every case random initial vectors iterated to the i.f.p. (see Table V).

This change of rule has a natural interpretation; effectively it allows non-commutativity, since *e.g.*, $x_i \oplus x_j$ may give two results, which we can interpret as the respective results of $x_i x_j$ and $x_j x_i$. Formally, this is an interesting generalization of the concept of a binary reaction system (as originally defined), but the convergence behavior does not appear to be startlingly different.

This same "non-commutative" generalization was tried on the system III.2.1.b; the results offered no surprises.

IX. A Specific Convergence Problem

1. The Exceptional Case I.2.j

As mentioned above, there are only two systems among all the 97 which exhibit an ambiguous convergence behavior. The most interesting of these has the form:

$$x_1' = x_1^2 + x_3^2 + 2x_1 x_3$$
$$x_2' = x_2^2 + 2x_1 x_2 \qquad \text{I.2.j}$$
$$x_3' = 2x_2 x_3$$

The system is degenerate, and has two n f.p.'s; in the following we shall ignore the behavior of the boundary points (since these present no problem) and only discuss the behavior under iteration of interior points.

By inspection, it is evident that the system possesses the i.f.p:

$$x_1 = x_3 = \frac{1}{4}$$
$$x_2 = \frac{1}{2} \qquad (45)$$

For three randomly chosen (interior) initial points, little convergence was evident even after some 85,000 iterations. In order to see better what was happening, we introduced new coordinates:

$$S = \frac{1 + x_1 - x_3}{2}$$
$$\alpha = \frac{1}{2} x_2 \qquad (46)$$

These are effectively Cartesian coordinates in the plane of the triangle formed by the three vertices $(1,0,0)$ $(0,1,0)$ $(0,0,1)$.

More specifically, in the original form of the transformation, the x_i's are constrained to move on the positive portion of plane $x_1 + x_2 + x_3 = 1$:

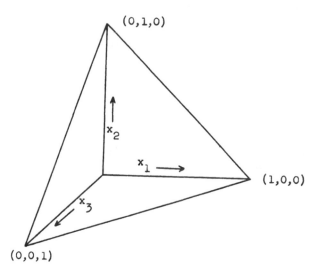

For algebraic convenience, we distort the triangle into a 45° triangle with base unity. The coordinates of a point in this triangle are then (S, α), as shown in the sketch below. Here

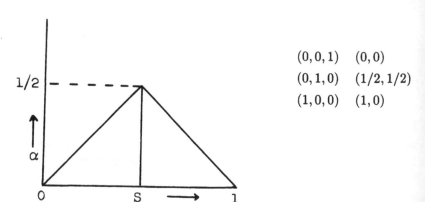

$(0,0,1)$	$(0,0)$
$(0,1,0)$	$(1/2,1/2)$
$(1,0,0)$	$(1,0)$

In terms of these new coordinates, the transformation takes the form:

$$S' = 1 - 4\alpha + 4\alpha^2 + 2\alpha S$$
$$\alpha' = 2\alpha S \tag{47}$$

and the i.f.p. is:

$$S = \frac{1}{2}$$
$$\alpha = \frac{1}{4} \tag{48}$$

Note that the Jacobian of (47) is exactly 1 at the f.p. (48). For future reference, we write down the inverse of (47):

$$S = \frac{\alpha'}{1 - \sqrt{S' - \alpha'}}$$
$$\alpha = \frac{1 - \sqrt{S' - \alpha'}}{2} \tag{49}$$

If we now make the further transformation:

$$x = S - \frac{1}{2}$$
$$y = \alpha - \frac{1}{4} \tag{50}$$

we get:

$$x' = -y + \frac{x}{2} + 4y^2 + 2xy$$
$$y' = y + \frac{x}{2} + 2xy \tag{51}$$

with i.f.p. $x = y = 0$.

Note that if we consider only *linear* terms:

$$x' = -y + \frac{x}{2}$$
$$y' = y + \frac{x}{2} \tag{52}$$

then an invariant ellipse exists:

$$x'^2 + x'y' + 2y'^2 = x^2 + xy + 2y^2 \tag{53}$$

Figure 1 shows a plot of the observed iterates in the x, y plane. The three curves show the behavior of successive iterates, the initial point of the sequence being taken respectively at $n = 1550$, $n = 10,101$, $n = 75,001$. Each curve is roughly an ellipse of the form:

$$x^2 + xy + 2y^2 = C \tag{54}$$

Numerically, at least, C appears to $\to 0$ as $n \to \infty$, or, to put it another way, the axes of the ellipse of reference are shrinking.

In order to convince ourselves that this apparent convergence was not simply the result of systematic round-off, we used the inverse transformation (48), numerically retracing our steps from $n = 10,101$ to $n = 1,550$. All coordinates agreed to 6 decimal figures, which pretty well precludes the possibility that the observed convergence is a numerical accident.

It should be noted, by the way, that no matter how close one is to the fixed point, the quadratic terms in (51) cannot be ignored, for according to (53), the linear terms by themselves will generate a sequence of iterates which will all lie on the ellipse of reference.

For the purpose of discussion, it is convenient to transform equation (51) so that the reference curve is a circle. The linear part of the transformation can be written:

$$\begin{pmatrix} x' \\ y' \end{pmatrix} = A \begin{pmatrix} x \\ y \end{pmatrix}, \quad A = \begin{pmatrix} \frac{1}{2} & -1 \\ \frac{1}{2} & 1 \end{pmatrix} \tag{55}$$

If we transform this with the matrix:

$$S = \begin{pmatrix} \sqrt{7} & 1 \\ 0 & -2 \end{pmatrix} \tag{56}$$

we find:

$$R = S^{-1} AS = \begin{pmatrix} \frac{3}{4} & \frac{\sqrt{7}}{4} \\ \frac{-\sqrt{7}}{4} & \frac{3}{4} \end{pmatrix} \tag{57}$$

216

which corresponds to a rotation through an angle $\theta = \cos^{-1}\frac{3}{4}$. The transform through S of the complete mapping (50) has, then, the form:

$$x' = \frac{3}{4}x + \frac{\sqrt{7}}{4}y + \frac{10}{\sqrt{7}}y^2 - 6xy$$
$$y' = \frac{-\sqrt{7}}{4}x + \frac{3}{4}y + 2y^2 + 2\sqrt{7}xy \tag{58}$$

2. The Asymptotic Behavior of the Angle of the Radius Vector Under Iteration

The transformations with which we are concerned do not preserve the ordinary measure (Lebesgue measure) of the space which is mapped into itself. Moreover, they are not one-one. In some cases they shrink a neighborhood of the fixed point into a proper part of itself and the limiting image of such a region may consist of this point alone. Obviously, an invariant measure, if it is to be constructed, would have to be of Lebesgue-Stieltjes type and assign positive values, in some cases, to sets consisting of single points.

One may be interested in the behavior of the angles (with a fixed direction issuing from a fixed point) of the vectors $T^i(\bar{x})$, where $\bar{x} = (x_1, x_2\, x_3)$; $i = 1, 2, \ldots$ more generally in the behavior of the points:

$$S_i = \frac{T^i(\bar{x})}{|T^i(\bar{x})|} \tag{59}$$

on the unit sphere. Something can be said about it even in cases where T does not transform any bounded region into itself. Thus, for example, if T is an arbitrary linear transformation of the n-dimensional Euclidean space into itself, $T(0) = 0$. Then the ergodic limit of the average of S_i exists. In other words, if C is an arbitrary "cone" of directions in space, *i.e.*, a sub-set of the unit sphere $f_C(s)$, its characteristic function being:

$$f_C(s) = 0 \text{ if } s \notin C; \quad f_C(s) = 1 \text{ if } s \in C$$

then for almost all s, the limit:

$$\lim_{N=\infty} \frac{1}{N} \sum_{i=1}^{N} f_C\left(T^i(s)\right) \tag{60}$$

217

exists. This follows for a general linear transformation T, from the well-known theorems giving, in fact, more precise information—in the two extreme cases: if T is an orthogonal transformation we have the Kronecker-Weyl theorem on equipartition; if T, considered as a matrix, has all coefficients positive, the Perron-Frobenius theorem asserts convergence to a unique direction. In the general case one obtains, by considering a decomposition of the space into sub-regions where one or the other behavior dominates, at least the *existence* of the ergodic limit.

Presumably, the theorem still holds true if T is a general homogeneous quadratic transformation of the n-dimensional space.

In our very special quadratic transformations of the plane, more can be said: The transformation of the previous section possesses the Knonecker-Weyl property: The angle described by the iterates of almost every point covers the circumference of the unit circle *densely* and *uniformly*.

We hope to show that if the linear part of a quadratic transformation Q, which has the origin as its fixed point, consists of a rotation through an irrational angle, then the iterates of Q converge to the origin provided one starts with points in a circle of sufficiently small radius.

A detailed discussion of these matters will be given in a subsequent report.

X. Further Generalizations

In view of the impracticability of studying all possible Binary Reaction Systems for any $N > 3$, we thought it worthwhile to generalize a few of our 3-variable systems to higher dimension by arbitrary but fixed rules. One such generalization is mentioned in SectionVII (see I.5.j — ext — 1 and subsequent discussion). Another essentially different way to generate "interesting" systems is to construct for any given 3-variable transformation the corresponding "super-system." This is constructively defined as follows:

We introduce nine variables, $y_1, y_2, \ldots y_9$, according to the prescription:

$$x_1 = y_1 + y_2 + y_3$$
$$x_2 = y_4 + y_5 + y_6 \tag{61}$$
$$x_3 = y_7 + y_8 + y_9$$

and substitute these in the original transformation. We then have three transformations for the three triads of variables $y_1 + y_2 + y_3$, $y_4 + y_5 + y_6$, $y_7 + y_8 + y_9$. Consider, *e.g.*, the system I.2.e. The last line of the transformation now reads:

$$y_7' + y_8' + y_9' = 2(y_1 + y_2 + y_3)(y_4 + y_5 + y_6)$$

In order to convert this into 3 separate expressions for y_7', y_8', y_9', we could, for example, *formally* identify the variables modulo 3, *i.e.*,

$$y_1 \sim y_4 \sim y_7$$
$$y_2 \sim y_5 \sim y_8 \tag{62}$$
$$y_3 \sim y_6 \sim y_9$$

Correspondingly, on the right-hand side, we could make the identification.

$$2y_1 y_4 \sim y_1^2$$
$$2y_2 y_5 \sim y_2^2$$
$$2y_2 y_6 + 2y_3 y_5 \sim 2y_2 y_3$$
$$2y_3 y_6 \sim y_3^2 \tag{63}$$
$$2y_1 y_6 + 2y_3 y_4 \sim 2y_1 y_3$$
$$2y_1 y_5 + 2y_2 y_4 \sim 2y_1 y_2$$

We can now write expressions for y_7', y_8', y_9' so that, with these formal identifications, the resulting sub-system will have the same form as the original 3-variable system, *i.e.*,

$$y_7' \sim y_1' = y_1^2 + y_2^2 + 2y_2 y_3 \sim 2y_1 y_4 + 2y_2 y_5 + 2y_2 y_6 + 2y_3 y_5$$
$$y_8' \sim y_2' = y_3^2 + 2y_1 y_3 \sim 2y_3 y_6 + 2y_1 y_6 + 2y_3 y_4 \tag{64}$$
$$y_9' \sim y_3' = 2y_1 y_2 \sim 2y_1 y_5 + 2y_2 y_4$$

219

In this way we (arbitrarily) obtain equations for y_7', y_8', y_9' in terms of the y_i's. When this is done for each triad, a 9-dimensional B.R.S. results.

In the present case, using the symbolic notation of Table II:

$(11) + (44) + 2(47) + (22) + (55) + 2(58) + 2(23) + 2(56) + 2(59) + 2(68)$

$(33) + (66) + 2(69) + 2(13) + 2(46) + 2(49) + 2(67)$

$2(12) + 2(45) + 2(48) + 2(57)$

$(77) + 2(17) + (88) + 2(28) + 2(29) + 2(38) + 2(89)$

$(99) + 2(39) + 2(79) + 2(19) + 2(37)$ (I.2.e — Super)

$2(78) + 2(18) + 2(27)$

$2(14) + 2(25) + 2(26) + 2(35)$

$2(36) + 2(16) + 2(34)$

$2(15) + 2(24)$

As a second example, we quote the result of treating system I.3.g in the same manner:

$(55) + 2(25) + (88) + (66) + (99) + 2(36) + 2(45) + 2(78) + 2(15) + 2(24)$

$2(46) + 2(79) + 2(16) + 2(34) + 2(56) + 2(89) + 2(26) + 2(35)$

$(44) + (77) + 2(14)$

$2(28) + 2(58) + 2(39) + 2(69) + 2(18) + 2(48) + 2(27) + 2(57)$

$2(19) + 2(37) + 2(49) + 2(67) + 2(29) + 2(38) + 2(59) + 2(68)$

$2(17) + 2(47)$ (I.3.g — Super)

$(22) + (33) + 2(12)$

$2(13) + 2(23)$

(11)

When these transformations were iterated for randomly chosen initial vectors, the sums:

$$x_1 = y_1 + y_2 + y_3, \quad x_2 = y_4 + y_5 + y_6, \quad x_3 = y_7 + y_8 + y_9$$

reached, of course, as they should, the same limiting configuration as was observed in the original 3-variable system. However, the actual values of the individual y_i varied with the initial configuration. The results are given in Table VI.

XI. Properties in the Large

1. Although it is, in general, possible to discuss the behavior of points under iteration in the neighborhood of a fixed point, the iteration behavior of such points over the whole domain (positive portion of the hyperplane) can, at present, be treated only experimentally. In what follows, we shall take the variables to be S, α, and the domain to be the corresponding 45° triangle with unit base (see Section IX.1 for a definition of this coordinate system).

As stated above, we have not found any general criteria for determining which of several possible limiting behaviors will be realized for a given 3-variable system, starting with a general point in the triangle. On the basis of a rather small sample ($\sim$ 3 random initial points for each system), it appears that, excluding boundary points, the limiting behavior is independent of the initial point for the large majority of systems. As shown in Table 1, however, there are (at least) 6 systems in which this limiting behavior depends on the initial point (we exclude from consideration the "pathological" systems II.1.d and II.1.f; see the discussions under these entries in Table II). Two of these, I.2.m and I.5.h, have been examined in greater detail. What was done was to look for boundaries which separate regions of different limiting behavior. This was accomplished by programming the computing machine to "search" the whole triangle in a systematic manner. On the first pass a crude net was used ($\Delta S = \Delta \alpha = .05$). Then, when the boundaries had been approximately located, a more refined interval was employed in the appropriate neighborhoods. For each trial point in the triangle, sufficient iterations had to be performed to identify the limiting behavior. Despite the apparent magnitude of the task (several hundred trial points had to be followed for some 70 iterations each), a complete search (first crude, then appropriately refined) takes only about 15 minutes of computing time per system.* The results for the two

---

* On the average, the machine will perform 50 iterations/sec. for a 3-variable Binary Reaction System. The systematic search of the triangle is somewhat slower for various reasons connected with input-output requirements.

systems studied are shown in Figures 2 and 3. In these, each calculated boundary point is determined to within an absolute error $< .0025$ *i.e.*, to within 1/4% of the length of the base of the triangle. In these figures, initial points lying in the region marked "OSC." will iterate to the appropriate boundary oscillation, while all points lying outside these regions will converge under iteration to the i.f.p. The boundaries appear to be complicated. We have not attempted to study what happens to points actually lying on the boundary curves; for this an analytical treatment is necessary.

In a few simple cases it is possible to give an analytical treatment of such boundary regions. As an example, we cite the system III.2.2.a. In the S, α coordinates (we use these to conform with our treatment of the other systems; the argument can be carried out in the original coordinates with equal ease) the transformation takes the form:

$$S' = 2S - S^2 + 3\alpha^2 - 4\alpha S$$
$$\alpha' = 2\alpha(1 - S) \tag{65}$$

There are 3 n.f.p.'s, namely, $(S = 0, \alpha = 0)$, $(S = 1, \alpha = 0)$, and $(S = 1/2, \alpha = 1/2)$, while the i.f.p. is $(S = 1/2, \alpha = 1/6)$. The boundaries $\alpha = 0$ and $S = \alpha$ are clearly transformed into themselves. In addition, there exists an invariant line:

$$S = 3\alpha .$$

All points lying on this line can easily be shown to iterate to the i.f.p.— with the exception of $S = \alpha = 0$, which is a non-attractive f.p. This line is shown on the diagram below. The curve $S = S^*$ is the locus of all points such that $S' = S$; its equation is:

$$S^* = \frac{1 - 4\alpha + \sqrt{(1 - 4\alpha)^2 + 12\alpha^2}}{2} \tag{66}$$

It is easily shown that a point lying in the region below the line $S = 3\alpha$, *i.e.*, such that $S > 3\alpha$, remains in this region under iteration; similarly for points lying above the line. Further, all points lying to the left of the curve $S = S^*$ remain to the left under iteration. It can be further shown that points lying to the right of $S = S^*$ and not situated at the corners of the triangle or on the line $S = 3\alpha$ will eventually cross

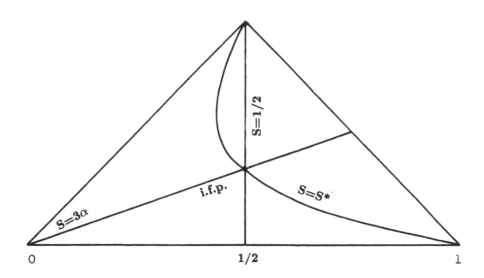

the curve. Furthermore, in the neighborhood of the i.f.p., the linear approximation to the transformation is ($x \equiv S - 1/2$, $y \equiv \alpha - 1/6$):

$$x' = \frac{x}{3} - y$$
$$y' = -\frac{x}{3} + y \tag{67}$$

from which it can be deduced that the i.f.p. is attractive only along the line $x = 3y$. Finally, using the fact that for $S < 1/2$, we always have $\alpha' - \alpha > 0$ and that correspondingly for $S > 1/2$, $\alpha' - \alpha < 0$, it can be shown that all points lying above the line $S = 3\alpha$ will iterate to $S = \alpha = 1/2$, while all points below the line will iterate to $S = 1$, $\alpha = 0$.

The reason why this system can be treated so simply is, of course, that the boundary curves are explicitly known. When this is not the case, it may be helpful to have before one a picture of the mapping (single iteration) of the entire triangle, as well as the curves along which α and S are stationary. A few interesting examples are given in Figures 4, 5, and 6.

2. There is one case which does not appear to converge either to a fixed point or to a finite oscillation. This is the system III.2.3.a:

$$x_1' = x_1^2 + 2x_1x_2$$
$$x_2' = x_2^2 + 2x_2x_3 \qquad\qquad (68)$$
$$x_3' = x_3^2 + 2x_1x_3$$

or, in terms of the coordinates, S, α:

$$S' = 2\alpha + S^2 - 3\alpha^2$$
$$\alpha' = 2\alpha(1 - S) \qquad\qquad (69)$$

The situation is illustrated in Figure 7. The i.f.p. is non-attractive and the 3 corners of the triangle are attractive only along the boundaries in a clockwise direction. Under iteration, points will spiral out, approaching arbitrarily close to the boundaries, but, *e.g.*, as the transformation $\alpha' = 2\alpha(1 - S)$ shows (for the bottom boundary), no point inside the triangle can ever reach the boundary. Thus a general point will continue to spiral indefinitely.

Numerically, a spurious convergence was observed owing to the fact that the several random initial points chosen rapidly iterated to within a distance less than 10^{-8} from one or another boundary line.

If one transforms the triangle into the unit circle in an appropriate manner, the situation can be viewed as follows: The center is a non-attractive fixed point, and all points lying in the circle spiral outwards towards the circumference. On the circumference itself, there are 3 fixed points located, say at $\theta = 0$, $2\pi/3$, $4\pi/3$, which in turn define 3 arcs. Any point lying on one of these arcs will move under iteration in a clockwise direction, ultimately converging to the fixed point which constitutes the right-hand boundary of the arc in question. Interior points, however, can never reach the boundary. In general, the sequence of iterates of any interior point (excluding the center) does not converge.

3. We have not studied in detail the *rate* of approach to the limiting configuration. In the neighborhood of a fixed point, this rate is usually easy to obtain, but for oscillating configurations the algebra is more difficult. For most of the cases studied, convergence (to 8 decimal places) was either attained within 100 iterations (some were much faster) or else not for many thousands of iterations. We shall not discuss this further except to remark that the path of approach to a fixed point may depend very critically on the initial point. As an example, we refer the reader to Figure 8. For this transformation, two of the corners of the triangle are non-attractive fixed points; initial points in their neighborhoods iterate smoothly to the i.f.p. In contrast, points in the neighborhood of the origin iterate to the i.f.p. in an oscillatory manner; the existence of a limiting line through the i.f.p. is clearly evident.

XII. Connection with Ordinary Differential Equations

No doubt it will have occurred to the reader that Binary Reaction Systems and their extensions have an obvious connection with systems of ordinary differential equations. Consider, for instance, the system of differential equations:

$$\frac{dx_1}{dt} = -x_1 + f_1(x_1 \ldots x_N)$$

$$\frac{dx_2}{dt} = -x_2 + f_2(x_1 \ldots x_N)$$

$$\cdot$$

$$\cdot \tag{70}$$

$$\cdot$$

$$\frac{dx_N}{dt} = -x_N + f_N(x_1 \ldots x_N)$$

where the $f_i(x_1 \ldots x_N)$ are homogeneous quadratic functions of the variables. A straightforward finite difference approximation to the system is:

$$x_1^{(n+1)} = (1 - \Delta t)x_1^{(n)} + \Delta t\, f_1^{(n)}$$

$$\cdot$$

$$\cdot \tag{71}$$

$$\cdot$$

$$x_N^{(n+1)} = (1 - \Delta t)x_N^{(n)} + \Delta t\, f_N^{(n)}$$

If we now restrict the f_i to be disjoint partial sums of the terms in $(x_1 + \ldots + x_N)^2$, such that $\sum_{i=1}^{N} f_i = (x_i + \ldots + x_N)^2$, i.e., just the terms that occur in our binary reaction transformation, then the above set of difference equations goes over into a Binary Reaction System for $\Delta t = 1$. With this restriction on the f_i, the system of difference equations has the property $\sum x_i^{(n)} = 1$ for all n if $\sum x_i^{(0)} = 1$, independent of the time-step Δt. It should be observed that for $\Delta t > 0$, the system has the same fixed points as the corresponding Binary Reaction System.

As a very simple example, consider the system:

$$x_1' = x_1^2 + x_2^2$$

$$\text{with i.f.p. } x_1 = x_2 = \frac{1}{2},$$

$$x_2' = 2x_1 x_2$$

$$\text{and b.f.p. } x_1 = 1,\ x_2 = 0 \ . \tag{72}$$

Any interior point $(x_1 \neq 1, 0)$ will iterate to the i.f.p. The corresponding differential equation is (eliminating x_2):

$$\frac{dx_1}{dt} = 2x_1^2 - 3x_1 + 1 \tag{73}$$

of which the solution is:

$$\left| \frac{1 - x_1}{1/2 - x_1} \right| = \left| \frac{1 - x_1^{(0)}}{1/2 - x_1^{(0)}} \right| e^t \ . \tag{74}$$

Clearly, as $t \to \infty$, $x_1 \to 1/2$, $x_1^{(0)} \neq 1, 1/2$. Thus, in this case, the asymptotic behavior of the differential equation is the same as that of the corresponding Binary Reaction System. This, however, is not generally so. For example, the transformation I.5.d (see Table II) possesses a boundary double point which is attained for a certain set of initial points. The corresponding differential equation system, when integrated according to the finite difference scheme (71) with $\Delta t = 2^{-6}$, converged to the n.f.p. $x_1 = 1$ regardless of the initial point.

In such cases the Binary Reaction System, viewed as a finite difference approximation to the corresponding system of differential equations, is clearly an "unstable" scheme. We hope to discuss the point further in a subsequent report.

Explanation of Graphs

1. Figure 1 is a plot of successive iterates of a point under the transformation equation (53) of the text. The outer curve shows the iterates from cycle $n = 1550$ (point i_1) to cycle $n = 1570$ (point f_1); the next curve goes from $n = 10,101$ (point i_2) to $n = 10,114$ (point f_2), while the innermost curve goes from $n = 75,001$ (point i_3) to $n = 75,011$ (point f_3).

2. Figures 2 and 3 show the (experimentally determined) boundaries between two types of limiting behaviors for two different Binary Reaction Systems (see discussion in Section XI). In terms of the S, α coordinates, these transformations are:

I.2.m (Figure 2):

$$S' = 1 + \frac{3S^2 - \alpha^2}{2} + 3\alpha S - 2S$$

$$\alpha' = \frac{S^2 - 3\alpha^2}{2} - 3\alpha S + 2\alpha$$

I.5.h (Figure 3):

$$S' = 1 + \frac{1 + \alpha^2 - 3S^2}{2} - \alpha + S + 3\alpha S$$

$$\alpha' = \frac{1 + S^2 - 3\alpha^2}{2} + \alpha - S - \alpha S$$

In each case, points in the inner region (marked "CONV.") will iterate to the fixed point, while points outside this region will reach an oscillatory final state (in I.5.h, all three outer regions are oscillatory, though only one is so marked).

3. Figures 4, 5, and 6. These figures give the S, α mapping explicitly for three different Binary Reaction Systems:

I.2.j (Figure 4):

$$S' = 1 - 4\alpha + 4\alpha^2 + 2\alpha S$$

(*cf.* equation (47) of the text)

$$\alpha' = 2\alpha S$$

I.5.h (Figure 5):

See 2. above.

III.2.1.a (Figure 6):

$$S' = S + 3\alpha + 6\alpha S$$
$$\alpha' = S - \alpha + 3\alpha^2 - S^2$$

In Figures 4 and 5, the numerically labelled lines are the transforms of lines of constant α; *e.g.*, in I.5.h the curve labelled .25 is the transform of the horizontal line $\alpha = .25$. The curves labelled $\alpha' = \alpha$ and $S' = S$ are lines of constant α and constant S, respectively.

Figure 6 is plotted in a 60°, 30° reference triangle, *i.e.*, in terms of the variables S and $t = \sqrt{3}\alpha$. The labelled curves are the transforms of lines of constant t, and the arrows indicate the order of the transformed points as the relevant lines of constant t are traversed from left to right (direction of increasing S). Note the change of direction for $t > \sqrt{3}/6$. (The vertical line with two arrows labelled 1/6 is the transform of $\alpha = 1/6$ or $t > \sqrt{3}/6$. This line is doubly covered, first upwards, then downwards, as S increases.)

4. Figure 7 illustrates the "non-convergent" case. See Section XI, equation (69), and the accompanying text.

5. Figure 8 illustrates different modes of convergence to a fixed point. See part 3 of Section XI for discussion.

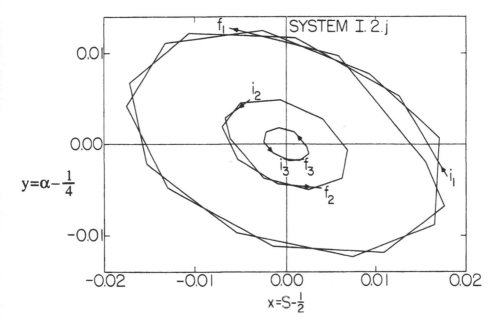

Fig. 1

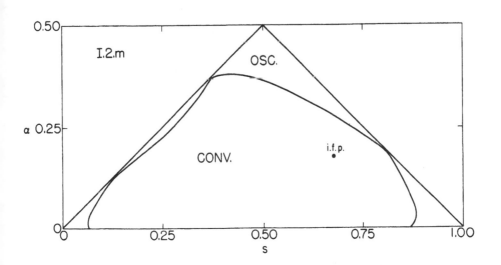

Fig. 2

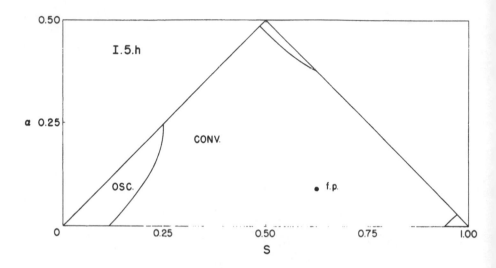

Fig. 3

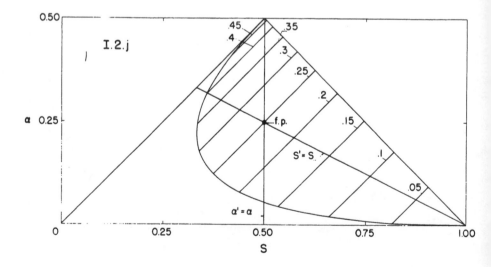

Fig. 4

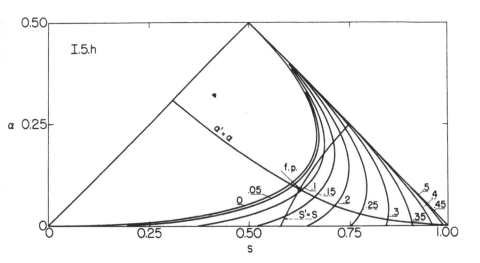

Fig. 5

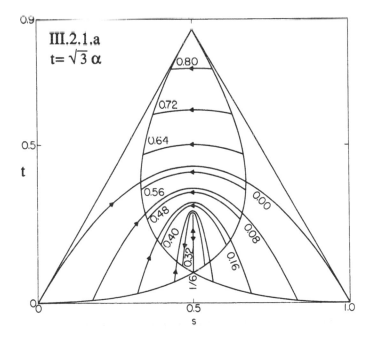

Fig. 6

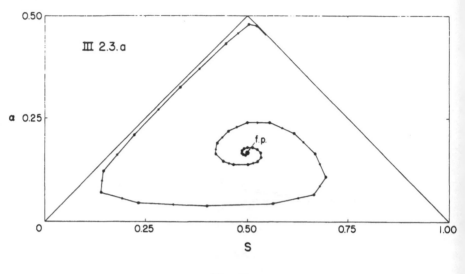

Fig. 7

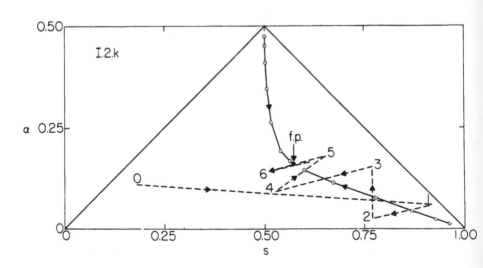

Fig. 8

Table I

Summary of Convergence Behavior of Three-Variable Binary Reaction Systems

1 n.f.p.	2 b.f.p.	3 i.f.p.	4 b.d.p.	5 b.t.p.	6 i.d.p.	Mixed Behavior
I.2.b	I.1.a	I.1.c	I.2.e	I.5.j	I.5.1	I.2.m(3,4)
I.2.c	I.1.b	I.2.a	I.2.n	I.5.r	I.5.p	I.2.p(3,4)
I.2.h	I.2.d	I.2.j	I.2.q	III.2.3.e	II.1.f	
I.2.i	I.2.f	I.2.k	I.2.r	III.2.3.f	III.1.f	I.f.h(3,5)
I.2.1	I.2.g	I.2.o	I.3.i			I.5.q(1,4)
I.3.a	I.3.b	I.3.e	I.4.d			III.2.2.a
I.3.c	I.4.g	I.3.f	I.4.e			(2 different
I.3.d	I.4.h	I.3.h				values of 1)
I.4.a	I.6.a	I.4.b	I.5.g			
I.4.c	II.1.a	I.4.f	I.5.i			III.2.3.b(4,1)
I.5.a	II.2.e	I.4.i	I.5.k			
I.5.b		I.5.f	I.5.m			
I.5.c	III.1.a	I.5.n	II.2.g			
I.5.d		I.5.o	II.2.h			
I.5.e		I.6.b	II.3.c			
II.1.b		I.6.c	III.2.2.b			
II.2.a		II.1.d				
II.2.c		II.2.b				
II.2.d		II.2.i				
II.2.f		II.3.d				
II.3.a		III.1.c				
II.3.b		III.1.d				
III.1.b		III.1.e				
		III.1.g				
		III.1.h				
		III.1.i				
		III.2.1.a				
		III.2.1.b				
		III.2.1.e				
		III.2.2.c				
		III.2.2.e				

Note: In addition, we have the following
  I.3.g   i.t.p.
  III.2.3.a non-convergent

Introduction to Table II

This table summarizes the properties of all 97 Binary Reaction Systems in 3 variables. The notation is explained in Section IV of the text. For a few systems, the behavior of arbitrary vectors under iteration can be predicted theoretically owing to the fact that the system reduces to a difference equation in a single variable (e.g., I.1.b, I.2.q, etc.).* We have not though it worthwhile to note these instances explicitly; anyone using the table will immediately discover them for himself.

For the 3-variable case, the coordinates of all fixed points could be written explicitly in terms of radicals, since the f.p. equation is at most of 4th degree. There are many interesting relationships between the roots of the various f.p. equations. We have not investigated these relationships systematically, although it could easily be done using standard tools (*cf.* L. E. Dickson).[6]

Table II
Three-Variable Binary Reaction Systems

I.1.a

$\left.\begin{array}{l} (11) + (22) + (33) \\ 2(12) + 2(13) \\ 2(23) \end{array}\right\}$ n.f.p.
  b.f.p.
  degenerate

no i.f.p.

$m \rightarrow$ b.f.p.

I.1.b

$\left.\begin{array}{l} (11) + (22) + (33) \\ 2(12) + 2(23) \\ 2(13) \end{array}\right\}$ n.f.p.
  2 b.f.p.'s
  doubly degenerate

no i.f.p.

$m \rightarrow$ b.f.p. $(x_3 = 0)$

---

* In some other cases, one may observe that one of the variables will obviously iterate to zero, in which case the limiting behavior is also evident.

Table II (cont.)

I.1.c

$$\left.\begin{array}{l}(11) + (22) + (33) \\ 2(32) + 2(23) \\ 2(12)\end{array}\right\} \begin{array}{l}\text{n.f.p.} \\ \text{b.f.p.} \\ \text{not degenerate}\end{array}$$

i.f.p. given by $(x = x_3)$

$$8x^3 - 20x^2 + 16x - 3 = 0 \qquad \left.\begin{array}{l} x_1 = .34116390 \\ x_2 = .39162172 \\ x_3 = .26721438 \end{array}\right\} \text{i.f.p.}$$

or $(y = 2x_1)$

$$y^3 + y - 1 = 0$$

$m \to$ i.f.p.

I.2.a

$$\left.\begin{array}{l}(11) + (22) + 2(12) \\ (33) + 2(23) \\ 2(13)\end{array}\right\} \begin{array}{l}\text{n.f.p.} \\ \text{degenerate}\end{array}$$

i.f.p. given by $(x = x_2)$

$$x^2 + x - 1/4 = 0$$

$m \to$ i.f.p.

$$\left.\begin{array}{l} x_1 = 1/2 \\ x_2 = \dfrac{\sqrt{2} - 1}{2} = .20710678 \\ x_3 = 1 - \dfrac{\sqrt{2}}{2} = .29289322 \end{array}\right\} \begin{array}{l} \text{i.f.p.} \\ (cf.\ \text{I.4.e}) \\ (cf.\ \text{III.1.g}) \end{array}$$

I.2.b

$$\left.\begin{array}{l}(11) + (22) + 2(12) \\ (33) + 2(13) \\ 2(23)\end{array}\right\} \begin{array}{l}\text{n.f.p.} \\ \text{degenerate}\end{array}$$

no i.f.p.

$m \to$ n.f.p.

Table II (cont.)

I.2.c

$$\left.\begin{array}{l}(11) + (22) + 2(13) \\ (33) + 2(23) \\ 2(12)\end{array}\right\}\begin{array}{l}\text{n.f.p.} \\ \text{not degenerate}\end{array}$$

no i.f.p.

$m \rightarrow$ n.f.p.

I.2.d

$$\left.\begin{array}{l}(11) + (22) + 2(13) \\ (33) + 2(12) \\ 2(23)\end{array}\right\}\begin{array}{l}\text{n.f.p.} \\ \text{b.f.p.} \\ \text{degenerate}\end{array}$$

no i.f.p.

$m \rightarrow$ b.f.p.

I.2.e

$$\left.\begin{array}{l}(11) + (22) + 2(23) \\ (33) + 2(13) \\ 2(12)\end{array}\right\}\begin{array}{l}\text{n.f.p.} \\ \text{b.d.p. (see below)} \\ \text{not degenerate}\end{array}$$

no i.f.p.

one coordinate of b.d.p. $(x = x_1)$ given by

$$\left.\begin{array}{lll}4x^3 - 4x^2 + 4x - 1 = 0 & x_1 = .56519772, & x_1 = .31944846 \\ x = .31944847 & x_2 = 0, & x_2 = .68055154 \\ m \rightarrow \text{b.d.p.} & x_3 = .43480228, & x_3 = 0\end{array}\right\}\text{b.d.p.}$$

Table II (cont.)

I.2.f

$$(11) + (22) + 2(13)$$ n.f.p.
$$(33) + 2(12)$$ b.f.p.
$$2(13)$$ degenerate

no i.f.p.

$m \to$ b.f.p.

I.2.g

$$(11) + (33) + 2(12)$$ 2 n.f.p.'s
$$(22) + 2(23)$$ b.f.p.
$$2(13)$$ doubly degenerate

no i.f.p.

$m \to$ b.f.p.

I.2.h

$$(11) + (33) + 2(12)$$
$$(22) + 2(13)$$ 2 n.f.p.'s   degenerate
$$2(23)$$

no i.f.p.

$m \to$ n.f.p. $(x_1 = 1)$

I.2.i

$$(11) + (33) + 2(13)$$ 2 n.f.p.'s $x_1 = 1/4$
$$(22) + 2(23)$$ degenerate $x_2 = 1/2$ i.f.p.
$$2(12)$$ $x_3 = 1/4$

$m \to$ n.f.p. $(x_1 = 1)$

Table II (cont.)

I.2.j

$$\left.\begin{array}{l}(11) + (33) + 2(13) \\ (22) + 2(12) \\ 2(23)\end{array}\right\} \begin{array}{l}\text{2 n.f.p.'s} \\ \text{degenerate}\end{array} \qquad \left.\begin{array}{l}x_1 = 1/4 \\ x_2 = 1/2 \\ x_3 = 1/4\end{array}\right\} \text{i.f.p.}$$

$m \to$ i.f.p. (See Section IX.1. of text for discussion)

I.2.k

$$\left.\begin{array}{l}(11) + (33) + 2(23) \\ (22) + 2(13) \\ 2(12)\end{array}\right\} \begin{array}{l}\text{2 n.f.p.'s} \\ \text{not degenerate}\end{array}$$

i.f.p. given by $(x = x_1)$

$8x^2 + 4x - 3 = 0$

$m \to$ i.f.p.

$$\left.\begin{array}{l}x_1 = \dfrac{\sqrt{7} - 1}{4} = .41143783 \\ x_2 = .32287566 \\ x_3 = .26568651\end{array}\right\} \text{i.f.p.}$$

I.2.l

$$\left.\begin{array}{l}(11) + (33) + 2(23) \\ (22) + 2(12) \\ 2(13)\end{array}\right\} \begin{array}{l}\text{2 n.f.p.'s} \\ \text{b.f.p.} \\ \text{doubly degenerate}\end{array}$$

no i.f.p.

$m \to$ n.f.p. $(x_2 = 1)$

I.2.m

$$\left.\begin{array}{l}(22) + (33) + 2(12) \\ (11) + 2(23) \\ 2(13)\end{array}\right\} \begin{array}{l}\text{b.d.p.} \\ \text{degenerate}\end{array} \qquad \begin{array}{l}x_1 = 1/2 \\ x_2 = \dfrac{1}{\sqrt{8}} = .35355339 \\ x_3 = \dfrac{2 - \sqrt{2}}{4} = .14644661\end{array}\left.\begin{array}{l}\\ \\ \end{array}\right\} \begin{array}{l}\text{i.f.p.} \\ (cf.\ \text{I.4.b}) \\ (cf.\ \text{III.1.d})\end{array}$$

Table II (cont.)

I.2.m (cont.)

    $m \to$ i.f.p. with initial vectors

$$x_1^{(0)} = 1/3 \qquad\qquad\qquad x_1^{(0)} = .10066270$$

$$x_2^{(0)} = 1/3 \qquad \text{and} \qquad x_2^{(0)} = .17982822$$

$$x_3^{(0)} = 1/3 \qquad\qquad\qquad x_3^{(0)} = .71950908$$

    $m \to$ b.d.p. with initial vector   (See Section XI for discussion)

$$x_1^{(0)} = .35613727$$

$$x_2^{(0)} = .62067802$$

$$x_3^{(0)} = .02318471$$

I.2.n

$$\left.\begin{array}{l} (22) + (33) + 2(12) \\[4pt] (11) + 2(13) \\[4pt] 2(23) \end{array}\right\} \begin{array}{l} \text{b.d.p.} \\[4pt] \text{degenerate} \end{array}$$

The only f.p. is the b.f.p., which is the i.f.p. of the 2-variable system

$$\text{System} \atop \text{S} \left\{\begin{array}{l} (22) + 2(12) \\ (11) \end{array}\right., \ i.e., \qquad \left.\begin{array}{l} x_1 = \dfrac{\sqrt{5}-1}{2} = .61803399 \\[10pt] x_2 = \dfrac{3-\sqrt{5}}{2} = .38196601 \\[10pt] (x_3 = 0) \end{array}\right\}$$

This f.p. is easily shown not to be attractive. And, experimentally $m \to$ b.d.p.

I.2.o

$$\left.\begin{array}{l} (22) + (33) + 2(13) \\[4pt] (11) + 2(23) \\[4pt] 2(12) \end{array}\right\} \begin{array}{l} \text{b.d.p.} \\[4pt] \text{not degenerate} \end{array}$$

    i.f.p. given by $(x = x_1)$ $\left.\begin{array}{l} x_1 = .39624265 \\ \end{array}\right.$

$$4x^4 + 8x^3 - 5x^2 + 3x - 1 = 0 \qquad \left.\begin{array}{l} x_2 = .33682695 \\ \end{array}\right\} \text{i.f.p.}$$

    $m \to$ i.f.p. $x_3 = .26693040$

Table II (cont.)

I.2.p

$$\left.\begin{array}{l}(22)+(33)+2(13)\\(11)+2(12)\\2(23)\end{array}\right\}\begin{array}{l}\text{b.f.p. }(cf.\ \text{I.2.n})\\\text{b.d.p.}\\\text{degenerate}\end{array}$$

i.f.p. given by $(x = x_1)$ $\quad x_1 = \dfrac{\sqrt{3}-1}{2} = .36602540$

$x^2 + x - 1/2 = 0$ $\qquad x_2 = 1/2$

$\qquad\qquad\qquad\qquad\quad x_3 = 1 - \dfrac{\sqrt{3}}{2} = .13397460$ $\left.\begin{array}{l}\\\\\\\end{array}\right\}\begin{array}{l}\text{i.f.p.}\\(cf.\ \text{I.3.h})\\(cf.\ \text{II.2.b})\end{array}$

$m \to$ i.f.p. with initial vectors

$x_1^{(0)} = 1/3$		$x_1^{(0)} = 5/16$		$x_1^{(0)} = .35613727$
$x_2^{(0)} = 1/3$	and	$x_2^{(0)} = 9/16$	and	$x_2^{(0)} = .62067802$
$x_3^{(0)} = 1/3$		$x_3^{(0)} = 2/16$		$x_3^{(0)} = .02318471$

$m \to$ b.d.p. with initial vector

$x_1^{(0)} = .04815165$

$x_2^{(0)} = .21990982$

$x_3^{(0)} = .73193853$

I.2.q

$$\left.\begin{array}{l}(22)+(33)+2(23)\\(11)+2(13)\\2(12)\end{array}\right\}\begin{array}{l}\text{b.d.p.}\\\text{not degenerate}\end{array}$$

i.f.p. given by $(x = x_1)$ $\quad x_1 = \dfrac{3-\sqrt{5}}{2} = .38196601$

$x^2 - 3x + 1 = 0$ $\qquad x_2 = \dfrac{\sqrt{5}-1}{8-2\sqrt{5}} = .35037291$ $\left.\begin{array}{l}\\\\\\\end{array}\right\}\text{i.f.p.}$

$m \to$ b.d.p. $\qquad\qquad\qquad\quad x_3 = .26766108$

Table II (cont.)

I.2.r

$$(22) + (33) + 2(23) \qquad \text{b.f.p.} \ (cf. \ \text{I.2.n})$$
$$(11) + 2(12) \qquad \qquad \text{b.d.p.}$$
$$2(13) \qquad \qquad \qquad \text{degenerate}$$

no i.f.p.

$m \to$ b.d.p.

I.3.a

$$(11) + (22) + 2(12)$$
$$2(13) + 2(23) \qquad \begin{matrix} 2 \ \text{n.f.p.'s} \\ \text{degenerate} \end{matrix}$$
$$(33)$$

no i.f.p.

$m \to$ n.f.p. $(x_1 = 1)$

I.3.b

$$(11) + (22) + 2(13) \qquad 2 \ \text{n.f.p.'s}$$
$$2(12) + 2(23) \qquad \qquad \text{b.f.p.}$$
$$(33) \qquad \qquad \qquad \text{doubly degenerate}$$

no i.f.p.

$m \to$ b.f.p.

I.3.c

$$(11) + (22) + 2(23) \qquad 2 \ \text{n.f.p.'s}$$
$$2(12) + 2(13) \qquad \qquad \text{b.f.p.}$$
$$(33) \qquad \qquad \qquad \text{not degenerate}$$

no i.f.p.

$m \to$ n.f.p. $(x = x_1)$

Table II (cont.)

I.3.d

$$\left.\begin{array}{l}(11) + (33) + 2(12) \\ 2(13) + 2(23) \\ (22)\end{array}\right\} \begin{array}{l}\text{n.f.p.} \\ \text{not degenerate}\end{array}$$

no i.f.p.

$m \to$ n.f.p.

I.3.e

$$\left.\begin{array}{l}(11) + (33) + 2(13) \\ 2(12) + 2(23) \\ (22)\end{array}\right\} \begin{array}{l}\text{n.f.p.} \\ \text{degenerate}\end{array} \qquad \left.\begin{array}{l}x_1 = 1/4 \\ x_2 = 1/2 \\ x_3 = 1/4\end{array}\right\} \text{i.f.p.}$$

i.f.p. exists

$m \to$ i.f.p.

I.3.f

$$\left.\begin{array}{l}(11) + (33) + 2(23) \\ 2(12) + 2(13) \\ (22)\end{array}\right\} \begin{array}{l}\text{n.f.p.} \\ \text{not degenerate}\end{array}$$

i.f.p. given by $(x = x_2)$
$\qquad 2x^3 + 4x2 - 1 = 0$
$m \to$ i.f.p.

$$\left.\begin{array}{l}x_1 = .34444609 \\ x_2 = .45160596 \\ x_3 = .20394795\end{array}\right\} \text{i.f.p.}$$

I.3.g

$$\left.\begin{array}{l}(22) + (33) + 2(12) \\ 2(12) + 2(13) \\ (11)\end{array}\right\} \begin{array}{l}\text{b.d.p.} \\ \text{not degenerate}\end{array} \qquad \left.\begin{array}{l}x_1 = .45823825 \\ x_2 = .33177946 \\ x_3 = .20998229\end{array}\right\} \text{i.f.p.}$$

Table II (cont.)

I.3.g (cont.)

    i.f.p. given by $(x = x_1)$

        $2x^4 - 3x^2 - x + 1 = 0$

      $m \to$ interior triple point

$$\left. \begin{aligned} x_1^{(n+3)} = x_1^{(n)} &= .73924369, \quad x_1^{(n+1)} = .08071790, \quad x_1^{(n+2)} = .49780563 \\ x_2^{(n+3)} = x_2^{(n)} &= .01294586, \quad x_2^{(n+1)} = .37280086, \quad x_2^{(n+2)} = .49567899 \\ x_3^{(n+3)} = x_3^{(n)} &= .24781045, \quad x_3^{(n+1)} = .54648124, \quad x_3^{(n+2)} = .00651538 \end{aligned} \right\} \text{i.t.p.}$$

I.3.h

$$\left. \begin{aligned} (22) + (33) &+ 2(13) \\ 2(12) &+ 2(23) \\ (11)& \end{aligned} \right\} \begin{aligned} &\text{b.f.p.} \quad (cf. \text{ I.2.n}) \\ &\text{b.d.p.} \\ &\text{degenerate} \end{aligned}$$

    i.f.p. given by $(x = x_1)$ $\left. \begin{aligned} x_1 &= \dfrac{\sqrt{3} - 1}{2} = .36602540 \\ x_2 &= 1/2 \\ x_3 &= 1 - \dfrac{\sqrt{3}}{2} = .13397460 \end{aligned} \right\} \begin{aligned} &\text{i.f.p.} \\ &(cf. \text{ I.2.p}) \end{aligned}$

        $x^2 + x - 1/2 = 0$

      $m \to$ i.f.p.

I.3.i

$$\left. \begin{aligned} (22) + (33) &+ 2(23) \\ 2(12) &+ 2(13) \\ (11)& \end{aligned} \right\} \begin{aligned} &\text{b.d.p.} \\ &\text{not degenerate} \end{aligned}$$

    i.f.p. given by $(x = x_1)$ $\left. \begin{aligned} x_1 &= \dfrac{3 - \sqrt{5}}{2} = .38196601 \\ x_2 &= 2\sqrt{5} - 4 = .47213596 \\ x_3 &= \dfrac{7 - 3\sqrt{5}}{2} = .14589803 \end{aligned} \right\} \begin{aligned} &\text{i.f.p.} \\ &(cf. \text{ I.5.f}) \end{aligned}$

        $x^2 - 3x + 1 = 0$

      $m \to$ b.d.p.

Table II (cont.)

I.4.a

$$\left.\begin{array}{l} (11) + 2(12) + 2(13) \\ (22) + (23) \\ 2(23) \end{array}\right\} \begin{array}{l} \text{2 n.f.p.'s} \\ \text{b.f.p.} \\ \text{doubly degenerate} \end{array}$$

no i.f.p.

$m \rightarrow$ n.f.p. $(x_1 = 1)$

I.4.b

$$\left.\begin{array}{l} (11) + 2(12) + 2(23) \\ (22) + (33) \\ 2(13) \end{array}\right\} \begin{array}{l} \text{2 n.f.p.'s} \\ \text{degenerate} \end{array}$$

$$\left.\begin{array}{ll} \text{i.f.p. given by } (x = x_2) & x_1 = 1/2 \\ \qquad 2x^2 - 2x + 1/4 = 0 & x_2 = \dfrac{2 - \sqrt{2}}{4} = .14644661 \\ \qquad m \rightarrow \text{i.f.p.} & x_3 = \dfrac{\sqrt{2}}{4} = .35355339 \end{array}\right\} \begin{array}{l} \text{i.f.p.} \\ (\textit{cf.} \text{ I.2.m}) \end{array}$$

I.4.c

$$\left.\begin{array}{l} (11) + 2(13) + 2(23) \\ (22) + (33) \\ 2(12) \end{array}\right\} \begin{array}{l} \text{2 n.f.p.'s} \\ \text{not degenerate} \end{array}$$

no i.f.p.

$m \rightarrow$ n.f.p. $(x_1 = 1)$

Table II (cont.)

I.4.d

$$
\left.\begin{array}{l}
(22) + 2(12) + 2(13) \\
(11) + (23) \\
2(23)
\end{array}\right\}
\begin{array}{l}
\text{b.f.p. } (cf.\ \text{I.2.n}) \\
\text{b.d.p.} \\
\text{degenerate}
\end{array}
$$

no i.f.p.

$m \to$ b.d.p.

I.4.e

$$
\left.\begin{array}{l}
(22) + 2(12) + 2(23) \\
(11) + (33) \\
2(13)
\end{array}\right\}
\begin{array}{l}
\text{b.f.p. } (cf.\ \text{I.2.n}) \\
\text{b.d.p.} \\
\text{degenerate}
\end{array}
$$

i.f.p. given by $(x = x_2)$

$x^2 - 2x + 1/2 = 0$

$m \to$ b.d.p.

$$
\left.\begin{array}{l}
x_1 = 1/2 \\
x_2 - 1 - \dfrac{\sqrt{2}}{2} = .29289322 \\
x_3 = \dfrac{\sqrt{2}-1}{2} = .20710678
\end{array}\right\}
\begin{array}{l}
\text{i.f.p.} \\
(cf.\ \text{I.2.a})
\end{array}
$$

I.4.f

$$
\left.\begin{array}{l}
(22) + 2(13) + 2(23) \\
(11) + (33) \\
2(12)
\end{array}\right\}
\begin{array}{l}
\text{b.d.p.} \\
\text{not degenerate}
\end{array}
$$

i.f.p. given by $(x = x_2)$

$4x^4 - 12x^3 + x^2 - 3x + 1 = 0$

$m \to$ i.f.p.

$$
\left.\begin{array}{l}
x_1 = .46164837 \\
x_2 = .27991085 \\
x_3 = .25844078
\end{array}\right\}
\text{i.f.p.}
$$

Table II (cont.)

I.4.g

$$\left.\begin{array}{l} (33) + 2(12) + 2(13) \\ (11) + (22) \\ 2(23) \end{array}\right\} \begin{array}{l} \text{n.f.p.} \\ \text{b.f.p.} \\ \text{degenerate} \end{array}$$

no i.f.p.

$m \rightarrow$ b.f.p.

I.4.h

$$\left.\begin{array}{l} (33) + 2(12) + 2(23) \\ (11) + (22) \\ 2(13) \end{array}\right\} \begin{array}{l} \text{n.f.p.} \\ \text{b.f.p.} \\ \text{degenerate} \end{array}$$

no i.f.p.

$m \rightarrow$ b.f.p.

I.4.i

$$\left.\begin{array}{l} (33) + 2(13) + 2(23) \\ (11) + (22) \\ 2(12) \end{array}\right\} \begin{array}{l} \text{n.f.p.} \\ \text{not degenerate} \end{array}$$

i.f.p. given by $(x = x_3)$ $\qquad\left.\begin{array}{l} x_1 = .45253634 \\ x_2 = .28737153 \\ x_3 = .26009213 \end{array}\right\}$
$\quad 2x^3 - 10x^2 + 14x - 3 = 0$
$\quad m \rightarrow$ i.f.p.

I.5.a

$$\left.\begin{array}{l} (11) + 2(12) + 2(13) \\ (22) + 2(23) \\ 2(33) \end{array}\right\} \begin{array}{l} \text{3 n.f.p.'s} \\ \text{triply degenerate} \end{array}$$

no i.f.p.

$m \rightarrow$ n.f.p. $(x_1 = 1)$

Table II (cont.)

I.5.b

$$\left.\begin{array}{l}(11) + 2(12) + 2(13) \\ (33) + 2(23) \\ (22)\end{array}\right\} \begin{array}{l}\text{n.f.p.} \\ \text{b.f.p. } (cf.\ \text{I.2.n}) \\ \text{b.d.p.} \\ \text{degenerate}\end{array}$$

no i.f.p.

$m \to$ n.f.p.

I.5.c

$$\left.\begin{array}{l}(11) + 2(12) + 2(23) \\ (22) + 2(13) \\ (33)\end{array}\right\} \begin{array}{l}\text{3 n.f.p.'s} \\ \text{degenerate}\end{array}$$

no i.f.p.

$m \to$ n.f.p. $(x_1 = 1)$

I.5.d

$$\left.\begin{array}{l}(11) + 2(12) + 2(23) \\ (33) + 2(13) \\ (22)\end{array}\right\} \begin{array}{l}\text{n.f.p.} \\ \text{b.d.p. (see below)} \\ \text{not degenerate}\end{array}$$

no i.f.p. In addition to the "trivial" b.d.p. $(0,1,0), (0,0,1)$, there is a b.d.p. which is a f.p. of the transformation

$$x_3'' = x_3^2 (2 - x_3)^2, \ viz.$$

$$\left.\begin{array}{ll} x_1 = \dfrac{\sqrt{5}-1}{2}, & x_1 = \dfrac{3-\sqrt{5}}{2} \\ x_2 = 0, & x_2 = \dfrac{\sqrt{5}-1}{2} \\ x_3 = \dfrac{3-\sqrt{5}}{2}, & x_3 = 0 \end{array}\right\} \text{b.d.p.}$$

This is not an attractive f.p. If $x_2^{(0)} = 0$, $x_3^{(0)} < (3 - \sqrt{5})/2$, the initial vector $\to x_1 = 1$. If $x_2^{(0)} = 0$, $x_3^{(0)} > (3 - \sqrt{5})/2$, the initial vector $\to$ the trivial b.d.p. This behavior has been verified numerically.

Table II (cont.)

I.5.e

$$\left.\begin{array}{l}(11) + 2(13) + 2(23) \\ (22) + 2(12) \\ (33)\end{array}\right\} \begin{array}{l}\text{3 n.f.p.'s} \\ \text{doubly degenerate}\end{array}$$

no i.f.p.

$m \rightarrow$ n.f.p. $(x_2 = 1)$

I.5.f

$$\left.\begin{array}{l}(11) + 2(13) + 2(23) \\ (33) + 2(12) \\ (22)\end{array}\right\} \begin{array}{l}\text{n.f.p.} \\ \text{b.d.p.} \\ \text{not degenerate}\end{array}$$

i.f.p. given by $(x = x_2)$

$$x^2 - 3x + 1 = 0$$

$m \rightarrow$ i.f.p.

$$\left.\begin{array}{l}x_1 = 2\sqrt{5} - 4 = .47213596 \\[2mm] x_2 = \dfrac{3 - \sqrt{5}}{2} = .38196601 \\[2mm] x_3 = \dfrac{7 - 3\sqrt{5}}{2} = .14589803\end{array}\right\} \begin{array}{l}\text{i.f.p.} \\ (cf.\ \text{I.3.i})\end{array}$$

I.5.g

$$\left.\begin{array}{l}(22) + 2(12) + 2(13) \\ (11) + 2(23) \\ (33)\end{array}\right\} \begin{array}{l}\text{n.f.p.} \\ \text{b.f.p.} \ (cf.\ \text{I.2.n}) \\ \text{b.d.p.} \\ \text{degenerate}\end{array}$$

no i.f.p.

$m \rightarrow$ b.d.p.

Table II (cont.)

I.5.h

$$(22) + 2(12) + 2(13)$$

$$(33) + 2(23) \left.\right\} \begin{array}{l} \text{b.t.p.} \\ \text{not degenerate} \end{array}$$

$$(11)$$

i.f.p. given by $(x = x_1)$

$$x^3 + 3x^2 - 1 = 0$$

$$\left. \begin{array}{l} x_1 = .53208889 \\ x_2 = .18479253 \\ x_3 = .28311858 \end{array} \right\} \text{i.f.p.}$$

$m \rightarrow$ i.f.p. with initial vectors

$$x_1^{(0)} = 1/3 \qquad x_1^{(0)} = .08662751$$
$$x_2^{(0)} = 1/3 \quad \text{and} \quad x_2^{(0)} = .32501609$$
$$x_3^{(0)} = 1/3 \qquad x_3^{(0)} = .58835640$$

$m \rightarrow$ b.t.p. with initial vector

$$x_1^{(0)} = .02476836$$
$$x_2^{(0)} = .01004474 \qquad \text{(See Section XI for discussion.)}$$
$$x_3^{(0)} = .96518690$$

I.5.i

$$(22) + 2(12) + 2(23) \left.\right\} \begin{array}{l} \text{n.f.p.} \\ \text{b.f.p. } (cf. \text{ I.2.n)} \\ \text{b.d.p.} \\ \text{degenerate} \end{array}$$

$$(11) + 2(13)$$

$$(33) \qquad .$$

no i.f.p.

$m \rightarrow$ b.d.p.

Table II (cont.)

I.5.j

$$\left.\begin{array}{l} (22) + 2(12) + 2(23) \\ (33) + 2(13) \\ (22) \end{array}\right\} \begin{array}{l} \text{b.d.p.} \\ \text{not degenerate} \end{array}$$

i.f.p. given by $(x = x_1)$ $\left.\begin{array}{l} x_1 = .48402830 \\ x_2 = .28168830 \\ x_3 = .23428340 \end{array}\right\}$ i.f.p.

$x^4 + 2x^3 + x^2 + x - 1 = 0$

$m \to$ b.t.p.

I.5.k

$$\left.\begin{array}{l} (22) + 2(13) + 2(23) \\ (11) + 2(12) \\ (33) \end{array}\right\} \begin{array}{l} \text{n.f.p.} \\ \text{b.f.p. } (cf.\ \text{I.2.n}) \\ \text{b.d.p.} \\ \text{degenerate} \end{array}$$

no i.f.p.

$m \to$ b.d.p.

I.5.l

$$\left.\begin{array}{l} (22) + 2(13) + 2(23) \\ (33) + 2(12) \\ (11) \end{array}\right\} \begin{array}{l} \text{b.t.p.} \\ \text{not degenerate} \end{array}$$
     $\left.\begin{array}{l} x_1 = .44504187 \\ x_2 = .35689587 \\ x_3 = .19806226 \end{array}\right\} \begin{array}{l} \text{i.f.p.} \\ (cf.\ \text{I.5.n for} \\ \text{value of } x_2) \\ (cf.\ \text{I.5.r}) \end{array}$

i.f.p. given by $(x = x_1)$

$x^3 - x^2 - 2x + 1 = 0$

$m \to$ i.d.p.

$$\left.\begin{array}{l} x_1^{(n+2)} = x_1^{(n)} = .67046846,\ x_1^{(n+1)} = .13146927 \\ x_2^{(n+2)} = x_2^{(n)} = .31224737,\ x_2^{(n+1)} = .41900277 \\ x_3^{(n+2)} = x_3^{(n)} = .01728417,\ x_3^{(n+1)} = .44952796 \end{array}\right\} \text{i.d.p.}$$

Table II (cont.)

I.5.m

$$\left.\begin{array}{l}(33) + 2(12) + 2(13)\\(22) + 2(23)\\(11)\end{array}\right\}\begin{array}{l}\text{n.f.p.}\\\text{b.d.p.}\\\text{b.f.p. } (cf.\ \text{I.2.n})\\\text{degenerate}\end{array}$$

no i.f.p.

$m \to$ b.d.p.

I.5.n

$$\left.\begin{array}{l}(33) + 2(12) + 2(13)\\(11) + 2(23)\\(22)\end{array}\right\}\begin{array}{l}\text{b.t.p.}\\\text{not degenerate}\end{array}$$

i.f.p. given by $(x - x_2)$ $x_1 = .51572947$
    $x^3 + 3x^2 - 4x + 1 = 0$ $x_2 = .35689587$ i.f.p.
    $m \to$ i.f.p. $x_3 = .12737466$ $(cf.\ \text{I.5.1 for}$
                                                       value of x_2)

I.5.o

$$\left.\begin{array}{l}(33) + 2(12) + 2(23)\\(22) + 2(13)\\(11)\end{array}\right\}\begin{array}{l}\text{n.f.p.}\\\text{b.d.p.}\\\text{not degenerate}\end{array}$$

i.f.p. given by $(x = x_1)$ $x_1 = .47283391$
    $x^3 + 4x^2 - 1 = 0$ $x_2 = .30359418$ i.f.p.
    $m \to$ i.f.p. $x_3 = .22357191$

I.5.p

$$\left.\begin{array}{l}(33) + 2(12) + 2(23)\\(11) + 2(13)\\(22)\end{array}\right\}\begin{array}{l}\text{b.t.p.}\\\text{not degenerate}\end{array}$$

Table II (cont.)

I.5.p. (cont.)

i.f.p. given by $(x = x_2)$

$$x^4 - 2x^3 + x^2 - 3x + 1 = 0$$

$$\left. \begin{array}{l} x_1 = .52645706 \\ x_2 = .35061327 \\ x_3 = .12292967 \end{array} \right\} \text{i.f.p.}$$

$m \to$ i.d.p.

$$\left. \begin{array}{ll} x_1^{(n+2)} = x_1^{(n)} = .42717428, & x_1^{(n+1)} = .49907610 \\ x_2^{(n+2)} = x_2^{(n)} = .53818709, & x_2^{(n+1)} = .21127856 \\ x_3^{(n+2)} = x_3^{(n)} = .04463863, & x_3^{(n+1)} = .28964534 \end{array} \right\} \text{i.d.p.}$$

I.5.q

$$\left. \begin{array}{l} (33) + 2(13) + 2(23) \\ (22) + 2(12) \\ (11) \end{array} \right\} \begin{array}{l} \text{n.f.p.} \\ \text{b.f.p. } (cf.\ \text{I.2.n}) \\ \text{b.d.p.} \\ \text{degenerate} \end{array}$$

no i.f.p.

$m \to$ n.f.p. with initial vectors

$$\begin{array}{ll} x_1^{(0)} = 1/3 & x_1^{(0)} = .28491108 \\ x_2^{(0)} = 1/3 \quad \text{and} & x_2^{(0)} = .47627166 \\ x_3^{(0)} = 1/3 & x_3^{(0)} = .23881726 \end{array}$$

$m \to$ b.d.p. with initial vector

$$x_1^{(0)} = .43634693$$
$$x_2^{(0)} = .16692265$$
$$x_3^{(0)} = .39673042$$

I.5.r

$$\left. \begin{array}{l} (33) + 2(13) + 2(23) \\ (11) + 2(12) \\ (22) \end{array} \right\} \begin{array}{l} \text{b.t.p.} \\ \text{not degenerate} \end{array}$$

i.f.p. given by $(x = x_2)$

$$x^3 - x^2 - 2x + 1 = 0$$

$$\left. \begin{array}{l} x_1 = .35689587 \\ x_2 = .44504187 \\ x_3 = .19806226 \end{array} \right\} \begin{array}{l} \text{i.f.p.} \\ (cf.\ \text{I.5.1}) \end{array}$$

$m \to$ b.t.p.

Table II (cont.)

I.6.a

$$\left.\begin{array}{l} 2(12)+2(13)+2(23) \\ (11)+(22) \\ (33) \end{array}\right\} \begin{array}{l} \text{2 n.f.p.'s} \\ \text{b.f.p.} \\ \text{degenerate} \end{array}$$

no i.f.p.

$m \to$ b.f.p.

I.6.b

$$\left.\begin{array}{l} 2(12)+2(13)+2(23) \\ (11)+(33) \\ (22) \end{array}\right\} \begin{array}{l} \text{b.d.p.} \\ \text{not degenerate} \end{array}$$

i.f.p. given by $(x = x_2)$ $\left.\begin{array}{l} x_1 = .56311573 \\ x_2 = .32878482 \\ x_3 = .10809945 \end{array}\right\}$ i.f.p.

$\quad$ $2x^4 + 2x^3 - x^2 - 3x + 1 = 0$

$m \to$ i.f.p.

I.6.c

$$\left.\begin{array}{l} 2(12)+2(13)+2(23) \\ (22)+(33) \\ (11) \end{array}\right\} \begin{array}{l} \text{n.f.p.} \\ \text{not degenerate} \end{array}$$

i.f.p. given by $(x = x_2)$ $\left.\begin{array}{l} x_1 = .56519772 \\ x_2 = .11535382 \\ x_3 = .31944846 \end{array}\right\}$ i.f.p.

$\quad$ $2x^3 + 2x^2 - 1 = 0$

$m \to$ i.f.p.

II.1.a

$$\left.\begin{array}{l} (11)+(22)+(33)+2(12) \\ 2(23) \\ 2(13) \end{array}\right\} \begin{array}{l} \text{n.f.p.} \\ \text{b.f.p.} \\ \text{doubly degenerate} \end{array}$$

no i.f.p.

$m \to$ b.f.p.

Table II (cont.)

II.1.b

$$\left.\begin{array}{l}(11)+(22)+(33)+2(13)\\2(23)\\2(12)\end{array}\right\} \begin{array}{l}\text{n.f.p.}\\\text{degenerate}\end{array}$$

no i.f.p.

$m \to$ n.f.p.

II.1.d

$$\left.\begin{array}{l}(11)+(22)+(33)+2(23)\\2(12)\\2(13)\end{array}\right\} \begin{array}{l}\text{n.f.p.}\\\text{2 b.f.p.'s}\\\text{doubly degenerate}\end{array}$$

There is a continuum of i.f.p.'s, since $\dfrac{x_2'}{x_3'} = \dfrac{x_2^{(0)}}{x_3^{(0)}} = \alpha.$
The i.f.p. corresponding to this α is

$$\left.\begin{array}{l}x_1 = 1/2\\x_2 = \dfrac{\alpha}{2(1+\alpha)}\\x_3 = \dfrac{1}{2(1+\alpha)}\end{array}\right\} \text{i.f.p.}$$

Note that if we write $x_2 + x_3 = y_2$, $x_1 = y_1$, then the system takes the form

$$\begin{array}{ll}y_1' = y_1'^2 + y_2^2 & \text{which, for a general point,}\\y_2' = 2y_1y_2 & \to y_1 = y_2 = 1/2\end{array}$$

$m \to$ appropriate i.f.p.

II.1.f

$$\left.\begin{array}{l}(11)+(22)+(33)+2(23)\\2(13)\\2(12)\end{array}\right\} \begin{array}{l}\text{n.f.p.}\\\text{not degenerate}\end{array}$$

There is a continuum of i.d.p.'s owing to the fact that
$\dfrac{x_2''}{x_3''} = \dfrac{x_2}{x_3}$. Thus, if $\dfrac{x_2^{(0)}}{x_3^{(0)}} = \alpha$, the i.d.p. is given by

Table II (cont.)

II.1.f (cont.)

$$x_1^{(n+2)} = x_1^{(n)} = 1/2, \qquad x_1^{(n+1)} = 1/2$$

$$\left. \begin{array}{ll} x_2^{(n+2)} = x_2^{(n)} = \dfrac{1}{2(1+\alpha)}, & x_2^{(n+1)} = \dfrac{\alpha}{2(1+\alpha)} \\[2mm] x_3^{(n+2)} = x_3^{(n)} = \dfrac{\alpha}{2(1+\alpha)}, & x_3^{(n+1)} = \dfrac{1}{2(1+\alpha)} \end{array} \right\} \begin{array}{l} \text{i.d.p.} \\ (\textit{cf. } \text{II.1.d}) \end{array}$$

This system reduces to the same 2-variable system as does II.1.c
$m \to$ appropriate i.d.p.

II.2.a

$$\left. \begin{array}{l} (11) + (22) + 2(12) + 2(13) \\[1mm] (33) \\[1mm] 2(23) \end{array} \right\} \begin{array}{l} \text{n.f.p.} \\ \text{degenerate} \end{array}$$

no i.f.p.

$m \to$ n.f.p.

II.2.b

$$\left. \begin{array}{l} (11) + (22) + 2(12) + 2(23) \\[1mm] (33) \\[1mm] 2(13) \end{array} \right\} \begin{array}{l} \text{n.f.p.} \\ \text{degenerate} \end{array}$$

i.f.p. given by $(x = x_3)$ $x_1 = 1/2$

$$x^2 + x - 1/2 = 0 \qquad \left. \begin{array}{l} x_2 = \dfrac{\sqrt{3}-1}{2} = .36602540 \\[3mm] x_3 = 1 - \dfrac{\sqrt{3}}{2} = .13397460 \end{array} \right\} \begin{array}{l} \text{i.f.p.} \\ (\textit{cf. } \text{I.2.p.}) \end{array}$$

$m \to$ i.f.p.

II.2.c

$$\left. \begin{array}{l} (11) + (22) + 2(13) + 2(23) \\[1mm] (33) \\[1mm] 2(12) \end{array} \right\} \begin{array}{l} \text{n.f.p.} \\ \text{not degenerate} \end{array}$$

no i.f.p.

$m \to$ n.f.p.

Table II (cont.)

II.2.d

$(11) + (33) + 2(12) + 2(13)$ ⎫
(22) ⎬ 2 n.f.p.'s
$2(23)$ ⎭ doubly degenerate

no i.f.p.

$m \rightarrow$ n.f.p. $(x_1 = 1)$

II.2.e

$(11) + (33) + 2(12) + 2(23)$ ⎫ 2 n.f.p.'s
(22) ⎬ b.f.p.
$2(13)$ ⎭ doubly degenerate

no i.f.p.

$m \rightarrow$ b.f.p.

II.2.f

$(11) + (33) + 2(13) + 2(23)$ ⎫
(22) ⎬ n.f.p.
$2(12)$ ⎭ doubly degenerate

no i.f.p.

$m \rightarrow$ n.f.p. $(x_1 = 1)$

II.2.g

$(22) + (33) + 2(12) + 2(13)$ ⎫ b.f.p. (*cf.* I.2.n)
(11) ⎬ b.d.p.
$2(23)$ ⎭ degenerate

no i.f.p.

$m \rightarrow$ b.d.p.

Table II (cont.)

II.2.h

$\left.\begin{array}{l}(22)+(33)+2(12)+2(23) \\ (11) \\ 2(13)\end{array}\right\}$ b.f.p. (*cf.* I.2.n)
b.d.p.
degenerate

i.f.p. exists

$m \rightarrow$ b.d.p.

$\left.\begin{array}{l}x_1 = 1/2 \\ x_2 = 1/4 \\ x_3 = 1/4\end{array}\right\}$ i.f.p.

II.2.i

$\left.\begin{array}{l}(22)+(33)+2(13)+2(23) \\ (11) \\ 2(12)\end{array}\right\}$ b.d.p.
not degenerate

i.f.p. exists

$m \rightarrow$ i.f.p.

$\left.\begin{array}{l}x_1 = 1/2 \\ x_2 = 1/4 \\ x_3 = 1/4\end{array}\right\}$ i.f.p. (*cf.* II.2.h)

II.3.a

$\left.\begin{array}{l}(11)+2(12)+2(13)+2(23) \\ (11) \\ (33)\end{array}\right\}$ 3 n.f.p.'s
doubly degenerate

no i.f.p.

$m \rightarrow$ n.f.p. $(x_1 = 1)$

II.3.b

$\left.\begin{array}{l}(11)+2(12)+2(13)+2(23) \\ (33) \\ (22)\end{array}\right\}$ n.f.p.
b.d.p.
not degenerate

no i.f.p.

$m \rightarrow$ n.f.p.

Table II (cont.)

II.3.c

$$\left.\begin{array}{l} (22) + 2(12) + 2(13) + 2(23) \\ (11) \\ (33) \end{array}\right\} \begin{array}{l} \text{n.f.p.} \\ \text{b.f.p. } (\textit{cf. } \text{I.2.n}) \\ \text{b.d.p.} \\ \text{degenerate} \end{array}$$

no i.f.p.

$m \to$ b.d.p.

II.3.d

$$\left.\begin{array}{l} (22) + 2(12) + 2(13) + 2(23) \\ (33) \\ (11) \end{array}\right\} \begin{array}{l} \text{b.t.p.} \\ \text{not degenerate} \end{array}$$

i.f.p. given by $(x = x_1)$

$\quad x^3 - x^2 + 2x - 1 = 0$

or (unreduced)

$\quad x^4 + x^2 + x - 1 = 0$ (*cf.* Eq. (21) of text)

or $(y = x_3)$

$\quad y^3 + 2y^2 + 2y - 1 = 0$

$m \to$ i.f.p.

$$\left.\begin{array}{l} x_1 = .56984029 \\ x_2 = .10544175 \\ x_3 = .32471796 \end{array}\right\} \text{i.f.p.}$$

III.1.a

$$\left.\begin{array}{l} (11) + (22) \\ 2(12) + 2(23) \\ 2(13) + (33) \end{array}\right\} \begin{array}{l} \text{2 n.f.p.'s} \\ \text{b.f.p.} \\ \text{doubly degenerate} \end{array}$$

no i.f.p.

$m \to$ b.f.p.

Table II (cont.)

III.1.b

$$\left.\begin{array}{l}(11)+(22)\\2(12)+2(13)\\2(23)+(33)\end{array}\right\}\begin{array}{l}\text{2 n.f.p.'s}\\\text{b.f.p.}\\\text{degenerate}\end{array}$$

no i.f.p.

$m \rightarrow$ n.f.p. $(x_3 = 1)$

III.1.c

$$\left.\begin{array}{l}(11)+(22)\\2(12)+2(23)\\2(13)+(33)\end{array}\right\}\begin{array}{l}\text{n.f.p.}\\\text{b.f.p.}\\\text{degenerate}\end{array}$$

i.f.p. exists

$m \rightarrow$ i.f.p.

$$\left.\begin{array}{l}x_1 = 1/4\\x_2 = \dfrac{\sqrt{3}}{4} = .43301270\\x_3 = \dfrac{3-\sqrt{3}}{4} = .31698730\end{array}\right\}\text{i.f.p.}$$

III.1.d

$$\left.\begin{array}{l}(11)+(33)\\2(12)+2(23)\\2(13)+(22)\end{array}\right\}\begin{array}{l}\text{n.f.p.}\\\text{b.f.p.}\\\text{degenerate}\end{array}$$

i.f.p. exists

$m \rightarrow$ i.f.p.

$$\left.\begin{array}{l}x_1 = \dfrac{2-\sqrt{2}}{4} = .14644661\\x_2 = 1/2\\x_3 = \dfrac{\sqrt{2}}{4} = .35355339\end{array}\right\}\begin{array}{l}\text{i.f.p.}\\(\mathit{cf.}\ \text{I.2.m})\end{array}$$

Table II (cont.)

III.1.e

$$\left.\begin{array}{l} (11) + (33) \\ 2(12) + 2(13) \\ 2(23) + (22) \end{array}\right\} \begin{array}{l} \text{n.f.p.} \\ \text{not degenerate} \end{array}$$

$$\begin{array}{ll} \text{i.f.p. given by} (x = x_1) & x_1 = .22815553 \\ \quad 4x^3 - 8x^2 + 6x - 1 = 0 & x_2 = .35220117 \\ \quad m \to \text{i.f.p.} & x_3 = .41964330 \end{array}\left.\begin{array}{l} \\ \\ \end{array}\right\} \text{i.f.p.}$$

III.1.f

$$\left.\begin{array}{l} (11) + (33) \\ 2(13) + 2(23) \\ 2(23) + (22) \end{array}\right\} \begin{array}{l} \text{n.f.p.} \\ \text{b.d.p. (see below)} \\ \text{not degenerate} \end{array}$$

$$\begin{array}{l} \text{i.f.p. given by } (x = x_3) \\ \quad 1 - 4(1 - x)^3 = 0 \end{array} \quad\quad \begin{array}{l} x_1 = .16374000 \\ x_2 = .46622052 \\ x_3 = 1 - \sqrt[3]{1/4} = .37003948 \end{array}\left.\begin{array}{l} \\ \\ \end{array}\right\} \text{i.f.p.}$$

A non-trivial b.d.p. exists, which is a f.p. of the transformation

$$x_3'' = 4x_3(1 - x_3)\left[1 - x_3(1 - x_3)\right]$$

This f.p. is given by the root of $(x = x_3)$

$$4x^3 - 8x^2 + 8x - 3 = 0$$

which leads to

$$x_1^{(n+2)} = x_1^{(n)} = .31944846, \; x_1^{(n+1)} = .56519772$$
$$x_2^{(n+2)} = x_2^{(n)} = 0 \quad\quad\quad , \; x_2^{(n+1)} = .43480228$$
$$x_3^{(n+2)} = x_3^{(n)} = .68055154, \; x_3^{(n+1)} = 0$$

Actually

$m \to$ i.d.p.

$$x_1^{(n+2)} = x_1^{(n)} = .17899745, \; x_1^{(n+1)} = .40604579$$
$$x_2^{(n+2)} = x_2^{(n)} = .20944248, \; x_2^{(n+1)} = .47510870 \left.\begin{array}{l} \\ \\ \end{array}\right\} \text{i.d.p.}$$
$$x_3^{(n+2)} = x_3^{(n)} = .61156007, \; x_3^{(n+1)} = .11884551$$

Table II (cont.)

III.1.g

$$\left.\begin{array}{l}(22) + (33) \\ 2(12) + 2(23) \\ 2(13) + (11)\end{array}\right\}\begin{array}{l}\text{b.f.p. (\textit{cf.} I.2.n)} \\ \text{b.d.p.} \\ \text{degenerate}\end{array}$$

i.f.p. exists

$m \to$ i.f.p.

$$\left.\begin{array}{l}x_1 = \dfrac{2 - \sqrt{2}}{2} = .29289322 \\[2mm] x_2 = 1/2 \\[2mm] x_3 = \dfrac{\sqrt{2} - 1}{2} = .20710678\end{array}\right\}\begin{array}{l}\text{i.f.p.} \\ (\textit{cf.} I.2.a)\end{array}$$

III.1.h

$$\left.\begin{array}{l}(22) + (33) \\ 2(12) + 2(13) \\ 2(23) + (11)\end{array}\right\}\begin{array}{l}\text{b.d.p.} \\ \text{not degenerate}\end{array}$$

i.f.p. given by $(x = x_1)$

$\qquad 8x^4 - 20x^3 + 17x^2 - 7x + 1 = 0$

$m \to$ i.f.p.

$$\left.\begin{array}{l}x_1 = .26882443 \\ x_2 = .39311571 \\ x_3 = .33805986\end{array}\right\}\text{i.f.p.}$$

III.1.i

$$\left.\begin{array}{l}(22) + (33) \\ 2(13) + 2(23) \\ 2(12) + (11)\end{array}\right\}\begin{array}{l}\text{b.d.p.} \\ \text{not degenerate}\end{array}$$

i.f.p. given by $(x = x_3)$

$\qquad 4x^4 - 8x^3 + 3x^2 + 3x - 1 = 0$

$m \to$ i.f.p.

$$\left.\begin{array}{l}x_1 = .27204695 \\ x_2 = .42351278 \\ x_3 = .30444027\end{array}\right\}\text{i.f.p.}$$

III.2.a

$$\left.\begin{array}{l}(11) + 2(23) \\ (22) + 2(13) \\ (33) + 2(12)\end{array}\right\}\begin{array}{l}\text{3 n.f.p.'s} \\ \text{not degenerate}\end{array}$$

i.f.p. $x_1 = x_2 = x_3 = 1/3$

$m \to$ i.f.p.

Table II (cont.)

III.2.1.b

$$\left.\begin{array}{l}(11) + 2(13) \\ (11) + 2(23) \\ (33) + 2(12)\end{array}\right\}\begin{array}{l}\text{n.f.p.} \\ \text{b.d.p.} \\ \text{not degenerate}\end{array}$$

i.f.p. $x_1 = x_2 = x_3 = 1/3$

$m \to$ i.f.p.

III.2.1.e

$$\left.\begin{array}{l}(33) + 2(12) \\ (11) + 2(23) \\ (22) + 2(13)\end{array}\right\}\begin{array}{l}\text{b.t.p.} \\ \text{not degenerate}\end{array}$$

i.f.p. $x_1 = x_2 = x_3 = 1/3$

$m \to$ i.f.p.

III.2.2.a

$$\left.\begin{array}{l}(11) + 2(13) \\ (22) + 2(23) \\ (33) + 2(12)\end{array}\right\}\begin{array}{l}\text{3 n.f.p.'s} \\ \text{doubly degenerate}\end{array}$$

i.f.p. $x_1 = x_2 = x_3 = 1/3$

$m \to$ i.f.p. $(x_1 = 1)$ with initial vectors

$x_1 = .37165698$ ___ $x_1 = .56881955$

$x_2 = .08887955$ and $x_2 = .00173087$

$x_3 = .53946347$ ___ $x_3 = .42944958$

$m \to$ n.f.p. $(x_2 = 1)$ with initial vector

$x_1 = .02489604$

$x_2 = .18806313$ (See Section XI for discussion)

$x_3 = .78704084$

Table II (cont.)

III.2.2.b

$$\left.\begin{array}{l}(22) + 2(23) \\ (11) + 2(13) \\ (33) + 2(12)\end{array}\right\} \begin{array}{l}\text{n.f.p.} \\ \text{b.d.p.} \\ \text{not degenerate}\end{array}$$

i.f.p. $x_1 = x_2 = x_3 = 1/3$

$m \to$ b.d.p.

III.2.2.c

$$\left.\begin{array}{l}(33) + 2(12) \\ (22) + 2(23) \\ (11) + 2(13)\end{array}\right\} \begin{array}{l}\text{n.f.p.} \\ \text{b.f.p. } (\textit{cf. } \text{I.2.n}) \\ \text{b.d.p.} \\ \text{degenerate}\end{array}$$

i.f.p. $x_1 = x_2 = x_3 = 1/3$

$m \to$ i.f.p.

III.2.2.e

$$\left.\begin{array}{l}(33) + 2(12) \\ (11) + 2(13) \\ (22) + 2(23)\end{array}\right\} \begin{array}{l}\text{b.t.p.} \\ \text{not degenerate}\end{array}$$

i.f.p. $x_1 = x_2 = x_3 = 1/3$

$m \to$ i.f.p.

III.2.3.a

$$\left.\begin{array}{l}(11) + 2(12) \\ (22) + 2(23) \\ (33) + 2(13)\end{array}\right\} \begin{array}{l}\text{3 n.f.p.'s} \\ \text{triply degenerate}\end{array}$$

i.f.p. $x_1 = x_2 = x_3 = 1/3$

$m \to$ no convergence (see Section XI for discussion)

Table II (cont.)

III.2.3.b

$\left.\begin{array}{l}(22) + 2(23) \\ (11) + 2(12) \\ (33) + 2(13)\end{array}\right\}$ n.f.p.
b.f.p. (*cf.* I.2.n)
b.d.p.
degenerate

i.f.p. $x_1 = x_2 = x_3 = 1/3$

$m \to$ n.f.p. with initial vectors

$x_1 = .29517934$ $x_1 = .05841813$

$x_2 = .10369878$ and $x_2 = .25242368$

$x_3 = .60112188$ $x_3 = .68915819$

$m \to$ b.d.p. with initial vector

$x_1 = .35613727$

$x_2 = .62067802$

$x_3 = .02318471$

III.2.3.e

$\left.\begin{array}{l}(33) + 2(13) \\ (11) + 2(12) \\ (22) + 2(23)\end{array}\right\}$ b.t.p.
not degenerate

i.f.p. $x_1 = x_2 = x_3 = 1/3$

$m \to$ b.t.p.

III.2.3.f

$\left.\begin{array}{l}(22) + 2(23) \\ (33) + 2(13) \\ (11) + 2(12)\end{array}\right\}$ b.t.p.
not degenerate

i.f.p. $x_1 = x_2 = x_3 = 1/3$

$m \to$ b.t.p.

Introduction to Table III

This table lists a few 4-variable generalizations (and two 5-variable ones) of selected 3-variable systems. The method of generalization is explained in the text (Section VII, I.5.j – ext – 1 and subsequent discussion). The basic notation is that of Table II, but degeneracy, b.f.p.'s, n.f.p.'s, *etc.*, are not noted.

Table III
Examples of Binary Reaction Systems for $N > 3$

I.1.c – ext – 1

$(11) + (22) + (44) + 2(14) + 2(24) + 2(34)$

$2(13) + 2(23)$

$2(12)$

(33)

i.f.p. given by $(x = x_1)$

$\quad 16x^3(1 + x) - 1 = 0$

$m \rightarrow$ i.f.p.

$\left.\begin{array}{l} x_1 = .35833637 \\ x_2 = .33933212 \\ x_3 = .24319009 \\ x_4 = .05914142 \end{array}\right\}$ i.f.p.

I.2.a – ext – 1

$(11) + (22) + 2(12) + 2(14) + 2(24) + 2(34)$

$(44) + 2(23)$

$2(13)$

(33)

i.f.p. given by $(x = x_3)$

$\quad x^4 - 2x^3 - x^2 + 2x - 1/2 = 0$

$m \rightarrow$ .i.f.p.

$\left.\begin{array}{l} x_1 = 1/2 \\ x_2 = .04258212 \\ x_3 = .34108137 \\ x_4 = .11633651 \end{array}\right\}$ i.f.p.

Table III (cont.)

I.2.k – ext – 1

$(11) + (44) + 2(23) + 2(14) + 2(24) + 2(34)$

$(22) + 2(13)$

$2(12)$

(33)

i.f.p. given by $(x = x_1)$

$\qquad 8x^2\left(8x^3 - 4x - 1\right) + 3 = 0$

$m \rightarrow$ i.f.p.

$\left.\begin{array}{l} x_1 = .42397159 \\ x_2 = .28099237 \\ x_3 = .23826556 \\ x_4 = .05677048 \end{array}\right\}$ i.f.p.

I.2.p – ext – 1

$(22) + (44) + 2(13) + 2(14) + 2(24) + 2(34)$

$(11) + 2(12)$

$2(23)$

(33)

$m \rightarrow$ b.d.p.

$(1,0,0,0),\ (0,1,0,0)$

$\left.\begin{array}{l} x_1 = \dfrac{\sqrt{3} - 1}{2} = .36602540 \\[2mm] x_2 = 1/2 \\[2mm] x_3 = \dfrac{-1 + \sqrt{5 - 2\sqrt{3}}}{2} = .11965684 \\[2mm] x_4 = x_3^2 = .01431776 \end{array}\right\}$ i.f.p.

I.3.g – ext – 1

$(22) + (44) + 2(12) + 2(14) + 2(24) + 2(34)$

$2(13) + 2(23)$

(11)

(33)

i.f.p. given by $(x = x_1)$

$\qquad 2x^5 + 2x^4 + 3x^3 + 3x^2 - 1 = 0$

$m \rightarrow$ b.t.p. $(1,0,0,0),\ (0,0,1,0),\ (0,0,0,1)$

$\left.\begin{array}{l} x_1 = .45003654 \\ x_2 = .30641100 \\ x_3 = .20253289 \\ x_4 = .04101957 \end{array}\right\}$ i.f.p.

Table III (cont.)

I.3.i – ext – 1

$(22) + (44) + 2(23) + 2(14) + 2(24) + 2(34)$

$2(12) + 2(13)$

(11)

(33)

i.f.p. given by $(x = x_1)$

$\qquad 2x^4 + x^3 + x^2 + 2x - 1 = 0$

$\left. \begin{array}{l} x_1 = .37973157 \\ x_2 = .45527986 \\ x_3 = .14419607 \\ x_4 = .02079250 \end{array} \right\}$ i.f.p.

$m \to$ i.f.p. with initial vectors:

$x_1^{(0)} = .23580752 \qquad x_1^{(0)} = .35656919$

$x_2^{(0)} = .45570511 \quad$ and $\quad x_2^{(0)} = .01450795$

$x_3^{(0)} = .16718123 \qquad x_3^{(0)} = .34834313$

$x_4^{(0)} = .14130614 \qquad x_4^{(0)} = .28057973$

$m \to$ b.t.p. with initial vector:

$x_1^{(0)} = .03633558$

$x_2^{(0)} - .15482236$

$x_3^{(0)} = .04547126$

$x_4^{(0)} = .76337080$

I.4.e – ext – 1

$(22) + 2(12) + 2(23) + 2(14) + 2(24) + 2(34)$

$(11) + (44)$

$2(13)$

(33)

i.f.p. given by $(x = x_3)$

$\qquad x^4 + x^2 + x - 1/4 = 0$

$m \to$ b.d.p.

$\left. \begin{array}{l} x_1 = 1/2 \\ x_2 = .25179510 \\ x_3 = .20583631 \\ x_4 = .04236859 \end{array} \right\}$ i.f.p.

Table III (cont.)

I.5.j – ext – 1

$(22) + 2(12) + 2(23) + 2(14) + 2(24) + 2(34)$

$(44) + 2(13)$

(11)

(33)

i.f.p. given by $(x = x_1)$

$x^8 + x^4 + 2x^3 + x^2 + x - 1 = 0$

$\left. \begin{array}{l} x_1 = .48325036 \\ x_2 = .22868204 \\ x_3 = .23353091 \\ x_4 = .05453669 \end{array} \right\}$ i.f.p.

$m \to$ interior periodic points of period 12:

Cycle	x_1	x_2	x_3	x_4
n	.66002500	.33689080	.00014037	.00294383
n+1	.56417301	.00019396	.43563301	.00000002
n+2	.00038792	.49154477	.31829119	.18977612
n+3	.86242865	.03626192	.00000015	.10130928
n+4	.24595298	.01026383	.74378319	$0(i.e. < 5 \times 10^{-9})$
n+5	.02042231	.36587139	.06049287	.55321343
n+6	.68740763	.30851591	.00041707	.00365939
n+7	.52688378	.00058679	.47252926	.00000017
n+8	.00117358	.49793601	.27760650	.22328391
n+9	.87242595	.05050729	.00000138	.07706538
n+10	.23293147	.00594148	.76112705	$0(i.e. < 5 \times 10^{-9})$
n+11	.01184765	.35458089	.05425707	.57931439

and $x_i^{(n+12)} = x_i^{(n)}$

Table III (cont.)

I.5.j – ext – 2

$(22) + 2(12) + 2(23) + 2(14) + 2(24) + 2(34) + 2(15) + 2(25)$
$+ 2(35) + 2(45)$

$(55) + 2(13)$

(11)

(33)

(44)

i.f.p. given by $(x = x_i)$

$x^{16} + x^8 + x^4 + 2x^3 + x^2 + x - 1 = 0$

$$\left.\begin{array}{l} x_1 = .48324807 \\ x_2 = .22571344 \\ x_3 = .23352870 \\ x_4 = .05453565 \\ x_5 = .00297414 \end{array}\right\} \text{i.f.p.}$$

$m \to$ interior periodic points of period 6:

Cycle	x_1	x_2	x_3	x_4	x_5
n	.83355924	.03412438	.01389264	.02095486	.09746888
n+1	.27188601	.03266085	.69482102	.00019301	.00043911
n+2	.06547728	.37782442	.07392200	.48277626	.00000004
n+3	.74749492	.00968042	.00428727	.00546440	.23307293
n+4	.38047069	.06073242	.55874865	.00001838	.00002986
n+5	.11787703	.42517497	.14475794	.31220006	$0(i.e. < 5 \times 10^{-9})$

and $x_i^{(n+6)} = x_i^{(n)}$

I.5.1 – ext – 1

$(22) + 2(13) + 2(23) + 2(14) + 2(24) + 2(34)$

$(44) + 2(12)$

(11)

(33)

i.f.p. given by $(x = x_1)$

$x^8 - 2x^5 + x^4 - 2x^3 - x^2 + 3x - 1 = 0$

$m \to$ i.f.p.

$$\left.\begin{array}{l} x_1 = .49177311 \\ x_2 = .20789913 \\ x_3 = .24184079 \\ x_4 = .05848697 \end{array}\right\} \text{i.f.p.}$$

Table III (cont.)

I.6.b – ext – 1

$2(12) + 2(13) + 2(14) + 2(23) + 2(24) + 2(34)$

$(11) + (44)$

(22)

(33)

i.f.p. given by $(x = x_2)$

$$2x^8 + 2x^6 + 2x^5 - x^4 + 2x^3 - x^2 - 3x + 1 = 0$$

$m \to$ i.f.p.

$\left.\begin{array}{l} x_1 = .56609505 \\ x_2 = .32057516 \\ x_3 = .10276844 \\ x_4 = .01056135 \end{array}\right\}$ i.f.p.

This i.f.p. is very close to that of II.3.d – ext – 1 (*q.v.*), as might be expected.

I.6.c – ext – 1

$2(12) + 2(13) + 2(23) + 2(14) + 2(24) + 2(34)$

$(22) + (44)$

(11)

(33)

i.f.p. given by $(x = x_1)$

$$x^7 + x^5 + x^4 + x^2 - 1/2 = 0$$

$m \to$ i.f.p.

(*cf.* I.6.b – ext – 1)

$\left.\begin{array}{l} x_1 = .56612299 \\ x_2 = .01066456 \\ x_3 = .32049525 \\ x_4 = .10271720 \end{array}\right\}$ i.f.p.

II.3.d – ext – 1

$(44) + 2(12) + 2(13) + 2(14) + 2(23) + 2(24) + 2(34)$

(33)

(11)

(22)

i.f.p. given by $(x = x_1)$

$$x + x^2 + x^4 + x^8 - 1 = 0$$

$m \to$ i.f.p.

(*cf.* Eq. (21) of the text)

$\left.\begin{array}{l} x_1 = .56616082 \\ x_2 = .10274465 \\ x_3 = .32053807 \\ x_4 = .01055646 \end{array}\right\}$ i.f.p.

Table III (cont.)

II.3.d – ext – 2

   $(55) + 2(12) + 2(13) + 2(14) + 2(15) + 2(23) + 2(24) + 2(25)$
      $+ 2(34) + 2(35) + 2(45)$

   (33)

   (11)

   (22)

   (44)

   i.f.p. given by $(x = x_1)$

   $x + x^2 + x^4 + x^8 + x^{16} - 1 = 0$

   (*cf.* Eq. (21) of the text)

   $m \rightarrow$ i.f.p.

$\left. \begin{array}{l} x_1 = .56612380 \\ x_2 = .10271779 \\ x_3 = .32049615 \\ x_4 = .01055094 \\ x_5 = .00011132 \end{array} \right\}$ i.f.p.

III.1.c – ext – 1

   $(11) + (22) + 2(14) + 2(24) + 2(34)$

   $2(13) + 2(23)$

   $2(12) + (44)$

   (33)

   i.f.p. given by $(x = x_3)$

   $1 - x^3 + 4(2x - 1)(1 - x - x^2)^2 = 0$

   $m \rightarrow$ i.f.p.

$\left. \begin{array}{l} x_1 = .35443555 \\ x_2 = .34053827 \\ x_3 = .24500079 \\ x_4 = .06002539 \end{array} \right\}$ i.f.p.

III.1.f – ext – 1

   $(11) + (44) + 2(14) + 2(24) + 2(34)$

   $2(13) + 2(23)$

   $2(12) + (22)$

   (33)

   i.f.p. given by $(x = x_3)$

   $4(1 - x - x^2)^2(x - 1) + 1 = 0$

   $m \rightarrow$ i.f.p.

$\left. \begin{array}{l} x_1 = .23278771 \\ x_2 = .36744075 \\ x_3 = .30608408 \\ x_4 = .09368746 \end{array} \right\}$ i.f.p.

Explanation of Tables IV and V

Each of these tables consists in a tabulation of the convergence behavior of random initial points under 26 transformations which are simple modifications of I.3.g. These 26 systems are generated as follows (see Section VIII.2 of text):

In Table IV, we retain the "skeleton"

$$x_1' = 1/2x_2^2 + 1/2x_3^2 + 2x_1x_2 + \dots$$
$$x_2' = 2x_1x_3 + 2x_2x_3 + \dots$$
$$x_3' = 1/2x_1^2 + \dots$$

To this we add in all possible ways the missing terms $1/2x_1^2$, $1/2x_2^2$, $1/2x_3^2$. Twenty-six new systems result (the 27th is identical with I.3.g). The notation is the same as in the previous tables, but we give only the numerical results, omitting comments and i.f.p. equations.

Table V is similar, except that the skeleton is

$$x_1' = x_2^2 + x_3^2 + x_1x_2 + \dots$$
$$x_2' = x_1x_3 + x_2x_3 + \dots$$
$$x_3' = x_1^2 + \dots$$

and the missing terms are x_1x_2, x_1x_3, x_2x_3.

Table IV
Modifications of System I.3.g.

Case 1.

$$1/2(11) + (22) + (33) + 2(12)$$
$$2(13) + 2(23)$$
$$1/2(11)$$
$$m \to \text{i.f.p.}$$

$$\left.\begin{array}{l} x_1 = .56804585 \\ x_2 = .27061611 \\ x_3 = .16133804 \end{array}\right\} \text{i.f.p.}$$

Case 2.

$$1/2(22) + 1/2(33) + 2(12)$$
$$2(13) + 2(23) + 1/2(11) + 1/2(22) + 1/2(33)$$
$$1/2(11)$$
$$m \to \text{i.f.p.}$$

$$\left.\begin{array}{l} x_1 = .47868722 \\ x_2 = .40674205 \\ x_3 = .11457073 \end{array}\right\} \text{i.f.p.}$$

Table IV (cont.)

Case 3.

$$1/2(22) + 1/2(33) + 2(12) \hspace{3cm} x_1 = .38326262$$
$$2(13) + 2(23) \hspace{3cm} x_2 = .37085176 \hspace{0.3cm} \} \, \text{i.f.p.}$$
$$(11) + 1/2(22) + 1/2(33) \hspace{3cm} x_3 = .24588562$$
$$m \rightarrow \text{i.f.p.}$$

Case 4.

$$1/2(11) + (22) + 1/2(33) + 2(12) \hspace{2cm} x_1 = .56267887$$
$$2(13) + 2(23) + 1/2(33) \hspace{2cm} x_2 = .27901737 \hspace{0.3cm} \} \, \text{i.f.p.}$$
$$1/2(11) \hspace{2cm} x_3 = .15830376$$
$$m \rightarrow \text{i.f.p.}$$

Case 5.

$$1/2(11) + 1/2(22) + (33) + 2(12) \hspace{2cm} x_1 = .54858377$$
$$2(13) + 2(23) + 1/2(22) \hspace{2cm} x_2 = .30094415 \hspace{0.3cm} \} \, \text{i.f.p.}$$
$$1/2(11) \hspace{2cm} x_3 = .15047208$$
$$m \rightarrow \text{i.f.p.}$$

Case 6.

$$(22) + (33) + 2(12) \hspace{2.5cm} x_1 = .51130605$$
$$2(13) + 2(23) + 1/2(11) \hspace{2.5cm} x_2 = .35797701 \hspace{0.3cm} \} \, \text{i.f.p.}$$
$$1/2(11) \hspace{2.5cm} x_3 = .13071694$$
$$m \rightarrow \text{i.f.p.}$$

Case 7.

$$1/2(11) + (22) + 1/2(33) + 2(12) \hspace{2cm} x_1 = .55386328$$
$$2(13) + 2(23) \hspace{2cm} x_2 = .27874434 \hspace{0.3cm} \} \, \text{i.f.p.}$$
$$1/2(11) + 1/2(33) \hspace{2cm} x_3 = .16739238$$
$$m \rightarrow \text{i.f.p.}$$

Table IV (cont.)

Case 8.

$$1/2(11) + 1/2(22) + (33) + 2(12)$$
$$2(13) + 2(23)$$
$$1/2(11) + 1/2(22)$$
$$m \rightarrow \text{i.f.p.}$$

$$\left. \begin{array}{l} x_1 = .52319512 \\ x_2 = .29610055 \\ x_3 = .18070433 \end{array} \right\} \text{i.f.p.}$$

Case 9.

$$1/2(22) + 1/2(33) + 2(12)$$
$$2(13) + 2(23) + 1/2(11) + 1/2(22)$$
$$1/2(11) + 1/2(33)$$
$$m \rightarrow \text{i.f.p.}$$

$$\left. \begin{array}{l} x_1 = .47452268 \\ x_2 = .40572059 \\ x_3 = .11975673 \end{array} \right\} \text{i.f.p.}$$

Case 10.

$$1/2(22) + 1/2(33) + 2(12)$$
$$2(13) + 2(23) + 1/2(11) + 1/2(33)$$
$$1/2(11) + 1/2(22)$$
$$m \rightarrow \text{i.f.p.}$$

$$\left. \begin{array}{l} x_1 = .43437288 \\ x_2 = .39376269 \\ x_3 = .17186443 \end{array} \right\} \text{i.f.p.}$$

Case 11.

$$1/2(22) + 1/2(33) + 2(12)$$
$$2(13) + 2(23) + 1/2(22) + 1/2(33)$$
$$(11)$$
$$m \rightarrow \text{i.f.p.}$$

$$\left. \begin{array}{l} x_1 = .42681726 \\ x_2 = .39100977 \\ x_3 = .18217297 \end{array} \right\} \text{i.f.p.}$$

Case 12.

$$1/2(22) + 1/2(33) + 2(12)$$
$$2(13) + 2(23) + 1/2(11) + 1/2(22)$$
$$1/2(11) + 1/2(33)$$
$$m \rightarrow \text{i.f.p.}$$

$$\left. \begin{array}{l} x_1 = .48181136 \\ x_2 = .40211755 \\ x_3 = .11607109 \end{array} \right\} \text{i.f.p.}$$

Table IV (cont.)

Case 13.

$(22) + 1/2(33) + 2(12)$ $x_1 = .50829830$

$2(13) + 2(23) + 1/2(11) + 1/2(33)$ $x_2 = .36251813$ } i.f.p.

$1/2(11)$ $x_3 = .12918357$

$m \rightarrow$ i.f.p.

Case 14.

$1/2(11) + 1/2(22) + 1/2(33) + 2(12)$ $x_1 = .54269064$

$2(13) + 2(23) + 1/2(22) + 1/2(33)$ $x_2 = .31005278$ } i.f.p.

$1/2(11)$ $x_3 = .14725658$

$m \rightarrow$ i.f.p.

Case 15.

$1/2(22) + 1/2(33) + 2(12)$ $x_1 = .39529957$

$2(13) + 2(23) + 1/2(33)$ $x_2 = .37727171$ } i.f.p.

$(11) + 1/2(22)$ $x_3 = .22742872$

$m \rightarrow$ i.f.p.

Case 16.

$1/2(22) + 1/2(33) + 2(12)$ $x_1 = .41841444$

$2(13) + 2(23) + 1/2(22)$ $x_2 = .38772371$ } i.f.p.

$(11) + 1/2(33)$ $x_3 = .19386185$

$m \rightarrow$ i.f.p.

Case 17.

$1/2(22) + 1/2(33) + 2(12)$ $x_1 = .42566975$

$2(13) + 2(23) + 1/2(11)$ $x_2 = .39057539$ } i.f.p.

$1/2(11) + 1/2(22) + 1/2(33)$ $x_3 = .18375486$

$m \rightarrow$ i.f.p.

Table IV (cont.)

Case 18.

$$1/2(22) + (33) + 2(12)$$
$$2(13) + 2(23)$$
$$(11) + 1/2(22)$$
$$m \to \text{i.f.p.}$$

$$\left.\begin{aligned} x_1 &= .41080271 \\ x_2 &= .35678917 \\ x_3 &= .23240812 \end{aligned}\right\} \text{i.f.p.}$$

Case 19.

$$(22) + 1/2(33) + 2(12)$$
$$2(13) + 2(23)$$
$$11 + 1/2(33)$$
$$m \to \text{i.f.p.}$$

$$\left.\begin{aligned} x_1 &= .44081956 \\ x_2 &= .34107310 \\ x_3 &= .21810734 \end{aligned}\right\} \text{i.f.p.}$$

(Very slow convergence; $|\text{Jacobian}| = .99411052$ at i.f.p.)

Case 20.

$$1/2(11) + 1/2(22) + 1/2(33) + 2(12)$$
$$2(13) + 2(23)$$
$$1/2(11) + 1/2(22) + 1/2(33)$$
$$m \to \text{i.f.p.}$$

$$\left.\begin{aligned} x_1 &= 1/2 \\ x_2 &= \frac{\sqrt{5}-1}{4} = .30901699 \\ x_3 &= \frac{3-\sqrt{5}}{4} = .19098301 \end{aligned}\right\} \text{i.f.p.}$$

Case 21.

$$1/2(11) + 1/2(22) + 1/2(33) + 2(12)$$
$$2(13) + 2(23) + 1/2(22)$$
$$1/2(11) + 1/2(33)$$
$$m \to \text{i.f.p.}$$

$$\left.\begin{aligned} x_1 &= .53484692 \\ x_2 &= .31010205 \\ x_3 &= .15505103 \end{aligned}\right\} \text{i.f.p.}$$

Case 22.

$$1/2(11) + 1/2(22) + 1/2(33) + 2(12)$$
$$2(13) + 2(23) + 1/2(33)$$
$$1/2(11) + 1/2(22)$$
$$m \to \text{i.f.p.}$$

$$\left.\begin{aligned} x_1 &= .51157583 \\ x_2 &= .30963296 \\ x_3 &= .17879121 \end{aligned}\right\} \text{i.f.p.}$$

Table IV (cont.)

Case 23.

$(22) + 1/2(33) + 2(12)$ $x_1 = .50306514$

$2(13) + 2(23) + 1/2(11)$ $x_2 = .36118335$ $\Big\}$ i.f.p.

$1/2(11) + 1/2(33)$ $x_3 = .13575151$

$m \rightarrow$ i.f.p.

Case 24.

$(22) + 1/2(33) + 2(12)$ $x_1 = .45127091$

$2(13) + 2(23) + 1/2(33)$ $x_2 = .34508366$ $\Big\}$ i.f.p.

(11) $x_3 = .20364543$

i.f.p. given by $(x = x_1)$

$$3x^4 - 6x^2 - 2x + 2 = 0$$

$m \rightarrow$ i.t.p.

$x_1^{(n+3)} = x_1^{(n)} = .87777286, \ x_1^{(n+1)} = .00766150, \ x_1^{(n+2)} = .34944206$

$x_2^{(n+3)} = x_2^{(n)} = .00011739, \ x_2^{(n+1)} = .22185332, \ x_2^{(n+2)} = .65049924$ $\Big\}$ i.t.p.

$x_3^{(n+3)} = x_3^{(n)} = .12210975, \ x_3^{(n+1)} = .77048518, \ x_3^{(n+2)} = .00005870$

Case 25.

$1/2(22) + (33) + 2(12)$ $x_1 = .44406835$

$2(13) + 2(23) + 1/2(11)$ $x_2 = .38371478$ $\Big\}$ i.f.p.

$1/2(11) + 1/2(22)$ $x_3 = .17221687$

$m \rightarrow$ i.f.p.

Case 26.

$1/2(22) + (33) + 2(12)$ $x_1 = .43425854$

$2(13) + 2(23) + 1/2(22)$ $x_2 = .37716097$ $\Big\}$ i.f.p.

(11) $x_3 = .18858049$

$m \rightarrow$ i.f.p.

Table V
Modifications of System I.3.g.

Case 1.

$$(22) + (33) + 2(12) + (13) + (23) \qquad \left. \begin{array}{l} x_1 = .52488860 \\ x_2 = .19960336 \\ x_3 = .27550804 \end{array} \right\} \text{i.f.p.}$$

(13) + (23)

(11)

$m \to$ i.f.p.

Case 2.

$$(22) + (33) + (12) \qquad \left. \begin{array}{l} x_1 = .39816095 \\ x_2 = .44330691 \\ x_3 = .15853214 \end{array} \right\} \text{i.f.p.}$$

2(13) + 2(23) + (12)

(11)

$m \to$ i.f.p.

Case 3.

$$(22) + (33) + (12) \qquad \left. \begin{array}{l} x_1 = .32471796 \\ x_2 = .24512233 \\ x_3 = .43015971 \end{array} \right\} \text{i.f.p.}$$

(13) + (23)

(11) + (12) + (13) + (23)

$m \to$ i.f.p.

Case 4.

$$(22) + (33) + 2(12) + (13) \qquad \left. \begin{array}{l} x_1 = 1/2 \\ x_2 = 1/4 \\ x_3 = 1/4 \end{array} \right\} \text{i.f.p.}$$

(13) + 2(23)

(11)

$m \to$ i.f.p.

Table V (cont.)

Case 5.

$$(22) + (33) + 2(12) + (23) \qquad\qquad x_1 = .48053382$$
$$2(13) + (23) \qquad\qquad\qquad\qquad\quad x_2 = .28855343 \quad \text{i.f.p.}$$
$$(11) \qquad\qquad\qquad\qquad\qquad\qquad x_3 = .23091275$$

$m \to$ i.f.p. (slow convergence)

Case 6.

$$(22) + (33) + (12) + (13) + (23) \qquad x_1 = .46557123$$
$$(13) + (23) + (12) \qquad\qquad\qquad\quad x_2 = .31767220$$
$$(11) \qquad\qquad\qquad\qquad\qquad\qquad x_3 = .21675657 \quad \text{i.f.p.}$$

$m \to$ i.f.p.   (*cf.* Case 9)

Case 7.

$$(22) + (33) + 2(12) + (13) \qquad\qquad x_1 = .48787213$$
$$(13) + (23) \qquad\qquad\qquad\qquad\qquad x_2 = .21060631 \quad \text{i.f.p.}$$
$$(11) + (23) \qquad\qquad\qquad\qquad\qquad x_3 = .30152156$$

$m \to$ i.f.p.

Case 8.

$$(22) + (33) + 2(12) + (23) \qquad\qquad x_1 = .43691113$$
$$(13) + (23) \qquad\qquad\qquad\qquad\qquad x_2 = .22408140 \quad \text{i.f.p.}$$
$$(11) + (13) \qquad\qquad\qquad\qquad\qquad x_3 = .33900747$$

$m \to$ i.f.p.

Case 9.

$$(22) + (33) + (12) + (13) + (23) \qquad x_1 = .46557123$$
$$(13) + (23) \qquad\qquad\qquad\qquad\qquad x_2 = .21675657$$
$$(11) + (12) \qquad\qquad\qquad\qquad\qquad x_3 = .31767220 \quad \text{i.f.p.}$$

$m \to$ i.f.p.   (*cf.* Case 6)

Table V (cont.)

Case 10.

$$(22) + (33) + (12)$$
$$2(13) + (23) + (12)$$
$$(11) + (23)$$
$$m \rightarrow \text{i.f.p.}$$

$$\left. \begin{aligned} x_1 &= .36632453 \\ x_2 &= .40727446 \\ x_3 &= .22640101 \end{aligned} \right\} \text{i.f.p.}$$

Case 11.

$$(22) + (33) + (12)$$
$$(13) + 2(23) + (12)$$
$$(11) + (13)$$
$$m \rightarrow \text{i.f.p.}$$

$$\left. \begin{aligned} x_1 &= .37008112 \\ x_2 &= .41249400 \\ x_3 &= .21742488 \end{aligned} \right\} \text{i.f.p.}$$

Case 12.

$$(22) + (33) + (12)$$
$$2(13) + 2(23)$$
$$(11) + (12)$$
$$m \rightarrow \text{i.f.p.}$$

$$\left. \begin{aligned} x_1 &= .35320996 \\ x_2 &= .38577366 \\ x_3 &= .26101638 \end{aligned} \right\} \text{i.f.p.}$$

Case 13.

$$(22) + (33) + (12) + (23)$$
$$2(13) + (23) + (12)$$
$$(11)$$
$$m \rightarrow \text{i.f.p.}$$

$$\left. \begin{aligned} x_1 &= .42578666 \\ x_2 &= .39291905 \\ x_3 &= .18129429 \end{aligned} \right\} \text{i.f.p.}$$

Case 14.

$$(22) + (33) + (12) + (13)$$
$$(13) + 2(23) + (12)$$
$$(11)$$
$$m \rightarrow \text{i.f.p.}$$

$$\left. \begin{aligned} x_1 &= .42878157 \\ x_2 &= .38736479 \\ x_3 &= .18385364 \end{aligned} \right\} \text{i.f.p.}$$

Table V (cont.)

Case 15.

$\quad$ (22) + (33) + (12) $\qquad\qquad\qquad$ $x_1 = x_2 = x_3 = 1/3$ i.f.p.

$\quad$ (13) + 2(23)

$\quad$ (11) + (12) + (13)

$\quad$ $m \to$ i.f.p.

Case 16.

$\quad$ (22) + (33) + (12) $\qquad\qquad\qquad$ $x_1 = x_2 = x_3 = 1/3$ i.f.p.

$\quad$ 2(13) + (23)

$\quad$ (11) + (12) + (23)

$\quad$ $m \to$ i.f.p.

Case 17.

$\quad$ (22) + (33) + (12) $\qquad\qquad\qquad$ $x_1 = x_2 = x_3 = 1/3$ i.f.p.

$\quad$ (13) + (23) + (12)

$\quad$ (11) + (13) + (23)

$\quad$ $m \to$ i.f.p.

Case 18.

$\quad$ (22) + (33) + (12) + (23) $\qquad\quad$ $\left. \begin{array}{l} x_1 = x_3 = .38196601 \\ x_2 = .23606798 \end{array} \right\}$ i.f.p.

$\quad$ (13) + (23) $\qquad\qquad\qquad\qquad\quad$

$\quad$ (11) + (12) + (13) $\qquad\qquad\qquad$ (*cf.* Case 20)

$\quad$ $m \to$ i.f.p.

Case 19.

$\quad$ (22) + (33) + (12) + (23) $\qquad\quad$ $\left. \begin{array}{l} x_1 = .41964338 \\ x_2 = .22815549 \\ x_3 = .35220113 \end{array} \right\}$ i.f.p.

$\quad$ (13) + (23)

$\quad$ (11) + (12) + (13)

$\quad$ $m \to$ i.f.p.

Table V (cont.)

Case 20.

$$(22) + (33) + 2(12)$$
$$(13) + (23)$$
$$(11) + (13) + (23)$$
$$m \rightarrow \text{i.f.p.}$$

$x_1 = x_3 = .38196601$
$x_2 = .23606798$ $\Big\}$ i.f.p.
(*cf.* Case 18)

Case 21.

$$(22) + (33) + 2(12)$$
$$2(13) + (23)$$
$$(11) + (23)$$
$$m \rightarrow \text{i.f.p.}$$

$x_1 = .42788476$
$x_2 = .30766792$ $\Big\}$ i.f.p.
$x_3 = .26444732$

Case 22.

$$(22) + (33) + 2(12)$$
$$(13) + 2(23)$$
$$(11) + (13)$$
$$m \rightarrow \text{i.f.p.}$$

$x_1 = .41421356$
$x_2 = x_3 = .29289322$ $\Big\}$ i.f.p.
(*cf.* Case 24)

Case 23.

$$(22) + (33) + (12) + (13)$$
$$(13) + (23) + (12)$$
$$(11) + (23)$$
$$m \rightarrow \text{i.f.p.}$$

$x_1 = .41594131$
$x_2 = .32699283$ $\Big\}$ i.f.p.
$x_3 = .25706586$

Case 24.

$$(22) + (33) + (12) + (13)$$
$$(13) + 2(23)$$
$$(11) + (12)$$
$$m \rightarrow \text{i.f.p.}$$

$x_1 = .41421356$
$x_2 = x_3 = .29289322$ $\Big\}$ i.f.p.
(*cf.* Case 22)

Table V (cont.)

Case 25.

$(22) + (33) + (12) + (23)$ $x_1 = .40157157$ ⎫

$(13) + (23) + (12)$ $x_2 = .32895639$ ⎬ i.f.p.

$(11) + (13)$ $x_3 = .26947204$ ⎭

$m \rightarrow$ i.f.p.

Case 26.

$(22) + (33) + (12) + (23)$ $x_1 = .39816095$ ⎫

$2(13) + (23)$ $x_2 = .31706428$ ⎬ i.f.p.

$(11) + (12)$ $x_3 = .28477477$ ⎭

$m \rightarrow$ i.f.p.

Table VI
Super-Systems

I.2.e — Super

a. Initial Configuration

$y_1 = .03348210$	$y_4 = .19438836$	$y_7 = .19222349$
$y_2 = .30728575$	$y_5 = .02148296$	$y_8 = .02173869$
$y_3 = .03472276$	$y_6 = .05711904$	$y_9 = .13761085$

b. Initial Configuration

$y_1 = .06578151$	$y_4 = .07528814$	$y_7 = .32057547$
$y_2 = .02301108$	$y_5 = .13873987$	$y_8 = .27225037$
$y_3 = .00731792$	$y_6 = .03090597$	$y_9 = .06612967$

Both of these gave the final periodic configuration $\left(y_i^{(n+2)} = y_i^{(n)}\right)$:

$y_1^{(n)} = .18055154$	$y_4^{(n)} = 0$	$y_7^{(n)} = .13889692$
$y_2^{(n)} = .38464618$	$y_5^{(n)} = 0$	$y_8^{(n)} = .29590536$
$y_3^{(n)} = 0$	$y_6^{(n)} = 0$	$y_9^{(n)} = 0$

$y_1^{(n+1)} = .18055154$	$y_4^{(n+1)} = .38464618$	$y_7^{(n+1)} = 0$
$y_2^{(n+1)} = 0$	$y_5^{(n+1)} = 0$	$y_8^{(n+1)} = 0$
$y_3^{(n+1)} = .13889692$	$y_6^{(n+1)} = .29590536$	$y_9^{(n+1)} = 0$

For the initial configuration

c.

$y_1 = .14398619$	$y_4 = .06422167$	$y_7 = .04324188$
$y_2 = .10409303$	$y_5 = .02141784$	$y_8 = .07707331$
$y_3 = .40338819$	$y_6 = .06973376$	$y_9 = .07284413,$

the final configuration was $\left(y_i^{(n+2)} = y_i^{(n)}\right)$:

$y_1^{(n)} = .10204732$	$y_4^{(n)} = .21740114$	$y_7^{(n)} = 0$
$y_2^{(n)} = .21740114$	$y_5^{(n)} = .46315040$	$y_8^{(n)} = 0$
$y_3^{(n)} = 0$	$y_6^{(n)} = 0$	$y_9^{(n)} = 0$

$y_1^{(n+1)} = .31944846$	$y_4^{(n+1)} = 0$	$y_7^{(n+1)} = .24574926$
$y_2^{(n+1)} = 0$	$y_5^{(n+1)} = 0$	$y_8^{(n+1)} = 0$
$y_3^{(n+1)} = .24574926$	$y_6^{(n+1)} = 0$	$y_9^{(n+1)} = .18905302$

Table VI (cont.)

I.3.g — Super

Initial Configurations:

a.

$y_1 = .01631507$	$y_4 = .11350670$	$y_7 = .24077161$
$y_2 = .06252989$	$y_5 = .05348809$	$y_8 = .05243129$
$y_3 = .01977135$	$y_6 = .03871014$	$y_9 = .40247586$

b.

$y_1 = .04649838$	$y_4 = .02351925$	$y_7 = .14210424$
$y_2 = .20513993$	$y_5 = .02478259$	$y_8 = .00038505$
$y_3 = .11967344$	$y_6 = .05513905$	$y_9 = .38275807$

These both gave the final configuration $\left(y_i^{(n+3)} = y_i^{(n)}\right)$:

$y_1^{(n)} = .24781045$	$y_4^{(n)} = .24675180$	$y_7^{(n)} = .00324338$
$y_2^{(n)} = .24675180$	$y_5^{(n)} = .24569765$	$y_8^{(n)} = .00322954$
$y_3^{(n)} = .00324338$	$y_6^{(n)} = .00322954$	$y_9^{(n)} = .00004246$
$y_1^{(n+1)} = .54648124$	$y_4^{(n+1)} = .00957014$	$y_7^{(n+1)} = .18319231$
$y_2^{(n+1)} = .00957014$	$y_5^{(n+1)} = .00016760$	$y_8^{(n+1)} = .00320812$
$y_3^{(n+1)} = .18319231$	$y_6^{(n+1)} = .00320812$	$y_9^{(n+1)} = .06141002$
$y_1^{(n+2)} = .00651538$	$y_4^{(n+2)} = .03009170$	$y_7^{(n+2)} = .04411082$
$y_2^{(n+2)} = .03009170$	$y_5^{(n+2)} = .13898048$	$y_8^{(n+2)} = .20372868$
$y_3^{(n+2)} = .04411082$	$y_6^{(n+2)} = .20372868$	$y_9^{(n+2)} = .29864174$

Initial Configuration:

c.

$y_1 = .11265988$	$y_4 = .09322974$	$y_7 = .03769313$
$y_2 = .25586952$	$y_5 = .00321861$	$y_8 = .00633159$
$y_3 = .16027686$	$y_6 = .09787061$	$y_9 = .23285006$

This gave the final configuration $\left(y_i^{(n+3)} = y_i^{(n)}\right)$:

$y_1^{(n)} = .04018182$	$y_4^{(n)} = .18558236$	$y_7^{(n)} = .27204143$
$y_2^{(n)} = .04001017$	$y_5^{(n)} = .18478956$	$y_8^{(n)} = .27087928$
$y_3^{(n)} = .00052591$	$y_6^{(n)} = .00242894$	$y_9^{(n)} = .00356053$
$y_1^{(n+1)} = .36799967$	$y_4^{(n+1)} = .36642757$	$y_7^{(n+1)} = .00481645$
$y_2^{(n+1)} = .00644452$	$y_5^{(n+1)} = .00641699$	$y_8^{(n+1)} = .00008435$
$y_3^{(n+1)} = .12336144$	$y_6^{(n+1)} = .12283443$	$y_9^{(n+1)} = .00161458$
$y_1^{(n+2)} = .05967020$	$y_4^{(n+2)} = .00104496$	$y_7^{(n+2)} = .02000274$
$y_2^{(n+2)} = .27559068$	$y_5^{(n+2)} = .00482623$	$y_8^{(n+2)} = .09238395$
$y_3^{(n+2)} = .40398281$	$y_6^{(n+2)} = .00707467$	$y_9^{(n+2)} = .13542376$

APPENDIX

1. In this appendix we summarize the results for a class of homogeneous quadratic transformations quite distinct from the class we have called Binary Reaction Systems. This class arises in a natural way in the study of a certain crude model of the "evolutionary process" which will be described below. It also has some mathematical interest, owing to the fact that the limiting behavior of all systems belonging to this class can be explicitly predicted.

2. **The Evolutionary Model**
    Consider a large population in which each distinct "type" of individual is labelled by an index pair (i,j), $i,j = 1,2,\ldots,N$. Let the fraction of the male population which is of type (i,j) be denoted by x_{ij}. We shall assume that $x_{ij} = x_{ji}$; also, we take the number of females of type (i,j) to be equal to the number of males of this type—hence there is no need to denote the fraction of females by a separate letter.

    We now impose a mating rule (random mating is assumed) which states, in effect, that if individuals of type (k,ℓ) mate with individuals of type (m,n), the progeny will be of all types (i,j) such that

    $$\begin{aligned} \min(k,m) \leq i \leq \max(k,m) \\ \min(\ell,n) \leq j \leq \max(\ell,n) \end{aligned} \tag{1}$$

Loosely speaking, we may call each index of a pair a "characteristic." A given mating will produce all possible children such that (1) is satisfied, the distribution of the two indices determining the progeny being the product of two identical distributions. The *number* of children will, of course, be proportional to the number of parents of each type. Mathematically:

$$x_{ij}^{(n+1)} = \sum_{k,\ell,m,n} \gamma_i^{km}\gamma_j^{\ell n}x_{k\ell}^{(n)}x_{mn}^{(n)} \tag{2}$$

The sum in (2) is to be carried out under the restriction (1). We specify the system further by postulating:

$$\begin{aligned} \gamma_i^{km} = \gamma_i^{mk} > 0, \ \min(k,m) \leq i \leq \max(k,m) \\ = 0 \text{ otherwise;} \end{aligned} \tag{3}$$

$$\sum_{i=m}^{k} \gamma_i^{km} = 1 \tag{4}$$

$$\sum_{i=m}^{k} i\gamma_i^{km} = \frac{m+k}{2} \tag{5}$$

In addition, we normalize by taking:

$$\sum_{i,j=1}^{N} x_{ij}^{(0)} = 1, \ 0 \leq x_{ij}^{(0)} \leq 1, \ \text{all } i,j \tag{6}$$

It is then evident that we have:

$$\sum_{i,j}^{N} x_{ij}^{(n)} = 1$$

for all n.

The class of systems defined above is the one we shall actually discuss. However, it may be of interest to at least mention the actual evolutionary model in connection with which equation (2) arises. Evolution is assumed to take place by mutation. A type (i,j) can give rise to two new types $(i+1,j)$ and $(i,j+1)$ with some small probability. When we include this (linear) effect, we get a series of equations:

$$x_{ij} = -\epsilon x_{ij} + \epsilon/2\big(x_{i-1,j} + x_{i,j-1}\big) + \sum \gamma_i^{km}\gamma_j^{\ell n} x_{k\ell} x_{mn} \tag{7}$$

Here ϵ is taken as some small number. We actually performed many numerical experiments on systems of the form (7) with special values of the γ_i^{km} satisfying (3), (4), (5). Two particularly convenient choices are:

$$\gamma_i^{jk} = \frac{1}{2^{|j-k|}}\begin{pmatrix} |j-k| \\ i - \min(j,k) \end{pmatrix} \tag{8}$$

and

$$\gamma_i^{jk} = \frac{1}{|j-k|+1}, \ \min(j,k) \leq i \leq \max(j,k) \\ = 0 \text{ otherwise} \tag{9}$$

Note that with our definition (3) we may take the sum in (2) as unrestricted, i.e., over-all $k, \ell, m, n = 1/2, \ldots, N$. We shall not discuss the behavior of the "mutating" system (7) in this report, but rather restrict ourselves to the pure "mating" system (2).*

---

* It may be objected that our mating rules have nothing to do with Mendel's Laws. This is intentionally the case. The Mendelian case has been treated in great generality in a series of papers by Hilda Geiringer;[7], See also C. C. Li:[8]

3. If we sum over one of the indices in (2), we obtain the system:

$$C_i^{(n+1)} = \sum_{k,m=1}^{N} \gamma_i^{km} C_k^{(n)} C_m^{(n)} \quad i = 1, \ldots N \tag{10}$$

$$C_i \equiv \sum_{j=1}^{N} x_{ij} \tag{11}$$

and, of course,

$$\sum_{i=1}^{N} C_i^{(n)} = 1 \quad \text{all } n. \tag{12}$$

By virtue of condition (5) on the γ_i^{km}, the system (10) possesses a *linear invariant* (distinct from $\sum C_i = 1$); in fact:

$$\sum_j j C_j^{(n+1)} = \sum_{\ell,m} C_\ell^{(n)} C_m^{(n)} \sum_j j \gamma_j^{\ell m} = \sum_{\ell,m} C_\ell^{(n)} C_m^{(n)} \left(\frac{\ell+m}{2}\right)$$

$$= \sum_\ell \ell C_\ell^{(n)} \tag{13}$$

The consequences of this property are very interesting. It turns out that the existence of this linear invariant enables one to predict explicitly the limiting behavior of any initial vector $(C_1^{(0)}, C_2^{(0)}, \ldots C_N^{(0)})$ when iterated according to (10). Using the fact that $\sum^N C_i = 1$ is also an invariant, we may define an invariant:

$$\sigma \equiv \sum_{i=1}^{N-1} (N-i) C_i \tag{14}$$

σ is, of course, explicitly determined by the initial vector. It can then be proved that every initial vector will converge to a definite fixed point which is determined as follows:

For the given value of σ, there is one index j such that:

$$N - j \geq \sigma > N - j - 1 \tag{15}$$

The f.p. is then explicitly given by:

$$\left.\begin{array}{l} C_j = \sigma - (N - j - 1) \\ C_{j+1} = N - j - \sigma \\ \text{all other } C_i = 0 \end{array}\right\} \tag{16}$$

The f.p. is independent of the actual values of the coefficients γ_i^{km} providing they satisfy (3), (4), and (5).*
These results can easily be referred back to the original variables x_{ij}. Defining a quantity α from (16) by:

$$C_j \equiv \frac{1}{1+\alpha} \tag{17}$$

we find that the corresponding symmetric tensor $(x_{mn}^{(0)})$ will converge to the final state:

$$\begin{aligned} x_{jj} &= \frac{1}{(1+\alpha)^2} \\ x_{j+1,j} = x_{j,i+1} &= \frac{\alpha}{(1+\alpha)^2} \\ x_{j+1,j+1} &= \frac{\alpha^2}{(1+\alpha)^2} \end{aligned} \tag{18}$$

The results can also be extended to the case of M "characteristics" $i_1, i_2, \ldots i_M$.
The fraction of the population of type $(i_1, \ldots, i_M)$ is then denoted by $x_{i_1 \ldots i_M}$. This is taken to be symmetric in all M indices; hence, there are:

$$\binom{N+M-1}{M}$$

types of individuals in the population. The corresponding mating rule is:

$$x_{i_1 \ldots i_M}^{(n+1)} = \sum_{S_1, S_2} \sum_{S_3, S_4} \cdots \gamma_{i_1}^{S_1 S_2} \gamma_{i_2}^{S_3 S_4} \cdots \gamma_{i_M}^{S_{2M-1} S_{2M}} \\ x_{S_1 S_3 \ldots S_{2M-1}}^{(n)} x_{S_2 S_4 \ldots S_{2M}}^{(n)} \tag{19}$$

If we sum over $M-1$ indices, we can again define:

---

* Of course the *rate* of convergence will, in general, depend on the actual values of the γ_i^{km}. In one case it is actually possible to solve explicitly for the iterates as functions of n. This is the case $N=2$ where we find:

$$x_{11}^{(n)} = \frac{1}{2^n}\left[x_{11}^{(0)} + \sigma^2(2^n - 1)\right]$$
$$\sigma = x_{11}^{(0)} + x_{12}^{(0)}.$$

This case can be viewed as a generalization of the Mendelian law for a single gene. The population has the same limiting configuration in both cases, but for Mendel's rules equilibrium is reached in a single step "(Hardy's Law)".

$$C_i \equiv \sum_{i_2,\ldots,i_M=1}^{N} x_{i_1\ldots i_M} \tag{20}$$

and we again get the system (10). In terms of the tensor $x_{i_1}\ldots i_M$, the final state is:

$$x_{j,j\ldots j} = \frac{1}{(1+\alpha)^M}$$

$$x_{j+1,j,j\ldots j} = \frac{\alpha}{(1+\alpha)^M}$$

$$\vdots \tag{21}$$

$$x_{j+1,j+1,\ldots j+1} = \frac{\alpha^M}{(1+\alpha)^M}$$

Here j and α are determined as in (16) and (17).

4. The explicit nature of these results is due to the existence of the linear invariant σ, which in turn is a consequence of the "mean-preserving" property (5) assumed for the coefficients γ_i^{jk}. However, the actual convergence properties are probably more closely connected with the "index-limiting" condition (3). (Our Binary Reaction systems do not, in general, have this property; the existence of oscillating final states may well be a consequence of allowing such mating rules as $j \oplus j \to k \neq j$.) Some cases have been investigated in which the conditions on the coefficients were relaxed. In particular, we have considered systems for which we no longer require:

$$\gamma_i^{jk} > 0 \quad \text{all } j,k .$$

As an example, we chose the case:

$$\sum_{i=1}^{N} \gamma_i^{jk} = \delta_{j,k} + \delta_{j+1,k} + \delta_{j-1,k} \tag{22}$$

This corresponds to a sort of "selective mating scheme" in which only individuals with "nearby" indices can mate. Of course we then no longer have $\sum C_i = 1$. In order to secure this property, we must "renormalize" at every step, i.e., set:

$$\frac{C_i'}{\sum C_i'} \to C_i'$$

If this is formally written out, the system is no longer quadratic. There are then many types of fixed points possible. So far it has not been possible to give an analytical treatment of such systems.*

References

1. See, e.g., R. Sauer, *Math. Ann.* **106**, 722, (1932), O. Baier, *Math. Ann.* **112**, 630.
2. Coddington and Levinson, *Theory of Ordinary Differential Equations*, Chapters 15, 16, McGraw-Hill, New York (1955).
3. V. Volterra, *Leçons sur la Theórie Mathématique de la Lutte Pour La Vie*, Gauthier-Villars, Paris (1939).
4. H. Weyl, *Math. Ann.* **77**, 313 (1916).
5. S. Ulam and J. von Neumann, *Bull. Am. Math. Soc.* **53**, 1120 (1947).
6. L. E. Dickson, *Modern Algebraic Theories*, Chapter II, Sanborn, Chicago, (1930).
7. Hilda Geiriger, *Ann. Math. Statistics*, **15**, 25–57 (1944), *Genetics*, **33**, 548–564, (1948).
8. C. C. Li, *Population Genetics*, Univ. of Chicago Press, Chicago (1955).

Note, however, that if we simply take $\gamma_i^{jk} = \frac{\delta_{ij} + \delta_{ik}}{2}$, we find that every initial vector is a fixed point.

Enrico Fermi

Johnny von Neumann with
11 year old Claire Ulam

Bob Richtmyer

David Hawkins

From left to right, Ulam, Mycielski, Bednarek

Bill Beyer

"Computers can also be used to investigate... and with less success, to study games of 'skill' like chess." (See page 302.) P. R. Stein playing the first game of chess he and others had programmed with Ulam in 1957 against the MANIAC.

11

NON-LINEAR
TRANSFORMATION STUDIES
ON ELECTRONIC COMPUTERS

With P. R. Stein
(LADC-5688, 1963)

This paper is a continuation of the study initiated and reported in the preceding chapter on Quadratic Transformations. Interactions of polynomial transformations, particularly cubic transformations in three variables, as well as the asymptotic and ergodic properties of the sequences of iterated points are considered. The theme of the computational study of difficult problems in pure mathematics is exemplified. (Eds.)

Introduction

This paper will deal with properties of certain non-linear transformations in Euclidean spaces—mostly in two or three dimensions. In the main they will be of very special and simple algebraic form. We shall be principally interested in the iteration of such transformations and in the asymptotic and ergodic properties of the sequence of iterated points. Very little seems to be known, even on the purely topological level, about the properties of specific non-linear transformations, even when these are bounded and continuous or analytic. The transformations we study in this paper are in fact bounded and continuous, but in general many-to-one, i.e., not necessarily homeomorphisms. In one dimension such transformations are simply functions with values lying in the domain of definition; for example, if $f(x)$ is continuous and non-negative in the interval $[0.1]$ and $\max[f(x)] \le 1$, then $x' = f(x)$ is a

293

transformation* of the type considered. Even in one dimension, however, nothing resembling a complete theory of the ergodic properties of the iterated transformation exists. On the algebraic side, we study in this paper the invariant points (fixed points), finite sets (periods)—and invariant subsets (*curves*) of these transformations—together with the means of obtaining them constructively. The topological properties of two (not necessarily one-dimensional) transformations $S(p)$, $T(p)$ are identical under a homeomorphism H: when $S(p) = H[T[H^{-1}(p)]]$. When S and T are themselves homeomorphisms—and for *one dimension*—necessary and sufficient conditions for conjugacy are known.[1] When S and T are one-dimensional, but not necessarily one-to-one, it is possible to give a set of necessary conditions for conjugacy; no meaningful sufficient conditions, however, are known.

For example, the set of fixed points of S has to be topologically equivalent to those of T. The same must hold for the set of periodic points, i.e., points such that the nth power of the transformation returns the point to its original position. The *attractive* and *repellent* fixed points must correspond, etc. These conditions are known from the corresponding study of homeomorphisms. For many-to-one transformations one may generalize these conditions by considering the *tree* of a point. For a given transformation T we define the *tree* of a point P as the smallest set Z of points such that:

a) P belongs to Z.
b) If a point Q belongs to Z, then $T(Q)$ belong to Z.
c) If Q belongs to Z, then all points of the form $T^{-1}(Q)$ belong to Z.

Obviously, for two transformations to be conjugate, the trees of corresponding points must be combinatorially equivalent and, in addition, their topological interrelations must be the same.**

The present study was initiated several years ago[2] with the consideration of certain homogeneous, quadratic transformations which we called *binary reaction systems*. A typical example is the following:

$$x_1' = x_2^2 + x_3^2 + 2x_1x_2 \,,$$
$$x_2' = 2x_1x_3 + 2x_2x_3 \,, \tag{1}$$
$$x_3' = x_1^2 \,,$$

---

* Here and throughout the paper a primed variable always represents the value obtained on the next iterative step. In a more explicit notation, the above equation would read: $x^{(n+1)} = f(x^{(n)})$.

** One-dimensional transformations are considered in more detail in Appendix I.

where we consider initial points P with coordinates x_1, x_2, x_3 satisfying:

$$0 \leq x_i \leq 1, \quad x_1 + x_2 + x_3 = 1 . \tag{2}$$

Since
$$x_1' + x_2' + x_3' = (x_1 + x_2 + x_3)^2 ,$$
the transformation (1) maps the two-dimensional region (2) into some sub-region of itself. The choice of these transformations was motivated by certain physical and biophysical considerations. For example, the set of equations (1) could be interpreted as determining the composition of a hypothetical population whose individuals are of three *types*, conventionally labeled 1, 2, and 3. The x_i would then represent the fraction of the total population which consists of individuals of type "*i*." The transformation can be thought of as a mathematical transcription of the *mating rule*:

> type 2 and type 2 produce type 1 ,
> type 3 and type 3 produce type 1 ,
> type 1 and type 2 produce type 1 ,
> type 1 and type 3 produce type 2 ,
> type 2 and type 3 produce type 2 ,
> type 1 and type 1 produce type 3 . (3)

For any assigned initial composition, i.e., any initial vector (x_1, x_2, x_3) satisfying (2), we may then ask: What is the final (or limiting) composition of the population after infinitely many "generations," that is, after infinitely many matings according to the scheme (3)?

In the present context, a *mating rule* can be defined as a system of three non-linear first-order difference equations of the form:

$$\begin{aligned} x_1' &= f_1(x_1, x_2, x_3) , \\ x_2' &= f_2(x_1, x_2, x_3) , \\ x_3' &= f_3(x_1, x_2, x_3) , \end{aligned} \tag{4}$$

where each f_i is the sum of some subset of the six homogeneous monomials x_1^2, x_2^2, x_3^2, $2x_1x_2$, $2x_1x_3$, $2x_2x_3$, and each such term must belong to one and only one f_i. Two transformations are called equivalent if they are conjugate under the (linear) transformation defined by a given permutation of the indices 1,2,3. (This is the only linear homeomorphism which preserves the homogeneous quadratic character of the transformation.) Under this definition of equivalence, it turns out that there are 97 inequivalent transformations of the above type. It quickly becomes apparent that, despite their formal simplicity, these transformations are very difficult to study analytically, particularly if one is interested in their iterative properties. For example, for most initial points in the region of definition, the sequence of iterates generated by repeated application of the transformation given by equation (1) converges to a set of three points:

$$p_2 = T(p_1) \, ,$$
$$p_3 = T(p_2) \, ,$$
$$p_1 = T(p_3) \, .$$
(5)

Using a standard terminology to be explained in detail below, we say that the "limit set" is a "period of order three." It is clear by inspection of transformation (1) that another limit set exists; if we write

$$p_1 = (x_1 = 1, \ x_2 = x_3 = 0), \quad p_2 = (x_1 = x_2 = 0, \ x_3 = 1) \, ,$$

then

$$p_2 = T(p_1), \quad p_1 = T(p_2) \, .$$
(6)

In addition there is the algebraic fixed point of (1):

$$p = T(p) \, .$$
(7)

The general initial vector, however, always leads to (5). Certain other quadratic transformations show an even more complicated behavior. An example is the transformation:

$$x_1' = x_2^2 + x_3^2 + 2x_1x_2, \quad x_2' = x_1^2 + 2x_2x_3, \quad x_3' = 2x_1x_3 \, .$$
(8)

This bears a close formal relationship to (1); in fact, they differ only by the exchange of a single term. The limit sets, however, are quite different. Transformation (8) has an attractive fixed point with coordinates:

$$x_1 = \frac{1}{2}, \ x_2 = \frac{\sqrt{2}}{4}, \ x_3 = \frac{2 - \sqrt{2}}{4} \, .$$
(9)

It also has a limit set of the type (6) with $p_1 = (1,0,0)$, $p_2 = (0,1,0)$. In this case, both limit sets are observed. It is found experimentally that the set of initial points leading to (9) is separated from those leading to the oscillatory limiting behavior (6) by a closed curve surrounding the fixed point (figure 2 of chapter 10). The analytical nature of this boundary curve remains unknown.

In view of the complicated behavior exhibited by these examples, we felt it would be useful to study these transformations numerically, making use of the powerful computational aid afforded by electronic computing machines. From one point of view our present paper may be looked on as an introduction, through our special problems, to modern techniques in *experimental* mathematics with the electronic computer as an essential tool. Over the past decade these machines have been extensively employed in solution of otherwise intractable problems arising in the physical sciences. In addition to solving the particular practical problem under consideration, this work has in some cases resulted in significant theoretical advances. Correspondingly, attempts to solve difficult physical problems have led to considerable improvements in the logical and technical design of computers themselves. In contrast, the use of electronic computers in pure mathematics has been relatively rare.* This may be partly due to a certain natural conservatism; in our opinion, however, the neglect of this important new research tool by many mathematicians is due simply to lack of information. In other words, the average mathematician does not yet realize what computers can do. It is our hope that the present paper will help to demonstrate the effectiveness of high-speed computational techniques in dealing with at least one class of difficult mathematical problems. With this end in mind, we have devoted the first section of our paper to a brief discussion of how computing machines can be used to study problems in pure mathematics. Much of this section is introductory in character, and is meant primarily for those readers who have had no firsthand experience in the use of computers. It also includes, however, a description of the numerical techniques used in this study; these may be of interest even to seasoned practitioners.

After our study of quadratic transformations in three variables,** we decided to investigate the iterative properties of other classes of polynomial transformations. As a natural generalization of the quadratics described above, we consider transformations of the form:

$$x'_i = f_i(x_1, \ldots, x_k) \quad (i = 1 \text{ to } k) , \tag{10}$$

where the f_i are disjoint sums of the homogeneous monomials which arise on expanding the expression:

$$F \equiv \left(\sum_{i=1}^{k} x_i \right)^m . \tag{11}$$

---

   * Perhaps the greatest computational effort has been expended on problems in number theory. See refs. 2 and 5.

   ** We shall not discuss this work here. Full details are contained in the above references. That report contains, in addition, some fragmentary results on a few particular quadratic transformations in higher dimensional spaces.

The number of such terms, each taken with its full multinomial coefficient, is

$$N_k^m = \binom{m+k-1}{k-1} .$$ (12)

By construction,

$$\sum_{i=l}^{k} f_i = F ,$$

so that if we take

$$\sum_{i=1}^{k} x_i = 1, \quad x_i \geq 0 ,$$ (13)

the (additive) normalization of the x_i is preserved. We are then dealing with positive transformations in a bounded portion of the Euclidean space of $k-1$ dimensions, i.e., just the hyperplane defined by (13). If $m = 2$, $k = 3$, these transformations are the 97 quadratics in three variables introduced above. The bulk of the present paper is devoted to the case $m = k = 3$, i.e., cubic transformations in three variables; there are 9370 independent transformations of this form. We have also examined the 34337 quadratic transformations in four variables, but our analysis of the results is not yet complete (January, 1963); for this case ($m = 2$, $k = 4$) we include only some statistical observations and a few interesting examples. These three cases—$m = 2$, $k = 3$; $m = 2$, $k = 4$; $m = 3$, $k = 3$—are the only ones for which an exhaustive survey is at present feasible. For other values of m and k the number of transformations to be studied is much too large.

The determination of the exact number T_k^m of inequivalent transformations for arbitrary m and k is an unsolved combinatorial problem. It can, of course, be reduced to enumerating those transformations which are invariant under one or more operations of the symmetric group on the k indices, but no convenient way of doing this is known. The problem, however, is not of much practical significance. A *lower limit* T_k^{*m} to the number T_k^m of inequivalent transformations is given by:

$$T_k^{*m} = S_k^{N_k^m} ,$$ (14)

where S_j^i is the Stirling number of the second kind. S_j^i is also the number of ways of putting i objects in j identical boxes, no box being left empty. This underestimates T_k^m by assuming in effect that each transformation has $k!$ non-identical copies, i.e., that no transformation is invariant under any permutation (except the identity). The following table illustrates the trend:

TABLE I

	N_k^m	T_k^{*m}	T_k^{*m}
$m = 2, \; k = 3$	6	90	97
$m = 3, \; k = 3$	10	9330	9370
$m = 2, \; k = 4$	10	34105	34337
$m = 4, \; k = 3$	15	2375101	—
$m = 2, \; k = 5$	15	210766920	—
$m = 3, \; k = 4$	20	45232115901	—

The T_k^m were obtained by direct enumeration—using, of course, all known shortcuts. For $m = 2$, $k = 4$, this enumeration was actually performed on a computer. In view of the huge values of the T_k^{*m} in the lower half of this table, it is unlikely that anyone will be interested in attempting a comprehensive numerical study of these transformations for values of m and k larger than those we have considered.

A general discussion of our results for the cubics in three variables and the quadratics in four variables is given in section II; the reader will also find there formal definitions of a few basic concepts and an explanation of the special terminology employed throughout the paper. Perhaps the most interesting result of this study is our discovery of limit sets of an extremely "pathological" appearance. The existence of such limit sets was quite unexpected,*—and is indeed rather surprising in view of the essential simplicity of the generating transformations. Sections III and IV are concerned with the effect—on the iterative properties of our transformations—of two types of structural generalization. Specifically, in section III we consider the one-parameter generalization—called by us the "Δt-modification"—which consists in replacing equation (10) by:

$$x_i' = (1 - \Delta t)x_i + \Delta t f_i(x_1, \ldots, x_k) \; (i = 1 \text{ to } k), \; 0 < \Delta t \leq 1 . \quad (12a)$$

This generalization has the special property of leaving the fixed points of the transformation invariant, although their character—i.e., whether they are attractive or repellent—may be altered. The detailed

---

* Quadratic transformations in three variables apparently do not exhibit similar pathologies.

discussion of the behavior of such transformations under variations of the parameter Δt is limited to the cubic case.

Section IV describes the result of introducing small variations in the coefficients of the monomials which make up the various f_i. Again we deal only with the cubic case, and indeed only with a few interesting examples chosen from our basic set of 9370 transformations. Let us denote the N_k^m monomials (e.g., x_1^3, $3x_2x_3^2$, $6x_1x_2x_3$, ...) in the expansion of (11) by the symbol M_j, $j = 1$ to N_k^m. The assignment of a particular index to a particular monomial is arbitrary.

Then we have

$$f_i = \sum_{j=1}^{N_k^m} d_{ij}M_j, \quad 1 \le i \le k,$$ (13a)

with

$$d_{ij} = 1 \text{ or } 0,$$ (14a)

$$\sum_{i=1}^{k} d_{ij} = 1.$$ (15)

The generalization then consists in relaxing the restriction (14). If this were done subject only to the condition that the d_{ij} all be non-negative, we should be dealing with a $(k - 1)N_k^m$ parameter family of positive, bounded, homogeneous polynomial transformations. At present nothing significant can be said about this class as a whole. As explained in section IV, our procedure has been to study one-parameter families of transformations which are in a certain sense "close" to some particular transformation of our original set.*

In section V we give a brief, heuristic discussion of the connection between our transformations—which are really first-order non-linear difference equations—and differential equations in the plane. Our conclusion is that the connection is not, in fact, very close, and that the techniques so far developed for treating non-linear differential equations do not seem suitable for handling the problems discussed in this paper.

---

* Some analogous but rather unsystematic investigations were carried out on quadratics in three variables, and are contained in the report cited in ref. 2. Subsequent to the appearance of that report we made some studies (unpublished) on quadratics with randomly chosen positive coefficients satisfying (15). For quadratics (at least in three variables), the conclusion seems to be that such randomly chosen transformations are most likely to lead under iteration to simple convergence for almost all initial points.

The final section of our main text—section VI—contains a description of a class of piece-wise linear transformations on the unit square. These transformations exhibit interesting analogies with our polynomial transformations in three variables. Relatively little work has been done on this "two-dimensional broken linear" case, but the preliminary results we report seem to indicate that a detailed study might prove worthwhile.

There are two appendices: Appendix I is largely devoted to an extended discussion of certain non-linear transformations in one dimension, on the unit interval. Some of these are special cases of our cubics in three variables; others originated independently of our principal study. It is perhaps rather surprising how little can be said theoretically even about this simple one-dimensional case. It turns out that some of the same phenomena are observed in one dimension as are found in the plane—e.g., the apparently discontinuous behavior of limit sets as a function of a monotonically varying parameter. Of course, the repeated iteration of a one-dimensional transformation is a much simpler matter than the corresponding process in several dimensions. However, as we soon discovered, great care must be taken to avoid the phenomenon of "spurious convergence." This point is discussed in some detail and a few—rather alarming—examples are given.

Appendix II contains the bulk of the photographic evidence— including the "pathology" of the limit sets—on which the discussion of sections III and IV is based. These pictures, together with others scattered throughout the main body of the text, constitute in a sense the unique contribution of this paper. In retrospect, it seems unlikely that our investigation could have been successfully carried out without the visual aid afforded by the oscilloscope and the polaroid camera. Put in the simplest terms, unless one knows precisely what one is looking for, mere lists of numbers are essentially useless. Automatic plotting devices however, such as the oscilloscope, allow one to tell at a glance what is happening. Very often the picture itself will suggest some change in the course of the investigation—for example, the variation of some hitherto neglected parameter. The indicated modification can often be effected in a few seconds and the result observed on the spot.*

Visual display is of very great value when one is in effect studying sets of points in the plane; when one passes to three dimensions automatic plotting ceases to be merely a convenience and becomes essential.

---

* This interaction of man and computing machine has sometimes been referred to as "synergesis."[3]

A glance at our pictures of three-dimensional limit sets—the result of iterating certain quadratic transformations in four variables—should convince even the most skeptical reader. In our opinion, it would be virtually impossible to make sense out of a mere numerical listing of coordinates of the points plotted in these photographs.

Of the many who have helped with this work, there are three to whom we are particularly indebted: Cerda Evans, Verna Gardiner, and Dorothy Williamson. These ladies did the actual coding and supervised all the machine calculations. Without their help this paper could not have been written.

I. The Role of the Computing Machine

1. The use of electronic computers for the solution of complicated or tedious problems, usually of practical origin, is by now familiar. Typical computer tasks are: the evaluation of integrals, the solution of large systems of linear equations, the solution of minimax problems (linear programming), the treatment of complicated boundary value or initial value problems, etc. One of the more impressive jobs that computers have done is to calculate the time history of immensely complicated physical systems (e.g., involving hydrodynamical motions, magnetic fields, etc.). Recently there has been considerable interest in using computers to attack problems of a less applied nature, for example those arising in combinatorial analysis[4] and number theory.[5] This work often takes on an experimental flavor; such experimentation has led to results of considerable interest, for example, the construction of certain types of mutually orthogonal latin squares.[6] Computers can also be used to investigate formal mathematical systems,[7] to reduce symbolic expressions,[8] and—with less success—to study games of "skill" like chess.[9]

The use of computing machines that we describe in the present paper differs in two respects from the examples just cited. On the one hand, our study is not essentially combinatorial in character, but falls rather in the domain of algebra and real variable function theory. On the other hand, we are not attempting to "solve" some well-defined problem; instead we investigate via repeated trials the asymptotic properties of certain non-linear transformations, usually without any advance knowledge of what we may find in a given case. Even "after the fact," so to speak, it is difficult to classify these asymptotic properties in a meaningful fashion; the broadness of the categories we employ for this classification* is merely a measure of our lack of insight into the structure of the observed limit sets.

---

* See section II.

Faced with this situation, one may ask the question: how does one recognize "convergence"—i.e., the existence of an invariant set—when one has no *a priori* numerical criteria to apply? We can only supply a partial answer to this question, but that answer has the advantage of simplicity, viz.: "use your eyes." The practical application of this "technique" involves, of course,* the use of automatic plotting devices.

2. Roughly speaking, computing machines are devices which perform the four elementary arithmetic operations on numbers in a certain—not necessarily simple—sequence. This sequence of operations is called the "program," and consists of a set of logical commands, both of the sequential ("do this and then do that") and of the branching ("if this holds, then do that") type. The program is composed by an investigator (the "programmer") and must therefore reflect his own limitations. Nevertheless, the machine may easily produce results quite unanticipated by the programmer, even if the program is essentially deterministic in nature.** A classic example—which happens to be relevant here—is the step-by-step application of some recurrence relation which generates a sequence whose trend the programmer cannot determine in advance. As an example, we may cite the following one-step recursion in a single variable:

$$y_{n+1} = w_n(3 - 3w_n + \sigma w_n^2), \quad w_n \equiv 3y_n(1 - y_n) . \tag{1}$$

Given some initial $0 < y_0 < 1$, we may ask: what is the result of applying the rule (1) N times, where N is some larger number, say 10^5? This particular transformation is discussed in detail in appendix I; here we quote three examples for the purpose of illustration.

(a) If $\sigma = 0.99004$, then for almost all y_0 the sequence of iterates produced by (1) converges (in $< 10^5$ steps) to a period of order 14.

(b) If $\sigma = 0.99005$, the corresponding limit set is a period of order 28.***

---

* Hand plotting is in general highly impractical, and clearly relinquishes the principal advantage of machine computation: SPEED.

** Strictly speaking, all programs used on digital computers are deterministic in nature: even when random numbers are employed, these are generated according to some fixed algorithm so that the sequence is in principle known.

*** These results were found by using IBM "STRETCH" computer. The periods are exact to within the accuracy of that machine, i.e., 48 binary digits (~15 decimals). See further in appendix I.

(c) If $\sigma = 0.99008$, no finite period is observed after $N = 5 \times 10^5$ steps.

So far as we are aware, this behavior could not be predicted by current analytical or algebraic techniques. Such phenomena are easy to study on a computer, however, because of the great speed with which it can carry out the (relatively simple) operations implied by an expression such as (1) above. In fact, 200000 iterations of this transformation takes slightly less than one minute on a really fast computer.*

3. As we mentioned in the introduction, the principal content of this paper is the study of the asymptotic properties of certain non-linear transformations of relatively simple form. This means that, if T is such a transformation, we examine the sequence

$$T(p), \ T^2(p), \ T^3(p), \ \ldots$$

for various initial points p lying in the domain of T. The mathematical object of interest to us is the set (or sets) of points to which these sequences converge. In the absence of any general analytical technique for calculating these "limit sets," we must have recourse to "brute-force" methods.

Some non-linear transformations which appear morphologically similar to those considered here can in fact be completely analyzed by elementary methods. We discovered one such case in the course of some earlier work on biological systems. It is described in our report on quadratics in three variables (see chapter 10). We restate these special results here:

Let

$$C_i' = \sum_{k,m=1}^{N} \gamma_i^{km} \, C_k C_m, \ 1 \le i \le N \ , \tag{2}$$

with coefficients satisfying

$$\gamma_i^{km} = \gamma_i^{mk} > 0, \ \min \, (k,m) \le i \le \max \, (k,m) \ , \\ \gamma_i^{km} = \gamma_i^{mk} > 0, \ \text{otherwise} \ , \tag{3}$$

$$\sum_{i=m}^{k} \gamma_i^{km} = 1 \ , \tag{4}$$

---

* This figure applies to STRETCH and includes all additional "diagnostic" operations such as checking for "convergence," etc.

$$\sum_{i=m}^{k} i\gamma_i^{km} = \frac{m+k}{2} \; . \tag{5}$$

We normalize the C_i by

$$0 \le C_i \le 1, \text{ all } i, \; \sum_{i=1}^{N} C_i = 1 \; . \tag{6}$$

This property is clearly preserved under iteration. With the coefficients defined as above, there exists a linear invariant:

$$\sigma \equiv \sum_{i=1}^{N-1} (N-i)C_i = \sum_{i=1}^{N-1} (N-i)C_i' \; . \tag{7}$$

Given an initial vector $(C_1^{(0)}, C_2^{(0)}, \ldots, C_N^{(0)})$ whose coordinates satisfy (6), σ is explicitly determined. It can then be shown that every initial vector satisfying (6) converges to a definite fixed point which is determined as follows. For the given value of σ, there is one value of the index j such that

$$N - j \ge \sigma > N - j - 1 \; . \tag{8}$$

The fixed point is then explicitly given by

$$C_j = \sigma - (N - j - 1), \; C_{j+1} - N - j - \sigma, \text{ all other } C_i - 0 \; . \tag{9}$$

Note that the fixed point is independent of the values of the coefficients γ_i^{km}.

As simple examples of coefficients satisfying (3), (4), and (5), we may mention

$$\gamma_i^{km} = \frac{1}{2^{|k-m|}} \binom{|k-m|}{i - \min(k,m)} \tag{10}$$

and

$$\gamma_i^{km} = \begin{cases} \frac{1}{|k-m|+1}, & \text{if } \min(k,m) \le i \le \max(k,m), \\ 0, & \text{otherwise.} \end{cases} \tag{11}$$

For a fuller discussion of this transformation and its possible applications, we refer the reader to the original report.

The term "brute-force" refers to the fact that, in order to determine the convergence properties of some transformation T belonging to our class, we must in general actually evaluate $T^k(p)$ for $k = 1, 2, \ldots, N$, where N is likely to be quite large, sufficiently large that is, so that we can observe convergence* to the limit set. To make matters clear, let us consider a specific example. We choose the cubic transformation

$$x_1' = x_3^3 + 3x_1x_3^2 + 3x_3x_1^2 + 6x_1x_2x_3 \ ,$$
$$x_2' = x_1^3 + 3x_2x_3^2 + 3x_3x_2^2 \ , \tag{12}$$
$$x_3' = x_2^3 + 3x_1x_2^2 + 3x_2x_1^2 \ .$$

We take some initial point $p = (x_1, x_2, x_3)$ whose coordinates satisfy:

$$x_1 + x_2 + x_3 = 1, \quad 0 \le x_i \le 1, \quad i = 1, 2, 3 \ . \tag{13}$$

The program then instructs the computing machine to evaluate the right hand side of (12), thus producing a new point $p' = (x_1', x_2', x_3')$; the coordinates of p' again of course satisfy (13). p' is then set to p, and the process is repeated. The iteration proceeds in this fashion until either some finite limit set is found** or an invariant set—presumably infinite—is "observed." The observation consists in looking at successive groups of consecutive iterates—in practice we have usually taken 900 points at a time—until no qualitative visual change is noted over a sample of several successive such groups of points. Since the transformation (12) is really two-dimensional, we may plot the successive points p in the plane. Accordingly, we define new coordinates S, a by the linear transformation***

$$S = \frac{1 + x_1 - x_3}{2}, \quad a = \frac{x_2}{2} \ . \tag{14}$$

The domain of the transformation is then the 45^0 isosceles triangle:

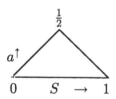

* "Convergence" must be of course understood in some approximate numerical sense. Our usual criteria are set forth in the next subsection.

** See the next subsection.

*** These are the coordinates employed in our earlier work on quadratics in three variables; we have retained them more for historical reasons than for any particular advantage they may possess.

In terms of these new variables, (12) takes the form:

$$S' = -\frac{1}{2}S^3 - \frac{15}{2}S^2a - \frac{3}{2}Sa^2 + \frac{3}{2}a^3 + 6Sa - 3a + 1 \equiv F(S,a) \,,$$
$$a' = \frac{1}{2}S^3 + \frac{3}{2}S^2a + \frac{3}{2}Sa^2 - \frac{7}{2}a^3 - 6Sa + 3a \equiv G(S,a) \,. \tag{15}$$

The computer is instructed to store 900 successive points

$$p(S^{(n)}, a^{(n)}), \ p(S^{(n+1)}), \ a^{(n+1)}), \ \dots \,,$$

and, when the last point has been calculated, to plot all 900 points on our oscilloscope screen.* If we choose, we may then photograph the resulting pattern with a polaroid camera. Such a photograph is shown in figure 1. Here one sees 900 successive high-order iterates ($n = 2700$ to 3600) of the initial point, $S = 1/2$, $a = 0.17$. For convenience, the triangle of reference is also shown.

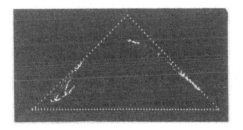

Fig. 1

This calculation—as well as all others which produced the photographs in this paper—was performed on the Los Alamos Laboratory's MANIAC II computer.** MANIAC II requires about 15 seconds to calculate 900 iterates of a point by repeated applications of

---

* The points are actually plotted in the order in which they are calculated, the whole pattern being replotted as many times as we wish. Actually, the plotting of 900 points is effectively instantaneous so far as the human eye is concerned. If we wish to *see* the points plotted in succession, we must introduce artificial time delays between the plotting of successive points.

** For the use of other computing machines in this work, see the next subsection.

a cubic transformation like (12) above. This figure includes the time spent in examining the successive points for simple convergence, as well as other "diagnostic" operations.* The actual numerical values of the coordinates may be printed out whenever desired by simply flipping a switch. On MANIAC II a decimal number is normally limited to eight significant figures. In the present paper, when there is occasion to quote numerical values obtained from MANIAC II print-outs, we shall generally reproduce them to seven figures without further specifying their accuracy.

Computer programs are, of course, not limited to generating sequences of numbers from an iterative formula such as (12). A considerable amount of sophistication can be incorporated into such a program so as to allow the machine to make "decisions" in the course of the calculation. It can, in fact, examine any property or any functional of the data that the programmer can describe in appropriate terms. One problem that is met with frequently in this work is to determine the points in a sequence of iterates that lie closest to some point, say within some chosen angle or set of angles. This sort of experiment is frequently of help in elucidating the local structure of a complicated limit set. Then again we may want to determine the average values of S and a, i.e., ergodic means, taken over the sequence. To achieve any sort of accuracy in such problems** we may be required to go to 50000 or even 100000 iterations. One saving feature is that several such diagnostic experiments can be carried out simultaneously. There are, however, special questions that must be dealt with by special programs. One such question arises in connection with our illustrative transformation (12). The complicated limit set shown in figure 1 is not the only one observed. This transformation has an attractive fixed point at:

$$S = F(S, a) = 0.6259977, \quad a = G(S, a) = 0.1107896 ; \quad (16)$$

indeed, the eigenvalues of the jacobian matrix*** evaluated at this point are complex, with $|\lambda|^2 = 0.4366967$. Consequently, there must be a neighborhood of this point in which all sequences will converge to it. The only way to find the boundary of this neighborhood is by

---

* For reasons of accuracy, the calculations are performed in the x_i coordinates; the transformation (14) to the S, a coordinates is carried out only for plotting purposes.

** More properly, to have confidence in the results. The accuracy cannot always be satisfactorily estimated.

*** The criterion for the nature of a fixed point is discussed in Section III.

trial and error. This is a time-consuming job, even for an electronic computer; if one picks a point close to the boundary of the region of convergence, several hundred—or even several thousand—iterations may be required before one can tell whether the chosen point lies inside or outside the region. Figure 2 shows the approximate boundary for the present case, drawn through 107 experimentally determined points. One of these is known to one part in 10^7, while the others have been determined only to 1 part in 10^4.*

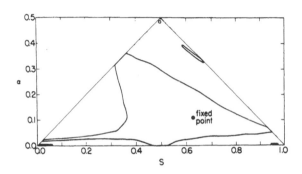

Fig. 2

4. General Procedure

a. *Cubic transformation in three variables.* Enough has been said above to make clear the necessity of using an electronic computer in such investigations. We must now say something about the systematic aspects of the study. All 9370 cubic transformations were initially studied on an IBM 7090.** First a complete list of inequivalent cubics was prepared—this was also done on the 7090—incidentally serving to check our original pencil and paper enumeration. Then by a completely automatic procedure, each transformation was taken in turn and four randomly-generated initial points were each taken as the start of an iterative sequence. For each point the iteration was continued until either convergence to a finite set of points was "observed" or 10000 iterations had been performed. By "observed" we mean that the machine sensed convergence to a fixed point or to a finite period of order

---

* The point $S = 1/2$, $a = 0.2952833$ lies in the region of convergence, while the point $S = 1/2$, $a = 0.2952834$ gives rise to a sequence which converges to the class IV limit set (see definition in section II).

** This computer is approximately five times as fast as MANIAC II.

≤ 300. More precisely, the computer was programmed to test whether the following conditions were satisfied

$$| x_i^{(n_1)} - x_i^{(n)} | < 10^{-7}, \ i = 1, 2, 3 . \tag{17}$$

If (17) is satisfied, a finite limit set has been reached to within the indicated accuracy. For $n = n_1 + 1$ this means convergence to a fixed point ("simple convergence"). Otherwise, the limit set is a period of order $n - n_1$. In practice, values of the x_i were stored at fixed time steps $n_1 = 300, \ 600, \ldots$, the test (17) being performed on each step. If "convergence" was found, the appropriate values of the x_i were printed out and the next random initial point was used, etc. If no such convergence was found after 10000 steps, the values of the iterates for the last few steps were printed, and the computer proceeded as before.

When all the cubic transformations had been studied in this fashion, the "interesting" cases—i.e., those in which no convergence was observed—were examined one by one on MANIAC II, where the visual oscilloscope display could be consulted. Many cases of apparent non-convergence turned out in fact to be convergent with the iteration carried further. It should be stressed that the restriction to 10000 iterations, which we imposed in the course of the systematic, fully automatic survey of all cubic transformations, was merely one of convenience; without some such reasonable limitation, the automatic survey would have taken too long. The same remark applies to the decision that only four randomly generated initial points be taken for each case. Past experience has shown that this last restriction is not unreasonable when a complete survey of transformations is contemplated. By this we mean that the behavior of an arbitrary transformation of our class is "likely" to be defined even if iterates of only four random points are studied. To be sure, in some cases the limit set depends in a very complicated way on the initial point; for such a transformation this crude sampling technique is not adequate. In these cases, however, the four random trials are likely to produce two difference limit sets; this in itself is an indication that the transformation in question should be studied in more detail.

For the detailed examination of a given transformation, many relatively sophisticated MANIAC programs are available. We may, in effect, study any properties of the transformation that seem of interest. Typically, these may include:

1. Determination of non-attractive fixed points (see section III).
2. Checking for periodicity.
3. Exhibiting some qualitative properties of the mapping, e.g., by showing the images under the transformation of a family of lines.

4. Determining the dimensions of the limit set.

5. Verifying that low-order periods are attractive (see section III).

6. Examining the dependence of the limit set on the initial point.

We cannot expatiate here on the actual procedures involved; sufficient to say that the use of visual display (i.e., the oscilloscope plot) is an essential tool in all this analysis.

b. *Quadratic in four variables.* All (34337) inequivalent transformations of this class were studied by the same fully automatic method as that used to study the cubics. For this purpose a faster machine than the IBM 7090 was clearly required; we were fortunate enough to have access to the IBM 7030 STRETCH computer, which is approximately 4 times as fast as the 7090 and 20 times as fast as MANIAC II. Only partial results are reported in section II, since our analysis of the STRETCH print-outs is not complete.

The detailed study of a given quadratic in four variables is more difficult than the corresponding analysis for the three variable cubics: the domain is three-dimensional, being in fact the tetrahedron defined by

$$\sum_{i=1}^{4} x_i = 1, \ 0 \leq x_i \leq 1, \ 1 \leq i \leq 4 \ .$$

Thus a meaningful visual display involves plotting some properly chosen projection of the three-dimensional limit set. In some cases it may require several trials before an appropriately "revealing" viewing angle is found; consequently it was not feasible to plot every potentially "interesting" limit set in this fashion, and some sort of selective procedure had to be resorted to. The method we chose was to look at three plane projections first—e.g., x_1 versus x_2, x_1 versus x_3, and x_2 versus x_3. It turns out that one soon develops a feeling for the "interesting" case even without being able to build up an image of the actual three-dimensional configuration from the plane "slices." More serious than this purely technical difficulty is that resulting from the generally more complicated dependence of the limit set on the initial point: it turns out that in these transformations one is much more likely to miss something by restricting one's self to a few randomly generated initial points. At the present time, lacking any local or structural criteria for the prediction of asymptotic behavior, we see no way to overcome this difficulty.

II. Limit Sets

1. **Abbreviated Notation for Transformations.** In order to have a convenient way of referring to a particular transformation without having to reproduce its explicit form, we introduce at this point a simple shorthand notation. As already noted in the introduction, our cubic transformations in three variables may be written in the form:

$$x_i' = \sum_{j=1}^{10} d_{ij} M_j, \ i = 1, 2, 3 , \tag{1}$$

with

$$d_{ij} = 0 \text{ or } 1, \text{ all } i, j , \tag{2}$$

and

$$\sum_{i=1}^{3} d_{ij} = 1, \text{ all } j , \tag{3}$$

where the M_j are the separate terms in the expansion of $(x_1 + x_2 + x_3)^3$. We now choose the following conventional ordering of the M_j.

$$M_1 = x_1^3, \ M_2 = x_2^3, \ M_3 = x_3^3, \ M_4 = 3x_1 x_2^2, \ M_5 = 3x_1 x_3^2,$$
$$M_6 = 3x_2 x_1^2, \ M_7 = 3x_2 x_3^2, \ M_8 = 3x_3 x_1^2, \tag{4}$$
$$M_9 = 3x_3 x_2^2, \ M_{10} = 6x_1 x_2 x_3 .$$

Any cubic transformation of our class is then completely determined by specifying which terms M_j or, equivalently, which indices j, appear in the first two lines of the schema (1). Let us call the set of indices belonging to the first line C_1 and those belonging to the second line C_2; C_3 is of course the complement of $C_1 + C_2$ with respect to the full set $\{1, 2, \ldots, 10\}$ and need not be written down. Thus, for example, the transformation:

$$x_1' = x_3^3 + 3x_1 x_2^2 + 3x_1 x_3^2 + 3x_2 x_3^2 + 3x_3 x_2^2 + 6x_1 x_2 x_3 ,$$
$$x_2' = x_1^3 + 3x_3 x_1^2 , \tag{5}$$
$$x_3' = x_2^3 + 3x_2 x_1^2 ,$$

would appear in the form:

$$T_{C_1 C_2} : \begin{array}{l} C_1 = \{3, 4, 5, 7, 9, 10\} \\ C_2 = \{1, 8\} . \end{array} \tag{6}$$

An analogous notation may be adopted *mutatis mutandis* for quadratics in four variables. Any such transformation can be written in the form:

$$x'_i = \sum_{j=1}^{10} d_{ij} F_j, \quad i = 1, 2, 3, 4 ,$$

$$d_{ij} = 0 \text{ or } 1, \text{ all } i, j , \tag{7}$$

$$\sum_{i=1}^{4} d_{ij} = 1, \text{ all } j .$$

Our conventional assignment of indices to the F_j is as follows:

$$F_1 = x_1^2, \ F_2 = x_2^2, \ F_3 = x_3^2, \ F_4 = x_4^2, \ F_5 = 2x_1x_2,$$
$$F_6 = 2x_1x_3, \ F_7 = 2x_1x_4, \ F_8 = 2x_2x_3, \ F_9 = 2x_2x_4, \tag{8}$$
$$F_{10} = 2x_3x_4 .$$

Let Q_k denote the set of indices belonging to the kth line of the schema (7). Then any such transformation is specified by writing down three of the four Q_k thus:

$$T_{Q_1 Q_2 Q_3} : \begin{array}{l} Q_1 = \{2, 8, 9\} , \\ Q_2 = \{3, 7, 10\} , \\ Q_3 = \{5, 6\} \end{array} \tag{9}$$

represents the transformation

$$x'_1 = x_2^2 + 2x_2x_3 + 2x_2, x_4 ,$$
$$x'_2 = x_3^2 + 2x_1x_4 + 2x_3, x_4 ,$$
$$x'_3 = 2x_1x_2 + 2x_1x_3 , \tag{10}$$
$$x'_4 = x_1^2 + x_4^2 .$$

This notation will be used extensively throughout the paper.

2. Limit Set Terminology. By a *limit set* $L_p(T)$ we shall mean the set of all points of the region of definition* which are limit points,

---

* For cubics this is the S, a triangle introduced in section I; for quadratics it is the tetrahedron

$$\sum_{i=1}^{4} x_i = 1, \ x_i \geq 0 .$$

in the ordinary sense, of the set $T^n(p)$, $n = 1, 2, \ldots$, for fixed p. It may happen that $L_p(T)$ is independent of the initial point p; $L_p(T) \equiv L(T)$ could then be called the limit set of T. In general, $L(T)$ will only be defined for interior points p, since points on the boundary frequently* behave in a rather special way.

Thus, for example, if p_0 is a unique fixed point of the transformation: $T(p_0) = p_0$, and if the iterated images $T^n(p)$ of all interior points p converge to p_0, then $L(T) \equiv \{p_0\}$. If $p_0, p_1, \ldots, p_{k-1}$ form a system S of k points such that $T(p_i) = p_{i+1}$, $i = 1, 2, \ldots$ (mod k), and if for all interior points p, $\lim_{n \to \infty} T^{nk}(p)$ is one of these points, then $L_p(T) \equiv S$.

It might happen that the interior points divide into a finite number of classes C_1, C_2, $\ldots$, C_r such that all points p belonging to the same class $L_p(T)$ forms the same set; we should then have a finite number of limit sets L_1, L_2, $\ldots$, L_r. Some of these may contain a finite number of points, others may be infinite. For convenience we shall usually refer to a finite limit set containing k distinct points as a *period of order k*.

Although a given finite limit set belonging to some transformation T may legitimately be considered a "property" of that transformation, it is in no sense characteristic; many different transformations of our type may possess the same limit set, even for the same set of initial points. It should also be stressed that not every set of points $S \equiv \{p_1, \ldots, p_k\}$ such that $T(p_i) = p_{i+1}$ (i mod k) is properly a limit set. Such a set of points, each of which is a solution of the equation $T^k(p) = p$, must have the additional property that there exists a set of initial points whose iterated images converge to S. Finite sets S which have this property are conventionally termed *attractive*. Thus, we should properly refer to a finite limit set of k points as an *attractive period of order k*. In the sequel we shall usually omit the word *attractive* when the context makes it clear that this is what is meant.

There is, of course, no *structure problem* so far as finite limit sets are concerned; they are completely described by giving the coordinates of their constituent points. For infinite limit sets, the situation is different. On the basis of our numerical work alone, we cannot say with certainty that our transformations have such limit sets; the sets may in fact be finite (with an enormous number of points in them), but the presumption that they are infinite is very strong. For any observed *infinite* limit set we can at most say that it is not a period of order less than some very large k. Granting, however, that we are dealing with infinite sets, and that we may infer some of their properties by

---

* i.e., often enough to make it worthwhile excluding them in the definition of $L(T)$.

examining a sufficiently large finite subset* we may attempt to classify them according to their macroscopic morphological properties.

3. Infinite Limit Sets for Cubics in Three Variables. On the basis of our empirical study of cubic transformations, we may make a rough division of infinite limit sets into four classes:

Class I. This includes all limit sets that appear to have the form of one or more closed curves. Figures 3 through 6 will serve as examples of this class. The detailed structure of these "curves" has been studied numerically in some cases, but there are as yet no theoretical arguments to the effect that these are really one-dimensional continua.

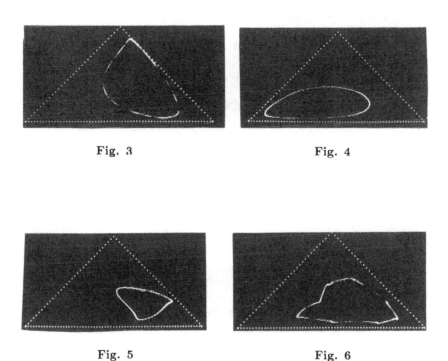

Fig. 3   Fig. 4

Fig. 5   Fig. 6

---

* This assumption underlies all our numerical work.

To illustrate one type of numerical study that we have carried out on these limit sets we cite the case of figure 3. This shows the "infinite" limit set $L(T)$ belonging to the transformations:

$$T_{C_1 C_2} : \begin{array}{l} C_1 = \{2, 5, 7, 9, 10\} \ , \\ C_2 = \{1, 3, 6, 8\} \ . \end{array} \tag{11}$$

In the S, a coordinates, this takes the form:

$$
\begin{aligned}
S' &= \frac{3}{2} S^3 - \frac{3}{2} S^2 a - \frac{15}{2} Sa^2 + \frac{23}{2} a^3 - 3S^2 - 3a^2 + \frac{3}{2} S + \frac{3}{2} a + \frac{1}{2} \ , \\
a' &= -\frac{3}{2} S^3 + \frac{3}{2} S^2 a - \frac{9}{2} Sa^2 + \frac{1}{2} a^3 + 3S^2 + 3a^2 - \frac{3}{2} S - \frac{3}{2} a + \frac{1}{2} \ .
\end{aligned} \tag{12}
$$

There is a (repellent) fixed point at:

$$S_0 = 0.6149341, \quad a_0 = 0.1943821 \ . \tag{13}$$

To six decimal places, the overall bounds on the curve are*

$$
\begin{aligned}
S_{\max} &= 0.816878 \text{ at } a = 0.058022 \ , \\
S_{\min} &= 0.411270 \text{ at } a = 0.204391 \ , \\
a_{\max} &= 0.435861 \text{ at } S = 0.552246 \ , \\
a_{\min} &= 0.017750 \text{ at } S = 0.728386 \ .
\end{aligned} \tag{14}
$$

To five decimal places, the average value of coordinates is found to be*

$$\bar{S} \equiv \frac{1}{N} \sum_{i=1}^{N} S^{(i)} = 0.62231, \quad \bar{a} \equiv \frac{1}{N} \sum_{i=1}^{N} a^{(i)} = 0.20772 \ . \tag{15}$$

This set $L(T)$ is the only infinite limit set the transformation seems to possess [the pair $(S = 1, a = 0)$, $(S = 1/2, a = 1/2)$ turns out to be an attractive period for this transformation.] For "most" initial points, the sequence of iterates converges to $L(T)$. If we choose as our initial point some $p \subset L(T)$, the curve will be traced out by successive images of p, though not in a continuous fashion. If, however, we look only at successive iterates of the 71st power of T: T^{71}, the curve is indeed generated in a relatively continuous fashion; the successive points $T^{(71n)}(p)$, $n = 1, 2, \ldots$, lie close to each other and trace the curve in a clockwise sense. This is illustrated in figure 7, where 246 successive values of $T^{(71)}(p)$ are plotted. It is striking that the nonuniform density of

---

* These results were obtained by carrying out $N = 9600$ iterations, starting from a point on the curve with coordinates: $S = 0.5841326$, $a = 0.4125823$.

points along the curve—as shown in figure 3—is reproduced by this sequence of iterates. We are thus led to the conjecture that $L(T)$ and $L(T^{(71)})$ coincide.

It is, of course, by no means generally true that $T^{(k)}$ and T will have the same limit set for an arbitrary T of our type (cf. the case of periodic limit sets where k is a multiple of the period). Further experiments have convinced us that $L(T) \equiv L(T^{(k)})$ for all k in this case. If this is so, the set is certainly infinite. That it is a continuum is also very probable.

The presumption that $L(T)$ is one-dimensional is supported by the following experiment. We choose a point p_0 which seems to lie, with all available precision, on a convex portion* of the curve, and obtain 100000 iterated images of it, keeping track of those iterates which lie closest to p_0. We find that the two points p_1 and p_2 of closest approach lie in opposite quadrants with respect to p_0, and that the slopes of the two line segments (p_1, p_0) and (p_2, p_0) are the same to within a fraction of a percent. This suggests (1) that the limit set is a curve, and (2) that the curve probably has a continuous derivative at p_0.**

Limit sets consisting of several separate curves (figures 8, 9, 10) may in principle be treated in the same manner, although it is then no longer true that $T^{(k)}$ will have the same limit set as T for all k. For example, if $L(T)$ consists of three separate curves, $L(T^{(3)})$ will coincide with only one of these—which curve depends, of course, on the initial point.

Class II. This class consists of those infinite limit sets all points of which lie on a pair of boundaries of the (S, a) triangle. Alternate iterates lie on alternate sides, hence the square of the transformation will have a limit set confined to one side of the triangle. $T^{(2)}(p)$ is then strictly one-dimensional for all p situated on one or the other of the two sides in question. The situation is illustrated in figures 11 and 12. There seem to be only a few such one-dimensional limit sets possible within our class of cubic transformations. Correspondingly, many different cubic transformations lead to the same pair of one-dimensional transformations when the set of initial points is restricted to a pair of sides of the (S, a) triangle.

---

* Overall convexity is rarely, if ever, a property of these limit sets.

** We do not conjecture that the derivative exists at every point, but we think it likely that the number of points where the derivative does not exist is at most a set of measure zero.

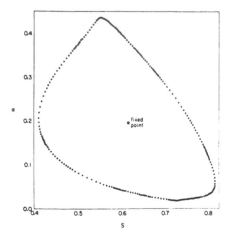

Fig. 7

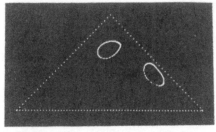

Fig. 8

Fig. 9

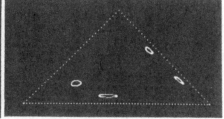

Fig. 10

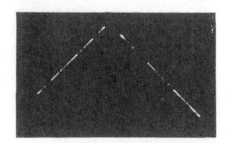

Fig. 11

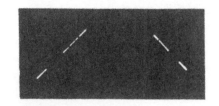

Fig. 12

For example, every transformation of the form:

$$x_1' = x_3^3 + 3a_1x_1x_3^2 + 3x_2x_3^2 + 3b_1x_3x_1^2 + 3x_3x_2^2 + 6c_1x_1x_2x_3 \ ,$$
$$x_2' = x_1^3 + x_2^3 + 3a_2x_1x_3^2 + 3b_2x_3x_1^2 + 6c_2x_1x_2x_3 \ , \qquad (16)$$
$$x_3' = 3x_1x_2^2 + 3a_3x_1x_3^2 + 3x_2x_1^2 + 3b_3x_3x_1^2 + 6c_3x_1x_2x_3 \ ,$$

with non-negative a_i, b_i, c_i satisfying:

$$\sum a_i = \sum b_i = \sum c_i = 1 \qquad (17)$$

will lead to the pair of one-variable polynomial transformations of 6th order:

$$y' = w[3 - 3w + w^2], \quad w \equiv 3y(1 - y) \ , \qquad (18)$$

$$u' = 3v(1 - v), \quad v \equiv u[3 - 3u + u^2] \ . \qquad (19)$$

In other words, transformations T of the form (16) have the property that $T^{(2)}$ transforms each of the lines $x_3 = 0$ and $x_i = 0$ into subsets of themselves (in the S, a coordinates, these lines are respectively the boundaries $S + a = 1$ and $S = a$). The study of such one-dimensional transformations is much easier than that of the original plane transformations, but there are certain serious computational pitfalls connected with high-order iteration (see appendix I).

Class III. The limit sets constituting this class will be referred to as *pseudo-periods*. They consist of relatively dense clusters of points localized at a finite set of *centers*, with a few scattered points in between (figure 13). Such limit sets have not been observed for our original cubic transformations with integer coefficients; they are, however, a prominent feature of the more general transformations discussed in sections III and IV.

Class IV. In this class we place all infinite limit sets not included in the first three classes. Viewed on the oscilloscope they appear as very complicated distributions of points with no recognizable orderly structure. Some examples are shown in figures 14 through 17. A few other examples will be discussed in detail in the following sections. For illustrative purposes, however, we include here a few remarks about figure 17.

This limit set belongs to the transformation:

$$T_{C_1C_2} : \begin{array}{l} C_1 = \{3, 4, 6, 7, 9, 10\} \ , \\ C_2 = \{5, 8\} \ . \end{array} \qquad (20)$$

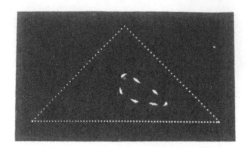

Fig. 13

Fig. 14

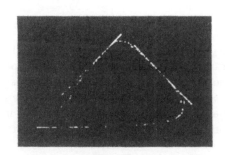

Fig. 15

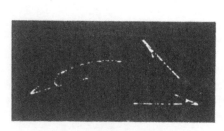

Fig. 16

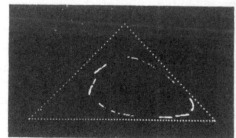

Fig. 17

As is evident from the photograph, it consists of seven separated pieces; each of these is invariant under the 7th power of the transformation. Extensive experimentation indicates that the gaps are *really there.* There appears to be no orderly structure within the separate pieces; in figure 18 we show about 385 consecutive images $T^{(7)}(p)$ (in the upper left-hand piece of the limit set) of some p lying in this subset.

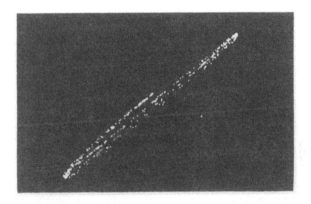

Fig. 18

4. Statistical Observations.

a. A large majority of our 9370 cubic transformations in three variables—some 75 per cent exhibited what might be called *simple* convergence for all initial points tried. For these the limit sets consist of a single point, i.e., a fixed point of the transformation. In many cases there are two such attractive fixed points, but we have not found a case in which both such points are interior to the (S, a) triangle.

We exclude here a few trivial cases such as the following. Consider

$$T_{C_1 C_2} : \quad \begin{aligned} C_1 &= \{1, 2, 3, 4, 5, 10\} , \\ C_2 &= \{6, 9\} . \end{aligned} \tag{21}$$

Explicitly, the second and third line read:

$$\begin{aligned} x_2' &= 3x_2(x_1^2 + x_2 x_3) , \\ x_3' &= 3x_3(x_1^2 + x_2 x_3) . \end{aligned} \tag{22}$$

Thus

$$\frac{x_2'}{x_3'} = \frac{x_2}{x_3} , \tag{23}$$

so that this ratio is fixed by the initial value, and we have a continuum of fixed points. Setting $x_2/x_3 \equiv r$, we find that the fixed point is given by

$$x_3 = \frac{1 + r - \sqrt{(1 + r^2)/3}}{1 + 3r + r^3} \tag{24}$$

with, of course,

$$x_2 = rx_3, \quad x_1 = 1 - (1 + r)x_3 . \tag{25}$$

If we consider the transformation derived from the above by interchanging the right-hand sides of (22), we shall have:

$$\frac{x_2''}{x_3''} = \frac{x_3'}{x_2'} = \frac{x_2}{x_3} , \tag{26}$$

yielding a corresponding continuum of limit sets which are periods of order two.

b. About 16.5 percent of the transformations seem to have only finite (periodic) limits sets; not surprisingly, most of these are of order two. More than half of the latter are of a *trivial* nature, that is, two vertices of the triangle permute under T. Less than 20 cases were found for which the limit set was a period of order $k > 3$. High-order periods are, however, frequently encountered in the study of the generalized transformations discussed in sections III and IV.

c. Some 5 percent of the cases were found to have several (i.e., two, rarely three) distinct finite limit sets of the types described above. For a given transformation it would in principle be possible to determine numerically the set of initial points whose iterated images converge to a particular one of the several limit sets; lack of time has prevented us from doing this except in a few cases. We only remark that there is in general no reason to suppose that the boundary of such a set of initial points is simple.

d. The remaining 3.5 percent, some 334 transformations, possess infinite limit sets. Most of these (roughly three-quarters of them) belong to class I, that is, they look like closed curves. Perhaps 5 percent of the rest belong to class II, the 20 percent residuum being of class IV type. As mentioned above, no examples of class III (pseudo-periods) limit sets were encountered in the study of our original group of cubic transformations (i.e., those with integer coefficients 1, 3, or 6).

e. No case has been found in which a transformation has two distinct class IV limit sets, although there are cases where one of several limit sets was of class IV type. One such has already been described in section I (page 16); a more complicated example will be mentioned in section IV below.

f. We can say very little about the rate of convergence of a sequence $T^{(n)}(p)$ to its $L_p(T)$. Sometimes it may be extremely rapid (10 to 20 iterations); in other cases many thousands of iterates may be required. If $L_p(T)$ consists of a single point, $L_p(T) \equiv \{p_0\}$, this rate can, of course, be calculated (for points sufficiently close to p_0) by solving the approximate, linear difference equations explicitly.

This is, however, not always sufficient. If the jacobian matrix, evaluated at p_0, has complex roots, and $|\lambda^2| = 1$, the linear difference equations may generate an invariant ellipse. Such a case was found in one of our quadratics in three variables, and is discussed in our report on that work. In the S, a coordinates, this transformation is

$$S' = 1 - 4a + 4a^2 + 2aS, a' = 2aS \qquad (27)$$

with fixed point at

$$S = \frac{1}{2}, \ a = \frac{1}{4} \ . \qquad (28)$$

Letting

$$x = S - \frac{1}{2}, \ y = a - \frac{1}{4} \ , \qquad (29)$$

the linear approximation is

$$x' = -y + \frac{1}{2}x, \ y' = y + \frac{1}{2}x \ . \qquad (30)$$

This then generates the invariant ellipse:

$$x'^2 + x'y' + 2y'^2 = x^2 + xy + 2y^2 \ . \qquad (31)$$

In fact, however, for the full (nonlinear) transformation, the fixed point is attractive.

5. Limit Sets for Quadratic Transformations in Four Variables.

All 34337 distinct systems of this type have been investigated on the STRETCH computer, as described in section I above.* A preliminary survey of the results indicates that only about 2 percent of

---

*   The computing time required for the whole study was only a fraction of what one would predict on the basis of 7 seconds for 10^5 iterations—an average for these recurrence relations as actually coded—because a large majority of cases "converged" in a few (~50) steps.

these transformations possess infinite limit sets. The finite limit sets need no special comment; they are of the same sort as those found in cubics in three variables—except, of course, that they are not in general plane sets. A few periods of rather high-order (more than 100 points!) were found, as well as a fair number of cases with 10 to 80 points. This probably should be expected in view of the greater variety of possible algebraic structures.

We are not yet in a position to classify the infinite limit sets as we have in the case of the cubics. Perhaps the closest analogy to sets of the class I type are those which appear to be closed curves in space. These are illustrated in figures 19 through 23. They are shown in convenient projections; the "coordinate system" in the center of the picture merely indicates the orientation relative to the viewer, who is conceived of as stationed at a certain distance from the origin along the y axis.* Figure 19 shows a limit set belonging to the transformation:

$$T_{Q_1 Q_2 Q_3} : \begin{matrix} Q_1 = \{1,3,4\} \,, \\ Q_2 = \{5,6,8\} \,, \\ Q_3 = \{7,10\} \,. \end{matrix} \qquad (32)$$

Presumably what one sees is a twisted space curve.

In figure 20, the limit set consists of two plane curves, one of which lies in the (x_1, x_3) plane, the other lying in a plane inclined at 45^0 with respect to the first. The corresponding transformation is

$$T_{Q_1 Q_2 Q_3} : \begin{matrix} Q_1 = \{1,2,9\} \,, \\ Q_2 = \{4,7,10\} \,, \\ Q_3 = \{3,5,6\} \,. \end{matrix} \qquad (33)$$

The observed limit set is at least consistent with the fact that (33) evidently transforms these planes into each other (so that the points lie alternately on the separate curves). More complicated twisted curves are possible (figure 21). We have also found quite implausible looking limit sets like that shown in figure 22. As a final example, we cite the transformation:

$$T_{Q_1 Q_2 Q_3} : \begin{matrix} Q_1 = \{1,7,9\} \,, \\ Q_2 = \{3,4,8\} \,, \\ Q_3 = \{2,5,10\} \,. \end{matrix} \qquad (34)$$

---

* The "reference system" (x, y, z) is parallel to, but displaced relative to, the actual coordinate system (x_1, x_2, x_3). The origin of the (x, y, z) system is in the (approximate) center of the picture; that of the (x_1, x_2, x_3) system is in the lower left-hand corner.

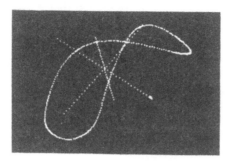

Fig. 19

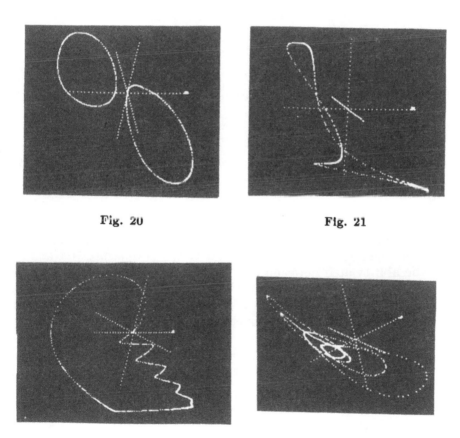

Fig. 20   Fig. 21

Fig. 22   Fig. 23

This has at least two infinite limit sets, one of which may be of class IV type (not shown); the other (figure 23) is a "curve" of unknown structure.

At the time of this writing (January, 1963) we are unable to say anything more specific about the limit sets for quadratics in four variables; to date, less than one-third of the seven hundred or so potentially "interesting" cases have been looked at on the oscilloscope.

III. The "Δt-Modification"

1. We discuss here a particular one-parameter generalization of our original cubic transformations in three variables which we have called the Δt-modification.* It consists in replacing the usual difference equations:**

$$x_i' = \sum_{j=1}^{10} d_{ij} M_j, \ i = 1, 2, 3 \ , \tag{1}$$

by

$$x_i' = (1 - \Delta t)x_i + \Delta t \sum_{j=1}^{10} d_{ij} M_j \ , \tag{2}$$

with

$$0 < \Delta t \le 1 \ . \tag{3}$$

If $\Delta t = 1$, we recover the original set (1); $\Delta t = 0$ is excluded, since the equations then become the trivial identity transformation.

The abbreviated notation of section II is extended in an obvious manner to cover this case. Thus, if (1) is represented symbolically by $T_{C_1 C_2}$, (2) may be symbolized by $T_{C_1 C_2 (\Delta t)}$. For a given p, the limit set will correspondingly be denoted by $L_p(T)$, $L_{p(\Delta t)}(T)$.

---

* This has already been mentioned in the Introduction, equation (12). The modification can, of course, be introduced for the general case [equations (10) through (13) of the Introduction.]

** The M_j are defined in section II, subsection 1, equation (4). Unless otherwise stated, equations (2) and (3) of section II are assumed to hold, as well as the condition

$$\sum x_i = 1, \ x_i \ge 0, \text{ for all initial points } p = (x_1, x_2, x_3) \ .$$

In the S, a coordinates, (1) appears in the form

$$S' = F(S, a), \quad a' = G(S, a) . \tag{4}$$

Correspondingly, (2) reads

$$S' = (1 - \Delta t)S + \Delta t F(S, a), \quad a' = (1 - \Delta t)a + \Delta t G(S, a) . \tag{5}$$

It is clear that the fixed points of (5) coincide with those of (4). As we shall see below, this fact enables us to find these fixed points by simple iteration, thus avoiding the unpleasant algebra involved in eliminating one variable from the pair of general cubics

$$S = F(S, a), \quad a = G(S, a) , \tag{6}$$

and then solving for the roots of the resulting high-order (≤ 9) polynomial. One can look upon (5) at the simplest (and most naive) finite difference scheme for approximating the first-order differential system

$$\frac{dS}{dt} = -S + F(S, a), \quad \frac{da}{dt} = -a + G(S, a) . \tag{7}$$

The analogy between (5) and (7) is not, however, very close;* consequently it is better to discuss (5) on its own merits. The effect of setting $\Delta t < 1$ (but > 0) on a single iteration is easy to see. Let us take a particular point S, a; the image produced by (4) will be denoted, as usual, by S', a', while we shall call the corresponding image under (5) S' mod, a' mod. Then

$$S' \text{ mod} - S' = \Delta t(S' - S), \quad a' \text{ mod} - a' = \Delta t(a' - a) . \tag{8}$$

In other words, the length of the iterative step is altered, while the direction remains the same. What happens on repeated iteration is, however, not all obvious. One expects that the limit set $L_{p(\Delta t)}(T)$ will in general have smaller diameter as we decrease Δt, but we cannot at present predict its structure as a function of Δt, even relative to the (observed) structure of $L_p(T)$. It is worthwhile illustrating this in a particular case. Consider the transformation.

$$T_{C_1 C_2} : \begin{array}{l} C_1 = \{3, 5, 7, 9, 10\} , \\ C_2 = \{1, 2, 8\} . \end{array} \tag{9}$$

---

*   For further discussion on this point see section V.

In the S, a coordinates, this reads explicitly:

$$T_A : \begin{array}{l} S' = S^3 - 6S^2a - 3Sa^2 + 4a^3 - \dfrac{3}{2}S^2 + 3Sa - \dfrac{3}{2}a^2 + 1 . \\[2mm] a' = -S^3 + 3Sa^2 + 2a^3 + \dfrac{3}{2}S^2 - 3Sa + \dfrac{3}{2}a^2 . \end{array} \qquad (A)$$

Since we shall refer to this transformation quite often in the sequel, we have given it the distinctive label (A). T_A has one interior* fixed point (repellent), whose coordinates are:

$$S_0 = 0.5885696, \quad a_0 = 0.1388662 . \qquad (10)$$

There are two infinite limit sets; these are shown in figure 24 and in figure 11 of section II. At the moment, we shall not be concerned with the limit set shown in figure 11; this is evidently of class II type and can therefore be studied in one-dimensional form.** The limit set shown in figure 24—which we shall henceforth refer to as $L(T_A)$—appears as an irregular pattern surrounding the fixed point (shown superimposed on the picture). Figure 25 again shows $L(T_A)$—this time enlarged by a factor 3, while figure 26 shows a portion of the upper left-hand corner*** enlarged about 14 times. Figure 24 shows 900 consecutive iterates, while figure 25 shows these same 900 points plus 1800 more. For comparison, in figure 27 we plot just 50 consecutive iterates. The approximate outer dimensions of $L(T_A)$ are[†]

$$\begin{array}{llll} S_{\max} & = & 0.754696 \text{ at } a & = & 0.077251 , \\ S_{\min} & = & 0.443911 \text{ at } a & = & 0.204610 , \\ a_{\max} & = & 0.277406 \text{ at } S & = & 0.491266 , \\ a_{\min} & = & 0.071196 \text{ at } S & = & 0.739170 . \end{array} \qquad (11)$$

We now contrast with $L(T_A)$ the limit sets $L_{(\Delta t)}(T_A)$ belonging to the generalized transformation $T_{A(\Delta t)}$. If we set $\Delta t = 0.9931$, we get a limit set entirely different from $L(T_A)$ (from the same initial point). This is shown in figure 28. It exhibits what we have called "pseudo-periodic" structure, that is, almost all the iterated images

---

* There is another repellent fixed point at a vertex of the triangle, namely $S = a = \frac{1}{2}$.

** See section II.

*** This is the region $0.455 \leq S \leq 0.525$, $0.225 \leq a \leq 0.278$.

[†] These results are based on a calculation with $N = 9600$ iterations.

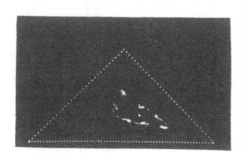

Fig. ~ ·

Fig. 26   Fig. 27

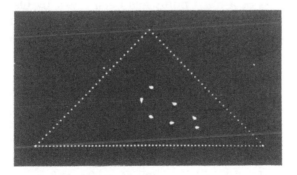

Fig. 28

of the initial point p are concentrated in the neighborhood of seven distinct "centers"—an example of class III limit set.*

With a very small change in Δt—namely, by setting $\Delta t = 0.9930$—we find instead a period of order 7. This is shown in figure 29. As we decrease Δt in small steps down to $\Delta t = 0.9772$ (figure 30), the corresponding $L_{(\Delta t)}(T_A)$ remains a period of order 7; the coordinates of the individual points appear to change continuously with Δt. For $\Delta t = 0.97713$, $L_{(\Delta t)}(T_A)$ is again a pseudo-period, and this character persists down to $\Delta t = 0.9770$ (figure 13 of section II). Below** $\Delta t = 0.9770$ $L_{(\Delta t)}(T_A)$ is a closed curve*** around the fixed point which shrinks in more or less continuous fashion as Δt is decreased. Figures 31, 32, and 33 illustrate, respectively, the limit sets for $\Delta t = 0.97$, 0.94, and 0.92. Finally, for $\Delta t < 0.9180154$ (see below), the limit set consists of a single point—the fixed point (10).† This peculiar behavior of $L_{(\Delta t)}(T_A)$ as a function of Δt is not an isolated instance, nor is it by any means among the most extreme examples we have encountered (see section IV for a considerably more "pathological" case). Within the class of cubic transformations we have studied, it seems to be an empirical rule that the more pathological the limit set looks for $\Delta t = 1$, the more complicated will be its behavior as Δt is decreased.

2. Attractive and Repellent Fixed Points. The fixed points of a cubic transformation in the standard form (4) are those real roots of the algebraic system (6) which lie in or on the boundary of the S, a triangle.†† We are interested both in finding the values of the coordinates of these fixed points and in determining whether the points are attractive or repellent. By attractive we mean as always that, for any point p in a sufficiently small neighborhood of the fixed point p_0, the sequence $T^{(n)}(p)$ will converge to p_0. A general criterion for the attractiveness of a fixed point has been given by Ostrowski,[10] viz: let $|\lambda_{max}|$ be the largest eigenvalue in absolute value of the jacobian matrix evaluated at the fixed point. Then if $|\lambda_{max}| < 1$, the point

---

* See section II for this classification scheme.

** We have not attempted to find the *critical* values of Δt with greater precision, though this could in principle be done to, say, 7 decimal places on MANIAC II.

*** This is, of course, only a conjecture. See the discussion in section II.

† In the language of functional analysis, $T_{A(\Delta t)}$ is a *shrinking operator* in this range of Δt values.

†† Brouwer's theorem assures us that there is at least one fixed point.

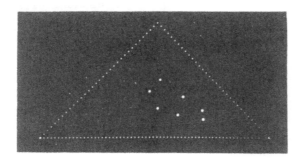

Fig. 29

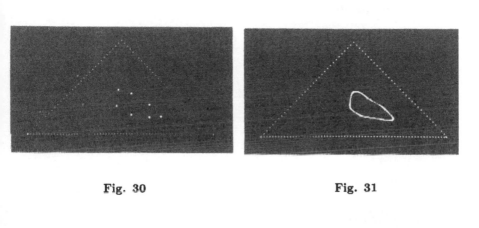

Fig. 30   Fig. 31

Fig. 32   Fig. 33

is attractive; if $|\lambda_{max}| > 1$ the point is repellent. The theorem says nothing about the case nor $|\lambda_{max}| = 1$, nor does it yield a method for determining theoretically the appropriate neighborhood. For the two-variable transformation (4), we may give the eigenvalues of the jacobian matrix explicitly:

$$\lambda = \frac{T_0 \pm \sqrt{T_0^2 - 4J_0}}{2} \ , \tag{12}$$

where T_0 is the trace:

$$T_0 = \frac{\partial F}{\partial S} + \frac{\partial G}{\partial a}\Big|_{\substack{S=S_0 \\ a=a_0}} \ , \tag{13}$$

and J_0 is the jacobian:

$$J_0 = \frac{\partial F}{\partial S}\frac{\partial G}{\partial a}\frac{\partial G}{\partial S}\frac{\partial F}{\partial a}\Big|_{\substack{S=S_0 \\ a=a_0}} . \tag{14}$$

For the modified system (5), we find correspondingly:

$$\lambda_{mod} = 1 - \Delta t + \frac{\Delta t}{2}\left(T_0 \pm \sqrt{T_0^2 - 4J_0}\right) . \tag{15}$$

If the roots are complex, i.e., if $T_0^2 < 4J_0$, we have

$$|\lambda_{mod}|^2 = 1 - \Delta t(2 - T_0) + \Delta t^2(1 - T_0 + J_0) . \tag{16}$$

Defining Δt_{lim} as the value of Δt for which $|\lambda_{mod}|^2 = 1$, we obtain

$$\Delta t_{lim} = \frac{2 - T_0}{1 - T_0 + J_0} \ . \tag{17}$$

Thus, for the case of complex roots, we may make the fixed point (S_0, a_0) attractive by choosing Δt such that

$$0 < \Delta t \leq \Delta t_{lim} . \tag{18}$$

Similarly, if λ_{max} is real and negative, and $|\lambda_{max}| > 1$,

$$\Delta t_{lim} = \frac{2}{1 + |\gamma_{max}|} \ . \tag{19}$$

It is clear that this artifice will not work if $\lambda_{\max} \geq 1$. Such a situation arises in the one-dimensional case:

$$x_1' = x_1^3 + 3x_1^2 x_2, \quad x_2' = x_2^3 + 3x_2^2 x_1, \quad x_1 + x_2 = 1 , \tag{20}$$

that is

$$x_1' = x_1^2(3 - 2x_i) . \tag{21}$$

The fixed points are $x_1 = 0$, $1/2$, 1; at these points the derivative $dx_1'|dx_1$, has the values 0, $3/2$, 0. Clearly both $x_1 = 0$ and $x_1 = 1$ are attractive fixed points; for all $x_1^{(0)} > 1/2$, $x_1^{(n)} \to 0$, while for all $x_1^{(0)} > 1/2$, $x_1^{(n)} \to 1$. The interior fixed point $x_1 = 1/2$ is repellent and cannot be made attractive by using the Δt-*modification*. The corresponding situation does not seem to occur for any of our cubic transformations in the plane.

In practice, all one has to do to obtain the numerical value of a repellent fixed point is to choose a sufficiently small Δt and iterate; on a computer, this calculation requires only a few seconds.*

3. Attractive Periods. The set of points constituting a period of order k are fixed points of $T^{(k)}$. Thus one may test whether a periodic limit set is attractive by applying Ostrowski's criterion to $T^{(k)}$. Let

$$J^{(n)} \equiv \begin{bmatrix} \frac{\partial S^{(n)}}{\partial S} & \frac{\partial S^{(n)}}{\partial a} \\ \frac{\partial a^{(n)}}{\partial S} & \frac{\partial a^{(n)}}{\partial a} \end{bmatrix}_{(S^0, \, a^0)} , \tag{22}$$

$$J_{n-1} \equiv \begin{bmatrix} \frac{\partial S^{(n)}}{\partial S^{(n-1)}} & \frac{\partial S^{(n)}}{\partial a^{(n-1)}} \\ \frac{\partial a^{(n)}}{\partial S^{(n-1)}} & \frac{\partial a^{(n)}}{\partial a^{(n-1)}} \end{bmatrix}_{S^{(n-1)}, \, a^{(n-1)}} , \tag{23}$$

where, e.g., $(S^0, \, a^0)$ is a fixed point of $T^{(k)}$.

Then, by the chain rule:

$$J^{(k)} = J_{k-1} \times J_{k-2} \times \ldots \times J_0 . \tag{24}$$

Thus $J^{(k)}$ is easily obtained by evaluating (24) over the periodic set in question; the application of Ostrowski's criterion is then immediate.

---

* Early in this investigation we made the "mistake" of taking $\Delta t > \Delta t_{\lim}$ in a few cases, and thereby discovered the interesting limit sets $L_{p(\Delta t)}(T)$.

We have often used this technique to convince ourselves that the periods are really limit sets and not the result of spurious or accidental convergence.*

IV. Modification of the Coefficients

1. In this section we present some results on the effect of modifying the original integer-valued coefficients of our cubic transformations in three variables. That is, we consider, as before, transformations of the form:

$$x_i' = \sum_{j=1}^{10} d_{ij} M_j , \tag{1}$$

$$\sum_{i=1}^{3} d_{ij} = 1 , \tag{2}$$

but we no longer require that the d_{ij} all be 1 or 0. As already remarked in the introduction, if we impose on the d_{ij} only the additional condition:

$$0 \le d_{ij} \le 1 \tag{3}$$

then (1), (2), and (3) define a class of cubic transformations depending on 20 parameters, e.g., d_{1j}, d_{2j} ($j = 1$ to 10).** Since we are unable to formulate a complete theory for the finite subclass of transformations characterized by the restrictions: $d_{ij} = 0$ or 1, all i, j, it is clear *a fortiori* that we do not have a theory for the infinite class.

In this paper we limit ourselves to showing how an experimental study of some special cases can help to throw light on the properties of our original cubic transformations.

In effect, what we do is study certain transformations which are "close to" some particular transformations of our basic type. A natural

---

* This technique has actually been used for periods with orders k as large as 148. For very large k the method might fail owing to round-off errors or other numerical inaccuracies.

** For the definition of the cubic monomials M_j, see equation (4) of section II. The domain of this class of transformations is again the region

$$\sum_{i=1}^{3} x_i = 1, \ 0 \le x_i \le 1 .$$

way to define a transformation close to some given $T_{C_1 C_2}$ would be to choose its coefficients as follows:

$$d_{ij} = 1 - \epsilon_{ij}, \quad j \subset C_i ,$$

$$d_{ij} = \epsilon_{ij}, \qquad j \not\subset C_i , \tag{4}$$

where the d_{ij} must satisfy (2) and (3), and the ϵ_{ij} are small. This class of transformations, defined with respect to some $T_{C_1 C_2}$, is still too extensive to study, even if the various ϵ_{ij} are restricted to a few discrete values. What we have actually done is to consider 20 such transformations, each of which depends on a single parameter ϵ. We denote these by the symbol:

$$T_{(r,s)\epsilon}, \ 1 \leq r \leq 10, \ 0 \leq \epsilon \leq 1, \ s = 0, 1 . \tag{5}$$

It is understood that these transformations are only defined relative to some $T_{C_1 C_2}$. For convenience we shall generally refer to the transformations $T_{(r,s)\epsilon}$ defined relative to some $T_{C_1 C_2}$ as *associated* transformations. The coefficients of the $T_{(r,s)\epsilon}$ are specified as follows:

For $j \neq r$:

$$\begin{aligned} d_{ij} &= 1, \quad j \subset C_i , \\ d_{ij} &= 0, \quad j \not\subset C_i . \end{aligned} \tag{6}$$

For $j = r$:

$$\begin{aligned} d_{ij} &= 1 - \epsilon, & r \subset C_1 , \\ d_{1r} &= (1-s)\epsilon, & r \not\subset C_1 , \\ d_{2r} &= 1 - \epsilon, & r \subset C_2 , \\ d_{2r} &= s\epsilon, & r \not\subset C_2 . \end{aligned} \tag{7}$$

In words: $T_{(r,s)\epsilon}$ is formed from $T_{C_1 C_2}$ by the replacement $M_r \rightarrow (1 - \epsilon)M_r$ wherever the term M_r occurs, and by adding ϵM_r to one of the other two lines of the three-line schema. As an example, consider the transformation T_A introduced in section III:

$$T_A : \begin{array}{l} C_1 = \{3, 5, 8, 9, 10\} , \\ C_2 = \{1, 2, 8\} . \end{array} \tag{8}$$

Relative to T_A, $T_{(5,1)\epsilon}$ would read:

$$\begin{aligned} x_1' &= M_3 + (1 - \epsilon)M_5 + M_7 + M_9 + M_{10}, \\ x_2' &= M_1 + M_2 + \epsilon M_5 + M_8 , \\ x_3' &= M_4 + M_6 , \end{aligned} \tag{9}$$

while $T_{(4,0)\epsilon}$ would take the form:

$$x_1' = M_3 + \epsilon M_4 + M_5 + M_7 + M_9 + M_{10} ,$$
$$x_2' = M_1 + M_2 + M_8 ,$$
$$x_3' = (1 - \epsilon)M_4 + M_6 ,$$

(10)

In the S, a coordinates, the $T_{(r,s)\epsilon}$ can be written:

$$S' = F(S,a) + \epsilon f_{rs}(S,a), \quad a' = G(S,a) + \epsilon g_{rs}(S,a) ;$$

(11)

the original $T_{C_1 C_2}$ is obtained from (11) by setting $\epsilon = 0$. For the two examples given above, we have:

$$T_{(5,1)\epsilon} : \quad \begin{aligned} f_{rs} &= f_{51} = -\frac{3}{2}(S - a)(1 - S - a)^2 , \\ g_{51} &= -f_{51} ; \end{aligned}$$

(12)

$$T_{(4,0)\epsilon} : \quad \begin{aligned} f_{rs} &= f_{40} = 12a^2(S - a) , \\ g_{40} &= 0 . \end{aligned}$$

(13)

It turns out that for these one-term modifications $T_{(r,s)\epsilon}$ we always have $g_{rs} = \pm f_{rs}$ or 0. f_{rs} can further be factored into a numerical coefficient c_{rs} and a function $M_r(S,a)$; the M_r are of course just the original cubic monomials expressed in terms of S and a. The c_{rs} and g_{rs} are determined as follows:

For $r \subset C_1$:
$$\begin{array}{llll} s = 0: & g_{rs} = 0, & c_{rs} = -1 , \\ s = 1: & g_{rs} = -f_{rs}, & c_{rs} = -\frac{1}{2} ; \end{array}$$

(14)

for $r \subset C_2$:
$$\begin{array}{llll} s = 0: & g_{rs} = -f_{rs}, & c_{rs} = \frac{1}{2} , \\ s = 1; & g_{rs} = f_{rs}, & c_{rs} = -\frac{1}{2} ; \end{array}$$

(15)

for $r \not\subset c_1, r \not\subset C_2$:
$$\begin{array}{llll} s = 0: & g_{rs} = 0, & c_{rs} = 1 , \\ s = 1: & g_{rs} = f_{rs}, & c_{rs} = \frac{1}{2} . \end{array}$$

(16)

2. We have studied the modified transformations $T_{(r,s)\epsilon}$ for a variety of our original cubic $T_{C_1 C_2}$ that happen to have infinite limit sets.

336

Our usual procedure has been to vary ϵ in steps of $1/100$ in the range $1/100 \leq \epsilon \leq 1/10$ for a given $T_{(r,s)\epsilon}$ relative to a given $T_{C_1 C_2}$, although on occasion intermediate values of ϵ have been used. Only for the transformation T_A [equation (8) above] have we looked at all 20 modified transformations. For a few other $T_{C_1 C_2}$ we have limited ourselves to selecting certain of the associated $T_{(r,s)\epsilon}$ for detailed study. Since this selection has generally been made on intuitive grounds, we cannot claim that the most "interesting" modifications of the original transformations have always been considered. Nevertheless, this part of our study has proved most revealing, especially as regards the structure of class IV limit sets.

Before describing the results, we insert a few remarks on the difference between the two types of generalizations we have considered, the Δt-*modification* of section III and the *associated* transformations $T_{(r,s)\epsilon}$. The Δt-*modification* is essentially nothing but the application of a technique frequently employed in the practical solution of nonlinear equations by iterative methods; it is, in fact, one way—perhaps the simplest—of introducing a linear convergence factor. Apart from our use of this device for obtaining the coordinates of the fixed point, our principal interest is in small convergence factors (Δt close to unity)— too small, in fact, to produce convergence to the fixed point. In view of the fact that the Δt-modified transformation $T_{C_1 C_2(\Delta t)}$ has precisely the same fixed point as the original transformation $T_{C_1 C_2}$, one might expect that there exists a close relationship between the corresponding limit sets $L_{p(\Delta t)}(T)$ and $L_p(T)$. In some sense this is true, as the examples given in section III show (see also below, subsection 4). We may express this more formally as follows:

We define a sequence of transformations $T_{C_1 C_2(\Delta t_i)} \equiv T_{\Delta t_i}$ with corresponding limit sets $L_{p(\Delta t_i)}(T)$ by some convenient rule:

$$\Delta t_0 = \Delta t_{\text{lim}}, \quad \Delta t_i = 1 - \frac{1 - \Delta t_0}{2^i} . \tag{17}$$

The sequence $T_{\Delta t_i}$, $i = 1, 2, \ldots$, clearly converges to $T_{C_1 C_2}$.

We then formulate the following conjecture:

Given a $T_{C_1 C_2}$ and a $\delta > 0$, then, for all p in the triangle, there exists an $N_{(p)}$ such that, for $i > N_{(p)}$ and for all $x \subset L_p(T)$, there exists a $y \subset L_{p(\Delta t_i)}(T)$ satisfying $| y - x | < \delta$.

The modification of $T_{C_1 C_2}$ defined by the associated transformations $T_{(r,s)\epsilon}$ differs from the Δt-*modification* in several respects. In the first place, the perturbation introduced is not linear. Furthermore, the fixed points of $T_{(r,s)\epsilon}$ are in general not the same as those of $T_{C_1 C_2}$ (fixed points on the boundary of the triangle may, of course, be common to $T_{(r,s)\epsilon}$ and $T_{C_1 C_2}$ for some pairs r, s).* Finally, each pair r, s must be treated separately; for fixed ϵ, perturbations of different terms of $T_{C_1 C_2}$ may lead to quite different limit sets. Nevertheless, a conjecture analogous to that formulated for the sequence $T_{\Delta t_i}$ would most probably turn out to be correct.

3. Limits Sets of the Transformations $T_{(r,s)\epsilon}$ Associated with T_A. Since we usually deal with values of ϵ of the form:

$$\epsilon_i = \frac{i}{100}, \ 1 \le i \le 10 , \tag{18}$$

we introduce a symbol to denote a set of such values:

$$I(i,j) \equiv \{\epsilon_n\}, \ i \le n \le j . \tag{19}$$

In addition, $R^k_{(r,s)}$ will denote the closed interval of ϵ:

$$R^k_{(r,s)} = [^- R^k_{(r,s)}, ^+ R^k_{(r,s)}], \ (^- R^k_{(r,s)} \le \epsilon \le ^+ R^k_{(r,s)} \tag{20}$$

for which the limit sets $L_{(r,s)\epsilon}$ of $T_{(r,s)\epsilon}$ are periods of order k. The photographs illustrating the examples that follow will be found, suitably labelled, in subsection 2 of appendix II.

There is one significant feature common to all the $T_{(r,s)\epsilon}$ associated with T_A; for every pair r, s at least one periodic limit set—that is, a period of order $k > 1$—was found in the range $I(1, 10)$. The order of periodicity of most frequent occurrence was $k = 7$. Thus, for example, for $(r, s) = (10,0)$, we found periodic limit sets with $k = 7$ over the range $I(3, 10)$, and the case $(r, s) = (6, 1)$ behaves in the same fashion over the same range. For both series of associated transformations, the limit sets for $\epsilon = 0.01$ are of class IV type and closely resemble $L(T_A)$. At $\epsilon = 0.02$, bright spots show up in the pattern (figure A-1); this usually indicates that one is *near* a period, i.e., that a relatively small change in ϵ will yield a transformation having a finite limit set. In the notation (20), this would be written: $-R^7_{(r,s)} - 0.02 \ll 1$. In

---

* The new fixed point $S_{(r,s)\epsilon} = S + \Delta s$, $a_{(r,s)\epsilon} = a + \Delta a$, calculated to first order in $(\Delta S, \Delta a)$, has both ΔS and Δa proportional to ϵ. The ration $\Delta S/\Delta a$ in this order is therefore independent of ϵ, though not of r, s. There are in fact, 6 possible directions of displacement, two for each of the three cases: $g_{rs} = f_{rs}$, $g_{rs} = -f_{rs}$, $g_{rs} = 0$, cf. equations (14), (15), and (16).

these two examples it happens that a period of order 7 is observed over the range $I(3, 10)$, that is: $I(3, 10) \subset R^7_{(r,s)}$. This is not generally the case. Thus for the case $(r, s) = (9, 0)$, $I(2, 9) \subset R^7_{(9,0)}$, whereas $L_{(9,0)0.01}$ and $L_{(9,0)0.10}$ are of class IV type and are morphologically similar to $L(T_A)$. It may be recalled (section III) that an analogous behavior was observed for the Δt-modified transformations $T_{A(\Delta t)}$, namely that $L_{\Delta t}(T_A)$ was found to be a period of order 7 for a particular range of values of Δt ($0.9930 \leq \Delta t \leq 0.9772$), and different in character (actually, of class III type) outside the range on both sides.

Periodic limit sets of order 7 have been found for some range $I(i, j)$ of ϵ in 9 out of the 20 possible cases. For one of these, $(r, s) = (2, 1)$, $I(4, 7) \subset R^7_{(2,1)}$, while $L_{(2,1)0.10}$ is periodic with $k = 28$ (figures A-2, A-3). In the transition region, i.e., for $^+R^7_{(2,1)} < \epsilon < {}^-R^{28}_{(2,1)}$, the limit sets are infinite. These are shown in figures A-4 and A-5 for the range $I(8, 9)$. They look like pseudo-periods, but, when suitably enlarged, they are seen to be of class I type (figures A-6, A-7). In these pictures one clearly sees with increasing ϵ the onset of instability—to use an expression from mechanics—and the eventual attainment of a different stable state. The transition region at the lower end of the range also contains infinite limit sets. Figures A-8 and A-9 show $L_{(2,1)0.03}$, first to normal scale, then enlarged. It is manifestly a class III limit set.

For other $T_{(r,s)\epsilon}$ periods of order $k > 7$ are found for certain ranges of the parameter ϵ, viz.: $k = 9, 16, 23, 30, 37, 46, 62, 148$. In two cases, two periods of relatively prime order are found in different sub-ranges of $I(1, 10)$. Thus $T_{(5,1)\epsilon}$ has two periodic limit sets, one with $k_1 = 23$ for $\epsilon = 0.01$ and one with $k_2 = 16$ over $I(9, 10)$. Similarly, $T_{(1,1)\epsilon}$ has a periodic limit set with $k_1 = 16$ for $\epsilon = 0.01$, and one with $k_2 = 9$ for $\epsilon = 1.10$. In these cases the dependence of the limit set on ϵ in the transition region $^+R^{k_1}_{(r,s)} < \epsilon < {}^-R^{k_2}_{(r,s)}$ is more complicated than that described above. For ϵ-values in this region and sufficiently close to the end-points we observe the expected pseudo-periodic limit sets. For values of ϵ not too close to either boundary the limit set may be either of class IV or of class I type. Figure A-10 shows $L_{(5,1)0.04}$ to normal scale; in figure A-11, a portion of the limit set is shown enlarged.

We conclude this subsection with two further examples. These illustrate a phenomenon previously mentioned in our general description of limit sets (section II), namely the coexistence of finite periods and class IV sets. Figures A-12 and A-13 show two distinct limit sets belonging to $T_{(3,1)0.01}$. One is a period of order $k = 23$, while the other is a class IV set closely resembling $L(T_A)$. The same phenomenon is perhaps more strikingly illustrated by the case of $T_{(5,0)0.02}$. Here we find both a class IV limit set and a period of order $k = 148$ (figures A-14 and A-15). We can say virtually nothing in this case about the dependence of the limit set on the initial point. Current computing facilities and techniques are not sufficiently powerful to effect an acceptably accurate determination of the respective regions of convergence without using prohibitive amounts of computing time. We have, however, carried out a few numerical experiments, the results of which certainly confirm our first impression that the geometrical structure of these regions is immensely complicated.

4. Study of the Associated Transformations for Other $T_{C_1 C_2}$. In this subsection we discuss a few additional examples to illustrate the dependence of infinite limit sets on the parameter ϵ. The relevant photographs and tables will be found in appendix II.

For our first example, we choose the transformation:

$$T_{C_1 C_2} \equiv T_B : \begin{array}{l} C_1 = \{2, 4, 6, 7, 9\}, \\ C_2 = \{5, 8, 10\} \end{array} \tag{21}$$

The class IV limit set $L(T_B)$ belonging to this transformation is shown in figure B-1. As is evident, it consists of three separate pieces. Each of these is, of course, a limit set for $T_B^{(3)}$. It is instructive to compare the limits sets $L_{\Delta t}(T_B)$ with those belonging to certain of the associated $T_{(r,s)\epsilon}$. In appendix II we list the results for only one case; $(r, s) = (1,0)$. The limit sets $L_{\Delta t}(T_B)$ and $L_{(1,0)\epsilon}$ are described in table B. There are (at least) three ranges of Δt values for which $L_{\Delta t}(T_B)$ is periodic; for Δt close to unity the behavior of $L_{\Delta t}(T_B)$ as a function of Δt is rather *wild*. As Δt approaches $\Delta t_{\lim} = 0.854320$ the (class I) limit set shrinks in a continuous manner. The behavior of $L_{(1,0)\epsilon}$ as ϵ is varied over $I(1, 10)$ is, if anything, more "pathological"; there are at least six different intervals $R_k^{(1,0)}$ for which the limit set is periodic, and each period has a different order. Note the similarity in appearance between the two class IV limit sets: $L_{\Delta t}(T_B)$ $(\Delta t = 0.994)$ and $L_{(1,0)0.01}$.

The next two examples may be taken together:

$$T_D : \begin{array}{l} C_1 = \{2, 7, 8, 9, 10\} \\ C_2 = \{4, 5, 6\} \; ; \end{array} \tag{22}$$

$$T_E : \begin{array}{l} C_1 = \{2, 5, 7, 8, 9\} \\ C_2 = \{4, 5, 10\} \; . \end{array} \tag{23}$$

The basic class IV limit sets $L(T_D)$ and $L(T_E)$ are shown in figures D-1 and E-1; their morphological resemblance is apparent. The behavior of the $L_{\Delta t}$ and $L_{(1,0)\epsilon}$ for these two cases is set forth in the tables and photographs of appendix II. Detailed comment is perhaps superfluous at this state of our knowledge; we limit ourselves to drawing attention to the following comparisons:

1. Compare $L_{\Delta t}(T_D)$ ($\Delta t = 0.97$) with $L_{(1,0)0.10}(T_D)$.
2. Compare $L_{\Delta t}(T_E)$ ($\Delta t = 0.97$) and $\Delta t = 0.96$) with $L_{(1,0)0.09}(T_E)$ and $L_{(1,0)0.10}(T_E)$.

5. The original transformations T_B, T_D, T_E are closely related from the point of view of formal structure. T_D and T_E differ by exchange of a single term between the defining sets C_1 and C_2, while each of these goes over into T_B under the simultaneous interchange of two terms between C_1 and C_2. A comparison of the associated limit sets for T_D and T_E shows that the initial similarity of $L(T_D)$ to $L(T_E)$ is roughly preserved under perturbation. This suggests the possibility that some meaningful classification based on algebraic form might be devised.* Of even greater interest is the correspondence, in these examples, between the $L_{\Delta t}$ and the $L_{(1,0)\epsilon}$ for some ranges of the respective parameters. We are not at present in a position to draw any significant conclusions from the existence of this correspondence; it seems likely, however, that a closer study of these examples would yield criteria enabling one to predict such behavior.

6. There is one property of these transformations which may safely be inferred from the data, namely, that they are close to transformations having periodic limit sets (for some common set of initial points), where *close* is to be interpreted with reference to some appropriate parameter space—e.g., a range of ϵ values of Δt values. Their limit sets are "close" to periods, not in the sense that pseudo-periods are, but rather by virtue of the fact that they contain points which lie close— perhaps arbitrarily close—to a set of algebraic solutions of $T^k(p) = p$. In other words, the Hausdorff distance between the set of period points and the limit set L is small. In this connection, the following piece

---

*   The difference in behavior of $L_{\Delta t}(T_B)$ on the one hand and $L_{\Delta t}(T_D)$, $L_{\Delta t}(T_E)$ on the other is undoubtedly due in part to the fact that in the first case the jacobian matrix has complex eigenvalues at the fixed point, while for T_D and T_E the eigenvalues are real; this is probably sufficient to explain the qualitative difference of behavior of the corresponding $L_{\Delta t}$ as $\Delta t \to \Delta t_{\text{lim}}$ for $\Delta t - \Delta t_{\text{lim}}$ sufficiently small.

of evidence may be presented. Consider the transformation $T_{(1.0)0.01}$ associated with T_E, for which we have observed that the sequence $T_{(1,0)0.01}^{(n)}(p)$, $n = 1, 2, \ldots$, converges to a period of order $k = 10$ for almost all p. Let us choose a p close to the fixed point. If we then examine the sequence for $n = 1, 2, \ldots, N$, where N is sufficiently large, we find that the images $T^{(n)}(p)$ of p have traced out a pattern which closely resembles the original class IV limit set $L(T_E)$ of figure E-1. This is shown in figure E-2. The bright spots are the points belonging to the periodic limit set $L_{1,0)0.01}$. Presumably this means that the effect of introducing a small perturbation into T_E, of the form specified by $T_{(1.0)0.01}$,* is to make the limit set $L(T_E)$ contract to 10 points. Alternatively, we could say that, as $\epsilon \to 0$, the periodic limit set $L_{(1,0)\epsilon}(T_E)$ *spreads out* until it becomes the class IV limit set $L(T_E)$.

This and other similar examples suggest that it might be useful to consider the periodic limit sets as fundamental, the hope being that one could develop an appropriate perturbation method, taking these periods as the *unperturbed states*. The effect of a small change of a parameter (in the direction of *instability*) is then simply to make the period non-attractive. This can in principle be studied by purely algebraic methods. Determining the structure of the resulting limit set—the *perturbed state*—is of course a more difficult matter.

In some cases this may amount to nothing more than the development of improved techniques for handling algebraic expressions of very high order. To clarify this statement, we offer one further example. Consider the Δt-modified transformations $T_{\Delta t}(T_A)$, where T_A is a transformation introduced in subsection 2 above. For $\Delta t = 0.99300$, $L_{\Delta t}(T_A)$ is a period of order 7. With a very small change in Δt—namely, for $\Delta t = 0.99301$—the limit set is of class III type, a pseudo-period. Rather than investigating $T_{A(\Delta t)}$ let us turn our attention to the seventh power of the modified transformation, $T_{A(\Delta t)}^7(p)$ ($\Delta t = 0.99301$). If we choose our initial point p sufficiently close to one of the (repellent) fixed points of $T_{A(\Delta t)}^7$** we find that the first 516 iterated

---

* If T_E is written in the form: $S' = F(S, a)$, $a' = G(S, a)$, then $T_{(1,0)\epsilon}$ is $S' = F(S, a) + \epsilon(S - a)^3$, $a' = G(S, a)$.

** The actual values are not known: we have not yet developed good techniques for finding the coordinates of the points of a non-attractive period. The initial point for this example was taken as: $S = 0.7034477$, $a = 0.1159449$, chosen on the basis of some simple numerical experimentation. It is close to one of the periodic points belonging to the limit set $L_{A(\Delta t)}$ ($\Delta t = 0.99300$), viz.: $S = 0.7037400$, $a = 0.1157123$.

images of p, $T^{(7n)}_{A(\Delta t)}$ (p), $n = 1, 2, \ldots, 516$, lie on an almost exact straight line in the S, a triangle. This is shown in figure A-16. The initial point p is at the lower right, and the successive images trace out the line continuously from right to left.* If we continue the iteration, we find that the later images deviate from the straight line, then oscillate in position, and finally settle down to generate another straight line with a different end-point—presumably very close to another fixed point of $T^{(7)}_{A(\Delta t)}$. It is clear that if one had powerful enough algebraic tools, one could calculate this linear behavior.

7. We close this section with two remarks:
1) A study of the $T_{\Delta t}$ and $T_{(r,s)\epsilon}$ associated with those $T_{C_1 C_2}$ which have only class I limit sets indicates that the latter are much more stable with respect to these one-parameter modifications than are the limit sets discussed above.
2) Even these *unstable* limit sets appear to be stable with respect to some one-parameter perturbations of the corresponding transformations. Thus, the transformations $T_{(3,0)\epsilon}$ associated with T_E have limit sets visually identical with $L(T_E)$ over the whole range $I(1, 10)$. Anomalies such as these make general pronouncements about absolute stability (or instability) impossible.

To illustrate: one might be tempted to explain the observed stability in this case as follows:
Explicitly, $T_{(3,0)\epsilon}$ has the form:

$$S' = F(S, a) + \epsilon(1 - S - a)^3, \quad a' = G(S, a) . \qquad (24)$$

Now the density of $L(T_E)$ is relatively large near the right-hand boundary of the triangle, $S + a = 1$. The perturbing term, however, vanishes on this line. Thus the transformation is *on the average* very little altered by the perturbation. But this "explanation" becomes less convincing when one looks at other transformations associated with T_E. $T_{(2,1)\epsilon}$, for example has the form:

$$S' = F(S, a) + \frac{1}{2}\epsilon(1 - S - a)^3, \quad a' = G(S, a) + \frac{1}{2}\epsilon(1 - S - a)^3 . \qquad (25)$$

One would expect the same argument to apply here, but in fact the limit sets only resemble $L(T_E)$ over the two ranges $I(1, 2)$ and $I(6, 10)$. In between, we get the familiar periodic and pseudo-periodic behavior.

---

* The final point plotted has coordinates: $S = 0.7030206$, $a = 0.11628136$, so the slope of the line is roughly $\Delta a / \Delta S \cong -0.713$. For this photograph, the scaling factor is approximately 2340.

V. Relation to the Theory of Differential Equations

1. As we remarked in section III, the non-linear transformations discussed in this paper exhibit certain analogies with systems of differential equations. In the following we confine ourselves to discussing the plane case.

An important study in the theory of differential equations, particularly as applied to non-linear mechanics, is that of so-called autonomous systems[11,12,13]

$$\frac{dx}{dt} = P(x,y), \; \frac{dy}{dt} = Q(x,y) \; . \tag{1}$$

The theory, initiated by Poincaré, seeks to determine the properties of the solutions of (1) under very general conditions, and to deduce such properties for particular cases without actually solving the equations explicitly (i.e., obtaining the general integral). In particular, the trajectories, given parametrically as a function of t:

$$x = x(t), \; y = y(t) \tag{2}$$

are investigated from a topological point of view. Fundamental is the classification of the singular points of the system (1), that is, the points x, y, where $P(x,y) = Q(x,y) = 0$. The behavior of trajectories in the neighborhood of singular points can be found by consideration of the linear approximation to (1); the real object of the theory, however, is to characterize and, where possible, predict behavior in the large. One of the most interesting phenomena connected with behavior in the large is the existence of closed trajectories, or *limit cycles*. The theorem of Poincaré and Bendixson[14] gives sufficient conditions for the existence of such. Unfortunately, the fulfillment of these conditions in particular cases is often hard to verify; to date no satisfactory theoretical method for dealing with an arbitrary given system has been found.*

2. If we write our general two-dimensional system of non-linear difference equations in the form:

$$\begin{aligned}
\frac{S^{(n)} - S^{(n-1)}}{\Delta t} &= -S^{(n-1)} + F(S^{(n-1)}, a^{(n-1)}) \; , \\
\frac{a^{(n)} - a^{(n-1)}}{\Delta t} &= -a^{(n-1)} + G(S^{(n-1)}, a^{(n-1)}) \; ,
\end{aligned} \tag{3}$$

---

* See reference 15. The practical applications are largely confined to stability theory. Also reference 13 and the literature there cited.

the analogy with (1) is evident. The fixed points of (3) correspond to the singular points of (1), and the behavior of solutions in the neighborhood of a fixed point can be investigated via the linear approximation; this procedure, in fact, yields Ostrowski's criterion (see section III). If the fixed point is attractive, the asymptotic solution in its neighborhood can of course be obtained. In the case of repellent fixed points (or if the initial point is outside the region of attraction of all attractive fixed points), the sequence of iterates sometimes converges to a limit set which appears to resemble a Poincaré limit cycle, i.e., a closed curve. In other cases, finite limit sets (periods) are obtained; on the other hand, one may observe limit sets of quite ambiguous geometrical, not to say topological, structure. These last two alternatives have no analogues in the case of differential equations.

In fact, the analogy between (3) and (1) is more apparent than real. The significant distinction lies, perhaps, in the fact that for our difference equations there is nothing corresponding to the trajectories of (1); successive iterates do not in general lie close to each other. This fact makes it difficult to use topological arguments to determine the character of the limit set. For sufficiently small Δt the sequence of iterates may resemble a trajectory to some extent, but the limit as $\Delta t \to 0$ is almost certain to be a single point.*

VI. Broken-Linear Transformations in Two-Dimensions

1. For certain special quadratic transformations in one dimension one can give an almost complete discussion of the iterative properties; this is possible because these transformations are conjugate to piecewise linear (*broken-linear*) mappings of the interval into itself. For example, the transformation: $x' = g(x)$, where $g(x) = 2x$, $0 \leq x \leq 1/2$; $g(x) = 2 - 2x$, $1/2 \leq x \leq 1$.** The iterative properties of the latter can be obtained from a study of the law of large numbers for the elementary case of Bernoulli. Stated differently, the behavior of iterates of this simple quadratic transformation turns out to depend on combinatorial rather than analytic properties of the function. With this in mind, we tried to see whether an analogous situation would obtain in two dimensions. Our non-linear, polynomial transformations of a triangle into itself might, we thought, be similar to suitably chosen broken-linear mappings of a square into itself, at least as regards their

---

* It may happen that some power $T^{(n)}$ of a transformation more closely resembles a trajectory; cf. the example cited in section II.

** See further in appendix I.

asymptotic behavior.

One simple generalization to two dimensions of broken-linear transformations in one variable is a mapping:

$$x' = f(x,y), \ y' = g(x,y) \ , \tag{1}$$

where each of the functions f and g is linear in regions of the plane. In other words, the graphs of these functions consist of planes fitted together to form pyramidal surfaces. The motivation for studying such transformations is the hope that their iterative properties will turn out to depend only on the folding of the plane along straight lines or, more specifically, on the combinatorics of the overlap of the various linear regions which is generated by the mapping. The simplest nontrivial case to investigate consists in taking $f(x,y)$ as a function defined by choosing a point in the square and making f maximum at this point, the function being linear in the four triangles into which the square is divided. $g(x,y)$ is defined in an analogous manner.

Each of the functions $f(x,y)$, $g(x,y)$ is thus made to depend on three parameters. Thus for f we choose a point x_1, y_1 in the square and erect a perpendicular of height $0 < d_1 \leq 1$ at this point; this defines a surface consisting of four intersecting planes. The transformation can then be given explicitly as follows:

$$\text{I}: x' = \frac{d_1}{x_1} \, x \ ,$$

$$\text{II}: x' = \frac{d_1}{1 - y_1} \, (1 - y) \ ,$$

$$\text{III}: x' = \frac{d_1}{1 - x_1} \, (1 - x) \ , \tag{2}$$

$$\text{IV}: x' = \frac{d_1}{y_1} \, y \ ,$$

where the regions I to IV are specified by the bounding lines:

$$L_1 : y = \frac{y_1}{x_1} \, x \ ,$$

$$L_2 : y = \frac{y_1 - 1}{x_1} \, x + 1 \ ,$$

$$L_3 : y = \frac{1 - y_1}{1 - x_1} \, x + \frac{y_1 - x_1}{1 - x_1} \ , \tag{3}$$

$$L_4 : y = \frac{y_1}{1 - x_1} \, (1 - x) \ ,$$

then

region I   is bounded by L_1, L_2, and $x = 0$,

region II is bounded by L_2, L_3, and $y = 1$,

region III is bounded by L_3, L_4, and $x = 1$,   (4)

region IV is bounded by L_4, L_1, and $y = 0$.

Analogous equations hold for $y' = g(x, y)$, with parameters x_2, y_2, d_2.

Of the several transformations of this type that we have studied numerically we mention only the following:

$$T_1 : \begin{array}{ll} x_1 = \dfrac{1}{3}, & y_1 = \dfrac{1}{3}, \quad d_1 = 0.95 \; , \\ x_2 = 0.6, & y_2 = 0.5, \quad d_2 = 0.95 \; , \end{array} \qquad (5)$$

$$T_2 : \begin{array}{l} x_1 = 0.5, \; y_1 = 0.9, \; d_1 = 1 \; , \\ x_2 = 0.3, \; y_2 = 0.7, \; d_2 = 0.8 \; , \end{array} \qquad (5)$$

and the one-parameter family:

$$T_z : \begin{array}{ll} x_1 = y_1 = z, & d_1 = 1 \; , \\ x_2 = y_2 = 1 - z, & d_2 = 1. \; , \end{array} \qquad (7)$$

The limit sets are shown in figures H-1, H-2, and H-3 through H-17 of appendix II.

$L(T_2)$ (figure II-2) is, in a sense, analogous to the class I limit sets we observed for some of our cubic transformations in three variables. In contrast, $L(T_1)$ (figure H-1) represents a new phenomenon—a connected "curve" (in this case, a collection of line segments) that does not close. More interesting, however, is the behavior of the limit sets $L(T_z)$ as z varies from 0.49 down to 0.01.* Initially $L(T_z)$ is an *open cycle* like

---

* The case $z = 1/2$ has not been studied. The reason for this is technical. Straightforward iteration will always produce sequences which degenerate to zero in a finite number of steps, owing to the fact that every iteration involves multiplication by 2. In a binary machine, this operation is a "shift" to the left. A sufficiently long chain of such left shifts will always result in zero. If one wants to study this case, one must replace multiplication by 2 by some arithmetically equivalent operations, e.g., multiplication by $C/(1/2C)$, where C is not a power of 2.

For this case ($z=1/2$), the problem becomes, of course, one-dimensional. The iterates remain on the line $x = y$, and the limit set is identical with that of the transformation $x' = g(x)$ introduced in subsection 1 above.

$L(T_1)$. With decreasing z, the limit set becomes more complex, until it resembles a class IV limit set (e.g., $z = 0.27$). With further decrease of z, the limit set appears to contract; the *tails* get shorter, and the points cover some sub-region of the square more and more densely. One interesting question—unanswered at the time of this writing—is: does T_z become ergodic for some range of z values? In figure H-10 we see 1000 (consecutive) points belonging to $L(T_z)$ for $z = 1/4$. Figure H-11 shows these same 1000 points together with the next 2000 points. It is evident that the region containing $L(T_z)$ is *filling in*. In our opinion, this is a strong indication of ergodicity.

For lower values of z, the same *ergodic* behavior is observed, until, at $z = 0.15$, the limit set splits into four disconnected pieces (figure H-15). As z is further decreased, these four pieces shrink; by the time we reach $z = 0.01$, the limit is *nearly* a single point.

2. It is clear that one can devise broken-linear transformations that are dense in the unit square; one may take, for example, product transformations with *independent* coefficients and use the one-dimensional result for each factor. For transformations of the type considered in this section, however, it is not easy to determine *a priori* what the limit set will be. It should be emphasized that there is no hope of demonstrating that our polynomial transformations in three variables are exactly conjugate to some two-dimensional broken-linear transformations. Presumably there is no such conjugacy. Nevertheless, one might hope that a somewhat weaker notion of equivalence than that of strict conjugacy could be introduced.

One suggestion along these lines is the following: define two transformations T and S to be asymptotically similar if for almost every initial point p the limit set $L_p(T)$ is topologically equivalent to $L_{p'}(S)$ for some suitably chosen p', and vice versa. Thus, for example: if for any two transformations T, S, the sets of iterates $T^{(n)}$, $S^{(n)}$ are dense in the (common) domain of definition for almost every initial point, then T and S are asymptotically similar in the above sense. Another special case in which two transformations, T and S, are asymptotically similar is when each transformation possesses just one attractive fixed point, the region of attraction being, in both cases, the whole space.

As we remarked above, in the case of broken-linear transformations the asymptotic behavior of iterates depends only on the combinatorial structure of the subdivisions of the fundamental regions (triangles) under repeated folding. Just how complicated this can be is shown by the behavior of the limit sets $L(T_z)$ belonging to the one-parameter family T_z discussed above. To date we have not managed to devise any good method for handling the Boolean algebra of these iterated intersections.

Appendix I

In this appendix we collect some general remarks about the process of iterating transformations, particularly in one dimension. We also discuss, in some detail, a few special one-dimensional transformations which we have had occasion to study.

1. One of the first simple transformations whose iterative properties were established is the following:

$$x' = f(x) \equiv 4x(1 - x) .$$

To obtain these properties we consider, instead of (1), the broken-linear transformation:*

$$x' = g(x), \quad g(x) = \begin{cases} 2x & \text{for } x \leq 1/2 , \\ 2(1 - x) & \text{for } x > 1/2 . \end{cases} \qquad (2)$$

The study of the iterates of this transformation is equivalent to investigating the iterates of a function $S(x)$ defined as follows:

$$\text{if } x = 0.a_1 a_2 a_3 \ldots a_n \ldots, \text{ where the } a_i \text{ are either 0 or 1}, \qquad (3)$$

$$\text{then } S(x) = 0.a_2 a_3 a_4 \ldots . \qquad (4)$$

In other words, $S(x)$ is merely a *left shift* of the binary word x by one place. The iterative properties of $S(x)$ are in turn deducible from the law of large numbers in the case of Bernoulli. In effect $S^{(i)}(x)$ falls into the first half of the interval if and only if $a_i = 0$. The *ergodic average*

$$\frac{1}{N} \sum_{i=1}^{N} F_{(I)}[S^{(i)}(x)]$$

is therefore the same as the fraction of ones among the a_i for $1 \leq i \leq N$. $F_{(I)}$ is the characteristic function of the interval $[0, 1/2]$.

The relation between (1) and (2) is that of conjugacy: there is a homeomorphism $h(x)$ of the interval $[0, 1]$ with itself such that

$$g(x) = h[f[h^{-1}(x)]] . \qquad (5)$$

---

* This transformation has already been mentioned in section VI.

Thus the study of the iterates of the quadratic transformation (1) reduces to the corresponding study for the broken-linear transformation (2). In this case, $h(x)$ can be written down explicitly:[15]

$$h(x) = \frac{2}{\pi} \, sin^{-1}(\sqrt{x}) \, . \tag{6}$$

2. The Set of Exceptional Points. In the case of the function $f(x) = 4x(1 - x)$, it is true, then, for almost every* initial point, that the sequence of iterated images will be everywhere dense in the interval, and what is more, the ergodic limit can be explicitly computed; it is positive for every sub-interval.

There exist, however, initial points x such that the sequence x, $f(x)$, $f[f(x)], \ldots$ is not dense in the whole interval $[0,1]$. Obviously, all periodic points, i.e., points such that, for some n, $f^{(n)}(x) = x$, are of this sort. It is interesting to notice, however, that there exist points x for which the sequence $f^{(i)}(x)$, $i = 1, 2, \ldots$, is infinite without being dense; there are, in fact, non-countably many such points. To show this we consider the equivalent problem of exhibiting such points for the function $S(x)$ introduced above. The construction then proceeds as follows. Consider a point $x = 0.a_1 a_2 \ldots a_n \ldots$ We define a set Z consisting of all those x's which have $a_n = a_{n+1} = a_{n+2}$ for all n of the form $n = 3i$. In other words, the set Z consists of points which have every binary digit repeated three times, the sequence being otherwise arbitrary. Consider now the transformation $x' = S(x)$, where $S(x)$, as defined in (4) above, is a shift of x one index to the left. We now look at the sub-interval from 0.010 to 0.011. Starting with any point in Z, it is clear that no iterated image will fall in this sub-interval; no three successive binary digits of a point in Z are of the form (010). It is easy to see that Z contains non-countably many points; in particular, it contains non-periodic points, so that the set of images $S^{(i)}(x)$, $i = 1, 2, \ldots$, is infinite, but not dense in $[0,1]$.

Presumably one can find points in Z for which the ergodic limit exists. The measure of the set Z is zero, but relative to Z, the set S of those points for which the *sojourn time* exists still form a *majority*. We may define *majority* either in the sense of Baire category or as follows. Take points $na \pmod 1$, $n = 1, 2, \ldots$, where a is an irrational constant. Consider the set N_1 of those n's for which $na \pmod 1$ belongs to Z, and also the set N_2 of those n's for which $na \pmod 1$ belongs to S. We then say that the points belonging to S form a *majority* of those points belonging to Z if the relative frequency of N_2 in N_1 is one.

---

* *Almost every* is to be understood in the sense of Lebesgue measure.

The behavior exhibited by the points belonging to Z is more general, in the sense of measure, for some other transformations of the interval [0,1]. It is possible to give examples of continuous functions such that, for almost every point, the iterated sequence will be nowhere dense in the interval, although the sequence does not converge to a fixed point.

3. A Remark on Conjugacy

Theorem. *Let $g(x)$ be the broken-linear function of equation (2) below, i.e.,*

$$g(x) = \begin{cases} 2x, & 0 \le x \le \frac{1}{2}, \\ 2(1-x), & \frac{1}{2} > x > 1. \end{cases} \tag{2}$$

Let $t(x)$ be a convex function on [0.1] which transforms the interval into itself, and such that $t(0) = t(1) = 0$. For some p in the interval, we must have $t(p) = 1$; by convexity, there is only one such point. Consider the lower tree* generated by the point 1. The necessary and sufficient condition that $t(x)$ be conjugate to $g(x)$ is that this tree be combinatorially the same as that generated by 1 under $g(x)$, and that the closure of this set of points be the whole interval, i.e., that the tree be dense in [0,1].

The condition is obviously necessary, since the point 1 generates a tree under $g(x)$ which is simply the set of binary rational points. Under any homeomorphism $h(x)$ which has to effect the conjugacy, the point 1/2 must go over into p, and our assertion follows.

To prove sufficiency, we construct $h(x)$ in the following manner.

We take $h(1/2) = p$ by definition. We next chose $h(1/4)$ to be the smaller of the two values of $t^{-1}(p)$; $h(3/4)$ is then by definition the larger of these two values. We then take $h(3/8)$ to be the smaller of the two values of $t^{-1}[h(3/4)]$, and so on. Proceeding in this fashion, we thus construct a function $h(x)$ defined for all binary rationals. It remains to prove that we can define it for all x by passage to the limit. This, however, follows from the assumption that these points are dense in [0,1] and that their order is preserved. The function $h(x)$ will obviously be monotonic, and, being continuous, will possess an inverse $h^{-1}(x)$. From our construction it then follows that $h[g(x)] = t[h(x)]$.

---

* By the *lower tree* of p (under $f(x)$) we understand the smallest set of all points with the following properties: (a) The set contains the given point p. (b) If a point belongs to the set, then so do all its counter-images under f.

4. Broken-Linear Transformations. In one dimension these are functions $f(x)$ that are continuous on $[0,1]$ and linear in sub-intervals of $[0,1]$. We assure that the graph of the function has a finite number of vertices, i.e., that $f(x)$ consists of a finite number of lines fitted together continuously. For these broken-linear transformations one certainly expects that the ergodic limit exists for almost every point. For example: if one considers the sequence of iterated images $T^{(n)}(p)$, then, for almost every initial point p, the time of sojourn should exist for all sub-intervals, i.e.,

$$\lim_{N \to \infty} \frac{1}{N} \sum_{i=1}^{N} f_R[T^{(i)}(p)]$$

should exist for almost all p and all measurable sets R; here f_R is the characteristic function of the set R.* The value of this limit may indeed depend on the initial point p; it is likely, however, that all the points of the interval can be divided into a finite number of classes such that, within each class, the value of the limit is the same.**

There is another *finiteness* property that these transformations may possess. Given n, consider all broken-linear transformations which have at most n pieces (i.e., the space divides into n regions, in each of which the transformation is linear). Then it may be conjectured that there are only a finite number of different types of such transformations, where any two transformations of the same type are asymptotically similar (in the sense defined above, section VI). In other words, according to this conjecture, the *type* (or class) that a given transformation belongs to does not depend on the precise numerical values of

---

* We should perhaps mention here a more general conjecture. Suppose T is a polynomial transformation of the sort described by equations (10) to (13) of the Introduction. We then conjecture that the sequence of iterated images $T^{(n)}(p)$ has the following property: Let C be any cone of directions in n-space, and let $f_C(p)$ be the characteristic function of this cone, i.e., $f_C(p) = 0$ if p does not belong to C, $f_C(p) = 1$ otherwise. Then, for almost every p,

$$\lim_{N \to \infty} \frac{1}{N} \sum_{i=1}^{N} f_C[T^{(i)}(p)]$$

exists.[16]

** See pages 71, 72 of S. Ulam, *A Collection of Mathematical Problems*, New York 1960. An analogous conjecture can be made concerning the actual limit sets $Lp(T)$ for four cubics in three variables.

the coordinates of those points where the derivative is undefined (*corner points*), but is determined only by the combinatorial structure of the subdivision of space into linear domains. In one dimension this means that the *type* of a transformation is determined by the number and interrelation of the nodes in the graph of the function, and not by their precise location.

5. Numerical Accuracy. The machines we use to compute the iteration process work with a fixed number of significant digits;* in MANIAC II, for example, this number is eight. It is therefore clear that any direct, single-step iterative process carried out on this computer will exhibit a period in not more than 10^8 steps. Given an algorithm which is iterative and of first order [i.e., the nth step depends only on the $(n-1)^{th}$], the process will, with great probability, exhibit a period which is much shorter. Statistically, one can reason as follows. If we assume a random distribution of, say, the last four digits of all computed numbers, then the probability that the cycle will close long before the full theoretical run of 10^8 steps is extremely close to one. Indeed, after producing numbers $A_1, A_2, \ldots, A_k$, the chance that A_{k+1} will be equal to one of the preceding numbers is of the order of $k/10^8$. If we continue the calculations up to A_{2k}, the chance that at least two numbers in the chain will coincide is approximately

$$1 - \left(1 - \frac{k}{10^8}\right)^k .$$

This is practically equal to

$$1 - \frac{1}{e} \text{ if } k \sim 10^4 .$$

Clearly, going to $3k$, $4k$, $\ldots$, the probability that the chain will be cyclic gets very close to one. The situation is quite different in an iterative process involving two or more variables, e.g., computing (A_{k+1}, B_{k+1}) from (A_k, B_k). On a probabilitic basis alone, one expects to encounter periods of length $\sim 10^8$ (the maximum possible being 10^{16}). In practice, this means that in two dimensions one does not expect to encounter periods of accidental or *false* periodicity unless one generates very long sequences.

---

* For any particular machine one can increase the accuracy by restoring to so-called *multiprecision arithmetic*. This can generally be done only at the cost of considerable loss in speed.

As the argument above shows, fortuitous periodicity can be a very real danger in one-dimensional iterative calculations. Indeed, we came across a striking example in the course of studying the asymptotic properties of the transformation.

$$y' = sin\ \pi y,\ 0 \leq y \leq 1\ .\tag{7}$$

This classical transformation is symmetric about $y = 1/2$, and maps the unit interval into itself. Furthermore, both fixed points—$y = 0$ and $y = 0.7364845$—are repellent. Now a simple argument shows that for such a function there cannot be any attractive periods of any finite order, that is, the derivative

$$\frac{d}{dy}\ T^{(k)}\ (y)\ |_{y=y_p}$$

evaluated at any periodic point $T^{(k)}(y_p) = y_p$ is always greater than one in absolute value.* Nevertheless, when we performed the iteration on MANIAC II, the limit set we observed was invariably one of two finite periods, the first of order 1578, the second of order 6168. These periods were exact to the last available binary digit! This *spurious convergence* is undoubtedly produced by the complicated interaction of several factors, e.g., the particular machine algorithms for multiplication and round-off, our choice of finite approximation to the function $\sin\pi y$, and so forth. When we repeated the calculation on the STRETCH computer (which works with about 15 significant figures), there was no observable tendency toward convergence to such periods.

If one is interested in obtaining the asymptotic distribution of iterates under some transformation like (7), one must either provide for more significant figures or resort to ingenious devices. One such *trick*—which appears to work well and is relatively convenient—consists essentially in computing the inverse transformation. Since the functions

---

* It is essential for this argument that the transformation maps the whole interval into itself. The number of distinct periods of order k can then be easily enumerated. The conclusion follows on noting that

$$\frac{d}{dy}\ T^{(k)}(y)$$

must be the same for all points y belonging to the same period. If, however, the maximum value of the function is less than one, $T^{(k)}(y)$ can have relative minima above the y axis; then, indeed, there may exist attractive periods, i.e., the line $y' \equiv T^{(k)}(y) = y$ may intersect the curve sufficiently close to an extremum so that the kth order fixed point is attractive.

under consideration are two-valued, this involves introducing a random choice at each step. Specifically, taking some initial point p_0, we compute the sequence

$$p_0, \ f^{-1)}(p_0), \ f^{-2}(p_0), \cdots .$$

Here the symbol $f^{-k}(p_0)$ implicitly contains the prescription that we choose one of the two values of the true inverse at random; thus the sequence

$$f^{-i}(p) = f[f^{-(i-1)}(p)], \ i = 1, 2, \ldots ,$$

implies a sequence of random decisions as to which counter-image to choose at each step. If the calculation is carried out in this fashion, the chance of falling into an exact period is vanishingly small until we reach a chain length in the neighborhood of the theoretical maximum. Once having obtained our inverse sequence, we can conceptually invert it, that is pretend that we started with the last point and proceeded to p_0 by direct iteration.*,**

6. In this sub-section we present some results of a study of a particular one-dimensional transformation:***

$$T_\sigma : \ y' = W(3 - 3W + \sigma W^2), \ W \equiv 3y(1 - y) . \qquad (8)$$

This arises in a natural manner from a certain sub-class of our generalized cubic transformations in three variables:

$$x_i' = \sum_{j=1}^{10} d_{ij} M_j , \qquad (9)$$

---

*,   One cannot, of course, actually reverse the calculation and expect to reproduce the sequence. If one could, there would be no need for this stochastic device.

**   This procedure presupposes a good method for generating random numbers. There are several of these which are well suited for use in automatic digital computers. One of the most common—and, in fact, the method used by us—is to generate the numbers by the chain:

$$R_{n+1} \equiv R_0 R_n (\text{mod } 2^k) ,$$

where k is the binary word length of the machine, and R_0 is some properly chosen constant. On MANIAC II, $R_0 = 5^{13}$. This chain closes, but its length is greater than 10^{12}. The lowest order binary digits are themselves not random, but this makes no difference in practice.

***   This transformation was introduced by the way of illustration in section I.

$$\sum_{i=1}^{3} d_{ij} = 1, \text{ all } j, \ 0 \le d_{ij} \le 1 \ . \tag{10}$$

Namely, we choose certain of the d_{ij} as follows:

$$d_{13} = \sigma, \ d_{17} = d_{19} = d_{34} = d_{36} = 1, \ d_{31} = 0 \ .$$

The rest are arbitrary, except that they must, of course satisfy (10).

If we restrict ourselves to the sub-class of initial points such that $x_3 = 0$ (i.e., the side $S + a = 1$ of our reference triangle), the second power of the transformation can be written in the form (8).

T_σ is symmetric about $y = 1/2$, but it does not map the whole interval $[0,1]$ into itself; its maximum value is

$$y'_{\text{max}} = T_\sigma \left(\frac{1}{2} \right) = \frac{9}{16} \left(1 + \frac{3\sigma}{4} \right) \tag{11}$$

so this is the right-hand boundary of the invariant sub-region. The left-hand boundary is then the image of y'_{max}. In the range $1 \ge \sigma \ge 0.9$, the fixed point varies from $y_0 = 0.8224922$ at $\sigma = 1$ down to $y_0 = 0.8193719$ at $\sigma = 0.90$. Over this range, the derivative at y_0 is negative and greater than 1 in absolute value; thus the fixed point is repellent.

We have studied this transformation as a function of the two parameters—the initial point y_0 and the coefficient σ—on the STRETCH computer. To study the asymptotic distribution, we divide the interval into 10000 equal parts and have the machine keep track of the number of points which fall into each sub-interval over the iteration history. Such a history was usually taken to be a sequence of length $n \times 10^5$, with $4 \le n \le 9$.* Should a period (of order $k \le 3 \times 10^5$) be detected by the machine, the calculation is automatically terminated. If the order of the period is not too great ($k \le 500$), the values of the periodic points are printed out, and the value of

$$\frac{d}{dy} T_\sigma^{(k)}(y)$$

is calculated over the period; if

$$\left| \frac{d}{dy} T_\sigma^{(k)}(y) \right| < 1 \ ,$$

---

* As we have already remarked (section I), the calculation of 2×10^5 iterates of this transformation requires about 1 minute on STRETCH.

this fact is strong evidence that the observed period is actually a limit set.

Our numerical investigation of (8) has been mostly restricted to the range $0.98 \leq \sigma \leq 1$.* Even in this restricted parameter range, the observed asymptotic behavior is of bewildering complexity. For $\sigma = 1$, the distribution of iterates in the interval is extremely non-uniform. There are large peaks at the end-points of the allowed sub-interval, with much complicated *fine-structure* in between, i.e., many relative maxima as well as sizeable intervals in which the distribution is locally uniform. This general behavior persists as σ is decreased down to $\sigma = 0.9902$. At $\sigma = 0.9901$, however, a dramatic change takes place; most of the points concentrate in a few small sub-intervals. The limit set is, in fact, a pseudo-period.** At $\sigma = 0.99009$, the pseudo-periodic behavior is still evident, but the occupied sub-intervals are larger. They contract, however, for $\sigma = 0.99008$, and at $\sigma = 0.990079$ a period of order $k = 42$ is found. Actually, for this particular value of the parameter σ, there appear to be two possible periodic limit sets, with $k = 42$ and $k = 84$ respectively. The dependence on the initial point is, of course, quite complicated.*** For example, as y_0 is varied over the values $y_0 = 0.11, 0.12, 0.13, \ldots, 0.21$, the period with $k = 42$ is found in all except three cases, namely $y_0 = 0.11, 0.16, 0.18$, which apparently lead to the periodic limit set with $k = 84$. Both periods are numerically well-attested; they are exact to the last binary digit and have

$$\left| \frac{d}{dy} T_\sigma^{(k)}(y) \right| < 1 \, .^\dagger$$

---

* Below $\sigma = 0.98$, various periodic limit sets of order $k \leq 24$ were found on MANIAC II. We have not studied this parameter range on STRETCH.

** It may actually be a period with order $k = 63049$. This is the result indicated by the machine. At present we have no way of verifying this.

*** Subsequent numerical work has shown that this y_0 dependence is spurious. The calculations were performed with multi-precision arithmetic; in some cases as many as its decimal places were retained!

† It is difficult to decide whether this behavior is real or not. Some supporting evidence for its *reality* is the observed behavior for values of σ very close to this *critical* value $\sigma = 0.990079$. We find that, with $\sigma = 0.9900789$, only a period with $k = 42$ is obtained, while on the other side, $\sigma = 0.9900791$, the limit set is always a period with $k = 84$.

As we continue down in σ, the limit sets are pseudo-periodic* until we reach $\sigma = 0.99007$, for which a period of order $k = 112$ is found. This appears to be the approximate right-hand boundary of a *periodic belt*; that is, in the range $0.98990 \leq \sigma \leq 0.99007$ there are only periodic limit sets for this transformation. All these have orders which are multiples of 14 (with $k \leq 112$), except for the (approximate) left-hand boundary of the σ-range ($\sigma = 0.9899$) where a period of order 7 exists. The dependence of these limit sets on the initial point is complicated, and will not be reproduced here.

At $\sigma = 0.98988$ there is another large discontinuity in behavior; we again find an asymptotic distribution which covers the whole allowable sub-interval in non-uniform fashion. No more periodic limit sets are found as σ is decreased to $\sigma = 0.9800$. The only phenomenon of note is the splitting of the limit set into two parts; this occurs somewhere between $\sigma = 0.986$ and $\sigma = 0.985$. The resulting gap—which contains the fixed point—continues to widen as $\sigma \rightarrow 0.980$.**

The results of this investigation—which is rather incidental and subordinate to the larger study reported in the present paper—clearly show that there is a great deal to be learned about the asymptotics of iterative processes, even in one dimension. It seems that "pathological" behavior is not a property of higher-dimensional systems alone. With regard to our study of the periodic limit sets, it may be argued that what we have really done is to investigate, in a rather indirect manner, the behavior of the roots of high-order algebraic equations as a function

---

* In this range, the machine *detected* some exact periods of huge order— e.g., $k = 295148$ ($\sigma = 0.990079$)! There seems to be no compelling reason to take this at face value.

** Such gaps have been observed in other cases. One interesting example is the transformation:

$$y' = W^2(3 - 2W), \quad W \equiv 3y(1 - y) ,$$

which is also a special case of one of our cubics in three variables. For this transformation, the limit set consists of four separate pieces, as follows: (I) $0.3455435 < y < 0.4086018$, (II) $0.4296986 < y < 0.5791385$, (III) $0.7562830 < y < 0.8146459$, (IV) $0.8220964 < y \leq 0.84375$.

In the present case, i.e., that of T_σ, the existence of the gap can easily be predicted. Let $y_0(\sigma)$ be the fixed point. There will clearly be a gap providing $T_\sigma^2(y_{max} > y_0(\sigma)$, since in this case only one of the two inverses of y_0 lies in the allowed interval. Of course, the determination of the critical value σ_c of σ from the equation $T_\sigma^2(y_{max}) = y_0(\sigma)$ would in any event have to be carried out numerically.

of their coefficients. This is certainly true, at least in part. It is therefore of interest to observe that the *iterative* method seems at present to be the only effective tool for treating this purely algebraic problem.

Appendix II

1. In this appendix we collect the photographs and tables illustrating the phenomena discussed in section IV and VI. The notation used has been described in sections II, III, IV, and VI. For convenience, some of the transformations are written out explicitly.

2. Modifications of the Transformation T_A. In shorthand notation, this transformation is

$$T_A : \begin{array}{l} C_1 = \{3, 5, 7, 9, 10\} , \\ C_2 = \{1, 2, 8\} . \end{array} \tag{1}$$

In the S, a coordinates, this reads:

$$S' = F(S, a) \equiv S^3 - 6S^2 a - 3Sa^2 + 4a^3 - \frac{3}{2}S^2 + 3Sa - \frac{3}{2}a^2 + 1 ,$$

$$a' = G(S, a) = -S^3 + 3Sa^2 + 2a^3 + \frac{3}{2}S^2 - 3Sa + \frac{3}{2}a^2 . \tag{2}$$

The (repellent) fixed point has coordinates:

$$S_0 = 0.5885696 ,$$
$$a_0 = 0.1388662 . \tag{3}$$

The *generalized* transformations based on T_A may be written:

$$S' = (1 - \Delta t)S + \Delta t\{F(S, a) + \epsilon f_{rs}(S, a)\} ,$$
$$a' = (1 - \Delta t)a + \Delta t\{G(S, a) + \epsilon g_{rs}(S, a)\} ; \tag{4}$$

the original transformation is recovered by setting $\Delta t = 1$, $\epsilon = 0$.

TABLE A				
Figure number	Δt	ϵ	r, s	Comments
A-1	1	0.02	6, 1	Class IV limit set
A-2	1	0.07	2, 1	Period: $k = 7$
A-3	1	0.10	2, 1	Period: $k = 28$
A-4	1	0.08	2, 1	Class I limit set
A-5	1	0.09	2, 1	Class I limit set
A-6	1	0.08	2, 1	Scaled plot of part of A-4. Scaling factor $\sim$77 ($0.472 \leq S \leq 0.485,\ 0.125 \leq a \leq 0.135$)
A-7	1	0.09	2, 1	Scaled plot of part of A-5. Scaling factor $\sim$37 ($0.465 \leq S \leq 0.492,\ 0.120 \leq a \leq 0.140$)
A-8	1	0.03	2, 1	Pseudo-period
A-9	1	0.03	2, 1	Scaled plot of part of A-8. Scaling factor $\sim$5.6 ($0.440 \leq S \leq 0.620,\ 0.175 \leq a \leq 0.265$)
A-10	1	0.04	5, 1	Class IV limit set
A-11	1	0.04	5, 1	Scaled plot of part of A-10. Scaling factor $\sim$5 ($0.44 \leq S \leq 0.64,\ 0.20 \leq a \leq 0.30$)
A-12	1	0.01	3, 1	Period: $k = 23$
A-13	1	0.01	3, 1	Class IV limit set
A-14	1	0.02	5, 0	Period: $k = 148$
A-15	1	0.02	5, 0	Class IV limit set
A-16	0.99301	1	—	514 successive iterates of $T_{A(\Delta t)}^{(7)}(p)$. Initial point: $S_0 = 0.7034477$, $a_0 = 0.1159449$. Scaling factor $\sim$909

3. Modifications of T_B. In shorthand notation, this transformation is

$$T_B : \begin{array}{l} C_1 = \{2, 4, 6, 7, 9\}\ , \\ C_2 = \{5, 8, 10\}\ . \end{array} \tag{5}$$

In the S, a coordinates, it takes the form

$$S' = F(S, a) \equiv 9S^2a - a^3 - \frac{3}{2}S^2 - 9Sa - \frac{3}{2}a^2 + \frac{3}{2}S + \frac{9}{2}a\ ,$$
$$a' = G(S, a) \equiv -S^2a + 3a^3 - \frac{3}{2}S^2 + 3Sa - \frac{3}{2}a^2 + \frac{3}{2}S - \frac{3}{2}a\ . \tag{6}$$

The coordinates of the fixed point are

$$S_0 = 0.6887703 ,$$
$$a_0 = 0.1592083 .$$

(7)

The only associated $T_{(r,s)\epsilon}$ discussed is the case $(r, s) = (1, 0)$; for this case, the generalized transformation, written in the form of equation (4), has

$$f_{rs} = f_{10} = (S - a)^3 ,$$
$$g_{rs} = g_{10} = 0 .$$

(8)

In table B below, all scaled plots show the region: $0.30 \leq S \leq 0.65$, $0.20 \leq a \leq 0.36$, i.e., the upper left-hand piece of the complete limit set (shown in figure B-1 for the unmodified transformation). The scale factor is ~ 2.9.

Figure number	Δt	ϵ	r, s	Comments
			TABLE B	
B-1	1	0	—	Class IV limit set (2700 points)
B-2	1	0	—	Scaled plot of part of figure B-1
B-3	0.996	0		Class IV
B-4	0.994	0	—	Class IV Compare figure B-8.
B-5	0.992	0	—	Class IV
B-6	0.990	0	—	Period: $k = 78$ (26 points in this piece)
B-7	0.980	0	—	Class IV
B-8	1	0.01	1, 0	Class IV: 12 separate pieces (4 shown here) Compare figure B-4
B-9	1	0.02	1, 0	Period: $k = 24$ (8 points shown here)
B-10	1	0.04	1, 0	Class IV
B-11	1	0.06	1, 0	Period: $k = 84$ (28 points shown here)
B-12	1	0.08	1, 0	Class IV
B-13	1	0.09	1, 0	Period: $k = 102$ (34 points shown here)
B-14	1	0.10	1, 0	Period: $k = 30$ (10 points shown here)
B-15	0.91	0	—	Class IV; shows "transition" from period with $k=3$ ($\Delta t = 0.92$)

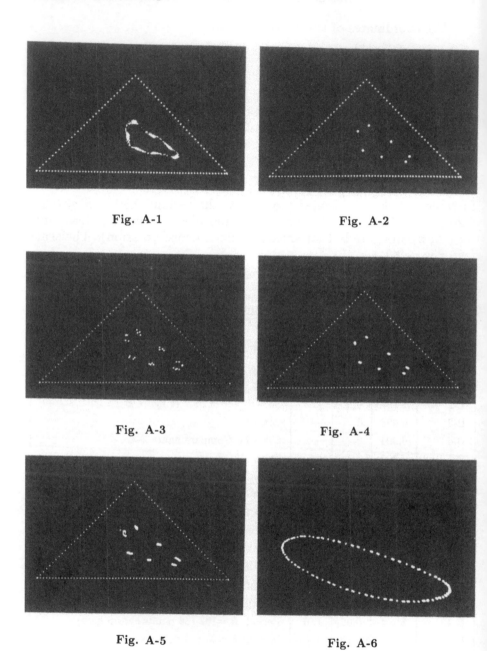

Fig. A-1

Fig. A-2

Fig. A-3

Fig. A-4

Fig. A-5

Fig. A-6

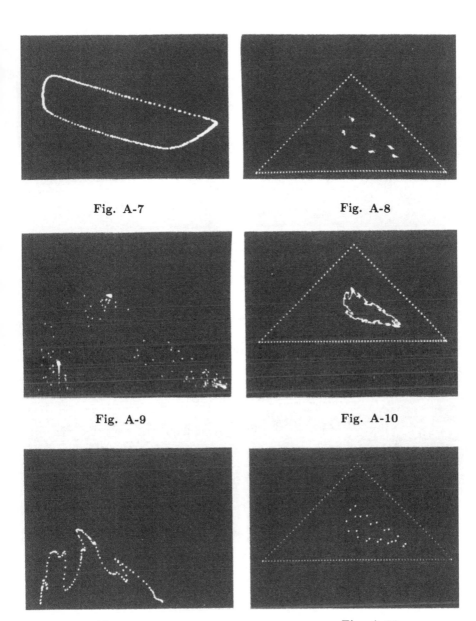

Fig. A-7

Fig. A-8

Fig. A-9

Fig. A-10

Fig. A-11

Fig. A-12

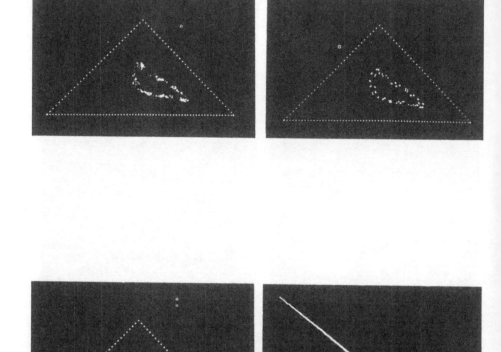

Fig. A-15   Fig. A-16

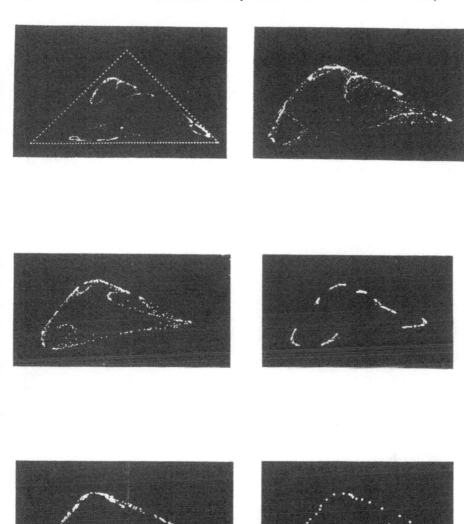

Fig. B-5   Fig. B-6

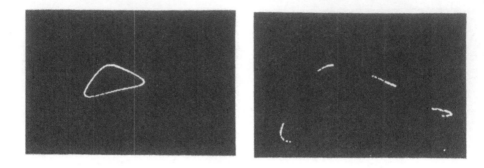

Fig. B-7   Fig. B-8

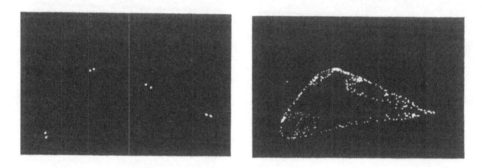

Fig. B-9   Fig. B-10

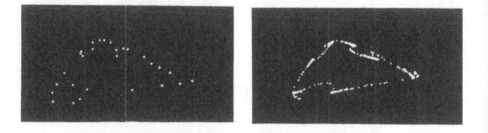

Fig. B-11   Fig. B-12

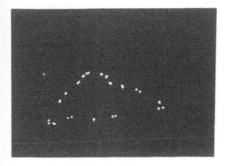

Fig. B-13

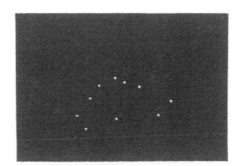

Fig. B-14

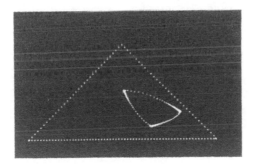

Fig. B-15

4. Modifications of T_D and T_E. These transformations are given by the schemes:

$$T_D : \begin{array}{l} C_1 = \{2,7,8,9,10\} , \\ C_2 = \{4,5,6\} , \end{array} \tag{9}$$

$$T_E : \begin{array}{l} C_1 = \{2,5,7,8,9\} , \\ C_2 = \{4,6,10\} . \end{array} \tag{10}$$

In the S, a coordinates, these read explicitly:

$$T_D : \begin{array}{l} S' = -\dfrac{3}{2}S^3 + \dfrac{3}{2}S^2 a + \dfrac{3}{2}Sa^2 + \dfrac{13}{2}a^3 - 6Sa - 6a^2 + \dfrac{3}{2}S + \dfrac{9}{2}a , \\[2mm] a' = \dfrac{3}{2}S^3 + \dfrac{9}{2}S^2 a - \dfrac{3}{2}Sa^2 - \dfrac{9}{2}a^3 - 3S^2 + 3a^2 + \dfrac{3}{2}S - \dfrac{3}{2}a , \end{array} \tag{11}$$

with fixed points:

$$\begin{array}{l} S_0 = 0.6525211 , \\ a_0 = 0.3056821 ; \end{array} \tag{12}$$

$$T_E : \begin{array}{l} S' = 9S^2 a - a^3 - 3S^2 - 12Sa + 3a^2 + 3S + 3a , \\ a' = -3S^2 a + 3a^3 + 6Sa - 6a^2 , \end{array} \tag{13}$$

with fixed point:

$$\begin{array}{l} S_0 = 0.6444612 , \\ a_0 = 0.3219578 . \end{array} \tag{14}$$

TABLE D				
Figure number	Δt	ϵ	r, s	Comments
D-1	1	0	—	
D-2	0.97	0	—	Compare figure D-6
D-3	0.96, 1. 0	0	—	Limit set for $\Delta t = 0.96$ superimposed on $L(T_D)$
D-4	0.83	0	—	
D-5	1	0.07	1, 0	
D-6	1	0.10	1, 0	Compare figure D-2

TABLE E				
Figure number	Δt	ϵ	r, s	Comments
E-1	1	0	—	
E 2	1	0.01	1, 0	Shows convergence to periodic limit (set $k = 10$) from initial point close to fixed point
E-3	0.97	0	—	Compare with E-5, E-6
E-4	0.96	0	—	Compare with E-5, E-6
E-5	1	0.09	1, 0	
E-6	1	0.10	1, 0	

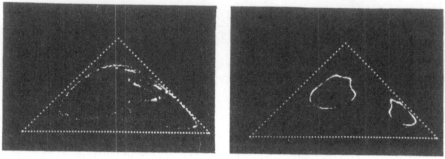

Fig. D-1

Fig. D-2

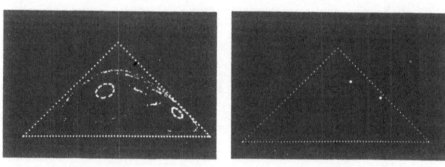

Fig. D-3

Fig. D-4

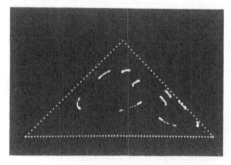

Fig. D-5

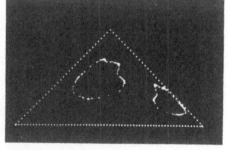

Fig. D-6

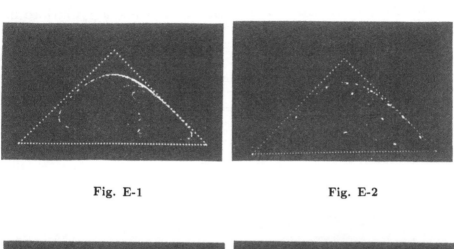

Fig. E-1   Fig. E-2

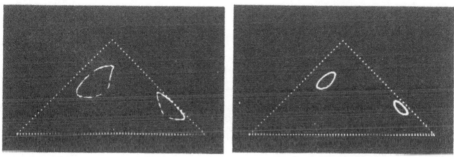

Fig. E-3   Fig. E-4

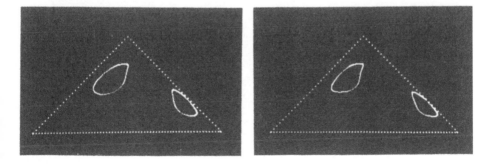

Fig. E-5   Fig. E-6

5. Broken Linear Transformations. These are described in section VI. Figures H-1 and H-2 show, respectively, 1000 points in the limit sets of T_1 and T_2. The latter are specified as follows:

$$T_1 : \begin{array}{l} x_1 = \dfrac{1}{3}, \quad y_1 = \dfrac{1}{3}, \quad d_1 = 0.95 , \\ x_2 = 0.6, \quad y_2 = 0.5, \quad d_2 = 0.95 , \end{array} \tag{15}$$

with fixed point:

$$x_0 = \frac{1}{2}, \; y_0 = \frac{2}{3} , \tag{16}$$

and

$$T_2 : \begin{array}{l} x_1 = 0.5, \quad y_1 = 0.9, \quad d_1 = 1 , \\ x_2 = 0.3, \quad y_2 = 0.7, \quad d_2 = 0.8 , \end{array} \tag{17}$$

with fixed point:

$$x_0 = \frac{80}{143}, \; y_0 = \frac{72}{143} . \tag{18}$$

The remaining figures, H-3 through H-17, show the limit sets belonging to the one-parameter family T_z:

$$T_z : \begin{array}{l} x_1 = y_1 = z, \quad\quad d_1 = 1 \\ x_2 = y_2 = 1 - z, \quad d_2 = 1 . \end{array} \tag{19}$$

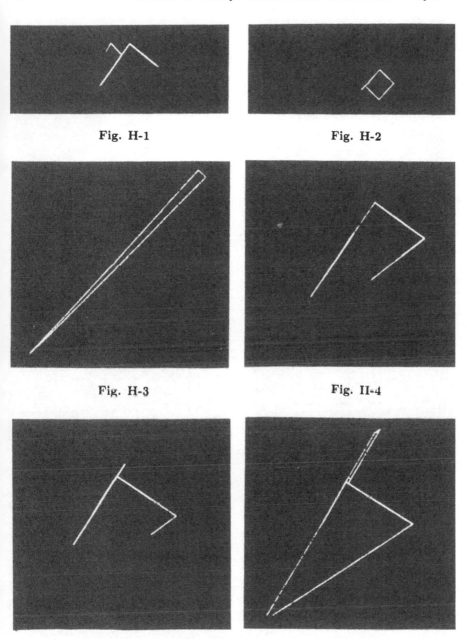

Fig. H-1   Fig. H-2

Fig. H-3   Fig. II-4

Fig. H-5   Fig. H-6

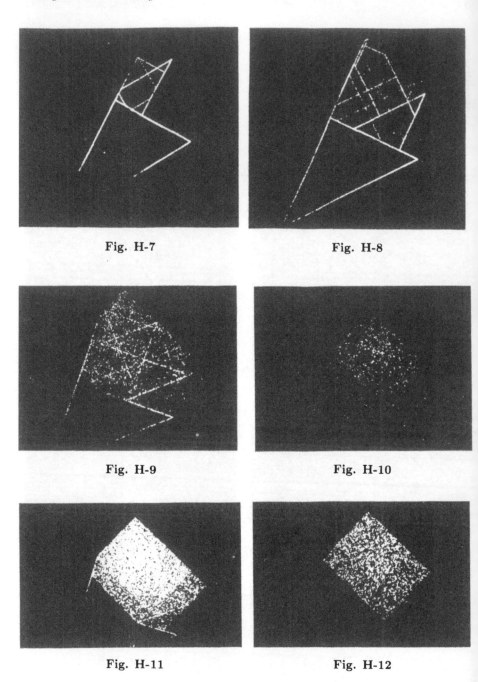

Fig. H-7

Fig. H-8

Fig. H-9

Fig. H-10

Fig. H-11

Fig. H-12

Fig. H-13   Fig. H-14

Fig. H-15   Fig. H-16

Fig. H-17

The identification is given by the following table (all figures except H-11 show 1000 points):

TABLE H		
Figure number	z	Comments
H-3	0.49	For $z - 0.5$, see the remarks in section VI, note 2
H-4	0.42	
H-5	0.40	Compare figure H-1
H-6	0.38	
H-7	0.36	
H-8	0.32	Like "class IV" limit set
H-9	0.27	Like "class IV" limit set
H-10	0.25	
H-11	0.25	Points of H-10 plus the next 2000 consecutive iterates
H-12	0.23	
H-13	0.19	
H-14	0.16	
H-15	0.15	
H-16	0.12	
H-17	0.03	Fixed point becomes attractive below $z = 0.01$

References

1. J. Schreier und S. Ulam, *Eine Bemerkung über die Gruppe der topologischen Abbildungen der Krieislinie auf sich selbst*, Studia Math. **5** (1935).
2. Menzel, Stein and Ulam, *Quadratic Transformations*, Part I, Los Alamos Report LA-2305, May, 1959 (Available from the Office of Technical Service, U.S. Dept. of Commerce, Washington, D.C.)
3. See for example, S. M. Ulam, *On some new possibilities ... computing machines*, I.B.M. Research Report 68, 1957.
4. See, e.g., Proceedings of Symposia on Applied Mathematics, Vol. X, American Mathematical Society (1958), or the article by Marshall Hall, Jr. in Surveys of Applied Mathematics, Vol. IV, 1958.

5. In the absence of a comprehensive reference, we refer the reader to the recent issues of Mathematics of Computation (1960-1962). See, e.g., Vol. 16, No. 80, October 1962, especially the article by D. H. Lehmer, et al.
6. See the article by E. T. Parker in Proceedings of Symposia in Pure Mathematics, Vol. VI, American Mathematical Society (1962). Parker's original construction supplied the final step in the disproof of a famous conjecture by Euler.
7. See H. Gelernter et al., Proceedings of the Western Joint Computer Conference, San Francisco, May 1960, pp. 143-149.
8. M. B. Wells, Proceedings of the IFIP Congress (1962).
9. C. Shannon, Philosophical Magazine 41, March, 1950. See also Kister, Stein, Ulam, Walden and Wells, Journal of the Association for Computing Machinery, Vol. 4, Number 2, April 1957.
10. A. Ostrowski, *Solution of Equations and Systems of Equations*, Chapter 18, (New York 1960).
11. A general reference is N. Minorsky, *Non-linear Oscillations*, Princeton 1962. More detail on theoretical points can be found in 12.
12. S. Lefschetz, *Differential Equations: Geometric Theory*, New York 1957.
13. Nemitsky and Stepanov, *Qualitative Theory of Differential Equations*, Princeton 1960.
14. See reference in footnote 11.
15. This result was first published by S. Ulam and J. von Neumann, American Mathematical Society Bulletin, Vol. 53 (1947), p. 1120, Abstract 403. See also the work of O. Rechard, Duke Mathematical Journal, Vol. 23 (1956), pp. 477-488.
16. Cf. the article by S. Ulam in *Modern Mathematics for the Engineer*, second series, edited by E. F. Beckenbach, New York 1961, p.280.

12

ON RECURSIVELY DEFINED GEOMETRICAL OBJECTS AND PATTERNS OF GROWTH

With R. G. Schrandt
(LA-3762, August 15, 1967)

This report formed a foundation of a now great ongoing effort involving many authors concerning the behavior and growth of cellular automata.

See, for instance, the proceedings of a March 1983 Los Alamos National Laboratory conference on cellular automata and reports written by Wolfram and others.

This report was reprinted in Essays on Cellular Automata *edited by A. Burks, published by the University of Illinois Press in 1970. (Author's note).**

ABSTRACT

Illustrations are given of computer-generated patterns exhibited by figures "growing" according to certain recursive rules. Examples of growing patterns in two- and three-dimensional space are given. Patterns are discussed in an infinite strip of a given width where periodic growth is observed. When modification of the rules of growth allows portions of the pattern to die out, configurations split into separate connected pieces, exhibiting the phenomena of both motion and some self-reproduction. A simple conflict rule together

---

* Also in *Science, Computers, and People,* by S. Ulam, Birkhäuser, 1986. (Eds.)

> with this modification allows a game of survival be-
> tween two systems growing in a finite portion of the
> plane. The examples show both the complexity and
> richness of forms obtained from starting with a simple
> geometrical element and the application of a simple
> recursive rule.

In this report we discuss briefly some empirical results obtained by experiments on computing machines. We continue the work described in a paper "On Some Mathematical Problems Connected With Patterns of Growth."[1]

Rules of Growth

Growth is in the plane subdivided into regular squares. The starting configuration may be any arbitrary set of (closed) squares. The growth proceeds by generations in discrete intervals of time. Only the squares of the last generation are "alive" and able to give rise to new squares. Given the nth generation, we define the $(n + 1)^{th}$ as follows: A square of the next generation is formed if

a) it is contiguous to one and only one square of the current generation, and

b) it touches no other previously occupied square except if the square should be its "grandparent." In addition:

c) of this set of prospective squares of the $(n + 1)^{th}$ generation satisfying the previous condition, we eliminate all squares that would touch each other. However, squares that have the same parent are allowed to touch.

In three dimensions the growth rules are the same. One merely replaces the squares by cubes and observes all three provisions.

We show an example of such a pattern growing on the infinite plane and then discuss patterns of growth in an infinite strip of a given width where a periodic growth is observed. We discuss, beyond the work mentioned in Ref.1, the behavior of figures growing according to our rules, with a new proviso: every element of the figure which is older than a specified number of generations, say two or three, "dies," i.e., is erased. This makes the figure move in the plane. We show some cases of such motion, with occasional splitting of the figures into separate connected pieces. In some cases these figures are similar to the original ones, thus exhibiting phenomena of both motion and of self-reproduction. As another amusement we tried on the computers the following game: starting, still in the plane with two separate initial elements, we let each grow according to our rule (including erasure or death of the "old" pieces); then when the two patterns approach each other we still apply the rule of a further growth of each figure with the proviso that the would-be grown new pieces are not put in if they should try to occupy the same square. This gives rise to a game for survival or a "fight" between two such systems—in some cases both figures die out.

Finally, we give an example of a similar process of growth in three-dimensional space (subdivided into regular cubes) with our rules for recursive addition of new elements.

Two-Dimensional Patterns

We present as examples of the planar type of pattern Figs. 1a and 1b. Our start was a single square. The patterns are plotted merely in one quadrant of the plane and show the result of 100 and 120 generations of growth, respectively. Fig. 1b shows the pattern on a large square with 100 units on a side; the portion of growth that extends beyond 100 units horizontally or vertically is not plotted. The figures are symmetric about the diagonal of the square, and the density of the occupied squares is about 0.44. There is no apparent periodicity in portions of this pattern. As shown in Ref. 1, the "stems" grow indefinitely on the sides of the quadrant, and the side branches split off from the stem. It is not known whether some of these side branches will grow to infinite length or whether they will all in turn be choked off by other side branches growing from the stem at later times.

In Fig. 2 we show a pattern grown from an initial configuration consisting of three noncontiguous squares at the vertices of an approximately equilateral triangle. One will note that the patterns in the subquadrants are both identical to those of Figs. 1a and 1b. The borders or strips between the subquadrants are due to interference between patterns generated by the individual starting squares. One of the strips reduces to a stem, since two of the starting squares generate patterns symmetric with respect to a 45° line through the center of the triangle.

By restricting in advance the growth of a pattern to an infinite strip of finite width in the plane, one observes a periodic growth. The proof that in a finite strip the pattern must be periodic is easily obtained. On inspection of our growth procedure one observes that the last generation is confined to a part of the strip which extends through its width and an equal distance in length back of the most forward square. There is only a finite number of possible patterns in such a square. Therefore, a configuration must repeat itself and from then on the whole process starts again. Figure 3 shows different patterns generated in strips of widths from 8 to 15 through 100 generations of growth. In each case the start is a single square in the upper left-hand corner of the strip. Table I gives the observed lengths of the periods for strips of widths 1 to 17. There seems to be no simple relation between the width of the strip and the length of the period.

Rules for Termination or "Death" in the Pattern

We have experimented with a rule for erasing, i.e., elimination of a part of the pattern after it is a fixed number of generations old. For this we have adopted a simpler rule of growth of the pattern by assuming only condition (a). Each square that is a certain fixed number k of generations old is erased or "dies" and becomes unoccupied. Later on, the pattern may grow back into these unoccupied positions. We took for k either the values 1, 2, or 3. For example, given a pattern, we grow the squares of the $(n+1)^{th}$ generation from those of the nth, and then erase those of the $(n-1)^{th}$. Under this rule the pattern will move and it may split up into disconnected pieces, as shown in Fig. 4a. It turns out that certain parts of the pattern replicate themselves in shape, and these repeat as subpatterns. One such subpattern consists of a straight strip of squares with two additional squares on each end. We call this rather frequent replication subpattern a "dog bone." (See Fig. 4b.)

Another construction concerns the behavior of such patterns in a finite portion of the plane. We have adopted a large square as the space for growth. Its boundary acted as an absorber so that each square which would possibly grow from a square on the boundary was not considered. This was studied under the simplified rule of growth mentioned above. Starting with an initial configuration, say a single small square, the pattern will grow and either eventually "die," or else will become periodic in time and continue indefinitely. In most cases the pattern eventually disappears or dies. This is because the death rule eliminates the old squares, and the simple conflict rule together with the boundary condition prevent any new squares from forming. For these problems we kept only the current generation, so $k = 1$. By this we mean that given a configuration, we produce the next one and then immediately erase the starting one. We have run a number of cases on a computer to ascertain either the period and its length, or the number of generations before the pattern terminates. This we have done in various sizes of the large square in which the game takes place. A sampling was obtained for sizes of the large square for 2×2 up to 8×8.

As an example, consider the square of size 6×6. There are, of course, 2^{36} possible initial configurations. Out of these we have chosen 132 such configurations at random, assuming that each of the 36 squares has 1/2 chance of being occupied initially. Each of these different starting configurations grew until it became periodic or died out. Let s be the number of states in each sequence and t the length of a period. The values of s ranged from 11 to 109 with an average of 33. The values of t were 1, 4, 6, 8, 12, and 24. Here $t = 1$ means the pattern died out. In our sample, 87 of the 132 cases had $t = 1$. The longest sequence has $s = 109$, with $t = 24$. In another experiment we tried 15 random starting configurations chosen in an 8×8 large square. The values of s ranged from 49 to 397, with t values of 1, 8, 12, and 16. Ten of our 15 experiments had $t = 1$.

We can formulate condition (u) of the rules of growth in another way if we keep only one generation before death. In this case the status of any square in the $(n+1)^{th}$ generation is determined only by the state of its four neighboring squares in the nth generation. Let us assign a 1 to an occupied square and a 0 to an empty one. We use the two operators (·) and (+), with the (+) modulo 2—that is, 1+1 = 0.

If a_n, b_n, c_n, d_n are the four neighbors of a square x_n and all four symbols have values 1 or 0, that is, they represent the states of the squares in the nth generation, then the state of the square x in the next generation is simply

$$x_{n+1} = \bar{c}_n \cdot \bar{d}_n \cdot (a_n + b_n) + \bar{a}_n \cdot \bar{b}_n (c_n + d_n)$$

where the bars above the symbols represent the complement (also modulo 2).

If the whole region in which the game is played is bounded, say, again by a large square, we will assume that the values on the boundary are always 0. The state of the configuration at time $(n + 1)$ is then obtainable by a fixed transformation from the state at time (n).

One of the interesting properties to determine is the existence of states which are self-replicating, that is, they reproduce themselves immediately. These are the fixed points of the transformation defined above. It is easily verified that there are none such (except those identically 0, which means the pattern dies out) for squares of size 2×2 and 3×3. There exists just one such state for the 4×4 square. This is given by

$$\begin{vmatrix} 0 & 1 & 1 & 0 \\ 1 & 0 & 0 & 1 \\ 1 & 0 & 0 & 1 \\ 0 & 1 & 1 & 0 \end{vmatrix}.$$

For the 5×5 square there are these two:

$$\begin{vmatrix} 1 & 0 & 1 & 0 & 1 \\ 1 & 0 & 1 & 0 & 1 \\ 0 & 0 & 0 & 0 & 0 \\ 1 & 0 & 1 & 0 & 1 \\ 1 & 0 & 1 & 0 & 1 \end{vmatrix} \text{ and } \begin{vmatrix} 1 & 1 & 0 & 1 & 1 \\ 0 & 0 & 0 & 0 & 0 \\ 1 & 1 & 0 & 1 & 1 \\ 0 & 0 & 0 & 0 & 0 \\ 1 & 1 & 0 & 1 & 1 \end{vmatrix}.$$

There are none of the 6×6 case. Here is an example of one for the 17×17 case. Let A be the second of the two 5×5 matrices. Let N_c and N_r be 5×1 and 1×5 matrices, respectively, with zero elements. Then the matrix

$$\begin{vmatrix} A & N_c & A & N_c & A \\ N_r & O & N_r & O & N_r \\ A & N_c & A & N_c & A \\ N_r & O & N_r & O & N_r \\ A & N_c & A & N_c & A \end{vmatrix}$$

is self-replicating.

Contests or Fights between Two Configurations

We may start, on a large finite square, with two different initial configurations each labeled, say, by a different color, so as to distinguish one set from the other. We let each grow according to condition a) of the rules of growth, plus the death rule. Now condition a) states that a square of the next generation is not formed if it is contiguous to two or more squares of the current generation. Two such squares of the current generation may be members of the same configuration or else one from each of the two different configurations. So the growth of these patterns is subject to restrictions for elements of the new generation within themselves separately, and when they are almost in contact, with the two taken together. One or both of these systems may then go to zero or one may survive, for some time or indefinitely.

Figures 5a, 5b, 5c, 5d illustrate one case of such a fight between two starting patterns in a 23×23 square. They show the situation at generations 16, 25, 32, and 33, respectively. We kept two generations before erasure for both patterns. We assumed as initial conditions for pattern A one square in the extreme lower left-hand corner and for pattern B one square placed one unit of distance off from the upper right-hand corner. After 33 generations (Fig. 5d) pattern B won, at which time the nth generation squares of pattern A were completely erased. (The $(n-1)^{th}$ generation squares of A will disappear the next generation.)

In another game we have started with two single squares in the same relative positions from the corners. For pattern A we kept one generation before erasure, and for pattern B two generations. In this case A won in 112 generations on the 23×23 board. There is no figure for this contest.

Three-Dimensional Model

We again used all three conditions of the rules of growth in forming a three-dimensional pattern. Figures 6a and 6b show two views of a model of such a pattern. Two-dimensional plots of the pattern were obtained from the computer after 30 generations of growth. The model was constructed from these plots and then photographed. The starting configuration was the single cube on the extreme left of Fig. 6a. This model represents the part grown in one octant of the space. In each octant there is a further threefold symmetry along the coordinate axes, of which we took the part $x > y$, $x > z$. There still remains a plane of symmetry at $45°$ to the x axis. The dark cubes represent the 30th generation elements.

Our examples show both the complexity and the richness of forms obtained from starting with a simple geometrical element and the application of a simple recursive rule. The amount of "information" contained in these objects is therefore quite small, despite their apparent complexity and unpredictability.

If one wanted to define a process of growth which is continuous rather than by discrete steps, the formulation would have to involve functional equations concerning partial derivatives.

It appears to us that a general study of the geometry of objects defined by recursions and iterative procedures deserves a general study— they produce a variety of sets different from those defined by explicit algebraic or analytical expressions or by the usual differential equations.

TABLE I

Width of Strip	Period of Pattern
1	1
2	2
3	3
4	5
5	5
6	8
7	13
8	13
9	13
10	26
11	13
12	91
13	13
14	106
15	106
16	75
17	93

Acknowledgments

The three-dimensional model was constructed by Barbara C. Powell and photographed by W. H. Regan.

Reference

1. S. Ulam in Proceedings of Symposia in Applied Mathematics, **XIV**, American Mathematical Society 1964, p. 215 to 224; see also J. C. Holladay, and S. Ulam, Notices of the American Mathematical Soc. **7** (1960), p. 234; and R. G. Schrandt, and S.Ulam, Notices of the American Mathematical Society **7** (1960), p. 642.

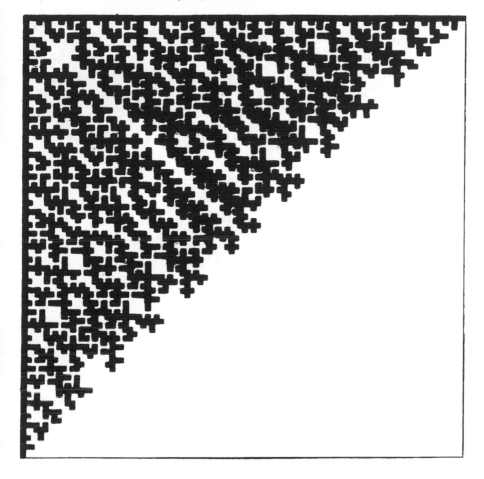

Fig. 1a.   Growth from a single starting square after 100 generations.

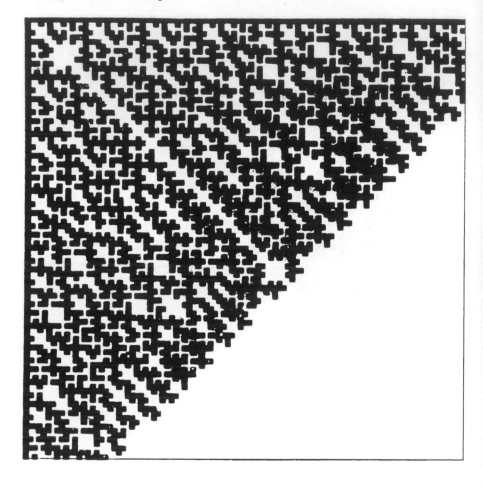

Fig. 1b. Same as Fig. 1a but after 120 generations.

Fig. 2. Growth from three noncontiguous starting squares.

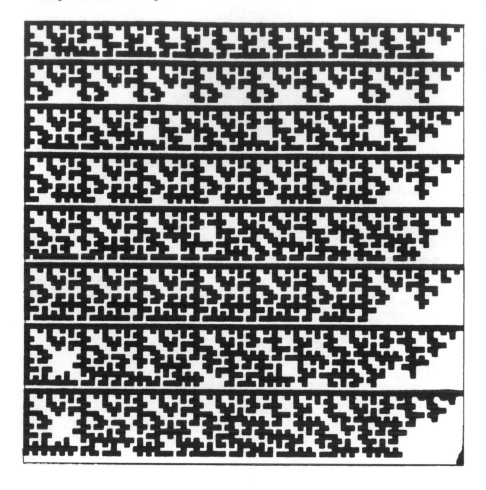

Fig. 3. Patterns generated in an infinite strip of widths 8 to 15, after 100 generations.

Fig. 4a. Growth from a single starting square with death rule, keeping two generations. The nth generation squares are cross hatched, the $(n-1)^{th}$ are blank. The integers at the top are the number of squares in the nth and $(n-1)^{th}$ generation, and the generation number.

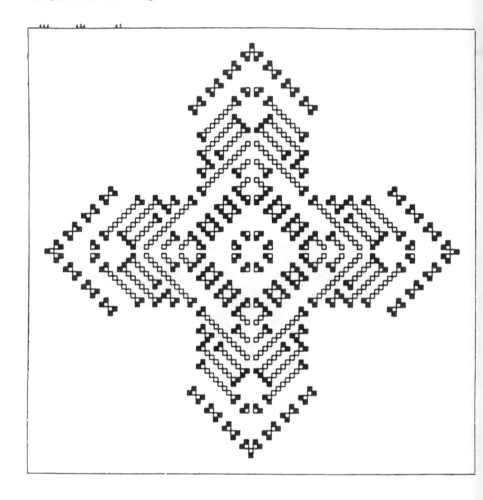

Fig. 4b. Same as Fig. 4a but after 45 generations.

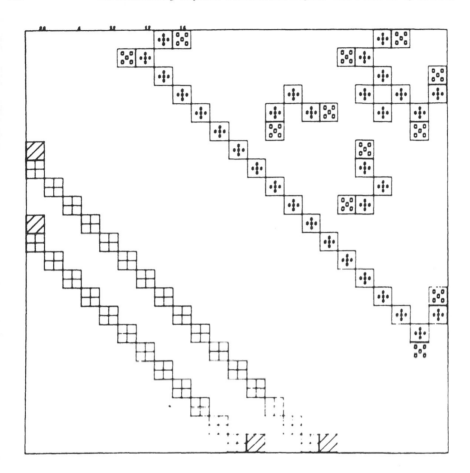

Fig. 5a.   Fight between two different patterns after 16 generations, keeping two generations before erasure. There are 26 $(n-1)^{th}$ generation squares of the lower pattern, and 4 n^{th} generation squares. The upper pattern has 32 $(n-1)^{th}$ generation squares, and 12 n^{th} generation squares.

Analogies between Analogies

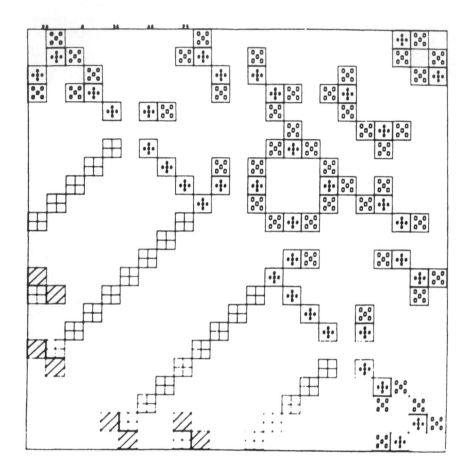

Fig. 5b. Same as Fig. 5a but after 25 generations.

394

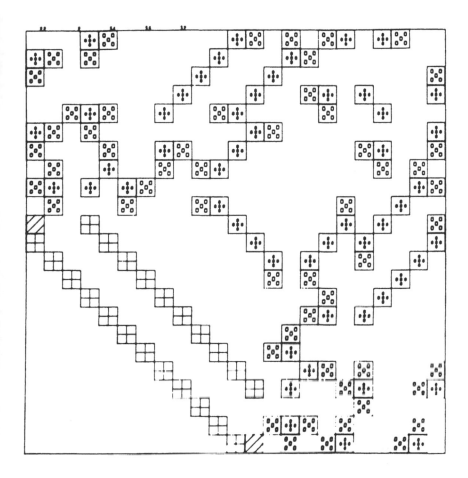

Fig. 5c. Same as Fig. 5a but after 32 generations.

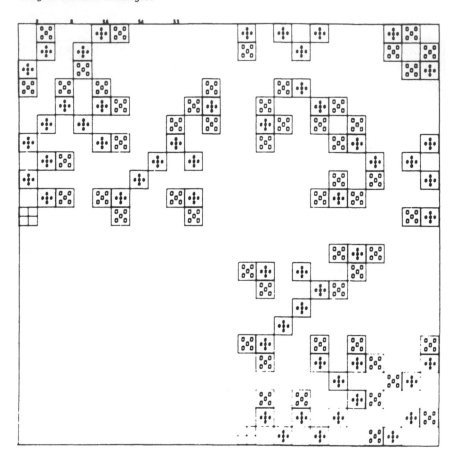

Fig. 5d. Same as Fig. 5a but after 33 generations. Lower pattern has been eliminated.

Fig. 6.   Model of three-dimensional pattern after 30 generations of growth. The starting configuration is the single cube on the extreme right.

13

COMPUTER STUDIES OF SOME HISTORY-DEPENDENT RANDOM PROCESSES

With W. A. Beyer and R. G. Schrandt
(LA-4246, October 28,1969)

This report is about a novel method for the study of several non-independent probability schemata with rather curious results on patterns of growth, iteration processes, and dependent random walks. (Author's note).

ABSTRACT

Various history-dependent random processes are investigated by computer and a few cases are investigated theoretically. These processes include history-dependent random walks, a combination of a birth process and a self-avoiding random walk, history-dependent randomly-generated increasing integer sequences, and randomly-generated integer sequences which might have prime-like densities. A possible random ergodic theorem for history-dependent processes is discussed.

I. Introduction

In this paper we consider some examples of random processes in which the probabilities of the outcome of the nth step depend upon the entire past history of the process. This, of course, means that they are non-Markovian processes. In contrast to Markovian processes, very little is known theoretically about these history-dependent processes. They are much more difficult to analyze. However, the real world abounds with examples of the latter.

Most of the examples are results of computer studies. In some instances, theoretical results are known and are reported here.

The computations were performed on a CDC-6600 machine. The random-number generator needed had the form $x_1 \equiv 5^{15}x_{i-1} \pmod{2^{35}}$ with $x_0 = 5^{15}$). The scales on the figures are linear.

II. Random Walk Examples

A. Self-Avoiding Random Walk

A process which cannot, in any reasonable way, be made into a Markov process is a self-avoiding walk, i.e., a walk on a lattice starting at some fixed point that is not permitted to visit any point more than once. A survey of this topic will be published by Domb.[1] Another version of a self-avoiding walk is a scattering process in which scattering is not permitted at a point that previously had been a scattering center. The physical idea is that the particle, which had been the scattering center, is moved by the scattering process.

B. History-Dependent Walk on the Line (Pólya)

An old example of a history-dependent process is the Pólya urn scheme described by Feller.[2] This has been used as a model of phenomena, such as contagious diseases, where an occurrence of a disease increases the probability of further occurrences.

As a special case of the Pólya scheme on two symbols, consider the following example. Let $a_0 = 0$, $a_1 = 1$ and $a_{n+1} = \widehat{a_i}$ $(i = 0, 1, \ldots, n)$ where $\widehat{a_i}$ means that one of the a_i $(i = 0, 1, \ldots, n)$ is selected at random with uniform probability. Let

$$Y_n = \frac{\sum_{i=0}^{n-1} a_i}{n} .$$

Then Prob. $[Y_n = k/n] = 1/n^{-1}$ for $n > 1$ and $k = 1, 2, \ldots, n-1$. As discussed by Feller in Ref. 3, p. 237, $Y = \lim_{n \to \infty} Y_n$ exists with probability 1, since $[Y_n]$ is a martingale and the distribution of Y is uniform on $[0,1]$; i.e.,

$$\text{Prob. } [0 \leq a < Y < b \leq 1] = b - a .$$

The process Y_n can be interpreted as a random walk on the horizontal line of integers with 0 interpreted as a step to the left and 1 as a step to the right.

C. History-Dependent Walk on a Plane Lattice

A two-dimensional version of the Pólya scheme on two symbols is a walk on the lattice of plane points with integer coordinates starting at the origin and is executed according to the following rule. The decision to execute either a horizontal step or a vertical step is made independently each time with Prob. [horizontal step (H)] = Prob.[vertical step (V)] = 1/2. After the decision H or V is made, a second decision is made. In the case of H, a decision is made to take the step right or left in accordance with the Pólya two-symbol game discussed previously. In a similar way, a decision is made to take a step up or down in the case of V. Figure 1 shows the terminal points of 10,000 walks of 64 steps each for walks made by these rules.

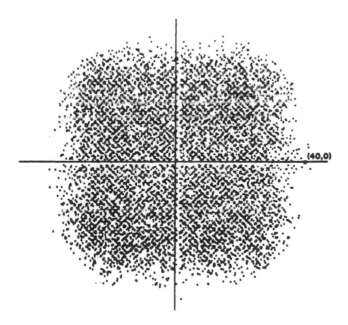

Fig. 1. End points of 10,000 random walks of 64 steps on the plane quadratic lattice starting at the origin in which the decision to execute a horizontal or vertical step is made independently with equal probability. The steps horizontally and vertically are made in accordance with the Pólya two-symbol game.

For comparison, the distribution of 10,000 walks of 64 steps in the case of classical Pólya walks is shown in Fig. 2. Classical Pólya walk is the walk on the points with integer coordinates starting at the origin and selecting one of the nearest neighbors with equal probability. The formula for the probability that the nth step takes the particle to (x, y) is

$$\frac{1}{4\pi^2 2^n} \int_{-\pi}^{\pi} \int_{-\pi}^{\pi} (\cos \alpha + \cos \beta)^n \cos x\alpha \cos y\beta \, d\alpha \, d\beta^* \ .$$

It seemed easier to generate the distribution of end points by a Monte Carlo procedure than to use this formula.

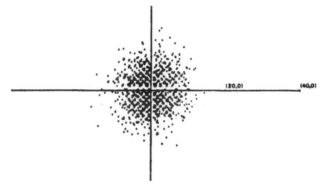

Fig. 2. End points of 1000 random walks of 64 steps on the plane quadratic lattice starting at the origin in the classical Pólya case.

A second two-dimensional version of the Pólya scheme on two symbols is as follows. The probability of a horizontal or vertical step is itself a Pólya scheme on two symbols. The remaining decisions are made as before and the resulting distribution is shown in Fig. 3. It is peaked about the coordinate axes in an approximately hyperbolic manner.

D. A History-Dependent Explosion

A plane configuration on the plane lattice of points with integer coordinates is generated. Let the origin be the first generation. Assuming that certain of the lattice points have been occupied by generations

---

* See ref. 2 p. 371.

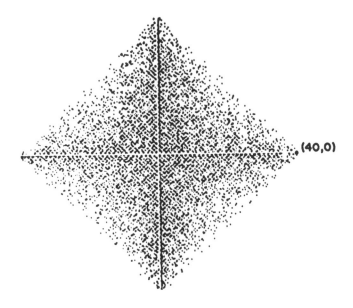

Fig. 3. End points of 1000 random walks of 64 steps on the plane quadratic lattice starting at the origin in which the sequence of decisions to execute a horizontal or vertical step forms a Pólya two-symbol game. The steps horizontally and vertically are made as in Fig. 1.

up to and including the nth generation, let the $(n+1)^{th}$ generation be determined as follows. For each point of the nth generation, two random numbers are selected to determine two neighbor points from four possible neighbor points for possible positions to be occupied in the $(n^{th}+1)^{th}$ generation. These two neighbor points are to be occupied provided they have not been previously occupied. As an example (see figure below), suppose the point (0) is the initial point. Assume that positions (1) and (2) are chosen by the random numbers as the new positions. Then, in the first generation, the walk is made from point (0) to points (1) and (2), because they are unoccupied. Now there are two terminal points, (1) and (2), from which to walk in the second generation. If the directions $(1) \to (0)$ or $(2) \to (0)$ are now chosen as one of the new directions, this walk is not executed, because point (0) is occupied. If both directions (1) → (3) and (2) → (3) are chosen, the first walk is executed, but the second is not. The walk terminates if all possible directions chosen for all terminal points lead to points that are occupied.

Two computer examples are given for this type of walk. In the first case, a maximum of two random numbers were chosen for the possible new directions of the walk for each terminal point. The same direction could occur twice if the two random numbers were in the same interval

$$\frac{k}{4} \le r_1, r_2 < \frac{k+1}{4}, \ k = 0, \ 1, \ 2, \ or \ 3 \ .$$

In this case, only one direction for the walk was allowed for this terminal point. In this run, the walk terminated after 108 generations, with 656 total points. This is plotted in Fig. 4.

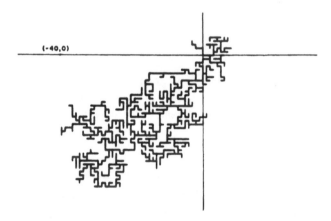

(-40,0)

Fig. 4. A history-dependent explosion. The rules for generation of this explosion are explained in the text.

In the second example, this situation was not allowed because the second random number was rejected and a new number was generated until two distinct new directions were obtained for each terminal point. This walk apparently does not stop. In Fig. 5 the walk is plotted through 65 generations, with 3126 points occupied. There are 100 active terminal points in the sixty-fifth generation.

The explosion is in some respects similar to a self-avoiding random walker, except the walker multiplies. Put in other terms, the explosion is in some respects similar to a branched polymer.[4]

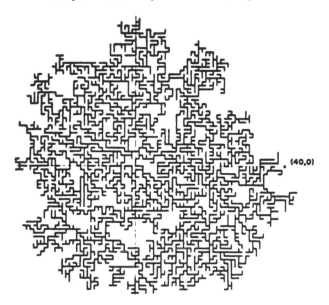

(40,0)

Fig. 5. A second history-dependent explosion with different rules for generation as explained in the text.

III. Integer Sequences Generated by History-Dependent Random Processes

In this section, sequences of integers obtained from the following three random processes are discussed.

(a) $a_{n+1}=a_n+ \widehat{a_i},$ $i = 1,\ldots, n$
(b) $a_{n+1}=\widehat{a_i} + \widehat{a_i},$ $i, j = 1,\ldots, n$
(c) $a_{n+1} = \widehat{a_i} - \widehat{a_j},$ $i, j = 1,\ldots, n, \; (i > j)$

where $\widehat{a_i}$ has the meaning given above. In (a), a_1 is given. In (b) and (c), a_1, a_2 are given.

For (a), we will only mention that Kac and Ulam have shown that the expected value of a_n, $E(a_n)$, is asymptotic for large n to $e^{\sqrt{n}}$.

In case (b), it is easy to show by induction that

$$E(a_n) = \frac{1}{3} \, (a_1 + a_2)n \text{ for } n > 2 \, .$$

One thousand sample sequences of case (b) were obtained by a Monte Carlo sampling with $a_1 = -1$, $a_2 = 1$. Let a_n^k be the nth member of the kth sequence with $1 \le n \le 100$ and $1 \le k \le 1000$. The averages

$$b_n = \frac{1}{1000} \sum_{k=1}^{1000} a_n^k$$

are plotted in Fig. 6 as a function of n. In this case, $E(a_n) = 0$, $n = 3, 4, \ldots$. It is seen that the averages increase with n.

Fig. 6. The averages over 1000 sequences generated by $a_{n+1} = \widehat{a_i} + \widehat{a_j}$, $ij = 1, \ldots, n$ with $a_1 = -1$ and a_2 as a function of n.

In Fig. 7, a second example is given with $a_1 = 0$ and $a_2 = 1$ and $1 \le k \le 5000$. Then $E(a_n) = 1/3n$. However, we plot the deviation

$$\frac{1}{5000} \sum_{k=1}^{5000} \left| a_n^k - \frac{1}{3}n \right|$$

as a function of n. The quantity apparently increases linearly with n.

In the case (c), it can be shown that the expected value of a_n, $E_n = E(a_n)$, satisfies the recurrence relation

$$E_{n+2} = \frac{2n}{n+1} E_{n+1} - \frac{n^2 - n + 2}{n(n+1)} E_n \quad (n = 1, 2, \ldots) .$$

The asymptotic value of E_n is discussed in the Appendix and it is shown that $E_n = 0(1/n)$. Figures 8 and 9 show a graph of E_n for a_1, $a_2 = 0$ and 1 and a_1, a_2, $= 1$, 0, respectively.

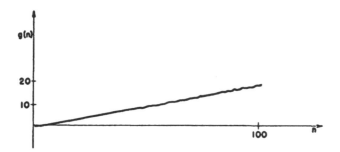

Fig. 7.   Graph of the function $g(n) = 1/5000 \sum_{k=1}^{5000} | a_n^k - E(a_n) |$ for the case where $a_{n+1} = \widehat{a_i + a_j}$ with $a_1 = 0$, $a_2 = 1$. $\{a_n^k\}_{k=1,\dots,5000}$ is a random sampling of 5000 sequences.

Fig. 8.   Graph of the function E_n defined by $E_{n+2} = [2n/(n+1)]E_{n+1} - [(n^2 - n+2)/n(n+1)] E_n$ which is the expected value of a_n where $a_{n+1} = \widehat{a_i - a_j}$, $i,j = 1, 2, \dots, n$. Here $a_1 = 0$ and $a_2 = 1$.

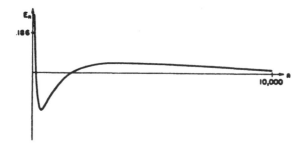

Fig. 9.   Same as Fig. 8, except that $a_1 = 1$, $a_2 = 0$.

407

Figure 10 shows a graph of the quantity

$$b_n = \frac{1}{1000} \sum_{k=1}^{1000} a_n^k, \; n = 1, \; \ldots, \; 100$$

where a^k is the nth member of the kth sample sequence for case (c) $a_1 = -1$, $a_2 = 1$. The average increases with n, but here n remains small.

Figure 11 is a graph of the function

$$b_n = \frac{1}{500} \sum_{k=1}^{500} \mid a_n^k - E(a_n) \mid$$

for $n = 1, \; \ldots, \; 1000$ with $a_1 = 0$, $a_2 = 1$. $[a_n^k]$ are 500 sample sequences of case (c). This function appears to be parabolic.

Fig. 10. The averages over 1000 sequences generated by $a_{n+1} = \widehat{a_i} - \widehat{a_j}, i, j = 1, \; \ldots, \; n$ with $a_1 = -1$, $a_2 = 1$ as a function of n.

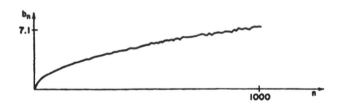

Fig. 11. The function $b_n = 1/500 \sum_{k=1}^{500} |a_n^k - E(a_n)|$ for $n = 1, 2, \ldots, 1000$ with $a_{n+1} = \widehat{a_i} - \widehat{a_j}$ and $a_1 = 0$, $a_2 = 1$. $\{a_n^k\}_{k=1,\ldots,500}$ is a random sampling of 500 sequences.

IV. Number Theoretical Games

In this section, we discuss sequences of positive integers generated by a history-dependent random process which might have densities like the primes.

Let d_1, d_2, ..., d_k be the sequence of differences of the first k primes. Let $d_{k+1} = \widehat{d_i}$ ($i = 1, \ldots, k$) and $d_{k+2} = \widehat{d_\ell} + 2$ ($\ell = 1, \ldots, k; \ell \neq i$). In general, let $d_{k+2j-1} = \widehat{d_i}$ ($i = 1, \ldots, k + 2j - 2$) and $d_{k+2j} = \widehat{d_\ell} + 2$ ($\ell = 1, \ldots, k + 2j - 2; \ell \neq i$) for $j = 1, 2, \ldots$. Let $s_0 = 1$, $s_i = s_{i-1} + d_i$ ($i = 1, \ldots, n$).

The following results are obtained.

1. For $k = 8$ and $d_1, \ldots, d_8$, the first eight real prime differences, i.e., 1, 2, 2, 4, 2, 4, 2, 4, and $n = 1000$, 200 "games" are played. The average of s_{1000} is 7986. The one-thousandth real prime is 7919. The average number of real primes in s_1, ..., s_{1000} is about 26%.

2. $k = 49$, d_1, ..., d_{49} are the first 49 real prime differences. Two hundred "games" are played and the average s_{1000} is 7643. The average number of real primes in s_1, ..., s_{1000} is about 30%. In 20 "games" it is found that the average $s_{10,000} = 101,356$, whereas the $10,000^{th}$ prime is 104,723. The average number of primes in s_1, ..., $s_{10,000}$ is 20%.

A second history-dependent random process for generating sequences with perhaps prime-like density is similar to the above. Again, let d_1, d_2, ..., d_k be the sequence of differences of the first k primes with k even. Let $d_{k+1} = \widehat{d_i}$ ($i = 1, 2, \ldots, k/2$) and $d_{k+1} = \widehat{d_i} + 2$ ($i = k/2 + 1, \ldots, k$), etc., as before. For $n = 1000$ with 200 "games," the average value of s_{1000} is 7892 and an average of 26% of the primes up to 7892 were obtained.

P. Stein[5] has used such sequences to generate minimal binary additive bases for the even integers.

V. A Problem

Starting with Ulam and von Neumann[6] and Pitt[7] there have been various versions of the random ergodic theorem stated and proved. One version is as follows.[8] Let $[X, \sum, \mu]$ be a measure space and $[\mathcal{J}, \Delta, p]$ be a probability measure space with $\mathcal{J}$ a set of measure-preserving transformations defined on X. Let $\mathcal{J}^* = \prod_{i=1}^{\infty} \mathcal{J}_i$ where $\mathcal{J}_i = \mathcal{J}$ for all i and $p^* = \prod_{i=1}^{\infty} p_i$ where $p_i = p$ for all i. Then if $F(x) \epsilon L^1(X, \mu)$, it follows that

$$p^* \left\{ \lim_{n \to \infty} \frac{1}{n} \sum_{k=1}^{n} f(T_k \ldots T_1 x) = f^*(x) \text{ for almost all } x \text{ in } X \right\} = 1$$

where $(T_1, T_2, \ldots)$ is a point of $\mathcal{J}^*$ and $f^*(x) \epsilon L^1(X, \mu)$.

In the above theorem, the transformations T_i are chosen independently. One can ask, "Suppose the T_i are not chosen independently, but are chosen to form a Markov chain, or are chosen in accordance with the rules governing a Pólya process (see above), does some random ergodic theorem hold?" One could specialize to the case of choosing from two transformations, each of which is a rotation of the circle.

Appendix

Theorem. $E(k) = O(1/k)$ where

$$k(k+1)E_{k+2} - 2k^2 E_{k+1} + (k^2 - k + 2)E_k = 0 \quad (k = 1, 2, \ldots) . \quad (1)$$

Proof. Define the vector

$$\mathcal{E}(k) \equiv \begin{bmatrix} e^1(k) \\ e^2(k) \end{bmatrix} \equiv \begin{bmatrix} e(k) \\ e(k+1) \end{bmatrix} .$$

Then Eq. (1) is equivalent to the vector equation

$$\mathcal{E}(k+1) = \begin{bmatrix} 0 & 1 \\ -\frac{k-1}{k+1} & -\frac{2}{k(k+1)} \quad \frac{2k}{k+1} \end{bmatrix} \mathcal{E}(k) . \quad (2)$$

We first show that any vector solution to Eq. (2) is bounded for all $k \geq 1$ in the norm $\|\mathcal{E}\| = |\mathcal{E}^1| + |\mathcal{E}^2|$. Write

$$B(k) \equiv \begin{bmatrix} 0 & 1 \\ -\frac{k-1}{k+1} & \frac{2k}{k+1} \end{bmatrix} ,$$

and

$$A(k) \equiv \begin{bmatrix} 0 & 0 \\ -\frac{2}{k(k+1)} & 0 \end{bmatrix} .$$

Consider

$$\bar{\mathcal{E}}(k+1) = B(k)\bar{\mathcal{E}}(k) . \tag{3}$$

A fundamental matrix set of solutions for Eq. (3) (Ref. 9) is given by

$$Y_k = \begin{bmatrix} 1 & \frac{1}{k-1} \\ 1 & \frac{1}{k} \end{bmatrix} \tag{4}$$

for $k = 2, 3, \ldots.$ From work shown in Ref. 9, any solution $\mathcal{E}(k)$ to Eq. (2) for $k \geq 2$, can be expressed by

$$\mathcal{E}(k) = \sum_{s=2}^{k-1} Y_{k-s+1} \, Y_2^{-1} \, A(s+1) \, \mathcal{E}(s) + Y_k Y_2^{-1} \mathcal{E}(2) . \tag{5}$$

Since

$$\sum_{k=1}^{\infty} A(k) < \infty ,$$

it follows from Lemma 3.2 of Ref. 9, p. 21, and Eq. (5) that $|\mathcal{E}(k)|$ is bounded for $k \geq 2$ and hence for $k \geq 1$. Now a computation gives

$$Y_{k-s+1} \, Y_2^{-1} \, A(s+1) \, \mathcal{E}(s) = - \begin{bmatrix} 1 & - & \frac{1}{k-s} \\ 1 & - & \frac{1}{k-s+1} \end{bmatrix} \frac{4\mathcal{E}^1(s)}{(s+1)(2+2)} ,$$

and

$$Y_k Y_2^{-1} \mathcal{E}(2) = \begin{bmatrix} -\mathcal{E}^1(2) + 2\mathcal{E}^2(2) + \frac{2}{k-1}\mathcal{E}^1(2) - \frac{2}{k-1}\mathcal{E}^2(2) \\ -\mathcal{E}^1(2) + 2\mathcal{E}^2(2) + \frac{2}{k}\mathcal{E}^1(2) - \frac{2}{k}\mathcal{E}^2(2) \end{bmatrix} .$$

Thus

$$\mathcal{E}(k) = - \sum_{s=2}^{k-1} \begin{bmatrix} 1 & - & \frac{1}{k-s} \\ 1 & - & \frac{1}{k-s+1} \end{bmatrix} \frac{4\mathcal{E}^1(s)}{(s+1)(s+2)}$$

$$+ \begin{bmatrix} -\mathcal{E}^1(2) + 2\mathcal{E}^2(2) + \frac{2}{k-1}\mathcal{E}^1(2) - \frac{2}{k-1}\mathcal{E}^2(s) \\ -\mathcal{E}^1(2) + 2\mathcal{E}^2(2) + \frac{2}{k}\mathcal{E}^1(2) - \frac{2}{k}\mathcal{E}^2(2) \end{bmatrix} \tag{6}$$

for $k \geq 2$. The summation can be written

$$-\sum_{s=2}^{k-1}\begin{bmatrix} 1 & - & \frac{1}{k-s} \\ 1 & - & \frac{1}{k-s+1} \end{bmatrix}\frac{4\mathcal{E}^1(s)}{(s+1)(s+2)} =$$

$$-4\sum_{s=2}^{k-1}\frac{\mathcal{E}^1(s)}{(s+1)(s+2)} + 4\sum_{s=2}^{k-1}\begin{bmatrix} \frac{1}{k-s} \\ \frac{1}{k-s+1} \end{bmatrix}\frac{\mathcal{E}^1(s)}{(s+1)(s+2)} . \qquad (7)$$

Since $\mathcal{E}^1(s)$ is bounded, it is seen that the first term on the right of Eq. (7) can be written

$$-4\sum_{s=2}^{k-1}\frac{\mathcal{E}^1(s)}{(s+1)(s+2)} = -4\sum_{s=2}^{\infty}\frac{\mathcal{E}^1(s)}{(s+1)(s+2)}$$

$$+4\sum_{s=k}^{\infty}\frac{\mathcal{E}^1(s)}{(s+1)(s+2)} = C + O\left(\frac{1}{k}\right), \qquad (8)$$

where

$$C = -4\sum_{s=2}^{\infty}\frac{\mathcal{E}^1(s)}{(s+1)(s+2)} \quad \text{is a constant.}$$

Now consider the second term on the right of Eq. (6). For $2 \leq s \leq k/2$

$$\frac{1}{k-s} \leq \frac{2}{k}$$

$$\frac{1}{k-s+1} \leq \frac{1}{(k/2+1)} = \frac{2}{k+2} = O\left(\frac{1}{k}\right),$$

and for $k/2 < s \leq k-1$

$$\frac{1}{(k-s)(s+1)(s+2)} \leq \frac{1}{(k/2+1)(k/2+2)} = \frac{4}{(k+2)(k+4)},$$

$$\frac{1}{(k-s+1)(s+1)(k+2)} \leq \frac{2}{(k+2)(k+4)} .$$

Thus, with [ ] denoting integer part, we have

$$\sum_{s=2}^{k-1} \frac{\mathcal{E}^1(s)}{(k-s)(s+1)(s+2)} = \sum_{s=2}^{[k/2]} \frac{\mathcal{E}^1(s)}{(k-s)(s+1)(s+2)}$$

(9)

$$+ \sum_{s=[k/2]+1}^{k-1} \frac{\mathcal{E}^1(s)}{(k-s)(s+1)(s+2)} = O(1/k) + kO(1/k^2) = O(1/k) \ ,$$

and similarly for

$$\sum_{s=2}^{k-1} \frac{\mathcal{E}^1(s)}{(k-s+1)(s+1)(s+2)} \ .$$

Hence, from Eqs. (6), (8), and (9), we have

$$\mathcal{E}(k) = \bar{\lambda} + O(1/k) \ ,$$

where $\bar{\lambda} = [\begin{smallmatrix}\lambda\\\lambda\end{smallmatrix}]$ is a constant vector. Therefore,

$$\mathcal{E}(k) = \lambda + O(1/k) \ .$$

Substituting this into Eq. (1), we obtain

$$[k(k+1) - 2k^2 + k^2 - k + 2] \ \lambda + [k(k+1) - 2k^2 + k^2 - k + 2]O(1/k)$$

$$= 2(\lambda + O(1/k)) = 0 \ .$$

Thus, $\lambda = 0$, which completes the proof.

An Approximating Differential Equation

If Eq. (1) is written in the form

$$\frac{E(k+2\Delta k) - 2E(k+\Delta k) + E(k)}{(\Delta k)^2} + \frac{2}{k+1} \frac{E(k+\Delta k) - E(k)}{\Delta k}$$

$$+ \frac{2}{k(k+1)} \ E(k) = 0 \ ,$$

with $\Delta k = 1$, then the differential equation

$$e'' + \frac{2}{k+1} e' + \frac{2}{k(k+1)} e = 0 , \qquad (10)$$

could be regarded as an approximation to Eq. (1).

In Eq. (10), k is regarded as a continuous variable. It can be shown that the general solution to Eq. (10) is a linear combination of the functions

$$F\left(\frac{1+i\sqrt{7}}{2} , \frac{1-i\sqrt{7}}{2} ; 2, -k\right) ,$$

and

$$F\left(\frac{1-i\sqrt{7}}{2} , \frac{1+i\sqrt{7}}{2} ; 2, 1+k\right) ,$$

where F denotes the Gauss hypergeometric function. Hence, for large k, the solution to Eq. (10) is a linear combination of the functions e_1 and e_2 where

$$e_1 = \frac{1}{\sqrt{k}} \cos\left[\frac{\sqrt{7}}{2} \log k\right] + O\left(\frac{1}{k^{3/2}}\right) ,$$

and

$$e_2 = \frac{1}{\sqrt{k}} \sin\left[\frac{\sqrt{7}}{2} \log k\right] + O\left(\frac{1}{k^{3/2}}\right) .$$

Regarding Eq. (10) as an approximation to Eq. (1), it can be shown from the theory of difference approximations to differential equations that if k_0 is fixed and $E(k_0) = e(k_0)$ as well as $E(k_0 + 1) = e(k_0 + 1)$, then for $k > k_0$ we have

$$\frac{|\,e(k) - E(k)\,|}{|\,e(k)\,|} \leq \frac{|\,e''(\xi)\,|}{4\,|\,e(k)\,|} \left[e^{k-k_o} - 1\right] , \qquad (11)$$

where $k_0 < \xi < k$. Thus, while it is true that $e''(\xi) \rightarrow \infty$, the exponential factor in Eq. (11) prevents Eq. (10) from being a very good approximation to Eq. (1) for arbitrarily large k. This difference equation can also be discussed, perhaps more successfully, by using generating functions.

References

1. C. Domb, "Self-avoiding Walks on Lattices," to be published in Advan. Chem. Phys.
2. W. Feller, *An Introduction to Probability Theory and Its Applications,* (John Wiley and Sons, New York 1968), Vol. 1, 3rd ed.
3. W. Feller, *An Introduction to Probability Theory and Its Applications,* John Wiley and Sons, New York 1966), Vol. 2.
4. L. V. Gallagher and S. Windmer, "Monte Carlo Study of Flexible Branched Macromolecules," J. Chem. Phys. **44**, 1139 (1966).
5. P. Stein, private communication.
6. S. M. Ulam and J. von Neumann, "Random Ergodic Theorem," Bull. Am. Math. Soc. **51**, 660 (1945).
7. H. R. Pitt, "Some Generalizations of the Ergodic Theorem," Proc. Cambridge Phil. Soc. **38**, 325 (1942).
8. P. Revesz, "A Random Ergodic Theorem and its Application in the Theory of Markov Chains," in *Ergodic Theory,* (Academic Press, New York, 1963) Fred. B. Wright, Ed.
9. K. S. Miller, *Linear Difference Equations,* (W. A. Benjamin, Inc., New York, 1968).

Everett and Ulam in Madison, Wisconsin, in 1941.

14

THE ENTROPY OF INTERACTING POPULATIONS

With C. J. Everett
(LA-4256, August 1969)

This report is a novel probabilistic approach to defining distributions of functionals of thermodynamical systems, including for instance, interactions between radiation and particles. (Author's note).

ABSTRACT

A study is made of interacting populations of "particles" which closely parallels the Boltzmann kinetic theory and the Planck-Einstein-Tolman treatment of radiation interacting with matter. The analogy is perhaps surprising since it appears that our postulates do *not* embody the physics of such systems, but are nevertheless quite reasonable, and applicable to similar situations. While we have deliberately used the language of physics for its intuitive appeal, one may well consider for example the implications of replacing "particles" by "people" and "energy" by "wealth." It is especially interesting that the "reversibility paradox" is excluded by confining the discussion to a scalar "energy" rather than a vector "velocity." As Boltzmann might have said, "Go ahead, reverse the energy."[1]

I. The System of "Particles"

1. The Interaction Postulates. We consider a system of "particles," of which there are $N(E, t)dE$ possessing "energy" on $(E, E+dE)$, $0 \leq E < \infty$, at time $t \geq 0$, and undergoing $c(E_1, E_2)N(E_1, t)dE_1 N$ $(E_2, t)dE_2$ "collisions," or binary E_1, E_2-interactions per unit time, between particles on the indicated ranges, where $c(E_1, E_2) > 0$ is a function of form $f(E_1 + E_2)$. We further assume that the (probable) number $\psi(E_1, E_2, E)dE$ of particles emerging on $(E, E + dE)$ from such a collision has the properties

ψ1. $\psi(E_1, E_2, E) > 0$; $0 < E < E_1 + E_2$, $0 < (E_1 \& E_2) < \infty$

ψ2. $\psi(E_1, E_2, E) = \psi(E_2, E_1, E)$

ψ3. $\psi(E_1, E_2, E) = \psi(E, E_1 + E_2 - E, E_1)$

ψ4. $\displaystyle\int_0^{E_1+E_2} \psi(E_1, E_2, E)dE = 2$.

From ψ2, ψ3 follows the symmetry

ψ5. $\psi(E_1, E_2, E) = \psi(E_1, E_2, E_1 + E_2, -E)$ of the ψ-distribution about its midpoint, and hence from ψ4 follows

ψ6. $\displaystyle\int_0^{E_1+E_2} E\psi(E_1, E_2, E)dE = E_1 + E_2$.

The uniform distribution $\psi(E_1, E_2, E) = 2/(E_1 + E_2)$ is by no means the only one having these properties. For example, $4(E_1 + E_2)^{-1} \sin^2 2\pi (E_1 + E_2)^{-1} (E_1 + E)$, $0 < (E_1 \& E) < (E_1 + E_2)/2$, symmetrically extended to the full interval $(0, E_1 + E_2)$, yields such a ψ-function (cf. Appendix I).

2. The Boltzmann Equation. The above assumptions imply the equation

$$\partial N(E)/\partial t = \int\int_{E_1+E_2 \geq E} \frac{1}{2} c(E_1, E_2)N(E_1)N(E_2)\psi(E_1, E_2, E)dE_1 dE_2$$
$$- \int_0^{\infty} c(E, E_2)N(E)N(E_2)dE_2 \equiv B_1 - B_2 \equiv B_3 \qquad (1)$$

for the change of $N(E,t)$ with time. (N.B. Hereafter *all* functions $N(E,t)$ are written $N(E)$ for brevity.)

For a solution of (1), we verify at once the conservation laws

$$\frac{d}{dt}\int_0^\infty N(E)dE = 0 = \frac{d}{dt}\int_0^\infty EN(E)dE$$

showing the values N_0, E_0 of these integrals to be constant in time. This rests on the fact that the properties of c and ψ imply

$$\int_0^\infty B_3 dE = 0 = \int_0^\infty EB_3(E)dE , \tag{2}$$

that is to say

$$\int_0^\infty B_1 dE = \int_0^\infty B_2 dE \text{ and } \int_0^\infty EB_1(E)dE = \int_0^\infty EB_2(E)dE . \tag{3}$$

For, these B_1 integrals may be written in the form

$$\int_{E_1=0}^\infty \int_{E_2=0}^\infty \int_{E=0}^{E_1+E_2} ,$$

and the relations in (3) follow from $\psi 4$, and from $\psi 6$.

We next throw (1) into the equivalent form

$$\partial N(E)/\partial t = \int_{E_2=0}^\infty \int_{E_1=0}^{E+E_2} \frac{1}{2}c(E,E_2)\psi(E,E_2,E_1) \tag{4}$$
$$\{N(E_1)N(E+E_2-E_1) - N(E)N(E_2)\}dE_1 dE_2 \equiv B_1 - B_2 \equiv B_3 .$$

To see this, one makes the transformation $E_1 = F_1$, $E_2 = F + F_2 - F_1$ in B_1, using $c(E_1, E_2) = f(E_1 + E_2)$ and $\psi 3$, and uses $\psi 4$ to express B_2 as a double integral.

3. The Boltzmann H-function. For a solution $N(E)$ of (1,4) we now define (ignoring the nice question of units)

$$H_N(t) = \int_0^\infty N(E) \log N(E)dE ,$$

for which

$$\frac{d}{dt}H_N(t) = \int_0^\infty \partial N(E)/\partial t \log N(E)dE = \int_0^\infty B_3 \log N(E)dE$$

(since N_0 is invariant). Using the form of B_3 in (4), and making the change of notation $(E, E_1) \rightarrow (E_1, E)$, the last integral becomes

$$\int_{E_1=0}^{\infty} \int_{E_2=0}^{\infty} \int_{E=0}^{E_1+E_2} \frac{1}{2} c(E_1, E_2,) \psi(E_1, E_2, E)$$

$$\{N(E)N(E_1 + E_2 - E) - N(E_1)N(E_2)\} \log N(E_1) dE_2 dE_1 .$$

Due to the symmetry in E_1, E_2, this is equal to the same integral with $N(E_2)$ replacing $N(E_1)$ in the log factor. Averaging the two results, we obtain

$$\int_0^{\infty} B_3 \log N dE = \int_{E_1=0}^{\infty} \int_{E_2=0}^{\infty} \int_{E=0}^{E_1+E_2} \frac{1}{4} c(E_1, E_2,) \psi(E_1, E_2, E).$$

$$\{N(E)N(E_1 + E_2, -E) - N(E_1)N(E_2)\} \log N(E_1)N(E_2) dE dE_2 dE_1 .$$

If we now make the transformation $E_1 = F$, $E_2 = F_1 + F_2 - F$, $E = F_1$, use $c(E_1, E_2) = f(E_1 + E_2)$ and $\psi 3$, and change notation $(F_1, F_2, F) \rightarrow (E_1, E_2, E)$ we obtain the same formula with the log factor replaced by $- \log N(E)N(E_1 + E_2 - E)$. Averaging once more, we obtain finally

$$H'_N(t) = \int_0^{\infty} \partial N(E)/\partial t \log N(E) dE = \int_0^{\infty} B_3 \log N(E) dE \equiv$$

$$- \int_0^{\infty} \int_0^{\infty} \int_0^{E_1+E_2} \frac{1}{8} c(E_1, E_2) \psi(E_1, E_2, E)$$

$$\{N(E)N(E_1 + E_2 - E) - N(E_1)N(E_2)\} \tag{5}$$

$$\log \frac{N(E)N(E_1 + E_2 - E)}{N(E_1)N(E_2)} dE dE_2 dE_1 \leq 0$$

with equality iff (at the time t in question),

$$\{N(E)N(E_1 + E_2 - E) - N(E_1)N(E_2)\} \equiv 0 ; $$
$$0 \leq E \leq E_1 + E_2 < \infty . \tag{6}$$

(For, $c > 0$, $\psi > 0$, and $x - y \leq (\geq)0$ as $\log x/y \leq (\geq)0$) .

Moreover, it is well known that a (continuous) function $N(E)$ satisfies (6) iff it is of form $\beta e^{-\alpha E}$ (cf. Appendix II).

4. The Steady State Solution. If $N_0(E)$ is a time *independent* function, the following conditions are now seen to be equivalent:

S1. $N_0(E)$ satisfies (4) (steady state solution)

S2. For $N = N_0(E)$, $B_3(E) \equiv 0$ in (4)

S3. For $N = N_0(E)$, $\int_0^\infty B_3(E) \log N(E)dE = 0$ in (5)

S4. For $N = N_0(E)$, $\{N(E)N(E_1 + E_2 - E) - N(E_1)N(E_2)\} \equiv 0$ in (6)

S5. $N_0(E) = \beta e^{-\alpha E}$ (with $\alpha = N_0/E_0$, $\beta = N_0\alpha$ if we stipulate the totals N_0, E_0).

It is here apparent that we do *not* have Boltzmann statistics, for which $N_0(E)dE = \beta E^{1/2}e^{-\alpha E}$, with $\alpha = (3/2)(N_0/E_0)$ and $\beta = 2N_0\alpha^{3/2}/\pi^{1/2}$ (see ref. 2).

5. Time Dependent Solutions. If $N(E,t)$ is a solution of (4), with invariants N_0, E_0, which is *not* the solution $N_0(E)$ in S5 at any t, then

$$H_N(t) = \int_0^\infty N(E) \log N(E)dE$$

is strictly decreasing, since $H_N'(t) < 0$, $t \geq 0$, in (5). Moreover, we shall now prove

$$H_N(t) > H_{N_0}; \ t \geq 0 \tag{7}$$

and hence the existence of $\lim_{t\to\infty} H_N(t) \equiv H_N^* \geq H_{N_0}$.

To verify (7), note first that, if $M(E)$ is *either* $N(E)$ or $N_0(E)$, then

$$\int_0^\infty M \log N_0 dE = \int_0^\infty M (\log \beta - \alpha E)dE = N_0 (\log \beta - 1) ,$$

because both have the same totals N_0, E_0. Since these integrals have the same value, we have

$$H_N(t) - H_{N_0} = \int_0^\infty N \log N dE - \int_0^\infty N_0 \log N_0 dE$$

$$= \int_0^\infty \{N \log N/N_0 + N_0 - N\} \, dE > 0 .$$

(For the integrand is of form $f(x) = x \log x/x_0 + x_0 - x$, with $f(x_0) = 0$, and $f'(x) = \log x/x_0 \leq (\geq) 0$ as $x \leq (\geq) x_0$. Thus $H_N(t) - H_{N_0} \geq 0$, and if equality held for any t_0, we should have $N(E, t_0) \equiv N_0(E)$.)

If, in addition, $N(E,t)$ is sufficiently well-behaved, with the limits $\lim\limits_{t\to\infty} H'_N(t) = 0$, and $\lim\limits_{t\to\infty} N(E,t) = N^*(E)$, we may conclude from (5) that $N^*(E) = N_0(E)$, i.e., the time dependent solution $N(E,t)$ of (4) approaches the steady state.

We do not investigate the existence of these limits; apparently they have not been established even in the simplest kinetic theory. (See however ref. 1.) It is clear that the second limit implies the first, since (5) then shows the *existence* of $\lim\limits_{t\to\infty} H'_N(t) \equiv C$, and by the theorem of the mean, $H'_N(\tau) = (H_N(t+\Delta t) - H_N(t))/\Delta t \to (H^*_N - H^*_N)/\Delta t = 0 = C$.

II. A Linked System of "Particles" and "Photons"

6. Interaction Assumptions. We now consider a system of "particles" and "photons," of which there are respectively $N(E,t)dE$ and $\widetilde{N}(E,t)dE$ on $(E, E+dE)$ at time $t \geq 0$, $0 < E < \infty$. Particles are subject to the rules of §1, while photon-particle interactions are governed by two positive functions $A(E_2, E_1)$, $B(E_2, E_1)$, $0 < E_1 < E_2 < \infty$, according to the Einstein postulates:

P1. $N(E_2, t)dE_2 B(E, E_2)\widetilde{N}(E - E_2, t)dE$ gives the number, per unit time, of $(E_2, E_2 + dE_2)$ particles *raised* to $(E, E+dE)$ by absorption of an $(E - E_2)$-photon; $0 < E_2 < E < \infty$.

P2. $N(E_2, t)dE_2 B(E_2, E)\widetilde{N}(E_2 - E, t)dE$ gives the corresponding number "induced" by the presence of $(E_2 - E)$-photons to *drop* to $(E, E+dE)$ with creation of such a photon, and

P3. $N(E_2, t)dE_2 A(E_2, E)dE$ gives the number of such particles spontaneously decaying to $(E, E+dE)$, with creation of an $(E_2 - E)$-photon, $0 < E < E_2 < \infty$.

P4. The functions A, B are related by the equation

$$A(E_2, E_1) \equiv B(E_2, E_1)R(E_2 - E_1)$$

where R is the function of the energy *difference*. (In the Planck-Einstein case, $R(F) = 8\pi(hc)^{-3}F^2$, and $N, \widetilde{N}$ are numbers per cm³.)

7. The "Boltzmann-Einstein Equation." The analogue of (1) in §2 is seen to be the linked system (cf. (4) for B_3)

$$\partial N(E)/\partial t = B_3 \; + \int_0^E N(E_2)B(E,E_2)\widetilde{N}(E-E_2)dE_2$$

$$+ \int_E^\infty N(E_2)\{B(E_2,E)\widetilde{N}(E_2-E)+A(E_2,E)\}dE_2$$

$$- \int_E^\infty N(E)B(E_2,E)\widetilde{N}(E_2-E)dE_2$$

$$- \int_0^E N(E)\{B(E,E_2)\widetilde{N}(E-E_2)+A(E,E_2)\}dE_2$$

$$\partial\widetilde{N}(F)/\partial t \;=\; -\int_0^\infty N(E_2)B(E_2+F,E_2)\widetilde{N}(F)dE_2$$

$$+ \int_F^\infty N(E_2)\{B(E_2,E_2-F)\widetilde{N}(F)+A(E_2,E_2-F)\}dE_2 \;.$$

Combining integrals, bringing all lower limits to 0, making an inversion on the proper integral resulting, and using P4, one can show that

$$\partial N(E)/\partial t =$$
$$B_3 - \int_0^E B(E,E-F)\{N(E)[\widetilde{N}(F)+R(F)]-N(E-F)\widetilde{N}(F)\}dF$$
$$+ \int_0^\infty B(E+F,E)\{N(E+F)[\widetilde{N}(F)+R(F)]-N(F)\widetilde{N}(F)\}dF$$
$$\equiv B_3 - B_4 + B_5 \tag{8}$$

$$\partial\widetilde{N}(F)/\partial t =$$
$$\int_0^\infty B(E+F,E)\{N(E+F)[\widetilde{N}(F)+R(F)]-N(E)\widetilde{N}(F)\}dF \equiv B_6 \;.$$

We next verify, for a solution $N, \widetilde{N}$ of (8), the relations

$$\frac{d}{dt}\int_0^\infty N(E)dE = 0 = \frac{d}{dt}\left\{\int_0^\infty EN(E)dE + \int_0^\infty F\widetilde{N}(F)dF\right\}$$

and hence the invariance of the particle number N_0, and of the *total* energy $\mathcal{E}_0 = E_0 + \widetilde{E}_0$. For the first, we have to prove

$$\int_0^\infty B_3 dE - \int_0^\infty B_4 dE + \int_0^\infty B_5 dE = 0 \;.$$

But the first integral is zero by (2), and making the transformation

$$E = E' + F, \; F = F' \tag{9}$$

on the second shows it equal to the third. For the second relation we must show that

$$\int_0^\infty E B_3 dE - \int_0^\infty E B_4 dE + \int_0^\infty E B_5 dE + \int_0^\infty F B_6 dF = 0 \ .$$

Here, the first is zero by (2), and the transformation (9) on the second makes the result clear.

8. The H-function. For a solution $N, \widetilde{N}$ of (8) we define[3]

$$H_{N,\widetilde{N}}(t) = \int_0^\infty N(E) \log N(E) dE$$

$$+ \int_0^\infty \{\widetilde{N}(F) \log \widetilde{N}(F) - [\widetilde{N}(F) + R(F)] \log [\widetilde{N}(F) + R(F)]\} dF$$

with derivative

$$\frac{d}{dt} H_{N,\widetilde{N}}(t) = \int_0^\infty \partial N/\partial t \ \log \ N dE + \int_0^\infty \partial \widetilde{N}/\partial t \ \log \frac{\widetilde{N}}{\widetilde{N} + R} dF$$

$$= \int_0^\infty B_3 \log \ N dE - \int_0^\infty B_4 \log \ N dE \tag{10}$$

$$+ \int_0^\infty B_5 \log \ N dE + \int_0^\infty B_6 \log \frac{\widetilde{N}}{\widetilde{N} + R} \ dF \ .$$

We know the first integral from (5), and making the transformation (9) on the second shows (10) to be

$$H'_{N,\widetilde{N}}(t) = \int_0^\infty B_3 \log \ N(E) dE$$

$$- \int_0^\infty \int_0^\infty B(E + F, E)\{N(E + F)[\widetilde{N}(F) + R(F)] \tag{11}$$

$$- N(E)\widetilde{N}(F)\} \log \frac{N(E + F)[\widetilde{N}(F) + R(F)]}{N(E)\widetilde{N}(F)} \ dF dE \le 0$$

with equality iff (at the time t in question), *both*

$$\{N(E)N(E_1 + E_2 - E) - N(E_1)N(E_2)\} \equiv 0; \ 0 \le E \le E_1 + E_2 < \infty$$

and

$$\{N(E + F)[\widetilde{N}(F) + R(F)] - N(E)\widetilde{N}(F)\} \equiv 0; \ 0 < (E \& F) < \infty \ . \tag{12}$$

424

(cf. §3.) As we have seen, the first of these implies $N = \beta e^{-\alpha E}$, and hence from the second follows $\widetilde{N} = R(F)(e^{\alpha F} - 1)^{-1}$. Conversely, such a pair satisfy (12).

9. The Steady State. For a pair of time *independent* functions $N_0, \widetilde{N}_0$, we now find that the following are equivalent:

$S'1$. $N_0, \widetilde{N}$ satisfy (8) (steady state solution)

$S'2$. For $N = N_0(E)$, $\widetilde{N} = \widetilde{N}_0(F)$,

$$B_3(E) - B_4(E) + B_5(E) \equiv 0 \equiv B_6(F) \text{ in (8)}.$$

$S'3$. For $N = N_0(E)$, $\widetilde{N} = \widetilde{N}_0(F)$,

$$\int_0^\infty \{B_3(E) - B_4(E) + B_5(E)\} \log N(E) dE$$
$$+ \int_0^\infty B_6(F) \log \frac{\widetilde{N}(F)}{\widetilde{N}(F) + R(F)} \, dF = 0 \text{ in (10), (11)}.$$

$S'4$. For $N = N_0(E)$, $\widetilde{N} = \widetilde{N}_0(F)$,

$$\{N(E)N(E_1 + E_2 - E) - N(E_1)N(E_2)\} \equiv 0 \text{ and}$$
$$\{N(E + F)[\widetilde{N}(F) + R(F)] - N(E)\widetilde{N}(F)\} \equiv 0 \text{ in (12)}.$$

$S'5$. $N_0(E) = \beta e^{-\alpha E}$, $\widetilde{N}_0(F) = R(F)(e^{\alpha F} - 1)^{-1}$.

Here, stipulation of the totals $N_0, \mathcal{E}_0$ determines α, β, and hence also $\widetilde{N}_0, E_0, \widetilde{E}_0$. If we take the special function $R(F) = 8\pi(hc)^{-3}F^2$, we find again $\alpha = N_0/E_0$, $\beta = \alpha N_0$, and $\widetilde{N}_0 = 16\pi\zeta(3)(hc)^{-3}\alpha^{-3}$, $\widetilde{E}_0 = 8/15 \, \pi^5(hc)^{-3}\alpha^{-4}$. Note that for us, $1/\alpha = E_0/N_0$ and *not* $(2/3)E_0/N_0$ $(= kT)$.

10. Time Dependent Solutions. If $N(E,t)$, $\widetilde{N}(E,t)$ is a solution of (8), with invariants $N_0, \mathcal{E}_0$, which is *not* the solution of $S'5$ at any time t, then $H_{N,N'}(t)$ is strictly decreasing, and bounded below by $H_{N_0,\tilde{N}_0}$, as we now show.

Note first that $\tilde{N}_0(F) = R(e^{\alpha F} - 1)^{-1}$ implies $\tilde{N}_0 + R = \tilde{N}_0 e^{\alpha F}$. Hence, if $M, \widetilde{M}$ is *either* the pair $N, \tilde{N}$ or the pair $N_0, \tilde{N}_0$ with the same two invariants, we find

$$\int_0^\infty M \log N_0 dE + \int_0^\infty \widetilde{M} \log \tilde{N}_0 dF - \int_0^\infty (\widetilde{M} + R) \log (\tilde{N}_0 + R) dF$$

$$= \int_0^\infty M (\log \beta - \alpha E) dE + \int_0^\infty \widetilde{M} \log \tilde{N}_0 dF - \int_0^\infty \widetilde{M} (\log \tilde{N}_0 + \alpha F)$$

$$- \int_0^\infty R \log (\tilde{N}_0 + R) dF = N_0 \log \beta - \alpha E_0 - \int_0^\infty R \log (\tilde{N}_0 + R) dF \ .$$

The integrals having this common value, we may write

$$H_{N,\tilde{N}}(t) - H_{N_0,\tilde{N}_0} \equiv$$

$$\int_0^\infty N \log N dE + \int_0^\infty \tilde{N} \log \tilde{N} dF - \int_0^\infty (\tilde{N} + R) \log (\tilde{N} + R) dF$$

$$- \left\{ \int_0^\infty N_0 \log N_0 dE + \int_0^\infty \tilde{N}_0 \log \tilde{N}_0 dF - \int_0^\infty (\tilde{N}_0 + R) \log (\tilde{N}_0 + R) dF \right\}$$

$$= \int_0^\infty N \log N/N_0 dE + \int_0^\infty \tilde{N} \log \tilde{N}/\tilde{N}_0 dF$$

$$- \int_0^\infty (\tilde{N} + R) \log \frac{\tilde{N} + R}{\tilde{N}_0 + R} dF$$

$$= \int_0^\infty \{N \log N/N_0 + N_0 - N\} dE$$

$$+ \int_0^\infty \left\{ \tilde{N} \log \tilde{N}/\tilde{N}_0 - (\tilde{N} + R) \log \frac{\tilde{N} + R}{\tilde{N}_0 + R} \right\} dF > 0 \ .$$

We have seen (§5) that the first integrand is non-negative. The second, of form

$$g(y) = y \log y/y_0 - (y + R) \log \frac{y + R}{y_0 + R} \ ,$$

has $g(y_0) = 0$, and $g'(y) = \log (1 + R/y_0) - \log (1 + R/y) \leq (\geq) 0$ as $y \cdot \leq (\geq) y_0$ and thus is also ≥ 0. Hence $H_{N,\tilde{N}}(t) > H_{N_0,\tilde{N}_0}$, by the argument of §5, and we have the existence of $\lim_{t \to \infty} H_{N,\tilde{N}}(t) \equiv H^*_{N,\tilde{N}} \geq H_{N_0,\tilde{N}_0}$.

If we assume the limits $\lim\limits_{t\to\infty} H'_{N,\tilde{N}}(t) = 0$, $\lim\limits_{t\to\infty} N(E,t) = N^*(E)$, $\lim\limits_{t\to\infty} \tilde{N}(F,t) = \tilde{N}^*(F)$, then from (11) follows

$$N^*(E) = N_0(E) \text{ and } \tilde{N}^*(F) = \tilde{N}_0(F)$$

as defined in $S'5$. The remark at the end of §5 applies here as well.

Appendix I

It is apparent that definition of a ψ-function $\psi(E_1, E_2, E)$ is equivalent to defining a function $f(E_1, S, E)$ such that

f1. $f(E_1, S, E) > 0;\ 0 < (E_1 \& E) < S < \infty$

f2. $f(E_1, S, E) = f(S - E_1, S, E)$

f3. $f(E_1, S, E) = f(E, S, E_1)$

f4. $\displaystyle\int_0^S f(E_1, S, E)dE = 2$.

For, $f(E_1, S, E) \equiv \psi(E_1, S - E_1, E)$ and $\psi(E_1, E_2, E) \equiv f(E_1, E_1 + E_2, F)$ serve to define each in terms of the other.

Moreover, given an f-function, the function $h(E_1, S, E) \equiv f(E_1, S, E)$ for E_1, E on $(0, S/2)$ satisfies

h1. $h(E_1, S, E) > 0$

h3. $h(E_1, S, E) = h(E, S, E_1)$

h4. $\displaystyle\int_0^{S/2} h(E_1, S, E)dE = 1$.

Conversely, given such an h-function defined for all E_1, S, E with $0 < (E_1 \& E) < S/2 < \infty$, we may extend h to an f-function by the consistent definitions

$$\begin{aligned}
f(E_1, S, E) = &h(E_1, S, E), & E_1 \epsilon(0, S/2),\ E\epsilon(0, S/2)\\
&h(S - E_1, S, E), & E_1 \epsilon(S/2, S),\ E\epsilon(0, S/2)\\
&h(E_1, S, S - E), & E_1 \epsilon(0, S/2),\ E\epsilon(S/2, S)\\
&h(S - E_1, S, S - E), & E_1 \epsilon(S/2, S),\ E\epsilon(S/2, S)\ .
\end{aligned}$$

The function $h(E_1, S, E) = 4/S \ sin^2 \ 2\pi(E_1+E)/S$ satisfies $h1, 3, 4$ and therefore defines a ψ-function, as indicated.

Appendix II

Obviously $N(E)$ in (7) satisfies (6). Conversely, setting $E = 0$ in (6), and $L(E) = \log (N(E)/N(0))$, we must have

$$L(E_1 + E_2) = L(E_1) + L(E_2); \ E_1, E_2 \geq 0 .$$

Then for integers $m, n \geq 1$,

$$L(m \cdot 1) = mL(1) ,$$

$$L(1) = L\left(n \cdot \frac{1}{n}\right) = nL\left(\frac{1}{n}\right) , \ \text{and}$$

$$L\left(\frac{m}{n}\right) = L\left(m \cdot \frac{1}{n}\right) = mL\left(\frac{1}{n}\right) = \frac{m}{n}L(1) .$$

If fractions $m/n \to E > 0$, we have

$$L(E) = L\left(\lim \frac{m}{n}\right) = \lim L\left(\frac{m}{n}\right) = \lim \frac{m}{n} L(1) = L(1)E$$
$$= \log (N(E)/N(0)) .$$

Thus $N(E) = N(0) \exp L(1) \cdot E = \beta e^{-\alpha E}$.

References

1. M. Kac, *Probability and Related Topics in Physical Sciences*, (1959) Ch. III, Interscience Publishers, Ltd., London.
2. E. H. Kennard, *Kinetic Theory of Gases*, (1938) Ch. II, McGraw-Hill Book Co., Inc., New York.
3. R. C. Tolman, *Statistical Mechanics*, (1927) pp. 198-203, Chemical Catalogue Co., Inc., New York.

15

SOME ELEMENTARY ATTEMPTS AT NUMERICAL MODELING OF PROBLEMS CONCERNING RATES OF EVOLUTIONARY PROCESSES

With R. Schrandt
(LA-4573-MS, December 1970)

This report is an attempt to study mathematically models of evolution showing the qualitative difference between the rates of development in sexual and non-sexual processes. A further number of papers were stimulated by this report.

See also the earlier Proceedings of a meeting at the Wistar Institute on "Mathematical Challenges to the Neo-Darwinian Interpretation of Evolution" held in April 1966 where a paper of mine called "How to formulate mathematically problems of rate of evolution" was published. (Wistar Institute's Symposium Monograph No. 5, pp. 21-33, 1967.)

The work of the above report, done with Schrandt, was performed before the Wistar meeting. (Author's note). *

ABSTRACT

An account of numerical work prepared on electronic computers—the problems concerned the ratio at which favorable mutations spread throughout the population

---

* This report is also reprinted in the proceedings of a Los Alamos conference dedicated to Stan Ulam, Evolution, Games and Learning that occurred May 20-24, 1985 in Los Alamos. (Eds.)

subject to a "survival of the fittest" mechanism. Models of asexual reproduction showed the expected linear growth in the number of improvements. The bisexual process greatly accelerated the average acquisition rate.

I. Introduction

In this report, we shall present an abbreviated account of calculations performed by us in the mid 1960's. These calculations were preliminary and intended merely as the zeroth approximation to the problem concerning rates of evolution—a process which we have here severely stylized and enormously oversimplified. A mention of the results of such calculations in progress at that time was made at a meeting in 1966 at the Wistar Institute in Philadelphia by one of us. The discussion there, as reported in the proceedings of the meeting, was rather frequently misunderstood and the impression might have been left that the results somehow make it extremely improbable that the standard version of the survival-of-the-fittest mechanism leads to much too slow a progress. What was really intended was indications from our computations—simple minded as they were—that a process involving only mitosis, in absence of sexual reproduction, would be indeed much too slow. However, and most biologists realize it anyway, the Darwinian mechanism together with mixing of genes accelerate enormously the rate of acquiring new "favorable" characteristics and leave the possibility of sufficiency of the orthodox ideas quite open. Numerous requests addressed to us for the elucidations and details of the numerical setup made us decide to give this account of our computations.

Perhaps the greatest uncertainties—the strongest objections to any calculation of the sort described in the pages that follow—must concern the values of the constants which are assumed initially or should indeed concern even their meaning in the interpretation we have chosen. We have tried to interpret the survivability of individuals by changes in the number of offspring which carry the species in time measured by a discrete succession of generations. The value of "favorable" mutations was mirrored in the increased proportion of offspring. The same, needless to say, is true of the *frequency* of favorable mutations. We have disregarded the lethal and the unfavorable mutations. We assumed a special form of the advantages which an individual holds relative to the rest of the population by comparing the number of his "improvements" with the average number present at that time in the population. We assumed a proportionality law, again arbitrarily. In some problems we have penalized the individuals whose score in the improvements was less than the average; in some of the problems we considered only the positive excess as leading to a greater number of offspring. Another debatable procedure is the way we have handled the growth of the population by normalizing periodically the total number of individuals to a constant figure. If the number of individuals holding a certain number of "advantages" after normalization dropped below 1, we summarily dismissed such representatives. It should be stressed here strongly that this procedure makes it very hard to find an analytic model equivalent to our numerical work. We do not have any clear idea of the necessary scaling laws concerning the effect of changing the constants alpha, kappa (defined below) and the size of the population. All this is true throughout all the problems. In the calculations involving combinations of genes from both parents, further assumptions were made of independent inheritance of the "improved" genes, etc. As will be seen in the description of the individual problems, we have chosen successively less unrealistic assumptions. Clearly in the counting of the new improved genes coming from the "father" and from the "mother", one has to take care not to count the same "improvements" from each twice. As will be seen, this precaution has the effect of slowing down the at first seemingly exponential growth into something more like a quadratic function of time (we have studied throughout the calculations the number of "improvements" present in the population as a function of time, that is to say of the generation number).

In order to get a feeling for the dependence of the results on the values of the constants, more such computations must be tried in the future and additional variables have to be considered—certainly the "kind" of the improved or favorable new gene has to be taken into account. A most important question concerns the existence of new genetic instructions involving perhaps logical prescriptions, that is to say recipes for operations and actions of the components rather than merely their chemical composition. An improvement in programming or interpretation of action by a gene or group of genes may be equivalent to a very large number of the "favorable changes" with which our computations have dealt so far.

Our first problem with the code name ADAM concerned asexual reproduction. We feel that the time scale to acquire a characteristic in an organism, such as the development of an eye, by a sequence of consecutive favorable mutations, is extremely long if one does not resort to something like sexual mating in the population. In the following rough and elementary estimate, the constants assumed are crude, but err toward "faster" evolution than what is to be expected.

Definitions: Let

T = time of existing life ($\sim 10^9$ yrs.);

τ = time for one generation say (~ 3 days);

G = the number of generations $= T/\tau = 10^{11}$;

N = the existing population size ($\sim 10^{11}$);

K = the total number of "favorable mutations" necessary to produce the desired characteristic ($\sim 10^6$);

α = the chance of a favorable mutation per individual per generation ($\sim 10^{-10}$);

γ = the "value" of a single favorable mutation expressed as a survival rate ($\sim 10^{-6}$). That is, an individual having this mutation would have (in expectation) $(k + \gamma)$ descendants, versus k descendants for an individual not having this mutation.

Therefore, in the first generation, the expected value of the population that could have one mutation is $N \cdot \alpha = 10$. In $1/\gamma = 10^6$ generations, a sizeable portion (approximately $1/e$) of the population would have this mutation, and in about 10^7 generations, most of the population would have it. But the time to acquire all the mutations would be about $K \cdot 10^7$, or 10^{13} generations, which is like the age of the universe.

II. ADAM

In order to study, on a computer, the rate at which a population can acquire a sequence of mutations we needed a set of more amenable parameters, which, it is hoped, could eventually be scaled down to the set given above. For the first problem, called ADAM, we used the following set: $N = 100$, $K = 100$, $\alpha = .02$, and $\gamma = .1$.

The method was as follows: In any one generation, each member of the population N had a probability α of acquiring one new independent mutation. Each individual then had one child, with a probability of extra children given by $\gamma(K_g - K_0)$, where K_g = the total number of different mutations possessed by this individual and, K_0 = the minimum number of mutations possessed by any individual in the population. (If $n/\gamma < K_g - K_0 \leq (n+1)/\gamma$, $n = 1, 2, 3, ...$, the individual had n extra children, and a probability of $\gamma(K_g - K_0 - n/\gamma)$ for an $(n+1)^{th}$ child.)

The children were then given the number of mutations possessed by their parent. The parent population then was assumed to have died, and the children formed the new population. The numbers of mutations in the population were recorded in categories by counts n_i, $i = 0, 1, 2, 3, \ldots$, where n_i = the number of individuals having a total of i mutations, and $\Sigma n_i = N$, the size of the new population. In the next generation, each member of this new population had the same probability α of acquiring another new mutation, and had children according to the above recipe. These children with their number of mutations recorded then became the population for the succeeding generation, etc.

It was necessary to renorm the population periodically, since the number of children increased in each generation. This was done by reducing the count n_i in each category by $1/2$, to the nearest smaller integer, when the population reached 200, which is double the initial population.

The weighted average number μ of mutations possessed by the population was computed for each generation from the categories n_i. This average was then plotted as ordinate against the generation time as abscissa. The slope of this curve is then the rate at which the population can acquire a sequence of mutations, as a function of the parameters α and γ.

This rate of acquiring mutations turned out indeed to be linear. For the parameters $\alpha = .02$ and $\gamma = .1$, the slope was about .1; or a majority of the population acquired an additional mutation every 10 generations. Several problems were run with smaller values of γ, that is, $\gamma' = f \cdot \gamma$, where f = a fraction. The graphs were all linear, with decreasing slopes, which decreased more closely with $\sqrt{f}$ than with f itself. There was no appreciable change in the slope by doubling the initial population to 200. Figure 1 shows a plot of 3 cases: $N = 100$, $\alpha = .01$, and $\gamma = .1$, .05 and .01 respectively.

A second version of ADAM was run with the K_0 in the probability recipe defined as the *average* number of mutations in the population, instead of the minimum number. Those individuals having fewer than the average number of mutations ($K_g < K_0$) had a probability of $\gamma(K_0 - K_g)$ of no children, (if $K_0 - K_g \geq 1/\gamma$, they had no children deterministically), and a probability of $1 - \gamma(K_0 - K_g)$ of one child. The individuals with more than the average number of mutations had the same probability for extra children as the one defined above. This version required fewer renormings of the population and it led to a somewhat greater slope than the $K_0 =$ min. recipe. The graph was still linear. Figure 2 illustrates the two versions with the same parameters $N = 100$, $\alpha = .02$, and $\gamma = .1$. The recipe with K_0 = average was used in all subsequent problems.

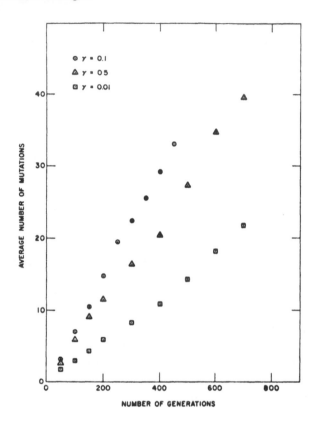

Fig. 1.

III. EVE

In the second class of problems, we introduced reproduction in the population, and also, what seems important, fluctuations in the manner in which the offspring received mutations from their parents. This problem was naturally called EVE. The initial parameters used were the same as in ADAM. The population acquired new mutations according to the probability α, as before. A random mating of the population N was then defined, resulting in $N/2$ pairs of individuals. (For N odd, the population was arbitrarily reduced by one to obtain $(N - 1)/2$ pairs.) Each pair then constituted a set of parents. The number of children from each set was again determined by the probability function $\alpha(K_g - K_0)$. Here K_g = the total number of mutations possessed by the set of parents, and K_0 = the average total number of mutations of all the pairs of this mating.

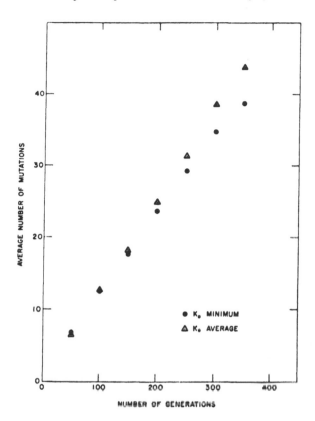

Fig. 2.

The children were produced in pairs. Thus if a set of parents had $K_g > K_0$, they had two children for certain, and with a probability given by $\gamma(K_g - K_0 - n/\gamma)$, $n = 1, 2, 3, ...$, where $n/\gamma < K_g - K_0 < (n + 1)/\gamma$, of $2n$ extra children. If for the set of parents, $K_g < K_0$, they had, with the probability $\gamma(K_0 - K_g)$ no children, and with the probability $1 - \gamma(K_0 - K_g)$ two. Again if $K_0 - K_g \geq 1/\gamma$, they had no children.

The number of mutations acquired by each child was obtained under a binomial distribution centered about the *average* number of mutations of the parents. If the parents had a count of $(x + y)$ mutations, the number for each child was obtained under the distribution centered at $(x + y)/2$, with its minimum at zero and maximum at $(x + y)$. The number of mutations for each child separately was determined under this distribution. Thus a child could possibly obtain as much as the sum of numbers of mutations of its parents. This recipe involving fluctuations in the inheritance would, we thought, speed up the rate of acquisition of mutations (compared to always giving the offspring $(x + y)/2$ favorable mutations).

The parents having died, the children became the new population. As before, they were classified in terms of counts of the number of individuals having each number of mutations. As before, the individuals had the chance α new mutations, and were mated at *random* in $N/2$ pairs. These pairs then had children whose mutations were again determined from the binomial distribution, and these children constituted the population for the following generation, etc.

In our random mating, the sex of the individual was not distinguishable. No attempt was made to keep members of the same "family" from mating. Their number of mutations was not necessarily the same, since it was determined separately for each child under the probability distribution. (The identity of the family was lost once the members were classified according to their mutations.)

The norming of the increasing population was done in the same way as in ADAM: All categories were halved when the population doubled.

The rate of acquiring mutations turned out to be much faster than in ADAM, and appeared to be exponential. In Fig. 3 we have plotted this rate on a semilog scale for four problems: $N = 100$, $\alpha = .02$, and $\gamma = .1, .05, .025$, and $.01$ respectively. The reduction in acquisition rate with γ was somewhat similar to that in ADAM.

In these problems no attempt was made to keep the histories of the different mutations. We define α to be the rate of acquiring new mutations, but we divide the population only in terms of numbers of individuals having a fixed number of mutations. The children too acquire mutations only under the distribution of the total count of the parent's mutations. Thus one might suspect that the exponential rate of acquisition could be due to a doubling of the identical mutations possessed by both parents.

A. EVE-PQ

To correct for this, we computed an expected number ν of mutations that the parents should have in common. This is given by

ν = $(T_1 \cdot T_2)/S$ where

T_1 = the total number of mutations possessed by one parent,

T_2 = the total number of mutations possessed by the other parent, and

S = the total number acquired by the entire population. (The

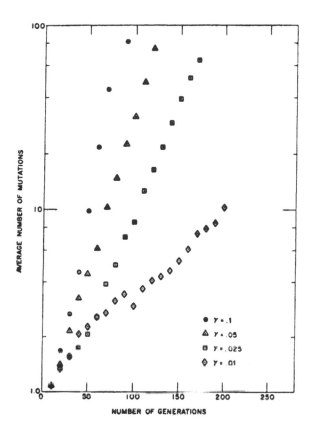

Fig. 3.

total S is accumulated as each individual in each generation has the probability α of acquiring a new mutation. If $\bar{N}$s is the average population in each generation, after k generations, S is approximately $\bar{N} \cdot k \cdot \alpha$.)

We allotted this number ν to the children for certain and then played a game of chance for additional mutations using the reduced binomial distribution centered at the midpoint of the total count of independent mutations possessed by the parents. In this manner, we count more correctly, the number in common only once.

This method leads to a slower rate of acquisition; it is still exponential in the beginning but tails off to something like a quadratic function, as more mutations must of necessity be held in common. Results from four problems are plotted in Fig. 4 on a semilog scale. The

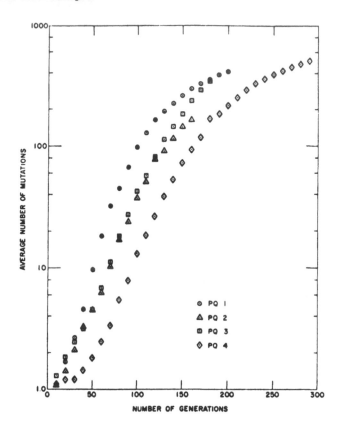

Fig. 4.

cases PQ1 and PQ2 have $N = 100$, $\alpha = .02$ and $\gamma = .1$ and $.05$ respectively. The case PQ3 is the same as PQ2 except that the sample size was doubled, $N = 200$. It had a somewhat higher value for μ, but the slopes are like those of PQ2. The last case, PQ4, had the same parameters as PQ3 except that α was cut to $.01$. This problem was run to 290 generations and showed a definite bending over of the curve to almost a linear rate of acquisition after 230 generations as the population had more and more mutations in common.

B. EVE-PM

In the two versions of EVE already discussed, the population mated at random uniformly. We next considered a version where the mating was not uniform, i.e., preferential in the following sense: We arbitrarily divided the population equally into three groups ranked according to their number of mutations, i.e., the first had the individuals with the greatest number of mutations. We specified that 3/4 of the population in each group mate at random uniformly within their own group. The remaining 1/4 would mate at random with equal probability from either of the two remaining groups. For example, if we name the groups A, B, and C, an individual from group A would have a 3/4 chance of acquiring a mate from group A, and a 1/8 chance of a mate from each of the groups B and C. We called this problem EVE-PM. The EVE-PQ method was used to estimate the mutations in common and to count them only once.

The rate of acquisition of mutations should reflect this preferential mating. A comparison of the curve PM1 of Fig. 5 with PQ1 of Fig. 4, (both with the same parameters), shows indeed that the initial acquisition rate of mutations under preferential mating is much higher than under random mating. But the curve PM1 tails off very rapidly after 100 generations, and at 200 generations the mutation rate is almost the same for the two problems. This indicates that at this point most individuals have acquired most of the mutations available in the total population, so that preferential mating has the same effect as the uniform mating. Our small sample size of 100 in part causes this phenomenon to occur so soon.

The computing time goes up rapidly with the initial population; one problem of preferential mating with an initial sample of 400 was run to 96 generations. The result is shown in the curve PM2 of Fig. 5, with the same parameters as PM1 (except for the population size). The preferential mating has a greater initial effect in the larger population, but the slope of this curve too is beginning to decrease.

The mutation acquisition rate in all the EVE problems of sexual reproduction can apparently be divided into three stages:

- An initial exponential rate, as few mutations are held in common (compared to the size of the population).
- A rate, roughly quadratic, as more mutations are held in common by the parents. This number in common is approximated by computing the expected intersection of the number possessed by each parent, assuming the parents had acquired them independently.
- A terminal rate, almost linear, as most of the mutations in the population are in common. If all the mutations were in common, the subsequent rate of acquisition must be linear, since new mutations are obtained only by the α rate of new acquisitions, which is a linear function.

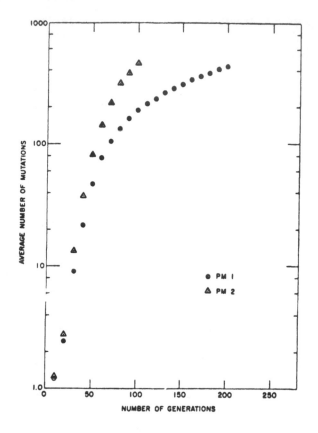

Fig. 5.

C. EVE-POS

In order to check these assumptions somewhat, a problem was run in which histories were kept of each new mutation. This was done by representing each new mutation as a bit position in a matrix of words in the computer. Each individual in the population and its children had their mutations recorded in such a matrix, that is to say mutation was specifically identified.

We called this problem EVE-POS. The sexual reproduction scheme was the same as before. The population was mated at random, and each set of parents had a probability of having extra children given by $\gamma(K_g - K_0)$. In both K_g and K_0 the actual mutations in common were counted only once. This number was known for each mating, so no approximation for the expected intersection was needed.

If both parents possessed the i-th mutation, it was then given to each of the children. If neither parent possessed this mutation, it was not acquired by any of their offspring. If only one parent possessed this i-th mutation, each child had a probability of $1/2$ of acquiring it.

The norming for this problem was different from before, namely, when the population doubled, each individual was given a half chance of surviving.

With this recipe for receiving mutations, any individual mutation can be lost to the population, since the parents die off in each generation. For example, the k-th mutation is initially acquired in the α recipe by one individual. If this individual and its mate have no children, the mutation is lost. If they have n children, there is a probability of $(1/2)^n$ that none of the children get it, in which case it is lost in the next generation. There is a chance that this particular mutation will be lost in each subsequent generation, although these probabilities are getting smaller. The k-th mutation is initially acquired only once, and by only one individual. The mutations that survived were "packed" in the bit positions of the matrices. This relieved the space limitation in the memory and allowed the problem to be run much further in time.

It was discovered that approximately 80% of the total of new mutations acquired by individuals were "lost" after matings in subsequent generations. Thus the parameter α in problem EVE-POS has a different meaning from that in EVE-PQ. In EVE-POS it denotes the probability of an individual acquiring a new mutation. In EVE-PQ it denotes the probability of acquiring a mutation that survives and will eventually be acquired by the entire population.

Figure 6 shows the mutation rate for two cases of EVE-POS, plotted on a log scale. The parameters are $\alpha = .02$, $\gamma = .1$, and $N = 100$ and 400, respectively. The case POS1 ($N = 100$) was run to 251 generations, and POS2 ($N = 400$) to 90 generations.

Note that the acquisition rate and the values of μ are considerably smaller in POS1 than in PQ1, which has the same parameters. This is because of the different interpretation of the parameter α. A reasonable comparison with POS1 would be to run PQ1 and compute our $\nu = (T_1 \cdot T_2)/S'$, where $S' = .2S$, since about 80% of the mutations are lost.

The acquisition rate for POS1 becomes linear after about 100 generations. The relatively small sample size is a contributing factor. Some statistics were compiled on the distribution of the available mutations. They are given in Table I.

An estimate was made of the expected number of mutations held in common compared to the actual number held by an average set of parents. The expected number was computed assuming that each

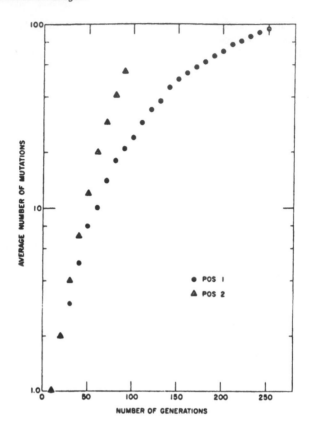

Fig. 6.

parent had the average number μ of mutations. Then $\nu = \mu^2/S'$, where S' = the total number of surviving mutations. The actual number held in common was about 1.3 times this expected number, after 251 generations.

In the problem with the larger sample size of 400 (POS2), the mutation acquisition rate remains approximately quadratic through 90 generations. At this point 18% of the mutations were held by at least 50% of the population, versus 36% for the population of 100. For this larger sample size, at 90 generations, separate distributions were kept of mutations acquired in the first 45 generations, versus those acquired in the last 45 generations. This data and some statistics on those mutations held in common are given in Table II.

TABLE I
Problem—EVE POS1-N = 100, $\alpha = .02$, $\gamma = .1$

After 90 generations, there were 55 surviving mutations out of 266 acquired mutations: 79% were lost.
Distribution of the 55 mutations:

    Min. - 15 (least number held by any individual)
    Aver.- 21 (average number held—this is the number μ that is plotted)
    Max. - 27 (greatest number held by an individual)
    36% of the mutations were held by at least 50% of the population
    1.8% of the mutations were held by the entire population.

After 251 generations, there were 130 surviving mutations out of 744 acquired: 85% were lost.
Distribution of the 130 mutations:

    Min. - 87
    Aver. - 94
    Max. - 101
    74% of the mutations were held by at least 50% of the population
    41% of the mutations were held by the entire population.

The actual number of mutations held in common was approximately 1.3 times the expected number.

TABLE II
Problem—EVE POS2 - $N = 400$, $\alpha = .02$, $\gamma = .1$

After 90 generations, there were 239 surviving mutations out of 1067 acquired: 78% were lost.
Distribution of the 239 total mutations:

    Min. - 42
    Aver. - 55
    Max. - 71
Distribution of mutations initially acquired in the first 45 generations (115):

    Min. - 35
    Aver. - 49
    Max. - 63
Distribution of mutations acquired in the last 45 generations (124):

    Min. - 1
    Aver. - 6.4
    Max. - 13
There were 18% of the mutations held by 50% of the population, 27% held by 40% of the population, and 32% held by at least 30% of the population. At 90 generations, the actual number of mutations held in common was approximately 2.4 times the expected number.

The above problems give some indication of the rate at which a population can acquire mutations in terms of a finite and rather small sample size, and in terms of the procedures which we adopted for acquiring and transmitting of the mutations. These methods involved using rather large values of the parameters alpha and gamma. The problem remains to find scaling factors. The computing time on the old IBM-7094 for the problem POS2 was over one hour, for 90 generations of growth. The computing time for the PQ code was much faster, in the order of minutes, but the parameter α was in effect much larger than the one in the POS code.

IV. SUMMARY

The problem ADAM with asexual reproduction gave a linear rate of acquisition of mutations. In reducing the parameter γ to $f \cdot \gamma$, where f is a fraction, the acquisition rate was reduced by a factor more like $\sqrt{f}$ than like f itself.

In problem EVE with sexual reproduction, the acquisition rate appeared to be exponential if the initial population were large enough. But with a small population, more of the *same* mutations were held in common by the parents. This caused the rate to change from an exponential to a "quadratic" and eventually to a linear one when most mutations were common to the majority of the population. The problem EVE-PQ involved approximating this number (our formula for ν).

The advantage of preferential mating over random mating gave an initially pronounced increase in the acquisition rate, but this was soon offset by the smallness of the population. In effect, as more mutations were held in common, the range of the distribution of mutations became narrow. After that the preferential mating was not much different from the uniform one.

The EVE-POS problem (where we kept a history of the mutations) gave us a measure of the distribution of mutations as a function of their age. It showed that most of the mutations initially acquired by one individual were lost in subsequent matings. This caused a redefinition of the probability α in computing the expected number of mutations held in common. This problem also showed that the actual number held in common was greater than the estimated number ν, by a factor of about 1.3 for the sample size of 100, and 2.4 for the sample size of 400. This is not too surprising, since the expected size of the intersection assumes independent sampling, whereas the mutations are acquired by something more like a Markov process.

16

THE NOTION OF COMPLEXITY

With W. A. Beyer and M. L. Stein
(LA-4822, December 1971)

This report is a study of complexity per se in certain algebraical systems. Much subsequent work seems to have been stimulated by these results. (Author's note.)

ABSTRACT

The notion of the arithmetic complexity $|n|$ of an integer n is defined in terms of the minimum number of additions, multiplications, and exponentiations required to combine 1's to form n. The value of $|n|$ is calculated for $n<2^{10}$. n is called complicated if $|n|>|n_1|$ for every $n_1<n$. Of the first 19 complicated numbers, 14 are prime. A conjecture about a relation between complexity and entropy is proposed. Some computations are presented to support this conjecture.

I. Introduction

In this report we discuss notions of complexity in some algebraic structures. These notions are also applicable to more general combinatorial situations that perhaps lack any algebraic pattern in the classical sense. We concentrate on a few special cases for which we studied and calculated a special notion of complexity. Essentially, we examined a special notion of complexity for ordinary integers with a little excursion on such a notion for integers *modulo a prime*.

445

The notion of complexity, in our view, is separate from, though associated with the idea of the amount of *information* or *entropy* of a system. We mention briefly a possible axiomatic approach to defining a real number called complexity for elements of a set or of a class on which certain *operations* are performed. These could be binary operations; our set could be a set of integers, and the operations could be addition, multiplication, and exponentiation, for example. It is this case that was examined on a computing machine and to which most of this report is devoted.

Another case would be a class of subsets of a given set, with allowed operations being the Boolean operations of *union* and *intersection* or *union* and *complementation*. One could add other operations, for example, the direct product of sets and also *projection*. This would correspond to allowing quantifiers in our theory. One can study a notion of complexity for vectors in a countable space or even in the continuum. An important study would be that of a *relative* complexity; that is to say, complexity of elements or "expressions" when the complexity of certain symbols is normalized to 1. In what has been sometimes called "speculation" on constants in physical theories, for example, the whole art seems to depend on the success of attempts to define some known important numbers, e.g., the dimensionless ratios

$$\frac{M_{proton}}{M_{electron}} = 1836.11 \ldots$$

and

$$\frac{e^2}{hc} = 137.1 \ldots$$

by use of only a few artificially introduced constants which should be as "simple" as possible. (cf. the attempts by Eddington[1] and some very recent ones by Good[2] and Wyler.[3])

Considered "genetically," a mathematical theory resembles a tree in that one obtains, from a given number of symbols corresponding to "variables" and from a number of allowed operations, expressions that elongate by branching. The simplifications and abbreviations may then reduce the length of the expressions.

One could try to define *complexity* in a mathematical structure by postulating certain of its properties, somewhat like postulating properties of a *measure*.

Let the structure, S, consist of elements x, y, ... It may be finite or infinite. We have in the set S a number of, say, binary operations R_1, R_2, ... R_n. We want to assign a number $c(x) \geq 0$ to each element

x of S and to each R_i $(i = 1 \ldots n)$ so that the following properties should hold.

a. If $z = R_i(x, y)$, then $c(z) = c(R_i(x, y)) \leq c(x) + c(y) + c(R_i)i = 1 \ldots n$.

b. For each element z, if $z = R_j(x, y)$, we should have for one case at *least* $c(z) = c(x) + c(y) + c(R_j)$.

c. $c(x_0) = c(x_1) = \ldots c(x_n) = 1$ for some preassigned elements $x_0 \ldots, x_n$ in S.

Needless to say, one can define analogous desiderata for the case in which the operations are more general than binary ones.

Obviously, in the case to which our exercise is devoted, these postulates are satisfied. Moreover, they define the complexity uniquely if, as must be the case in general, the complexity was normalized for some elements. (In our case, we assume the complexity of the integer 1 to be equal to 0.) We hope to study this notion more thoroughly for the more general case and also to perform experiments to determine complexity functions for the case in which S is a class of sets. Ultimately, one would wish to discuss the complexity of genetic codes and biological organisms quantitatively.

("Integer" always means a positive integer.)

II. Arithmetic Complexity of Integers

The arithmetic complexity $|n|$ of an integer n is defined as the fewest number of operators: $+$, $\times$, $\times\times$ (addition, multiplication, and exponentiation) which combine 1's to form n. Thus, $|1| = 0$; $|2| = 1$ since $2 = 1 + 1$; and $|5| = 4$ since $5 = (1 + 1) \times\times (1 + 1) + 1$ and not fewer than four operators with 1's will form five. Obviously, for a and b integers, $|a + b|$, $|ab|$, and $|a^b|$ are each not more than $|a| + |b| + 1$. For an infinity of integers n, the relation $|n + 1| = |n| + 1$ holds.

For the purpose of calculating the complexity of some integers, all correct formulas (up to some number of operators) involving $+$, $\times$, $\times\times$, and the number 1 were enumerated using parenthesis-free notation on a computer. It required one hour of computer time to enumerate the integers with complexity ≤ 6. Ralph Cooper made the following observation. Each correct formula involving $n(>0)$ operators is the composition of two formulas, one formula with n_1 operators and one formula with n_2 operators such that $n = n_1 + n_2 + 1$. One generates the integers of complexity n by first generating tables of integers of complexity $<n$. One partitions $n - 1$ into $n_1 + n_2$ in all ways and combines the integers of complexity n_1 with the integers of complexity n_2 to produce integers of complexity not larger than n. This method is considerably more efficient than the previous method. Table I lists the complexity of all integers $< 2^{10}$.

From the above construction, one sees that an upper bound $\ell^1(k)$ to $\ell(k)$, the number of integers of complexity k, is given by the solution of

$$\ell^1(k+1) = \sum_{j=0}^{k} \ell^1(j)\ell^1(k-j) \, ,$$

with $\ell^1(0) = 1$. The solution to this equation is given by

$$\ell^1(k) = \frac{1}{k+1}\binom{2k}{k}2^{-k} \, ,$$

which implies that

$$\ell(k) \leq \frac{2^k}{k\sqrt{\pi k}} + O(2^k k^{-5/2}) \, .$$

Two additional forms of complexity have been considered and calculated.

a. Complement Complexity. To make complexity symmetric in 0's and 1's, we introduce a slightly different complexity, the complement complexity $\bar{K}(y|n)$. Define the complement operation C by $C(x|n) = 2^n - 1 - x$. $\bar{K}(y|n)$ is defined as the fewest operations of addition, multiplication, exponentiation, and complementation that combine 1's to form y. In the count of operations, the first three are given the value 1 and the last is given the value 0. Thus $\bar{K}(y|n) = \bar{K}(2^n - 1 - y|n)$. Table II gives the values of $\bar{K}(y|n)$ for $y < 2^{10}$ and $n = 10$.

b. Modulo a prime p Complexity. In addition to the operations of $+$, $\times$, and $\times\times$, the operation of mod_p is allowed and is defined by $mod_p(x) = x - p[x/p]$ where p is a fixed prime and $[\ ]$ denotes the greatest integer. Table III gives the *modulo prime $p = 137$* complexity for integers < 137. Table IV gives the *modulo prime $p = 1009$* complexity for integers < 1009.

TABLE I. Complexity of Integers $< 2^{10}$

Complexity	Integer
0	1 1022 1023
0	1
1	2
2	3
3	4
4	5 6 8 9
5	7 10 16 27
6	11 12 17 18 25 28 32 36 64 81 256 512
7	13 14 15 19 20 24 29 33 37 49 54 65 82 100 125 128 216 243 257 513 729 1024
8	21 22 30 34 38 48 50 55 56 66 72 83 101 121 126 129 144 162 217 244 258 289 324 343 514 625 730 784 1000
9	23 31 35 39 40 45 51 52 57 58 67 73 74 75 84 96 98 102 108 122 127 130 145 163 164 169 192 196 200 218 225 245 250 259 290 325 344 361 400 432 486 515 576 626 676 731 768 785 841 1001

TABLE I. (cont.)

Group	41	42	44	46	53	59	60	63	68	76	78	80	85	87	90	97	99	103	109	110
10	41	42	44	46	53	59	60	63	68	76	78	80	85	87	90	97	99	103	109	110
	111	112	123	131	132	135	146	147	165	166	170	193	195	197	201	202	219	226	242	246
	251	252	260	288	291	300	326	345	362	375	384	401	433	434	441	484	487	488	516	577
	578	627	648	677	686	732	769	771	786	842	900	1002								
11	43	47	61	62	69	70	77	79	86	88	89	91	104	113	114	116	124	133	134	136
	140	148	150	153	160	167	168	171	180	189	194	198	203	204	220	224	227	247	249	253
	254	261	262	264	265	270	292	301	303	320	327	328	338	346	363	376	378	385	387	392
	402	405	435	436	442	450	485	489	490	500	517	518	520	521	529	579	580	628	649	650
	651	678	687	688	722	733	770	772	774	787	800	843	864	867	901	961	972	1003		
12	71	92	93	95	105	106	115	117	118	119	120	137	141	149	151	152	154	156	161	172
	174	175	176	181	185	190	199	205	206	208	221	222	228	232	234	248	255	263	266	271
	272	280	283	293	294	296	297	302	304	306	321	329	330	332	333	339	340	347	360	364
	366	377	379	381	386	388	390	393	394	403	404	406	410	437	438	443	448	451	452	459
	491	492	501	502	504	507	519	522	528	530	539	567	581	582	585	588	600	629	640	652
	654	656	675	679	689	690	723	724	734	735	737	738	750	756	773	775	777	788	801	802
	810	844	865	866	868	870	882	902	962	968	973	974	975	976	1004					
13	94	107	138	142	155	157	158	159	173	177	178	182	186	187	191	207	209	223	229	231
	233	235	240	267	268	273	274	275	281	284	295	298	305	307	308	309	322	331	334	336
	337	341	342	348	349	351	352	365	367	369	370	380	382	389	391	395	396	407	408	411
	415	416	425	439	440	444	449	453	454	455	460	464	476	493	494	495	498	503	505	506
	508	510	523	524	531	537	540	544	548	568	574	583	584	586	589	591	592	593	594	601

TABLE I. (cont.)

Grp	602	603	605	606	612	630	631	633	634	641	645	653	655	657	664	680	691	692	700	702
	704	720	725	726	736	739	745	747	751	752	753	757	776	778	780	783	789	790	792	793
	803	804	808	811	820	845	869	871	872	873	875	883	884	891	896	903	909	963	969	970
	977	978	980	999	1005	1006	1008	1009												
14	139	143	179	183	184	188	210	212	230	236	237	238	241	269	276	279	282	285	286	299
	310	312	315	316	319	323	335	350	353	356	359	368	371	372	383	397	398	399	409	412
	417	420	426	445	456	461	462	465	468	472	475	477	480	496	499	509	511	525	526	527
	532	536	538	541	542	545	549	550	560	561	566	569	575	587	590	595	604	607	608	609
	610	613	632	635	637	642	646	658	660	665	666	672	681	682	684	685	693	694	701	703
	705	707	715	721	727	728	740	741	746	748	754	755	758	759	761	762	765	779	781	791
	794	795	805	806	809	812	815	816	821	825	830	832	833	846	847	849	850	874	876	880
	885	886	892	897	904	910	918	924	925	928	936	960	964	971	979	981	982	984	985	1007
	1010	1014	1016																	
15	211	213	214	239	277	278	287	311	313	314	317	318	354	355	357	373	374	413	414	418
	421	423	424	427	429	446	447	457	458	463	466	469	470	473	478	481	483	497	533	534
	543	546	551	555	562	570	596	597	599	611	614	615	616	618	621	624	636	638	643	644
	647	659	661	662	663	667	668	670	673	674	683	695	696	698	706	708	714	716	742	743
	744	749	760	763	764	766	782	796	798	807	813	814	817	822	824	826	829	831	834	836
	837	840	848	851	854	855	857	853	877	878	879	881	887	888	889	893	898	905	906	908
	911	912	913	919	920	926	927	929	931	935	937	945	950	952	957	965	983	986	987	988
	990	996	1011	1012	1015	1017	1018	1020												

TABLE I. (cont.)

16	215	358	419	422	428	430	467	471	474	479	482	535	547	552	556	557	558	559	563	564
	565	571	572	573	598	617	619	620	622	639	669	671	697	699	709	711	712	713	717	718
	767	797	799	818	819	823	827	828	835	838	852	853	856	859	861	890	894	899	907	914
	915	916	917	921	922	930	932	938	944	946	951	953	954	958	966	989	991	992	993	
	997	998	1013	1019	1021	1022	1023													
17	431	553	554	623	710	719	839	860	862	895	923	933	939	940	941	942	947	948	949	955
	956	959	994	995																

TABLE II. Complement Complexity of Integers $< 2^{10}$

Complement Complexity	Integer																			
0	0	1																		
	1022	1023																		
1	2																			
	1021																			
2	3																			
	1020																			
3	4																			
	1019																			
4	5	6	8	9																
	1014	1015	1017	1018																
5	7	10	16	27																
	996	1007	1013	1016																
6	11	12	15	17	18	25	26	28	32	36	64	81	256	511	512	767	942	959	987	991
	995	997	998	1005	1006	1008	1011	1012												
7	13	14	19	20	24	29	31	33	35	37	49	54	63	65	80	82	100	125	128	216
	243	255	257	294	510	513	729	766	768	780	807	895	898	923	941	943	958	960	969	974
	986	988	990	992	994	999	1003	1004	1009	1010										
8	21	22	23	30	34	38	48	50	52	53	55	56	62	66	72	79	83	99	101	121
	124	126	127	129	144	162	215	217	225	239	242	244	254	258	289	293	295	324	343	347
	398	509	514	625	676	680	699	728	730	734	765	769	779	781	784	798	806	808	861	879
	894	896	897	899	902	922	924	940	944	951	957	961	967	968	970	971	973	975	985	989
	993	1000	1001	1002																

TABLE II. (cont.)

9	39	40	45	47	51	57	58	61	67	70	71	73	74	75	78	84	96	98	102	108
	120	122	123	130	143	145	160	161	163	164	169	182	192	196	200	214	218	224	226	238
	240	241	245	250	253	259	288	290	292	296	323	325	342	344	346	348	361	397	399	400
	432	435	447	486	508	515	537	576	588	591	623	624	626	662	675	677	679	681	698	700
	727	731	733	735	764	770	773	778	782	783	785	797	799	805	809	823	827	831	841	854
	859	860	862	863	878	880	893	900	901	903	915	921	925	927	939	945	948	949	950	952
	953	956	962	965	966	972	976	978	983	984										

10	41	42	44	46	59	60	68	69	76	77	85	87	90	93	95	97	103	104	105	106
	107	109	110	111	112	119	131	132	135	141	142	146	147	158	159	165	166	168	170	181
	183	189	191	193	195	197	198	199	201	202	213	219	223	227	237	246	248	249	251	252
	260	287	291	297	300	322	326	329	337	341	345	349	360	362	375	384	396	401	430	431
	433	434	436	437	441	445	446	448	450	478	484	485	487	488	494	507	516	529	535	536
	538	539	545	573	575	577	578	582	586	587	589	590	592	593	622	627	639	648	661	663
	674	678	682	686	694	697	701	723	726	732	736	763	771	772	774	775	777	786	796	800
	804	810	821	822	824	825	826	828	830	832	834	840	842	853	855	857	858	864	865	876
	877	881	882	888	891	892	904	911	912	913	914	916	917	918	919	920	926	928	930	933
	936	938	946	947	954	955	963	964	977	979	981	982								

11	43	86	88	89	91	92	94	113	114	116	118	133	134	136	138	140	148	150	153	156
	157	167	171	180	184	186	188	190	194	203	204	208	212	220	222	228	229	234	236	247
	261	262	264	265	270	286	298	299	301	303	306	320	321	327	328	330	331	335	336	338
	339	340	350	359	363	364	372	373	374	376	377	378	381	383	385	387	392	395	402	405
	428	429	438	439	440	442	443	444	449	451	452	476	477	479	480	482	483	489	490	493

TABLE II. (cont.)

							12									13					14	
543	638	695	801	885			209	284	371	456	564	644	738	813	908	356	460	599	706		562	
541	636	693	795	883			207	283	369	454	549	643	721	812	906	355	458	598	671		561	
540	631	692	794	875			206	282	367	453	548	641	719	802	886	354	457	568	670		557	
534	628	688	789	873			205	280	366	427	542	637	718	793	884	353	455	566	669		556	
533	621	687	787	870			187	279	365	426	532	635	716	792	874	352	425	565	668		554	
530	618	685	776	867			185	273	358	423	531	633	714	791	872	317	424	563	667		553	
528	595	684	762	866	980		179	272	351	416	527	632	708	790	871	314	422	560	666	850	552	
523	594	683	761	856	937		176	271	334	410	525	630	707	788	869	312	421	559	655	846	471	
521	585	673	759	852	935		175	269	333	406	524	629	705	760	868	310	420	558	653	845	470	
520	584	664	758	843	934		174	267	332	404	522	620	704	757	851	308	418	555	634	755	469	
519	583	660	753	839	932		172	266	319	403	501	619	691	756	849	281	417	551	616	749	467	
518	581	659	737	837	931		155	263	318	394	499	617	690	754	848	278	415	550	615	748	466	
517	580	651	725	835	929		154	235	316	393	498	613	689	752	847	277	411	526	614	747	462	
506	579	650	724	833	910		152	233	315	391	496	607	672	751	844	276	409	497	612	746	461	
505	574	649	722	829	909		151	232	309	390	492	600	665	750	838	275	408	473	608	745	419	712
504	572	647	720	820	907		149	231	307	388	491	597	658	744	836	274	407	472	606	742	414	710
503	571	646	717	819	905		139	230	305	386	481	596	657	743	818	268	389	468	605	715	413	611
502	547	645	703	815	890		137	221	304	382	475	570	656	741	817	178	370	465	603	713	412	610
500	546	642	702	811	889		117	211	302	380	474	569	654	740	816	177	368	464	602	711	313	609
495	544	640	696	803	887		115	210	285	379	459	567	652	739	814	173	357	463	601	709	311	604

TABLE III. Modulo Prime $p = 137$ Complexity of Integers < 137

Complexity	Integer
0	1
1	2
2	3
3	4
4	5, 6, 8, 9
5	7, 10, 16, 27
6	11, 12, 17, 18, 25, 28, 32, 36, 64, 81, 101, 119
7	13, 14, 15, 19, 20, 24, 26, 29, 33, 37, 44, 49, 50, 54, 61, 65, 79, 82, 92, 100
	102, 106, 120, 122, 125, 128
8	21, 22, 30, 34, 38, 41, 45, 48, 51, 55, 56, 60, 62, 63, 66, 68, 69, 72, 77, 80
	83, 88, 93, 99, 103, 107, 109, 117, 118, 121, 123, 126, 129, 130, 132, 133
9	23, 31, 35, 39, 40, 42, 46, 47, 52, 53, 57, 58, 59, 67, 70, 73, 74, 75, 76, 78
	84, 87, 89, 94, 96, 98, 104, 108, 110, 111, 112, 113, 115, 124, 127, 131, 134, 136
10	0, 43, 71, 85, 86, 90, 95, 97, 105, 114, 116, 135
11	91

TABLE IV. Modulo Prime $p = 1009$ Complexity of Integers < 1009

Complexity	Integer																			
0	1																			
1	2																			
2	3																			
3	4																			
4	5	6	8	9																
5	7	10	16	27																
6	11	12	17	18	25	28	32	36	64	81	256	512								
7	13	14	15	19	20	24	26	29	33	37	49	54	65	82	100	125	128	216	243	257
	507	513	548	729	960															
8	21	22	30	34	38	48	50	55	56	60	66	72	74	83	87	101	121	126	129	137
	144	162	169	217	244	258	287	289	324	343	383	384	508	514	527	549	625	710	730	763
	784	813	911	961	993	1000														
9	23	31	35	39	40	45	51	52	57	58	61	67	73	75	80	84	88	96	98	102
	106	120	122	127	130	138	142	145	148	163	164	170	173	174	189	192	196	200	218	225
	240	242	245	250	259	270	271	274	288	290	322	325	344	360	361	385	400	411	432	449
	464	480	486	490	509	515	528	538	550	572	573	576	617	626	631	635	640	670	676	707
	711	713	719	731	764	766	768	782	785	787	808	814	829	841	859	877	893	898	912	919
	928	962	977	985	994	1001														

TABLE IV. (cont.)

10

41	42	44	46	53	59	62	63	68	76	78	85	89	90	97	99	103	105	109	110
111	112	123	131	132	135	139	143	146	147	149	165	166	171	175	177	179	180	185	186
190	193	195	197	201	202	203	219	222	226	241	246	251	252	253	254	260	272	275	280
284	286	291	296	300	309	320	323	326	328	338	345	348	362	375	386	394	401	404	412
421	423	431	433	434	435	441	443	450	451	454	465	481	482	484	487	488	491	497	506
510	516	517	519	523	529	530	539	540	551	555	556	559	574	577	578	605	607	609	618
622	627	632	636	641	648	663	671	675	677	686	708	709	712	714	715	720	726	732	741
755	765	767	769	771	777	783	786	788	791	805	809	815	822	824	830	835	842	847	860
861	862	878	881	882	886	894	896	899	900	906	913	920	922	927	929	935	937	942	945
955	963	972	978	979	986	991	995	999	1002	1006	1007								

11

43	47	69	70	71	77	79	86	91	104	106	113	114	116	124	133	134	136	140	150
153	160	167	168	172	176	178	181	187	191	194	198	199	204	205	206	209	210	211	212
213	220	223	224	227	247	249	255	261	262	264	265	268	269	273	276	281	283	285	292
297	301	302	303	310	313	314	321	327	329	331	332	334	335	336	337	339	340	346	349
353	355	363	374	376	378	379	382	387	392	395	398	402	405	406	409	410	413	417	418
422	424	429	436	442	444	448	452	453	455	456	466	479	483	485	489	492	494	498	500
501	511	518	520	521	524	531	533	541	542	545	546	552	557	558	560	561	565	568	575
579	580	583	591	592	597	599	600	606	608	610	611	615	619	620	623	628	633	634	637
638	642	644	649	650	651	654	656	661	662	664	666	668	672	673	674	678	681	685	687
688	689	692	693	694	696	706	716	721	722	727	733	735	738	742	745	748	753	756	758
759	762	770	772	774	778	789	792	793	800	803	804	806	810	816	823	825	831	836	840
843	845	848	852	863	864	865	866	867	870	875	879	883	884	887	895	897	901	904	907
908	914	915	921	923	930	934	936	938	940	943	946	949	951	953	956	959	964	967	968
973	980	981	982	987	989	992	996	1003	1008										

TABLE IV. (cont.)

	0	92	93	95	107	115	117	118	119	141	151	152	154	156	157	161	182	183	188	207
12	206	214	221	226	232	234	235	248	263	266	277	278	282	293	294	295	298	299	304	306
	311	315	317	319	330	333	341	342	347	350	354	356	357	358	364	366	369	370	372	377
	380	381	388	390	393	396	397	399	403	407	414	415	419	420	425	426	430	437	438	445
	457	459	460	461	463	467	469	471	472	473	493	495	499	502	503	504	505	522	525	532
	534	536	543	544	547	553	554	562	563	566	567	569	571	581	582	584	585	588	593	598
	601	603	604	612	613	616	621	624	629	630	639	643	645	646	652	655	657	665	667	669
	679	682	690	695	697	700	717	718	723	724	725	728	734	736	737	739	743	746	747	749
	750	754	757	760	773	775	779	790	794	797	799	801	802	807	811	817	818	820	826	832
	837	844	846	849	853	858	868	869	871	872	876	880	885	888	889	902	905	909	916	917
	924	925	931	939	941	944	947	950	952	954	957	965	966	969	974	975	976	983	988	990
	997	1004	1005																	
13	94	155	158	159	184	215	229	231	233	236	237	267	279	305	307	308	312	316	318	351
	352	359	365	367	368	371	373	389	391	408	416	427	428	439	440	446	447	458	462	468
	470	474	475	476	496	526	535	537	554	570	586	587	589	590	594	602	614	647	653	658
	659	680	683	691	698	701	702	704	740	744	751	752	761	776	780	781	795	796	798	812
	819	821	827	833	834	838	850	851	854	857	873	874	890	891	903	910	918	926	932	948
	958	970	984	998																
14	230	236	239	477	478	595	596	660	684	699	703	705	828	839	855	892	933	971		
15	856																			

III. Complicated Numbers

One defines n to be a complicated number if $|n| > |n_1|$ for every $n_1 < n$. The complicated numbers $< 2^{10}$ are 1, **2**, **3**, 4, **5**, **7**, **11**, **13**, 21, **23**, **41**, **43**, **71**, 94, **139**, **211**, 215, **431**, and **863**. (Those in bold are also prime.) Obviously, there is an infinity of complicated numbers. We propose the following conjectures.

a. There exists K such that all complicated numbers $K_1 > K$ are prime.

b. Every sufficiently large integer n is the sum of $k < \log n$ complicated integers.

c. There exists c such that every sufficiently large n satisfies $|n| < c + \sqrt{\log n}$.

IV. Complexity and Entropy

Kolmogorov[4,5] has introduced the notion of complexity of a finite string over a given alphabet. For simplicity, suppose the alphabet to be $\{0, 1\}$. Let A be an algorithm that transforms finite binary sequences into binary sequences. By an algorithm is meant any of the various equivalent concepts used in logic. For a binary string x, one defines the complexity by

$$K_A(x) = \begin{cases} \underset{A(p,x)=y}{\text{Min}} \ \ell(p) \\ \infty \text{ if no } p \text{ exists such that } A(p) = x \ , \end{cases}$$

where $\ell(p)$ denotes the length of the binary string p. Analogously, one defines conditional complexity. Let $A(p, x)$ be an algorithm defined from pairs of binary strings to binary strings. Put

$$K_A(x) = \begin{cases} \underset{A(p)=x}{\text{Min}} \ \ell(p) \\ \infty \text{ if no } p \text{ exists such that } A(p) = x \ , \end{cases}$$

$K_A(y|x)$ is called the conditional complexity of y with respect to x. Kolmogorov regards complexity as analogous to entropy. We make the following conjecture.

Conjecture. Let a discrete binary information source S in the sense of Shannon[6] be given with entropy $H = -p \log p - (1-p) \log(1-p)$ where probability $(0) = p$ and probability $(1) = 1-p$; $0 < p < 1$. Let $\{x_1, x_2, \ldots, x_{2^n}\}$ be the set of all binary strings of length n arranged in order of decreasing probability. Let $k(n)$ be the least integer so that $\sum_{i=1}^{k(n)} \text{prob}(x_i) > r$ where $1/2 < r < 1$. Then asymptotically for large n,

$$H \approx \frac{1}{k(n)} \sum_{i=1}^{k(n)} K_A(x_i \mid n) . \tag{1}$$

$\Big($ In Eq. (1), K_A should be normalized so that when $p = 1/2$,

$$\frac{1}{k(n)} \sum_{i=1}^{k(n)} K_A(x_i \mid n) = 1 .\Big)$$

In other words, the most likely sequences from A have complexity approximately equal to the entropy of S.

In order to test the conjecture expressed in Eq. (1), we replaced $K_A(x_i|n)$ by $\lambda = \bar{K}(y|n)$, where λ is selected so that when $p = 1/2$,

$$\frac{1}{k(n)} \sum_{i=1}^{k(n)} \lambda = \bar{K}(x_i \mid n) = 1 .$$

Graphs of $H_1 = -p \log p - (1-p) \log(1-p)$ and

$$H_2 = \frac{1}{k(n)} \sum_{i=1}^{n} \lambda \bar{K}(x_i \mid n)$$

when $n = 10$ and $r = .75$ are shown in Fig. 1.

V. Complexity of N-Tuples of Integers

Matijasevič[7] has proved the following theorem. There exists a fifth-degree polynomial $Q(y_1, \ldots, y_k; z)$ with integer coefficients such that any enumerable set m of natural numbers (for example, the set of prime numbers) coincides with the set of natural values of the polynomial $Q(y_1, \ldots, y_k; a_m)$ where a_m is a certain number effectively constructed for the set m. From the result it follows that if one could

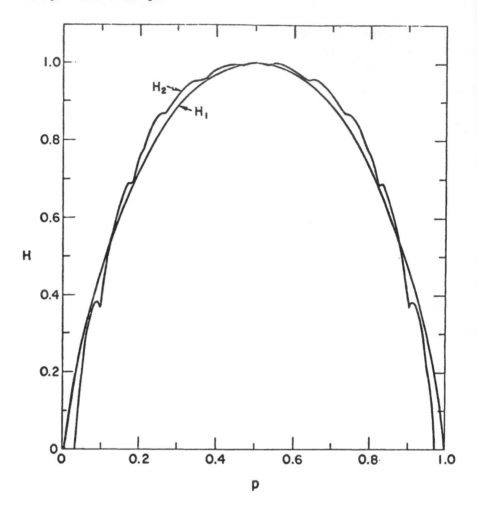

Fig. 1. Comparison of entropy $H_1 = -\sum p_i \log p_i$ and complement complexity H_2 as defined and discussed in text.

discuss complexity of n-tuples of integers, then one could discuss the complexity of enumerable sets of natural numbers by equating such complexity to the complexity of the associated polynomial Q.

References

1. A. S. Eddington, *Fundamental Theory* (Cambridge University Press, 1946).
2. I. J. Good, "The Proton and Neutron Masses and a Conjecture for the Gravitational Constant," Phys. Let. **33A**, 383–384 (1970).
3. A. Wyler, "Les Groupes des Potentiels de Coulomb et de Yukawa," Compt. Rend. Acad. Sci. Paris, **271**, 186–188 (1971).
4. A. Kolmogorov, "Three Approaches for Defining the Concept of Informationb Quantity," Problems of Information Transmission **1**, 3–11 (1965).
5. A. Kolmogorov, "Logical Basis for Information Theory and Probability Theory," IEEE Trans. Information Theory **IT-14**, 662–664 (1968).
6. C. E. Shannon and Warren Weaver, *The Mathematical Theory of Communication* (The University of Illinois Press, Urbana, 1949).
7. Ju V. Matijasevič, "Enumerable Sets are Diophantine," Soviet Math. Dokl. **11**, 354–358 (1970).

Additional References Not Used in Text

1. E. L. Lawler, "The Complexity of Combinatorial Computations: A Survey," Proceedings of 1971 Polytechnic Institute of Brooklyn Symposium on Computers and Automata.
2. D. W. Loveland, "On Minimal Program Complexity Measures," ACM Symp. Theory of Computing, Marina del Rey, California, May 5–7, 1969.
3. P. Young, "A Note on Dense and Nondense Families of Complexity Classes," Math. Systems Theory **5**, 66–70 (1971).

17

METRICS IN BIOLOGY, AN INTRODUCTION

With W. A. Beyer, M. L. Stein, and Temple-Smith*
(LA-4973, August 1972)

The role of measures of similarity, or dissimilarity, is basic to the development of taxonomical structures. Such measures, including the more restrictive ones, namely metrics having relevance for biological phenomena, are considered in this report. Of particular concern are non-traditional metrics of potential utility in recognition and taxonomy—particularly in molecular taxonomy. (Eds.)

ABSTRACT

The use of metrics in biology is discussed. Attention is given to metrics in pattern recognition, taxonomy, and especially molecular taxonomy. Various ways of constructing metrics between sequences for use in molecular taxonomy are discussed.

I. Introduction

To compare quantitatively different organisms, complex molecules, or biological entities in general, a measure of dissimilarity is required. More generally, all objects which form the elements of study

---

*Supported in part by a National Institutes of Health postdoctoral fellowship, grant HD 42801, Northern Michigan University.

in the natural sciences can be compared as to the degree of their differences. The notion of distance, in mathematics, is not directly or easily applicable to such studies, although intuitively any useful measure of the degree of difference between objects would seem to convey a measure of distance between them. The notion of distance between two sets of points in a metric space or between functions defined on some space (e.g., on the real line) is usually considered by comparing the values at each point separately. The differences are then either added in absolute value or integrated in the case of a continuum, and one may take, instead of sums of absolute values, the square root of sums of squares of the differences, etc. This, however, refers to numeric objects (sets, functions, operators) as rather fixed or rigid entities and does not in general involve moving or transforming one or both in order to obtain as close proximity of "fit" as possible. Obviously if one wants to compare two given different organisms, even purely geometrical organisms only, one tries to place them in positions where the comparison is made with real "corresponding" parts.[1] Mathematically, this means that one looks for the distance between two sets, modulo a class of transformations. This class can, but need not necessarily, form a group.

In this report we shall recall the elementary notions about metric spaces; we shall then redefine distances and "pseudo distances" between sets or, what is particularly important, between those *classes* of sets defining the entities one deals with in the natural sciences. The stimulus for this general discussion stems in part from the interesting work of Fitch and Margoliash[2] and others[3-6] on reconstruction of evolutionary trees from the data on the amino acid sequence of certain proteins. It also comes from certain unrelated studies of the general formulations of the problems of "recognition of patterns" and "artificial intelligence studies" which were undertaken by S. Ulam, R. Schrandt, and J. Mycielski.

In biological taxonomy there are attempts to define such mathematical concepts as subspaces and neighborhoods within the space of all organisms, yet the direct application of the more general concepts of metric spaces to this apparent mathematical area of natural science has met with only limited success.[7] The work of Sokal and Sneath[8] exemplifies one of the more successful attempts to extract evolutionary measures or distances from numeric or phenotypic taxonomy. These studies were often concerned more with statistical analysis of the variables being analyzed than with more fundamental mathematical considerations.

The advent of modern molecular biology, and with it the availability of comparative protein sequence data, has renewed interest in numerical taxonomy on the part of both biologists and mathematicians.[9-11] This is largely because the data are simple enough to permit tractable evolutionary distance calculations, although the simple nature of the 23-element protein space or four-element DNA genetic space may be misleading. The biological interpretations of this new molecular taxonomy have raised controversies about our understanding of evolution.[12,13]

Because of the current interest in molecular taxonomy and morphological distances in general we have outlined in this report the mathematical concepts of metric spaces and distances which may be applicable to these areas. This outline is then followed by a discussion of the protein sequence problem.

II. Dissimilarity Coefficients and Metrics

Let P be a set of objects. Following Jardine and Sibson (Ref. 14, pp. 77-78), one says that a function ρ from $P \times P$ to the real line is a dissimilarity coefficient if it satisfies the following requirements:

1. $\rho(p,q) \geq 0$ for all p, $q \epsilon P$,
2. $\rho(p,p) = 0$ for all $p \epsilon P$,
3. $\rho(p,q) = \rho(q,p)$ for all $pq P$.

Sometimes one requires

4. $\rho(p,q) = 0$ implies $\rho(p,r) = \rho(q,r)$ for all $r \epsilon P$ (evenness) ,

or

5. $\rho(p,q) = 0$ implies $p = q$ for all p, $q \epsilon P$ (definiteness) .

A dissimilarity coefficient which also satisfies

6. $\rho(p,r) \leq \rho(p,q) + \rho(q,r)$ for $p,q,r \epsilon P$ (triangle property)

in addition to properties 1-5 is called a metric. Intuitively, it would seem that any measure of distance should satisfy the triangle property. The triangle property is essential for relating meaningful topological notions to properties defined by distance. We should note here, however, that given any assignment of a "semidistance" satisfying only the first five properties one can obtain a metric from it by the following procedure: given two points, p, q, one considers all possible finite chains from p to q, continuing for example through p, x_1, x_2, ... x_n, q, and defining the distance from p to q as the minimum sum of the lengths of the chains:

$$\rho'(p,q) = \operatorname*{Min}_{[n,x_1,...,x_n]} \rho(p,x_1) + \rho(x_n,q) + \sum_{i=1}^{n-1} \rho(x_i,x_{i+1}) . \qquad (1)$$

Sometimes it is useful to require that an additional property be satisfied:

7. $\rho(p,q) \le \operatorname{Max}(\rho(p,r), \rho(q,r))$ for all p, q, $r \epsilon P$ (ultrametric inequality).

The ultrametric inequality is important in the theory of p-adic numbers and valuation theory.[15] Its relevance to biology is brought out by the following.[14]

"The strongest assumption about evolutionary rates which can be made is that they are constant. On this assumption the dissimilarities between present-day populations would be monotone with the times since their divergence. They would therefore be ultrametric, since the times of divergence of populations in an evolutionary tree form an ultrametric. The fact that the dissimilarities between present-day populations are rarely ultrametric refutes the hypothesis of constancy of evolutionary rates in terms of known measures of dissimilarity."

The following geometric interpretation can be given to the ultrametric inequality: every triangle is isosceles and its base has length less than or equal to that of the equal sides.

III. Metrics in the Space of Closed Sets, Hausdorff Distance, and Applications

One of the more general metrics is the Hausdorff distance. (See Ref. 16, pp. 166-172; Ref. 17, pp. 214-224; and Ref. 18, pp.20-32). This is a definition of distance between closed sets of points in a metric space. We assume here compactness of the underlying space P (this means that given any sequence of points x_n of P there exists a subsequence of these points converging to some point of P). Given two closed sets, A and B, one defines:

$$\rho(A,B) = \operatorname*{Max}_{x \epsilon A} \operatorname*{Min}_{y \epsilon B} \rho(x,y) + \operatorname*{Max}_{x \epsilon B} \operatorname*{Min}_{y \epsilon A} \rho(x,y), \qquad (2)$$

This distance satisfies the triangle property. Under this definition of distance the class of all closed sets in P becomes itself a compact metric space that is denoted by 2^P. One can now iterate our definition and consider sets in the space 2^{2^P}. This means that we consider classes of sets. Again, one can define a distance between any two of these classes with the use of the Hausdorff formula. This will be important in the sequel because when we speak of properties of sets we really consider classes of sets having a given property. So, for example, when we speak of sets of points on a screen "looking" like a letter A, we mean the aggregate or the class of such sets, distinguishable from the class of sets which "looks" like a letter B. In this way when we define a distance between objects independently of their size and orientation, for example, we have to consider, given a set, the class of all sets obtained from it by translations and rotations and also by changing of scale; then, given two different objects we are led to two classes of sets. The degree of their similarity or a quantitative measure of their differences should take into account possible changes of scale and position. If we now consider these two classes of sets and take the Hausdorff distance between them we do in essence the following. Given a set of the first class we look at the set in the second class that is as close to it as possible; we then take a maximum of this with respect to all choices of the first set. We then perform it symmetrically the other way around by taking a set from the second class, etc. The sum (or the maximum) of these two numbers gives a measure of distance between the "letter A" and the "letter B."

In unpublished notes W. A. Beyer and S. Ulam have compiled possible methods of measuring distances between sets and certain theorems which should be proved about the distances.

IV. Metrics for Molecular Taxonomy

In molecular taxonomy one considers sequences of amino acids defining the same protein* in various species. This means, mathematically, a class of codes each consisting of a sequence of symbols or words with 23 possible symbols. The encoded information gives the physical, chemical, and structural properties of the protein. The length of the sentences for the protein cytochrome c, e.g., is about one hundred words, and the first task is to define a distance function between any two such sequences of symbols, each assuming values from 0 to 22.[2,4]

---

* The phrase "same protein" means a class of proteins all of which perform the same biochemical function, and are by implication evolutionarily related.

This distance would then give a measure of dissimilarity. We shall, in this section, discuss the problem somewhat more generally. We may assume, for simplicity, that the symbols assume only two values: 0 and 1. We thus have a space of all sequences of this sort, of variable length, and we try to define a notion of distance between them under various postulates as to the equivalence or indistinguishability between some sequences. In other words, we shall assume, given a sequence, a class of other sequences "equivalent" to it and give definitions of distance between such classes. This is analogous with the illustration given above on classes of objects in the plane.

In mathematical studies, given two sequences of 0's and 1's, one may define the distance variously. For example, given $\alpha = [x_1, x_2, \ldots x_n]$ and $\beta = [y_1, y_2, \ldots y_n]$ as:

$$\rho(\alpha, \beta) = \sum_{i=1}^{n} |x_i - y_i|, \qquad (3)$$

or

$$\rho(\alpha, \beta) = \sum_{i=1}^{n} (x_i - y_i)^2. \qquad (4)$$

Still another way, suggested for coding problems by Hamming,[19] is:

$$\rho(\alpha, \beta) = \sum_{i=1}^{n} [1 - \delta(x_1, y_i)], \qquad (5)$$

where $\delta(x, y)$ is the Kronecker delta function. We note that this metric is equal to that defined by Eq. (3) only for binary sequences. This will be of value later when considering the problem of protein sequences which are formed from the 23-symbol amino acid space.

We might, however, assume that the given sequences of 0's and 1's are written not linearly but on a circumference of a circle, and we can arbitrarily rotate this circle rigidly so that each sequence of a given length n is equivalent to n-1 other sequences. Definition of distance, then, would concern a distance between classes of equivalents.

This is quite a general situation. In mathematics one defines a distance between two functions $f(x)$ and $g(x)$, for example, as follows:

$$\rho(f, g) = \text{Max } |f(x) - g(x)|, \qquad (6a)$$

$$\rho(f, g) = \int |f(x) - g(x)| \, dx, \qquad (6b)$$

or

$$\rho(f,g) = \left\{ \int (f(x) - g(x))^2 \, dx \right\}^{1/2}, \tag{6c}$$

etc. We may however, wish not to distinguish between functions which are obtained by shifting one from another. In this case we have to define distance between classes of functions, perhaps using the Hausdorff metric. Also consult the work of Marczewski and Steinhaus.[20]

We shall now consider still different definitions of distance between two sequences of 0's and 1's. One definition could be

$$\rho(\alpha,\beta) = \min_{n,m,n',m'} (n + m + n' + m') \tag{7a}$$

where n, m, n', and m' are defined by

$$(T_1)^n \, (T_2)^m \, \alpha = (T_1)^{n'} \, (T_2)^{m'} \, \beta . \tag{7b}$$

Here we allow two types of transformation or two kinds of "steps." T_1 consists of changing a 0 into 1 or vice versa. T_2 consists of a deletion of a symbol anywhere in the sequence and subsequent contraction of the rest, to close the gap. Given two sequences, one may define as a distance the minimum total number of steps performed on one or both of these sequences so as to bring them into identical form. As an example, let $\alpha = [010101010101]$ and $\beta = 101010101010]$. Then by Eq. (3) or (5) we would obtain

$$\rho(\alpha,\beta) = 12 \tag{8a}$$

since all places have different values, while by Eq. (7) we would have

$$\rho(\alpha,\beta) = 2 \tag{8b}$$

since by deleting the first symbol in α and the last in β one obtains identical sequences.

If one considers mutations in a chain of DNA and if amino acids defining a protein are considered to be the special types defined above, then the distance as constructed above may correspond to the number of necessary mutations to transform one sequence into the other or both into an ancestral one. This mutational-transformation approach has been applied by Reichert and Wong[21] to protein and RNA sequences. Mathematically, this function of a pair of sequences is of certain combinatorial interest. It is not obvious *a priori* whether given at random two such sequences of n binaries the distance between them as computed by the algorithm above will be, *on the average*, a linear function of n. It is clear that this average will be less than $n/2$. It is also of interest to consider infinite sequences and diverse definitions of distances between them.

One can allow a number of transformations of sequences which lead from a given one to sequences which we consider equivalent among themselves. Given such a division into classes, one can define the distance function between classes à la Hausdorff indicated above, starting with a given notion of distance between individual sequences.

Another distance-type function which is applicable to the sequence problem can be defined as follows: given two sequences, we consider the number of 1's in each. We take the absolute value of the difference between them. Next we consider the number of 1's followed by 0's, compare that number in the two sequences, and take the difference. We do the same for 0's followed by 1's, then 1's followed by two 0's, etc. We then add the numbers. This type of "Markov distance" gives us perhaps an idea of the "visual" distance between the two given sequences.

The last two distances have considerable appeal for the molecular sequence problem. The sequence transformation metric defined by Eq. (7) would seem to have a direct biochemical interpretation, as pointed out above. However, for the nonbinary sequences (such as proteins defined in the 23-symbol amino acid space) the direct interpretations of the different transformation operators as the analogs of the physical mutations become more difficult. This is partly because the physical events take place in 4-symbol RNA space and are not always simple* functions of these noncommuting operators.[21]

The "Markov distance" is also of considerable interest inasmuch as it appears to be a measure of the overall visual similarity of the sequences. This may be what is needed since in itself the sequence is not the object of ultimate interest but rather, as in the case of proteins, it is the three-dimensional structure which is, or chemical properties which are, encoded in the sequence.

It was in the light of the above considerations that a new sequence metric was defined. It began with an idea of Fitch.[9] Fitch's original proposal for detecting sequence homology was defined as follows. Let $\alpha = (X_1, \ldots, X_N)$ and $\beta = (Y_1, \ldots, Y_N)$ be two sequences of amino acids each of length N. Let $\eta(X, Y)$ be some measure of the distance from amino acid X to amino acid Y. Put

$$
\rho_n(\alpha, \beta) = \sum_{\ell=0}^{N-n} \sum_{k=0}^{N-n} \sum_{j=\ell+1}^{n+\ell+1} \eta(X_j, Y_{k+j})
$$
$$
- np(N - n + 1)^2, \ (1 \leq n \leq N) \, .
$$

(9)

---

* For example, genetic duplications, inversions, and the frame shifts as viewed in the protein space.

Here n is what is thought to be a statistically important subsequence size. The second term is the expected value of the measure assuming nonhomologous or random sequences with an average element probability of p. Our new sequence metric can be defined in a related manner. Put

$$\gamma(\alpha, \beta, n) = \sum_{\ell=0}^{N-n} \underset{0 \le k \le N-n}{\text{Min}} \sum_{j=\ell+1}^{n+\ell+1} \eta(X_j, \, y_{k+j-\ell}) \qquad (10a)$$

and

$$\rho'(\alpha, \beta) = \left[\sum_{n=1}^{N} \left(\frac{\gamma(\alpha, \beta, n)}{N-n+1}\right)^2\right]^{1/2} \left[\sum_{n=1}^{N} n^2\right]^{-1/2}. \qquad (10b)$$

Then the metric is

$$\rho(\alpha, \beta) = \text{Max}\left\{\rho'(\alpha, \beta); \; \rho'(\beta, \alpha)\right\}. \qquad (10c)$$

This metric has a number of potential advantages including the fact that it can be applied to sequences of varying length although no proof exists as to the triangle inequality for such cases. It also can give a measure as to the degree of redundance (subsequence duplications within or among the sequences). In a subsequent[22] paper we shall study this metric in greater detail.

V. Remarks

a. One of the major uses of distances in biology is in cluster analysis and evolutionary tree construction. J. A. Hartigan (unpublished notes) has pointed out the following objection: pairwise distances are a more sophisticated form of dissimilarity judgment than clusters, and so it may be inappropriate to use them to compute clusters. However, in tree construction where one wants to estimate length of branches, the distance concept is useful. There are other situations where one wants an estimate of the distance between clusters, and the distance concept is useful.

b. Another quantity which might be used in place of distance is a quantity σ to be thought of as relating to the probability of transition from p to q. $\sigma\,(p, q)$ should be a mapping from $P \times P$ to $[0,1]$ which satisfies

1. $0 \leq \sigma\ (p, q) \leq 1$

2. $\sigma\ (p, q) \geq \sigma\ (p, r)\ \sigma\ (r, q)$

for all $p, q, r \epsilon P$. Development of a theory of such a function might be worthwhile.

References

1. E. S. Smirnov, "Mathematische Studien über Individuelle und Kongregationen Variabilität," Verh. 5 Intern. Kong. Vererbungs-wiss. **2**, 1373-1392 (1927).
2. W. M. Fitch and E. Margoliash, "Construction of Phylogenetic Trees," Science **155**, 279-284 (1967).
3. T. H. Jukes, *Molecules and Evolution* (Columbia Univ. Press, New York, 1966).
4. M. O. Dayhoff, *Atlas of Protein Sequence and Structure* (National Biomedical Research Foundation, Silver Spring, Md., 1969).
5. D. E. Kohne, "Evolution of Higher Organisms DNA," Quant. Rev. Biophys. **3**, 327-381 (1970).
6. M. Goodman, J. Barrabas, G. Matsuda, and G. W. Moore, "Molecular Evolution and the Descent of Man," Nature **233**, 604-613 (1971).
7. E. Mayr, *Principles of Systematic Zoology* (McGraw-Hill, New York, 1969).
8. R. R. Sokal and P. H. A. Sneath, *Principles of Numerical Taxonomy* (W. H. Freeman and Co., San Francisco, 1963).
9. W. M. Fitch, "An Improved Method of Testing for Evolutionary Homology," J. Mol. Biol. **16**, 9-16 (1966).
10. T. Uzzell and K. W. Corbin, "Fitting Discrete Probability Distributions to Evolutionary Events," Science **172**, 1089-1096 (1971).
11. M. Kimura, "Evolutionary Rate at the Molecular Level," Nature **217**, 624-626 (1968).
12. B. Clarke, "Selective Constraints on Amino-Acid Substitutions During the Evolution of Proteins," Nature **228**, 159-160 (1970).
13. T. H. Jukes and J. L. King, "Deleterious Mutations and Neutral Substitutions," Nature **231**, 114-115 (1971).
14. N. Jardine and R. Sibson, *Mathematical Taxonomy* (John Wiley and Sons, New York, 1971).
15. G. Bachman, *Introduction to p-adic Numbers and Valuation Theory* (Academic Press, New York, 1964).
16. F. Hausdorff, *Set Theory* (Chelsea Publishing Co., New York, 1957).

17. K. Kuratowski, *Topology, Volume I* (Academic Press, New York, 1966).
18. K. Kuratowski, *Topologie, Volume II* (Polish Scientific Publishers, Warsaw, 1961).
19. R. W. Hamming, "Error Detecting and Error Correcting Codes," Bell System Tech. J. **29**, 147-160 (1950).
20. E. Marczewski and H. Steinhaus, "On a Certain Distance of Sets and the Corresponding Distance of Functions," Colloq. Mathematicum **6**, 319-327 (1958).
21. T. A. Reichert and A. K. C. Wong, "An Application of Information Theory to Genetic Mutations and Matching of Polypeptide Sequences," to appear in J. Theor. Biol.
22. W. A. Beyer, T. Smith, M. L. Stein, and S. M. Ulam, "A Molecular Sequence Metric and Evolutionary Trees," submitted to *Nature*.

18

ON THE THEORY OF RELATIONAL STRUCTURES AND SCHEMATA FOR PARALLEL COMPUTATION

With A. R. Bednarek
(LA-6734-MS, May 1977)

This report is about a combinatorial study of relations between compositions of transformations and "projective algebra" forming foundations of mathematical logic. It also contains proposals to utilize such ideas for parallel computation machines. (Author's note.)

ABSTRACT

This report will outline an area of work which, so far theoretical, will indicate some applications for construction and operation of computers which should be able to perform simultaneous operations, i.e., computations in parallel, particularly as these concern the composition and/or iteration of functions or, more generally, relations.

I. Introduction

This report will deal with indications of usefulness of parallel and series machines for several areas of mathematical constructions. [For example, such is the case in the Monte Carlo method, where a great number of independent samplings with low or medium arithmetical precision (not requiring very many decimals) may considerably speed up and enlarge the area of this application. In the Monte Carlo procedures the results are gathered not by a small number of very precise (many digit accuracy) calculations, but rather by obtaining statistics of very many independent or semi-independent histories of typical patterns or processes (to be developed in subsequent paragraphs). Thus, in the study of parabolic differential equations, a method of operating simultaneously on many channels would be useful. Similarly, in the discussion of hyperbolic differential equations or integral equations leading to or equivalent to such, we encounter a similar situation: Outside the cone of influence there is an independent or semi-independent development of disturbances and this can be studied all at once on a one-, three-, or multidimensional grid.]

Given two functions f and g, it would be nice to obtain the function $f(g)$, all at once by a single order, including the case when the two functions are given numerically, graphically, or tabulated. The present machines compose them serially, point by point, with very great speed to be sure. If we had at our disposal a machine combining the digital and analogue features, we could envisage this most important and powerful algorithm of analysis to be effected as fast as possible with only one computer. After all, it is the facility of differentiating a composite function that increases the power of the infinitesimal calculus.

In particular, we outline a method of *composing* functions and relations, using a novel, parallel, i.e., "all at once" computing scheme. We concern ourselves with a two-track theoretical investigation, namely: (1) studies of the role of composition, particularly as far as its simultaneous or parallel realization is concerned, in mathematical schemata. Here our concern is to encompass the standard operations of algebra and analysis in terms of composing point or set-valued functions (relations); (2) assuming a mechanism for effecting composition in a nonserial manner, we will study the advantages of such in computer organization, the modelling of neurological like phenomena such as iterative searches, memory building, "pattern recognition," processing of operations of the visual retina, etc.

Preliminary engineering studies (at the University of Florida) have established the feasibility of the fabrication of a computational element effecting instantaneous composition using existing solid state technology or some combination of optical and acoustical phenomena. However, our study is restricted to the theoretical implications of the existence of such a computer component.

II. Functional and Relational Composition

We have observed that in some early work of A. Tarski[27] there was given a set-theoretic formula for the composition $R \circ S$ of two relations which involved projections. In particular,

$$R \circ S = \pi\{(R \times X) \cap (X \times S)\}$$

where π is the projection $\pi(x, y, z) = (x, z)$. Technically speaking there should be included in the formula the morphisms $((x, y), z) \to (x, y, z)$ and $(x, (y, z)) \to (x, y, z)$. The following schematic should aid in the visualization of this formula.

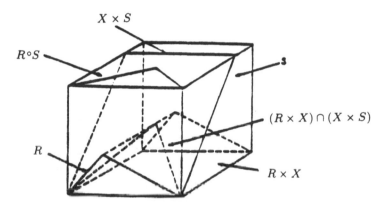

Subsequent discussions with electrical engineers and materials scientists have established the physical realizability of this operation employing existing solid state technology or some combination of optical and acoustical phenomena in a three-dimensional homogeneous medium. Some such schemata will be discussed in Appendix B.

Motivated by the above, our aim is to study mathematical operations and algorithms which would utilize parallel work and simultaneous operations on computing machines—be it on a single large computer with such facility, or on a number of smaller computers connecting together and operating simultaneously and largely independently— as far as arithmetical or Boolean operations are concerned. In particular, assuming the facility to effect the operations of *composition* of functions and relations in a parallel ("all at once") fashion, our concern is to encompass the standard operations of algebra and analysis in terms of composing point or set-valued functions.

In the following section we describe some specific mathematical investigations now under way. In Sec. IV we illustrate some of the possible implications and/or applications of the anticipated results.

III. Some Studies of Relational Structures

There are basically two directions in our study. The interconnection between composition and projection suggests the first of those, namely, the algebraization of the operations of projection and the construction of product sets. This corresponds to a (finite) model of the logical quantifiers; that is, the "there exists" and the "for all" operators. These, in addition to the usual Boolean and arithmetical orders on present day computers, would enlarge, even though modestly at first, the scope of theoretical mathematical investigations amenable to study on electronic machines. The second direction concerns the operation of composition of functions and transformations and, more generally, the composition of relations. In addition, we propose to study the interrelation of those two approaches—the algebraic and combinatorial properties of these.

Delineated below are some specific researches which we have undertaken.

(a) In Ref. 12 there was initiated the study of projective algebras—an initial step in the development of algebraic versions of logic from which have evolved the cylindric and polyadic algebras. Extending the work in Ref. 12, McKinsey[20] showed that every projective algebra was isomorphic to a certain kind of class of binary relations. There are many open problems (see, for example, Ref. 28) concerning these algebras, but in particular we are concerned with those that are isomorphic to *specific* relation algebras. Recently it has been observed that the projective algebras B_N of all binary relations on a finite set can be generated by a single element. We have characterized those algebras corresponding to B_N. The "projection" and "direct product" operators in the axiomatization of projective algebras can be replaced by a multiplication (composition). Is it possible to characterize those algebras that may be isomorphic to some class of topological relations; e.g., under suitable conditions, the "clopen" relations on a compact, totally disconnected space? Do there exist "Stone-type" theorems identifying spaces whose suitably associated algebras are isomorphic? Given a class of subsets, say, for example, in the plane, under what condition can this family be *projectively generated*? Recently V. Faber and J. Larson[13] established the existence of a countable collection of sets in the plane that could not be embedded in a finitely generated algebra.

McKinsey[20] gave postulates for a calculus of binary relations. We have been able to show that every such structure is a complete atomic projective algebra and, conversely, that there can be introduced into particular, complete atomic projective algebras a multiplication making these algebras of binary relations. This replacement of projections by a multiplication may result in a more amenable (for machine purposes) model of the first-order predicate calculus.

(b) At the other end of the algebraic spectrum we have started to investigate various semigroups (under composition) of functions or relations. The object of this line of research is to see how much information is carried by the multiplicative structure of the algebras in question; that is, to what extent can the Boolean operations on the objects be supplanted by the multiplicative substitute for the logical quantifiers? The results reported in Refs. 1 and 2 show, for example, that for some classes of spaces, the spaces are determined by these semigroups. In this connection we propose to investigate and extend the results in Refs. 4 and 26 concerning the *finite* generation of these semigroups. Connected with these problems are questions of the most economical and shortest expressions involving the minimum number of compositions of the base elements to produce all given ones. That such extensions are possible has been confirmed by our recently noticed theorem that all closed relations on the unit interval can be approximated by the composition of three fixed relations.

The applications of such algorithms may be interesting for the study of models of memory building (see Refs. 9 and 21) by constructing trees (with some loops) or coding visual and aural impressions by a process of composing existing codes in the nervous system or memory storage.

(c) Focusing on the operation of composition there is a class of problems concerning *fixed points* of transformations as well as properties of their iterates.* Given the possibility of simultaneous or parallel effectuation of composition there is open an entire avenue of investigation concerning such questions not only for transformations but for multivalued transformations; that is, relations. As observed in Ref. 28, computations on existing computers are particularly well suited for the display of asymptotic properties of iterates. However, to tabulate only the interesting points it would be preferable to have a visual evaluation of the *iterated* properties of the many (or all) points. We propose to study the implications of the possibility of nonserial composition for investigations, qualitative as well as quantitative, of fixed-point problems, mixing phenomena in hydrodynamics, topological features of force fields, etc.

(d) Two relations R and S on a set X are *product-isomorphic* if there exists a bijection $\psi : X \to X$ such that the transformation

---

* Schemata for such were discussed with John Pasta some years ago.

$(x, y) \rightarrow (\psi(x), \psi(y))$ takes R onto S. Employing composition this is equivalent to $R = \psi \circ S \circ \psi^{-1}$. Of course the original definition can be extended to n-ary relations. Is there an analogous compositional identity? It was noted in Ref. 28 that the notion of product isomorphism has an interesting relation to isomorphism in certain algebraic structures. In particular, if one associates with a group G the ternary relation $\{(x, y, z) \mid xy = z\}$ then two groups are isomorphic if and only if their associated relations are product isomorphic. We propose to investigate the possibility of associating with structures, such as groups, a *binary* relation (or relations) so that these representations would be product isomorphic. Such representations coupled with the ability to compute rapidly via the compositional identity above would provide a useful tool.

The concept of *weakly product-isomorphic*[28] for binary relations is equivalent to the assertion that $R = \psi \circ S \circ \lambda^{-1}$ for a *pair* of bijections. Numerous extensions of the above will be investigated. Recent investigations by A. R. Bednarek and Gian-Carlo Rota have revealed a connection between the concepts of weakly product-isomorphic functions or relations and *flows*.

A more extensive discussion of these concepts and problems is contained in Appendix A.

(e) Another problem to be investigated is that of the classification of "complexity" of sets under the usual operations of the type mentioned earlier; that is, Boolean, projection, and composition. Here we mean "complexity" in the sense defined in the paper by Beyer, Stein, and Ulam.[5] Also of concern is the relation of this complexity to "entropy" of such sets and the connected combinatorial problems.

(f) As mentioned earlier we have proved recently that all of the closed relations on the unit interval can be "approximated" by three such. Of course this approximation is dependent on the existence of an appropriate *metric*. In the case in question we used the Hausdorff metric. We are studying the question of possible metrics for various spaces of relations. For example, when there is an underlying metric, as there was above, on the space on which the relations are defined, then the Hausdorff metric is a natural choice. If, however, the subsets (relations) are subsets of a measure space then the Steinhaus distance

$$\frac{m(\Delta(A, B))}{m(A) + m(B)},$$

where Δ denotes the symmetric difference between sets, may be more natural. The particularization of the above (or possible alternatives) in the case when the underlying space is finite and the relations are therefore directed graphs will be studied.

IV. Possible Applications of Results on Relational Structures

In this section we wish to indicate some of the directions for possible applications of results realized from the theoretical investigations outlined in Sec. III.

For example, in the last-mentioned item in the preceding section we mentioned the study of possible metrics. In the "recognition" of figures, some kind of "distance" must be used to examine *similarity* or *analogy* between them. The nervous-system conversion behind the retina *might* operate using some schemata involving such abstract distances; see, for example, Ref. 21.

In speculating on how nonserial (parallel) composition might play a role in effecting some of the standard operations of analysis we make the following observation.
Suppose $X = [a, b]$ is a real interval

Let $H = \{(x, y) \mid x \geq y\} \subset X \times X \subset R \times R$. Now if $f : X \to X$, then the composition $f \circ H$ is the set $f \circ H = \{(x, y,) \mid f(x) \geq y\}$. Schematically:

Given instant composition and, for example, optical display of the same, if the luminosity of $A \subset X \times X$ is $\lambda(A)$, then

$$\int_a^b f \, dx = \frac{\lambda(f \circ H)}{\lambda(X \times X)} \; .$$

Studies concerning the further role of composition in integration (summation) are under way.

In another direction, one can explore the relation of the facility to compose in a nonserial fashion to, among other things, the recent work of N. Metropolis and Gian-Carlo Rota[18] on "significance arithmetic."

We speculate on some of the possible applications to problems of "information retrieval." To make the situation more concrete we employ a simple situation basic to some existing retrieval systems.

First, we assume the existence of a finite collection $W = \{w_k, w_2, \ldots, w_N\}$ of *words* and a finite collection $\mathcal{D} = \{d_1, d_2, \ldots, d_k\}$ of *documents*, where each document d_i is a vertex of the N-dimensional cube; that is, d_i is a sequence of length N with its jth entry 1 or 0, indicating that word w_j occurs or does not occur in the document. Of course we are not here concerned with how such a determination is made. A typical example is "key word in title indexing" of some collection of journal articles with the totality of "key words" constituting the collection W.

Given this situation, immediately there comes to mind two *relations* D, W, on the sets of documents $\mathcal{D}$ and words W, respectively. Namely $D \subset \mathcal{D} \times \mathcal{D}$ where $(d_i, d_j) \in D$ iff d_i and d_j have a common non-zero entry and $W \subset W \times W$ where $(w_i, w_j) \in W$ iff w_i, w_j occur together in some document.

Of course the composition $D \circ D [W \circ W]$ yields all those pairs of documents [words] that are correlated to a third document [word]; that is, for example, $(d, d') \in D \circ D$ iff there exists a document d^* such that $(d, d^*) \in D$ and $(d^*, d') \in D$. The process can be iterated with $D^{[1]} = D$, $D^{[2]} = D \circ D$ and $D^{[n]} = D^{[n-1]} \circ D$.

Given a word $w^* \in W$, consider the composition $W \circ (w^*, w^*) \circ W$ which consists of all pairs (w, w') related through w^*; that is $(w, w') \in W \circ (w^*, w^*) \circ W$ iff $(w, w^*) \in W$ and $(w^*, w') \in W$. In general, given a subset $W_0 \subset W$, we let $\Delta_{W_0} = \{(w, w) \mid w \in W_0\}$. Then $W \circ \Delta_{W_0} \circ W = \{(w, w') \mid$ there exists a $w_0 \in W$ such that $(w, w_0), (w_0, w') \in W\}$.

Given the relations D and W there is much interest in *clusters*; that is in the maximal complete subgraphs of D and W. Many algorithms exist for the determination of these (Refs. 3, 6, and 23). To our knowledge none of these focuses on composition. What is suggested is, given a reflexive-symmetric relation R on a finite set X, can one find more efficiently those subsets C of X maximal with respect to $C \times C \subset R$? Similarly, can searches in more sophisticated retrieval systems be conducted more efficiently given the facility to compose relations all at once? What role, if any, does relational composition play in the design of an *adaptive* retrieval system, that is, one of the form $Q \times G \to G$ where the "inputs" are questions and G the "states"— are admissible file organizations, with the transition resulting in file reorganizations *that take into account* the "information" carried by the queries presented?

It appears that some of the lines of research delineated above make contact with the recent work of L. N. Cooper (Refs. 9 and 21) concerning development of feature-detecting cells in visual cortex as well as possible organization of animal memory and learning. These connections were discussed by S. M. Ulam and L. N. Cooper at a recent meeting on Quantum Biology and it is expected that our researches will result in a strengthening of these connections.

We include here also an account of some work concerning simple preliminary experiments of pattern recognition. These were performed in 1970–71 by Robert Schrandt and Ulam in Los Alamos on one of the Los Alamos Scientific Laboratory computers.

It is proposed to extend these studies on a more general basis using variety of different notions of a metric distance between two sets of points in the plane or between two classes of such sets (or "pictures") on a screen.

It is speculated that analogous schemata may be operating in the nervous system, specifically, in the visual brain, on the layers behind the retina.

As will be seen later, the operation of *composition* of sets may play a role in the production of "examples" which are variations of *one* or *a few* impressions stored in the brain, to be compared with a new impression confronted by the nervous system (or by the computer).

A subsequent report will contain a discussion of distances to be used in purely mathematical investigations concerning the *degree* of a quantitative measure of analogy or similarity between mathematical structures, i.e., between relation sets, i.e., graphs, groups, and some other mathematical structures. The behavior of such distances as regards relational composition has been studied by A. Bednarek in his work on relation theory.

The following account summarizes work done in 1970–71 in Los Alamos by Robert Schrandt and Ulam.* The computations were done on the CDC 6600. Specifically we experimented with schemata for an automatic recognition by an electronic computer of visual data, i.e., pictures such as presented on a screen or on the retina of an optical system. The idea involves "teaching by examples." The simplest instance considered by us was to present the computer with data describing a handwritten letter "a" in a large number of examples, then present the machine with a number of examples of the letter "b." After this, a letter was written again and the machine was to decide whether it was "a" or "b." The visual system of a living organism contains ways to abstract from size, from translation and from rotations through small angles, at least. We should stress right away that rotations through larger angles, say, 90°, are not yielding equivalent impressions, so for example, N and Z are equivalent by a 90° rotation, but are perceived as different objects. The invariance then obtains

---

* We want to acknowledge also the interest and helpful suggestions of Professor Jan Mycielski.

only for "group neighborhood" small-angle rotations. A larger class of transformations which does not affect the notion of the same object is, of course, small deformations. [The notion of an object to be perceived, that is to say, a class of two-dimensional sets must involve abstraction from small changes, for the problem of recognition amounts to ways of distinguishing classes of two-dimensional sets when it comes to establishing visual memory (one-dimensional sequences for auditory impression, perhaps three-dimension ones for tactile ones)]. The main tool in our approach is the definition of a distance or, really, several distances between classes of sets. We shall indicate some plausible such metrics and speculate on the way these are coded and registered in the nervous system and the brain. Later on we shall speculate on iteration of such procedures, that is to say, families of classes which we can call ideas or notions of the first order for coding and registering those, etc.

In our simple-minded experiments we attempted to imitate the various versions of a handwritten letter as follows. An example of the letter a was written by hand and the points forming it were registered on a 64×64 grid in the computer in such a way that a unit square was touching the extremities of the set. Instead of writing a number of other examples by hand and putting the resulting pictures into the memory of the computer, which would be laborious and time-consuming, we write down once and for all, two transformations of the unitsquare into itself, call them S and T, which are a little different from identity transformations, and we apply in succession various *compositions* of these two, obtaining a number of transformations of the given first example. In our case if we wanted, say, 128 examples, we would take all possible compositions of seven transformations S and T, e.g., $STSSTST$, $TTSSTST$, etc., each giving an image of the original set. Clearly, if one wanted, say, 1024 examples, we would take compositions of 10 of these transformations. If T and S differ sufficiently little from the identity, the resulting transformations would be still "small" deformations. We have taken, for T and S, the following:

$$T : \begin{aligned} x' &= x + \epsilon_1 4y(1 - y) & \epsilon_1 = \frac{1}{10} \text{ or } \frac{1}{16} \\ y' &= y + \epsilon_1 4x(1 - x) \end{aligned}$$

$$S : \begin{aligned} x' &= x + \epsilon_2 y^2 & \epsilon_2 = \frac{1}{8} \text{ or } \frac{1}{12} \\ y' &= y + \epsilon_2 x^3 \end{aligned}$$

It was rather surprising to watch the images of the letter a (and similarly for b) as they appeared after these transformations, they gave the impression of being handwritten, too, sometimes by a shaking hand showing a variety of different styles.

We now want to discuss a way to code numerically each of the resulting sets of points and explain our criteria for comparing a new example with our two sets of examples to decide whether the new picture should belong to the first or to the second set. The unit square is divided successively into four, then 16, then 64...subsquares. We denote these by, Q_1, Q_2, Q_3, Q_4, then by Q_{11}, Q_{12}, Q_{13}, Q_{14}, $Q_{21}Q_{24}, Q_{23}$, Q_{24}, ..., and so on. Let Z_1, Z_2, ..., Z_N be the sets of points corresponding to our examples where N was, say, 128. Given a set Z, we associate with it a sequence of numbers which are binaries in 0's and 1's as follows:

In the intersections of Z and Q_1 we will write a 1 or a 0 depending on whether Z has points in common with this set or not. We continue, in this way, obtaining a sequence of "the characteristic functions of a given set Z."

We shall now consider weights assigned to the successive coordinates of the sequence, giving more importance to the large squares and diminishing it successively for the ones corresponding to the smaller squares. This is to stress the "global" properties of the set Z more than the "smaller details." In our procedure we gave to the four first squares, $Q_1 \ldots Q_4$, a weight of $1/4$ each, to the next batch a weight of $1/16$, and so on. We now assign a distance, ρ, between two sequences coding the two sets Z_i and Z_j as follows: If the sequences are

$$\bar{x} = (x_1, x_2, \ldots, x_{128})$$

and

$$\bar{y} = (y_1, y_2, \ldots, y_{128})$$

we put

$$\rho_1(\bar{x}, \bar{y}) = \frac{|x_1 - y_1|}{4} + \ldots \frac{|x_4 - y_4|}{4} + \ldots \frac{|x_5 - y_5|}{16} + \ldots \frac{|x_{128} - y_{128}|}{128}.$$

In this fashion we obtain two finite sets of points A and B corresponding to the examples of the letter a and the letter b, respectively, each a set of 128 points. Given now a "problem," that is to say, a new example of a letter written by hand, we have a single new point p in this metric space.

We can now compute the distances from p to all the points of A and the distance from p to the points of B. If the sum of the distance from p to the points of A is smaller than the sum of the distance from

p to the points of B, the computer decides that "the point p is more like the letter a."

In our experiments we have produced, as problems for the machine, by actual writing, a variety of letters a and b, imitating different styles of handwriting, and obtaining points $p_1 \ldots p_k$. In each case the computer has made a decision whether the letters were a or b. The results were over 80% correct!

The first experiment was based on the rather crude metric ρ_1, defined above.

A more suitable metric, which we shall call ρ_2, would be based on a more precise way of coding the visual picture by a numerical sequence. Instead of merely noting the presence (1) or absence (0) of the given set in any of our squares of the subdivision of the screen, one could write down a real number, S_1, $0 \leq S_1 \leq 1$, describing the *proportion* of the number of points in the given set S to the total number of points in the subsquare in question. In this fashion a picture or set Z would be coded by a sequence of real numbers S_1, $S_2, \ldots, S_N$. Again, one should use *weights* for the sets of squares—larger ones for the more "global" or "gross" features of the set.

Still another distance, ρ_3, could be defined to correspond to the "Hausdorff" distance between sets. This distance, ρ_H, introduced by Hausdorff, between closed sets A, B in a compact metric space E is defined by

$$\rho_H = \underset{x \epsilon B}{\text{Max}} \; \underset{y \epsilon A}{\text{Min}} \rho_E(x, y) \;\; + \;\; \underset{x \epsilon A}{\text{Max}} \; \underset{y \epsilon B}{\text{Min}} \rho_E(x, y),$$

ρ_E being the metric in E.

In our case ρ_E would be the ordinary (euclidean) distance between the points in our unit square. ρ_3 would be defined by computing the distances between the sets of our two pictures in each of the squares of the subdivision, again giving *weights* to each square, larger ones to the larger squares, diminishing for the "small detail squares."

Still another distance, ρ_4, would be defined analogously but starting with the "Steinhaus distance" ρ_S instead of the Hausdorff one. The Steinhaus distance between sets A, B is based on the *measure* (in our case simply the *number* of points, of course)

$$\rho_S(A, B) = \frac{m(A \triangle B)}{m(A) + m(B)} \, ,$$

$\triangle$ denoting the symmetric difference between the sets A, B.

488

To obtain ρ_4 we would again compute the distances between the sets in each square of the subdivision separately, keeping track of the weights of the squares as above.

We plan to undertake the corresponding experiments with these "more precise" distances.

Our crude beginning attempts referred to two letters only. For "reading" a text, the computer would have to possess in its memory sets of examples for each letter, of course.

One can now speculate about the building of memory in a living organism—in the nervous system connected to retinal impressions. There might exist a "wiring" system of nerves and their connections behind the retina, allowing us to abstract from orientation, size, translation of the picture presented. In addition, we might perhaps venture to postulate a system of "deformation" of a given picture (object) by a mechanism not unlike our composition of two fixed transformations providing a set of equivalent impressions.

The way to compute a *distance* or degree of analogy could be based not on our numerical procedure but could perhaps utilize a system of finding an analogous set in a collection of sets by a method similar to the one which astronomers use to scan a set of photographs of a region of the sky to discover a new or variable star by flipping a number of these in succession in that the eye, disregarding the rejective pictures, reacts to a point which "jumps out" on a background of constant impressions.

This could be imagined as follows: The more "global" parts of the code evoke from the memory places where these "trunks" of the nervous connections go. A set of pictures which are then compared *modulo* 2 with the impression received, that is, the *overlap* of the pictures is "dark"—when the total intensity of the symmetric difference is sufficiently small—the picture is examined in more detail until it is finally recognized or else put in the memory in the *new place* (together with all the "small deformations" of the *new* picture).

A more economical and efficient method of operating by the nervous system and the brain should be, of course, to keep in the memory just one example (or a few examples). Presented with a letter as a "problem" the brain can obtain very quickly from any example it has many deformed ones and compare the one to be recognized with this whole class in the manner described above. This avoids the loading of memory with unnecessary variants of a given "picture."

The next step is to devise, in an analogous manner the recognition of *classes* of sets of pictures which are not variants of one of them. The following may serve as a case of the general problem. Given are, say, 10 pictures representing various animals—a dog, a cat, a horse, etc. Given is a set of points representing another animal. One is to decide whether this new picture belongs to this class K or whether it represents, say, a tree, of which we also have in the memory a number of examples, forming a class. The problem deals obviously with variables of a higher class—families of sets—instead of sets individually.

The sets of each class have only some *properties* in common; they are not variables of just one set as before. The problem of recognition of a class would be attacked by considering functionals of sets. One set of functionals can be the Fourier, or rather Rademacher or Walsh co-efficients, describing a two-dimensional set. These, by the way, should be weighted, as in our previous discussion. But there must be several or many other sets of functionals, including one dealing with individual points, for example. The functionals whose sequence is put in the memory must overdetermine the sets which we consider and which our experience builds up gradually from childhood. Perhaps a hundred or more collections of sets of functionals are gradually acquired.

We shall now tentatively describe a possible mechanism of discerning whether a single set S should be put in the class of sets R or in the class U. For simplicity of explanation let us assume that the class R consists of two sets R_1 and R_2 and the class U of sets U_1 and U_2.

We examine functions $F_1 \ldots F_N$ for the sets U_1 and U_2 and find out which among them are the same or have close values for the sets R_1 and R_2. This is a collection forming a subset of the sequence $F_1 \ldots F_N$, call it $F^1_{i_1} \ldots F^1_{i_k}$. Analogously for sets U_1 and U_2. these might be $F^2_{j_1,} \ldots F^2_{j_\ell}$. For our "problem" presented by the set S we look at these functionals for the set S and find out whether the values of S are closer to the F^1 collection or else to the F^2 collection. We make our decision accordingly. (If none is sufficiently close we will decide that the set S has neither the "property" R nor U.)

In our next report we shall examine a possible way to scan this procedure, which, in the case where R and U contain a sizeable number of sets each, will be more "economical."

As a final example of a situation in which the type of computational facility we have alluded to might play a role, we outline a very general formulization which we label the "graph of mathematics." The iterative searches in this graph that might recognize "deep theorems," given the complexity of the graph, would involve the ability to make rapid computation on "parallel" searches over a great multitude of paths.

The idea is to consider a formal system of mathematics as a collection of groups of signs and symbols starting with a list of symbols, then selecting a system of axioms, for example Gödel's, (really for the first attempt taking a much more restricted and simpler one), we start with axioms which are groups of signs. We have rules of combination of these and rules of deduction which are formalized. These are not written down, but given a single group of formulae we consider joining them to others by the rules of procedure which include the symbols for Boolean operators, *and, or,* quantifiers, and substitutions. Therefore our graph has a form of a "pair tree," that is to say, from pairs of such formulae or from pairs of collections of symbols, we get new ones. In some cases, from a single one by "mitosis," we get another one (by omission or restriction, or some suitable "erasures" allowed by our procedures). We obtain a graph which bears a resemblance to "genealogical" graphs. These graphs can be characterized by the property that, going "backwards" from single points, or single individuals considered as vertices, one obtains two parents, ancestors of the individual. A single person has thus at most two, sometimes only one ancestor. (Allowing procreation by "mitosis.") These graphs form a natural generalization of graphs which are called trees, for which a considerable theory exists. Much less is known about graphs defined in the above way. A theory of such "pair trees" in analogy to the existing body of knowledge about tree graphs remains to be developed.

There are many problems which have been considered for trees. For example the reconstruction problem which has been solved for trees could be studied for pair trees. (A list of such problems will be appended later on.)

The graph of mathematics as we call it, is not the same as a genealogical tree since a given formula or a theorem can be derived by our procedures from different pairs of ancestors. (It could be considered as a pair tree with colors.)

It is possible to attempt to define the "value" or "interest" or "importance" or "depth" of a theorem, or a formula. This, roughly speaking, should have the following characteristics: some of the formulae are obtained only by a long train of successive pairs of ancestors and deductions. If they happen to be relatively short we consider them the more valuable. The really interesting ones are such which in their "neighborhoods"* have many others easily deducible from them. For example, the fixed point theorem of Brouwer has this property (the consequences of the "deep theorem" need not easily lead back to the "important" or "deep" formula or theorem.) One could try to define

---

* Having a pair tree, one can define a *distance* between any two formulae or two theorems by the length of the shortest connection between them through the chain of "ancestors." This will always be a *metric*. In this metric it makes sense to define "proximity" of a vertex to other vertices and to define what a neighborhood means.

what might be called a surprising formula or a surprising theorem. These could be such that can be deduced by a perhaps complicated sequence of operations from the starting list of symbols and axioms but are such that the path or a band of paths leading to them comes from a rather remote, "sidewise" located collection upstairs.

V. Relation Algebras: Some Examples

In the next to the last application of the preceding section attention was focused on a particular "distance" between classes of planar patterns (relations). While we have given several possible metrics for such objects, no systematic investigation of such spaces appears to have been made. At the algebraic end of the spectrum though, some attention has been paid to relational structures (pattern algebras); for example, the earlier mentioned calculus of relations of McKinsey,[20] the relation algebras and algebras of relations of Tarski[27] and others. To illustrate the character of these investigations, we summarize here results obtained relative to some of the more general of the relational structures, namely, semigroups of relations. Our focus throughout is on those questions of interest to anyone concerned with algebras of patterns, particularly those algebras including the iteration or composition of these patterns as one of the basic operations.

The family of all binary relations on a nonempty set X is a semigroup under the operation of composition. We denote this semigroup by $\mathcal{B}_X$. There has been increased interest in these semigroups due in part to a resolution, by Montague and Plemmons,[19] of a problem of Schwarz[25] concerning the character of the maximal subgroups of $\mathcal{B}_X$ and due further, to the interesting results obtained in a topological setting by K. D. Magill, Jr. and his students.

In particular, Montague and Plemmons[19] showed that every finite group is isomorphic to a maximal subgroup of some $\mathcal{B}_X$. Later Plemmons and Schein[24] (see Ref. 7 for a particularly elegant argument) extended this result to arbitrary groups.

In another direction attention has been focused on the morphisms of $\mathcal{B}_X$. In 1964, Crestey[10] showed that every automorphism of $\mathcal{B}_X$ which preserves finite unions is inner. One year later Zareckii[36] showed that every automorphism of the subsemigroup of $\mathcal{B}_X$ consisting of all relations with domain and range equal to X are inner. Finally, Magill,[16] and independently, Gluskin,[14] showed that every automorphism of $\mathcal{B}_X$ is inner. Beginning with the work of Clifford and Miller[8] and pursued by Magill and his students, in the topological case, investigations of the endomorphisms of $\mathcal{B}_X$ are under way.

Obviously, there are many interesting questions one may ask about $\mathcal{B}_X$, but the sampling of results above is intended to only indicate some of the possibilities. Furthermore, one could argue that all of this could be subsumed within the framework of *transformation semigroups* since $\mathcal{B}_X$ is isomorphic to the semigroup of all additive set functions of X that fix the empty set (define $\psi:\mathcal{B}_X \to \mathrm{Hom}(2^X, 2^X)$ by $\psi(R) = F_R$, where $F_R(A) = RA = \{x \mid (x,a) \epsilon R$ for some $a \epsilon A\}$ for all $A \epsilon 2^X$). This argument, however, loses some of its attraction particularly when one attempts to choose a proper topology to insure, for example, that the subsemigroup of all closed relations on some "nice" space is isomorphic to the semigroup of all continuous selfmaps of 2^X.

At the other end of the spectrum; that is, when X is finite, too little attention has been paid to $\mathcal{B}_X$. It appears that aside from some characterizations (Magill[17]) no systematic investigation of $\mathcal{B}_X$ has been undertaken and some very basic questions remain unanswered. (By way of illustration we include the multiplication table for $\mathcal{B}_2$; that is, the semigroup of all binary relations on a two-element set. We are grateful to M. Stein of the Los Alamos Scientific Laboratory for the computation of this table.)

Multiplication Table for $\mathcal{B}_2$

	R_1	R_2	R_3	R_4	R_5	R_6	R_7	R_8	R_9	R_{10}	R_{11}	R_{12}	R_{13}	R_{14}	R_{15}	R_{16}
R_1	1	1	1	1	1	6	1	6	1	1	11	11	1	6	11	16
R_2	1	1	1	1	2	6	2	6	3	3	11	11	4	8	12	16
R_3	1	2	3	4	1	6	3	8	1	2	11	12	1	6	11	16
R_4	4	4	4	4	4	8	4	8	4	4	12	12	4	8	12	16
R_5	1	1	1	1	5	6	5	6	9	9	11	11	13	14	15	16
R_6	1	1	1	1	6	6	6	6	11	11	11	11	16	16	16	16
R_7	1	2	3	4	5	6	7	8	9	10	11	12	13	14	15	16
R_8	4	4	4	4	8	8	8	8	12	12	12	12	16	16	16	16
R_9	1	5	9	13	1	6	9	14	1	5	11	15	1	6	11	16
R_{10}	1	5	9	13	2	6	10	14	3	7	11	15	4	8	12	16
R_{11}	1	6	11	16	1	6	11	16	1	6	11	16	1	6	11	16
R_{12}	4	8	12	16	4	8	12	16	4	8	12	16	4	8	12	16
R_{13}	13	13	13	13	13	14	13	14	13	13	15	15	13	14	15	16
R_{14}	13	13	13	13	14	14	14	14	15	15	15	15	16	16	16	16
R_{15}	13	14	15	16	13	14	15	16	13	14	15	16	13	14	15	16
R_{16}	16	16	16	16	16	16	16	16	16	16	16	16	16	16	16	16

$a=(0,0)$	$R_1=\{a,b,c,d\}$	$R_5=\{a,c,d\}$	$R_9=\{b,c,d\}$	$R_{13}=\{c,d\}$
$b=(0,1)$	$R_2=\{a,b,c\}$	$R_6=\{a,c\}$	$R_{10}=\{b,c\}$	$R_{14}=\{c\}$
$c=(1,0)$	$R_3=\{a,b,d\}$	$R_7=\{a,d\}$	$R_{11}=\{b,d\}$	$R_{15}=\{d\}$
$d=(1,1)$	$R_4=\{a,b\}$	$R_8=\{a\}$	$R_{12}=\{d\}$	$R_{16}=\phi$

We do not mean to imply by the above that the *semigroup* of relations on a set is the proper relation or pattern algebra to study. Quite the contrary, while the calculus of relations of McKinsey may represent too stringent an axiomatization, the disregard of the Boolean operations, particularly the natural order induced by them, errs in the other direction. This judgment is confirmed to some extent by an examination of some of the characterizations of semigroups of relations (Magill[17]) which reveal the presence of an *order* defined in terms of the multiplication. Although considerable insights can result from a study of semigroups of relations we feel that the multiplicative structure must be complemented by some additional structure; for example, some order respected by the multiplication. A variety of such algebras will be investigated.

In yet another direction, almost no attention has been paid to topological relational structures; that is, to relational algebras topologized in such a way as to insure the continuity of the operations involved.

Of fundamental interest in any algebra of patterns are the questions of generation; that is, the construction of the patterns from basic units using admissible operations. We illustrate these ideas with some recent results on generators and embeddings for semigroups of relations. These results, most of which have not appeared in the literature, are suggestive of questions that may be asked for any structure purporting to be a "relational structure."

Recently E. Howorka has proved the following:

Theorem. (Howorka). If X is a set then a relation R on X is of the form $f^{-1} \circ g$, where f and g are functions iff card $R \leq$ card X.

As an immediate corollary we have that *every* binary relation on an infinite set is of the form $f^{-1} \circ g$; that is, every relation is the composition of the inverse of a function and of a function. In addition, Howorka[15] observed, extending the results in Ref. 4, that the calculus of relations $\mathcal{B}_N$, that is, the calculus of all relations on a finite set of card N could be generated by a *single element*. This followed from the proposition.

Theorem. (Howorka). Let $\mathcal{B}_n$ denote the collection of all binary relations on the set $X = \{1, 2, \ldots, n\}$. Let $R = \{(1, 2), (2, 3), \ldots, (n-1, n)\}$ and $S = \{(n, 1)\}$. The subsemigroup $[R, S]$ of $\mathcal{B}_n$ generated by R and S under relational composition contains all of the atoms of $\mathcal{B}_n$.

By observing that $S = [R^2 \cup (R \circ R^c) \cup (R^c \circ R)]^c$, where R^c denotes the complement of R in $X \times X$, we then know that the relation R generates all of $\mathcal{B}_n$ under the Boolean operations and composition.

Very little was known until recently about generators of finite semi-groups and, in particular, about generators of $\mathcal{B}_N$. Recently E. Norris[22] proved the following:

Theorem. (Norris). The semigroup $\mathcal{B}_X$ satisfies ($*$) below iff X is finite.
($*$) For all $R, S \epsilon \mathcal{B}_X$, $R \circ S = \Delta$ implies $R^{-1} = S$ where S is a permutation on X.

This implies that $\mathcal{B}_N$ cannot be generated by a pair of generators. It is not known how many generators are required for $\mathcal{B}_N$.

There follows an example of a result similar to those suggested in Ref. 28 for projective algebras.

T. Evans[11] showed that any countable semigroup can be embedded in a semigroup generated by two elements. S. Subbiah[26] gave an elegant constructive proof of the fact that any countable collection of transformations on the set of natural numbers N was contained in a subsemigroup of τ_N, the full transformation semigroup on the natural numbers, generated by two such transformations. Since any countable semigroup can be embedded in τ_N, Evan's result was an immediate consequence of Subbiah's theorem.

A minor modification of Subbiah's proof permits the following extension of her theorem to relations on the set of natural numbers.

Theorem. If $\{R_n\}$ is any countable family of relations on the natural numbers N, then there exist two relations on N that generate a semigroup containing $\{R_n\}$.

Proof. As in Ref. 26 we let

$$A_n = \{2^n(2m - 1) - 1\}_{m=1}^{\infty} \text{ for } n = 1, 2, 3, \ldots .$$

Now let ϕ and T be relations on N defined as follows:

$$\phi = \{(x, 2x) \mid x \epsilon N\}, \text{ and}$$

$$T = \{(x, x - 1) \mid x \text{ even}\} \cup$$

$$\bigcup_{n=1}^{\infty} \{x \times (\tfrac{1}{2^{n+1}}(x + 1) + \tfrac{1}{2})R_n \mid x \epsilon A_n\}$$

where we note that, in general, for any relation R, by xR we mean the set $xR = \{y \mid (x, y) \epsilon R\}$.

Assert that $R_n = \phi T \phi^n T^2$, where juxtaposition denotes relational composition.

Suppose $(x, y)\epsilon R_n$, then $(x, 2x)\epsilon\phi(2x, 2x-1)\epsilon T, (2x-1, 2^n(2x-1))$ $\epsilon\phi^n$, $(2^n(2x-1), 2^n(2x-1)-1)\epsilon T$. Note $2^n(2x-1)-1\epsilon A_n$ so that $(2^n(2x-1)-1) \times (1/2^{n+1}((2^n(2x-1)-1+1)+1/2)R_n$ is contained in T. But

$$\frac{1}{2^{n+1}}\left(2^n(2x-1)-1+1)+\frac{1}{2}\right)R_n = xR_n \text{ , so that}$$

$$(2^n(2x-1)-1, y) \ \epsilon \ T \ .$$

Thus $R_n \subset \phi T \phi^n T^2$.

Suppose $(x, y)\epsilon\phi T \phi^n T^2$, then $(x, a)\epsilon\phi$, $(a, b)\epsilon T$, $(b, c)\epsilon\phi^n$, $(c, d)\epsilon T$ and $(d, y)\epsilon T$, for some $a, b, c, d\epsilon N$. So $a = 2x$, $b = 2x-1$, $c = 2^n(2x-1)$, $d = 2^n(2x-1)-1$. Since $d = 2^n(2x-1)-1\epsilon A_n$, we have

$$y \ \epsilon \ \left(\frac{1}{2^{n+1}} \ (d+1) + \frac{1}{2}\right)R_n = xR_n \ .$$

Hence
$$\phi T \phi^n T^2 \subset R_n \text{ , and therefore}$$

$$R_n = \phi T \phi^n T^2 \ .$$

If the family $\{R_n\}$ is a family of functions then the relation T is a function so that the result particularizes to that of Ref. 26.

This completes our illustration of the character of the results and problems that might be of concern to investigators of "algebras" purporting to be algebras of patterns, particularly two-dimensional patterns. Generalizations to higher dimensions of some concepts and problems considered are given in Appendix A.

Appendix A

Examples of Combinatorial and Set-Theoretical Problems Concerning the Operation of Forming Product Sets

The theory of relations, the theory of graphs, also the theory of projective or cylindric algebras have as their bases the following set-theoretical setup:

Let E be an abstract set, finite or infinite. By E^n we understand the set of ordered n-tuples: $E^n = E \times E \times \ldots,\ E$, the elements being $(e_1,\ e_2,\ \ldots,\ e_n)$ where $e_1 \in E$ for $i = 1 \ldots n$. E^∞ denotes the set of infinite (countable) sequences $(e_1,\ e_2,\ \ldots,\ e_n \ldots)$.

For the study of binary *relations*, we consider E^2 and associate with the given relation R between pairs of elements x, y, a set of edges; the edges correspond to certain pairs of elements of E, namely those pairs which are R-related.

In this way a given relation R, or a graph G, can be associated with a subset, call it A of E^2.

(More generally, in a study of other algebraical or combinatorial notions, for example in ternary relations, say a group H we might associate with it a subset of a higher direct product. Thus for a group operation, defined on an abstract set E, we might consider in E^3 the set of such triplets (e_1, e_2, e_3) such that $e_1 \circ e_2 = e_3$ where $\circ$ denotes the group operation. Analogously when we deal for example with rings, we could consider in E^6 a subset of sixtuples $(e_1,\ e_2,\ e_3,\ e_4,\ e_5,\ e_6)$ where $e_1 + e_2 = e_3$, $e_4 \circ e_5 = e_6$ where $+$ and $\circ$ denote the two arithmetic operations defining the ring. For infinite operations, i.e., in some topological theories, say, in considering Fréchet spaces where the fundamental notion is that of convergence of a sequence: $(e_1,\ e_2,\ \ldots,\ e_n,\ \ldots)$ to an element e_0, we may "represent" the topological space by a subset of E^∞ consisting of the sequences: $(e_0,\ e_1,\ e_2,\ \ldots,\ e_n,\ \ldots)$.

We introduce now a general notion of *product isomorphism* and *product homomorphism* which subsume the notions of isomorphism, and homomorphism used in these theories.

Two subsets A, B contained in E^2 are called product isomorphic if there exists a one-one mapping $T(E)$ and the mapping T^2 defined by $T^2(x, y) = (T(x),\ T(y))$ maps the set A onto B: $T^2(A) = B$. Analogously a product homomorphism $S^2(E)$ is defined by a not necessarily one-one mapping $S(E) = E$ by $(x, y) \rightarrow (S(x), S(y))$.

In complete analogy one defines a product isomorphism T^n of E^n onto itself by $T^n(E^n) = E^n$ through $(x_1,\ x_2, \ldots,\ x_n) \rightarrow (T(x_1),\ T(x_2),\ \ldots,\ T(x_n))$ and two subsets A, B of E_n are called product isomorphic if there exists a T and $T^n(A) = B$.

Similarly we define the product homomorphism S^n. B is called product homomorphic to A if there exists an S such that $S^n(A) = B$.

n can be any finite integer or $n = \infty$.

It is obvious now that two relation sets R_1 and R_2, or two graphs G_1 and G_2, are isomorphic to each other in the usual sense if and only if the "representations" of them as defined above are product isomorphic in our sense.

Similarly, the isomorphism of two groups H_1 and H_2 is equivalent to the product isomorphism of the subsets "representing" them in our sense. Analogously, homeomorphism of two topological spaces is obviously equivalent to the product isomorphism of their "representations" in E^∞. It is equally clear that homomorphisms, algebraically or topologically translate into homomorphisms of the representations.

We shall now present a few examples of combinatorial problems arising in the theory of properties of the product isomorphisms.

The first questions concern the enumerations.

(1) How many nonproduct isomorphic sets can one have in E^n or equivalently, of course, how many classes of product isomorphic sets are there? (Product isomorphism is a transitive relation, so the classes are disjoint.) The asymptotic estimates are known for E^2—that is to say for the number of nonisomorphic relations or nonisomorphic graphs, in case of E, finite. (For E countably infinite the number is of the power of the continuum.) Much less if known for $n = 3$ or higher, i.e., the number of non-isomorphic general ternary relations, for example.

(It is of the order of $2^{\binom{N}{3}}$, N being the cardinality of E; it is obviously $= c$ for cardinality of E equal $\aleph$.)

(2) Analogous problems concerning product homeomorphisms. Less is known for this case. The question is how many sets can one have in E^n so that none of them is a homomorphic image of the other? (Recently, E. Howorka and, independently, J. Mycielski showed a continuum set of graphs such that none is a homomorphic image of any other in the set.)

(3) Questions concerning the existence of "universal" countable subsets of E^n, i.e., such sets A in E^n that every countable subset X in E^n should be isomorphic to a subset of A; this, of course, in the case when the cardinality of E is that of a continuum.

(4) Problems concerning the existence of "economical" bases for a given class (e.g., a countable one of subsets A_1, A_2, ..., A_m, ... of E^n). This means, for instance, furnishing finite number of sets B_1, B_2, ..., B_j such that by the operations of Boolean algebra and of projection and direct product formation one can obtain, among others, all the sets A_1, ..., A_n... (There is a whole class of problems of this sort, mainly still unresolved.) For the case of K finite can the base be formed with just 2 sets? The problem is of interest when the cardinality of E is say of the power of continuum or higher—the class of given sets may then be uncountable.

(5) Problems of the above sort become more difficult when their analogs are stated for the notion of "equivalence" of two sets A, B through a decomposition.

We illustrate this notion for subsets of E^2. Two subsets A, B of E_2 are called *equivalent by finite decomposition* if:

$$A = A_1 \cup A_2 \cup \ldots \cup A_k$$

$$B = B_1 \cup B_2 \cup \ldots \cup B_k$$

$$A_i \cap A_j = \emptyset \text{ for } i \neq j$$

$$B_i \cap B_j = \emptyset \text{ for } i \neq j$$

and each A_i is product isomorphic to the corresponding B_i—the product isomorphism depending in general on i that is $A = T_i^2(B)$: through a transformation T_i.

In the case when the cardinality of E is finite the problem concerns for example the number of sets nonequivalent by a decomposition into two subsets. If E is infinite, e.g., of the power $\aleph$ of the continuum or higher, one can ask about the possible number of sets nonequivalent by decomposition even into a countable number of sets:

$$A = \bigcup_{n=1}^{\infty} A_i \, , \quad B = \bigcup_{n=1}^{\infty} B_i \, .$$

(6) A notion of "weak" product isomorphism: e.g., for $n = 2$. For subsets $A, B \subset E_2$, we call these weakly product-isomorphic if there exist two transformations T_1 and T_2 of E onto itself. Such that the transformation $T_{1,2}(E^2) \to E_2$

defined by $(x, y) \to (T_1(x), T_2(y))$

carries A onto $B : T_{1,2}(A) = B$.

(7) A definition of "iteration" of subset $A \subset E^n$. This will be a subset ^2A of E^{2n} defined as follows:

^2A consists of all sequences of elements $(e_1, e_2, \ldots, e_n, f_1, f_2, \ldots, f_n)$, where $(e_1, \ldots, e_n)$ and $(f_1, \ldots, f_n)$ were arbitrary elements of A.

Suppose ^2A is product isomorphic to ^2B; under what conditions can one assert that A is product isomorphic to B? In all generality this is not true. For example, Fox constructed, in an answer to a problem posed by one of us, an example of two topological spaces S_1 and S_2, and the S_1^2, and S_2^2 are homomorphic (in the usual topological sense) but S_1 and S_2 are not. For algebraic or combinatorial structures, the answers are in general negative—but true for special classes, e.g., finite groups, graphs of special character, etc. When can we assert that if $^2A_2$ is a homomorphic product image of $^2A_1$ then A_2 is a homomorphic image of A_1?

The above will merely serve as isolated specimens of a great class of problems formulable in the spirit of combinatorial set theory from a unified point of view.

Appendix B

Physical Realizations of Nonserial Compositions

In this appendix we consider the possibilities for the physical realization of nonserial composition.* In particular, in A there is described the preliminary engineering design of a device that effects this operation. One could consider the problem of more efficacious designs along these lines with the objective of possible fabrication of a pilot model to serve as a heuristic aid in early theoretical studies, but the interconnection problems preclude a system of even modest capability.

In part B we describe possible lines of investigation of three-dimensional computing in a homogeneous medium. While presenting many interesting possibilities, particularly in the elimination of the size restriction inherent in A, this endeavor faces many unsolved problems in the area of materials science. What is considered here is an initial assessment of physical phenomena that might be exploited to effect the realization of the principal processes of cylindrification, intersection, and projection, the processes basic to nonserial composition.

A. Discrete Systems

Here we are concerned with the preliminary engineering design of a discrete composition machine, in particular the logic, memory, input-output, control, and size. The cube of Fig. 1 is referred to in the description of this proposed hardware.

---

* We are particularly grateful to Derek Dove, Professor of Materials Science and Engineering, University of Florida, for generous discussion of the schemata that follow.

This machine is described as a cube with three relevant working surfaces, A, B, and C, where pairs of indices (i, j, k) describe locations on each surface as indicated in Fig. 1, with each index having n possible values. In this description, surfaces A and B are thought of as being the input mappings to be composed, with the results projected on surface C.

The logic of the composition is described as a two-level AND-OR array. A typical cell on surface A is associated by logic AND with each of the n cells on the same row of surface B, with the result projected to a row of internal points. For instance, typical internal point c_{ijk} holds the results of the logical AND of cells a_{ik} and b_{jk}, as

$$c_{ijk} = a_{ik} \wedge b_{jk} .$$

Now the projection upward onto the C surface is taken as follows. Each cell on the C surface,

$$C_{ij} = \bigvee_{k=0}^{2^n - 1} c_{ijk}$$

where C_{ij} is a typical cell in the C surface and where the recursion operator indicates a logical OR in the k direction of all the c_{ij} elements below it.

A possible realization to illustrate this idea is as follows. Each of the n^3 AND gates can be realized by a transistor such that transistor conduction indicates a logic 1, occurring if the voltage on the base is high AND, and the voltage on the emitter is low. As shown in Fig. 1, variable ik is a bus bar which supplies base current from surface A to any number up to n transistors. Variable jk is a similar bus bar from the B surface. Each bar is programmed, at surfaces A or B, by connection to a voltage supply.

The recursion OR, operating in the k direction, can be considered as a bus bar connecting together all the collectors of the transistors in each level below it which have the same (i, j) coordinates. The display cell on each cell of the C surface is a lamp, which is on if one or more transistors below it are turned on by the (ik, jk) coincidence.

One of the system requirements is that a result of a composition displayed on surface C needs to be mapped back onto surface A in order to provide the capability of several iterative compositions with a fixed mapping on B. This can be accomplished with an $n \times n$ array of flip-flops on each of the two surfaces A and C, where one of the surfaces would required dual rank flip-flops.

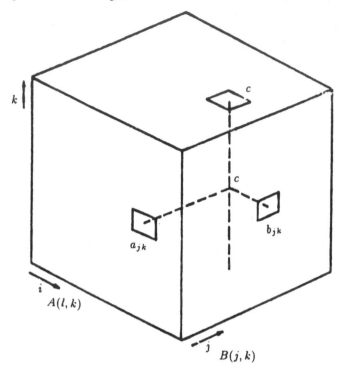

Logical AND at each $c_{ijk} = a_{ik} \bigwedge b_{jk}$

Logical OR vertically $C_{ij} = C_{ij} = \bigvee\limits_{k=0}^{2^n-1} c_{ijk}$

A possible cell realization:

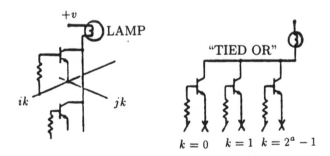

Fig. 1. Schematic for a discrete composition machine.

B. Homogeneous and Partially Parallel, Partially Serial Systems

Here we consider physical phenomena that might be possibly exploited to effect a realization of the processes of projection and intersection.

Several major types of systems may be envisaged. The projection path may be delineated by some physical structure such as an electrical conductor, and intersection may be realized by the use of discrete components as in A. We may refer to such a system as a *discrete system*. Alternatively, projection may be carried out by the motion of an advancing wave front, while intersection in such a case may involve the interaction of wave fronts in a *homogeneous* medium. Evidently these two examples represent the extremes of completely discrete and completely homogeneous systems.

The present status of such arrangements may be summarized briefly by noting that while the technology exists for the construction of an entirely discrete system, such a system suffers from an inefficient use of elements and presents interconnection problems of extreme difficulty for a system of even modest capability. Modern high-density fabrication techniques lend themselves naturally to two-dimensional organization.

Three-dimensional computing in a homogeneous medium, while an exciting possibility, presents unsolved problems in both mechanisms of defining a vector or scalar wave property over a large region of space and in choice of wave/wave interaction phenomena.

This part of the appendix is concerned with the exploration of the implications of function composition for parallel computing schemes, and with the preliminary assessment of the potentialities and limitations of homogeneous medium interactions. Finally a study is proposed of partially parallel, partially serial systems, based upon modern high density solid state optical sensor arrays.

1. Homogeneous Medium. Wave front delineation: The spatial limitations imposed upon a wave front by diffraction effects are, of course, well known; the resolution with which a desired image may be reproduced may be readily evaluated from a knowledge of wave length, system observations, and effective aperture.

The process of projecting a two-dimensional image "intact" through a region of space can only be carried out with a resolution far inferior to the two-dimensional resolution capabilities of the system. In optical terms an extreme depth of field requires a very small beam divergence which in turn limits lateral resolution.

One would need to examine the information packing density that may be "projected" using optical, holographic, acoustic, and electron beam techniques.

Interaction in Medium. In order to carry out the intersection process utilizing the influence of two projected wave fronts, it is necessary that the intersecting wave fronts produce an effect that may be detected by a third orthogonal wave front and that may in addition be distinguished above the background "noise" level due to the nonintersecting wave fronts.

Particular attention needs to be given to an evaluation of known optical and acoustic phenomena for their potentialities for the present applications. Examples are (i) Photochromic effects in which a beam of one wavelength produces a coloration in a crystalline or glassy medium, a second beam of different wavelength produces a bleaching or other change of the colored region. (ii) Quenched fluorescence, a phenomena where the fluorescence produced by light of a particular wavelength may be locally quenched by light of another wavelength. (iii) Acoustic beams, that may give rise to a stress-induced birefringence. By suitable summation of local stress fields, a change may be made that is detectable by a polarized light beam. Evidently each phenomenon offers certain potentialities and limitations.

The phenomena need to be evaluated for signal/noise ratio, information packing density, and the inherent natural lifetimes associated with them.

2. Parallel/Serial Image Processing. A promising compromise approach to a machine permitting functional composition is one employing solid state imaging technology. Such a system might consist of two planar input optical sensor arrays, a cathode ray tube output display, and line and frame scan circuitry.

Each sensor array consists of n rows of sensors, each row containing m sensors connected in a charge transfer chain. Thus the charges produced upon a line of sensors may be stepped sideways simultaneously by simple scan circuitry. As the charges reach the end of the row they activate further circuitry controlling, for example, the intensity of an electron beam being scanned synchronously across a phosphor screen. Thus in a solid state television system, the sensor array is exposed to an optical image, giving rise to greater or lesser charge buildup in the individual sensors. The charges are then read out by stepping the charges in each line along each row in sequence. This information is used to modulate a scanning electron beam and so the image is regenerated upon a phosphor screen.

504

In the composing system an input image is presented to a line-organized sensor array, and a second image is presented to a similar second array. In order to form an image of the mapped intersection of the projection of the two arrays as described earlier, the information content of the first array is stepped along a row by one unit, all rows stepping simultaneously. The row output signals enter a chain-of-logic elements. The second sensor rows are now stepped continuously in synchronism with a scanning electron beam and are compared with the first column of signals of panel 1. The row output signals enter the chain-of-logic elements and, if an output signal is encountered on any corresponding rows of the two arrays, then the intensity of the electron beam is modulated.

The information of the first array is now stepped again, the electron beam is displaced one unit sideways, the second array is re-exposed if necessary and a new line is scanned.

In this way an image is built up on a display screen having the required properties. By using a storage display tube this image may be retained and used to expose the first sensor array, light output and optical sensitivities being quite compatible in this regard.

The two sensor arrays, scanning circuitry, and logic chain may be fabricated upon one silicon single-crystal wafer, using the state-of-the-art photolithograph techniques. It is not unreasonable to contemplate arrays of 500 × 500 elements.

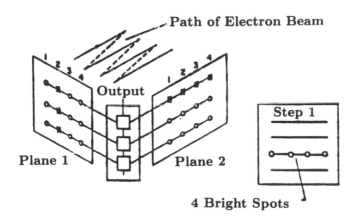

The logic chain gives a signal if and only if a signal is present on both input terminals. The information in plane 1 is stepped sideways; i.e., change in column 2 is moved to column 3 one step at a time. At each step an electron beam sweeps along a line and plane 2 is stepped through 1 to 5. If a signal is present on both inputs of the chain of comparators at some step, then the intensity of the electron beam is increased.

In a simpler scanning arrangement the two images may be mechanically scanned, one slowly, one rapidly and repetitively, across two columns of photosensors, for example, by rotating or oscillating mirrors. The scan rate of the images is synchronized to the faster scan of the electron beam of a display tube as before. Signals are compared as before and coincident signals give rise to a brief increase in intensity of the electron beam. Such devices are fully within the present state of the art of silicon microelectronic technology.

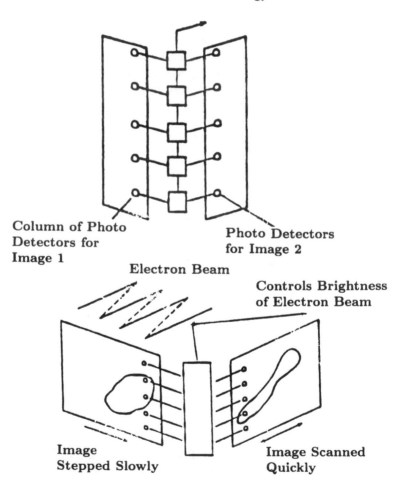

Column of Photo
Detectors for
Image 1

Photo Detectors
for Image 2

Electron Beam

Controls Brightness
of Electron Beam

Image
Stepped Slowly

Image Scanned
Quickly

References and Related Publications

1. A. R. Bednarek and K. D. Magill, Jr., "Q-Composable Properties, Semigroups and CM-Homomorphisms," Trans. Am. Math. Soc. **171** (1972), 383–398.
2. A. R. Bednarek and K. D. Magill, Jr., "Semigroups of Clopen Relations," submitted.
3. A. R. Bednarek and O. E. Taulbee, "On Maximal Chains," Rev. Roum. Math. Pures Appl. **11** (1966), 23–25.
4. A. R. Bednarek and S. M. Ulam, "Generators for Algebras of Relations," Bull. Amer. Math. Soc. **82** (1976) 781–782.
5. W. A. Beyer, M. L. Stein, and S. M. Ulam, "The Notion of Complexity," Los Alamos Scientific Laboratory report LA-4822 (December 1971).
6. R. E. Bonner, "On Some Clustering Techniques," IBM J. Res. Dev. **8** (1964).
7. A. H. Clifford, "A Proof of the Montague-Plemmons-Schein Theorem on Maximal Subgroups of the Semigroup of Binary Relations," Semigroup Forum **1** (1970), 303–314.
8. A. H. Clifford and D. D. Miller, "Union and Symmetry Preserving Endomorphisms of the Semigroup of All Binary Relations on a Set," Czech. Math. Journ. **20** (95) (1970), 303–314.
9. L. N. Cooper, "A Possible Organization of Animal Memory and Learning," Nobel Symposia **24** (1973), 252–264.
10. M. Crestey, "Applications Multiformes Partielles," Collectanea Mathematica, Univ. of Barcelona **16** (1964), 111–126.
11. T. Evans, "Embeddings for Multiplicative Systems and Projective Geometries," Proc. Amer. Math. Soc. **3** (1952), 614–620.
12. C. J. Everett and S. M. Ulam, "Projective Algebra I." Am. J. Math. **68** (1946), 77–88.
13. B. Faber and J. Larsen, Private communication.
14. L. M. Gluskin, "Automorphisms of Semigroups of Binary Relations." Ural. Gos. Univ. Mat. Zap. 6, tetrad **1** (1967), 44–54.
15. E. Howorka, "Generators for Algebras of Relations. Preliminary Report," Notices Am. Math. Soc. **24** (1977), A-4-A-5.
16. K. D. Magill, Jr., "Automorphisms of the Semigroup of All Relations on a Set," Canadian Math. Bull. **9** (1966), 73–77.
17. K. D. Magill, Jr., "Isomorphisms of Triform Semigroups," J. Australian Math. Soc. **10** (1969), 185–193.
18. N. Metropolis and Gian-Carlo Rota, "Significance Arithmetic on the Algebra Binary Strings," in *Studies in Numerical Analysis*, B. K. P. Scaife, Ed., (Academic Press, New York and London, 1974) pp. 241–252.

19. J. S. Montague and R. J. Plemmons, "Maximal Subgroups of the Semigroup of Relations," J. Algebra **13** (1969), 575–587.
20. J.C.C. McKinsey, "Postulates for the Calculus of Binary Relations," The Journal of Symbolic Logic **2** (1940), 85–97.
21. M.M. Nass, and L. N. Cooper, "A Theory for the Development of Feature Detecting Cells in Visual Cortex," Report, Center for Neural Studies, Brown University.
22. E. N. Norris, "A Characterization of Finiteness," unpublished.
23. R. E. Osteen, "Clique Detection Algorithms Based on Line Addition and Line Removal," Siam J. Appl. Math. **26** (1974), 126–135.
24. R. J. Plemmons and B. M. Schein, "Groups of Binary Relations," Semigroup Forum **1** (1970).
25. S. Schwarz, "On the Semigroup of Binary Relations on a Finite Set," Czech. Math. J. **20** (1970), 632–679.
26. S. Subbiah, "Another Proof of a Theorem of Evans," Semigroup Forum **6** (1973), 93–94.
27. A. Tarski, "On the Calculus of Relations," The Journal of Symbolic Logic **3** (1941), 73–89.
28. S. M. Ulam, *A Collection of Mathematical Problems* (Interscience, New York, 1960).
29. S. M. Ulam, "Computers," Sci. Am., Vol. 211, No. **3** (1964), 203–216.
30. S. M. Ulam, "Computations on Certain Binary Branching Processes," in *Computers in Mathematics Research,* (North Holland Publishing Company, Amsterdam, 1968).
31. S. M. Ulam, "Electronic Computers and Scientific Research, (Parts I and II)," Comput. Autom., Vol. 12, No. 8, 20–4 (Pt. I); Vol. 12, No. 9, 35–40 (Pt. II), 1963.
32. S. M. Ulam, "General Formulations of Simulation and Model Construction " in *Prospects for Simulation and Simulators of Dynamic Systems,* (Macmillan & Co., Ltd., London, New York, 1967).
33. S. M. Ulam, "Generalizations of Product Isomorphisms (Résumé)", in *Recent Trends in Graph Theory* **186**, Lecture Notes in Mathematics, Springer, 1971.
34. S. M. Ulam, "On Some Possibilities in the Organization and Use of Computing Machines," International Business Machines report IBM-RC-68, 1957.
35. S. M. Ulam, "Some Ideas and Prospects in Biomathematics," Annu. Rev. of Biophys. Bioeng., **1** (1972), 227–292.
36. K. A. Zareckii, "The Semigroup of Completely Effective Binary Relations," *Theory of Semigroups and Appl. I* (Russian), Izdat., Saratov. Univ., Saratov (1965), 238–250.

19

THE SCOTTISH BOOK
A LASL Monograph

(LA-6832, 1977)

The "Scottish Book" played a vital role in the stimulation of mathematical research—particularly in topology and in real and functional analysis. It is a collection of problems posed by the Lwów mathematicians before World War II, to which Ulam was one of the principal contributors, and which he translated, had typed (at his own expense), and circulated privately in 1957.

Its first printed version appeared in 1977 as a LASL Monograph.

More recently a new version under the title "The Scottish Book, Mathematics from the Scottish Café", edited by R. Daniel Mauldin, (Birkhäuser-Boston, 1982) delineates the evolution of this collection, lists the problems, and provides commentaries on the status of most of them.

While it would be redundant to reproduce the entire LASL Monograph here, Ulam's preface to his 1957 translation of the original collection is of historical interest and is thus included. (Eds.)

Preface to Monograph

Numerous requests for copies of this document, addressed to Los Alamos Scientific Laboratory or to me, appear to make it worthwhile (after a lapse of some 20 yrs) to reprint, with some corrections, this collection of problems.

This project was made possible through the interest and active help of Robert Krohn of the Laboratory.

It is a pleasure to give special thanks to Dr. Bill Beyer for his perspicacious review of the changes and the revised version of some formulations. Thanks are due to Martha Lee DeLanoy for editorial work.

Stan Ulam
Los Alamos, NM
May 1977

Preface

The enclosed collection of mathematical problems has its origin in a notebook which was started in Lwów, in Poland in 1935. If I remember correctly, it was S. Banach who suggested keeping track of some of the problems occupying the group of mathematicians there. The mathematical life was very intense in Lwów. Some of us met practically every day, informally in small groups, at all times of the day to discuss problems of common interest, communicating to each other the latest work and results. Apart from the more official meetings of the local sections of the Mathematical Society (which took place Saturday evenings, almost every week!), there were frequent informal discussions mostly held in one of the coffee houses located near the University building—one of them a coffee house named "Roma," and the other "The Scottish Coffee House." This explains the name of the collection. A large notebook was purchased by Banach and deposited with the headwaiter of the Scottish Coffee House, who, upon demand, would bring it out of some secure hiding place, leave it at the table, and after the guests departed, return it to its secret location.

Many of the problems date from years before 1935. They were discussed a great deal among the persons whose names are included in the text, and then gradually inscribed into the "book" in ink. Most of the questions proposed were supposed to have had considerable attention devoted to them before an "official" inclusion into the "book" was considered. As the reader will see, this general rule could not guarantee against an occasional question to which the answer was quite simple or even trivial.

In several instances, the problems were solved, right on the spot or within a short time, and the answers were inscribed, perhaps some time after the first formulation of the problem under question.

As most readers will realize, the city of Lwów, and with it the "Scottish Book," was fated to have a very stormy history within a few years of the book's inception. A few weeks after the outbreak of World War II, the city was occupied by the Russians. From items at the end of this collection, it is seen that some Russian mathematicians must have visited the town; they left several problems (and prizes for their solutions). The last date figuring in the book is May 31, 1941. Item Number 193 contains a rather cryptic set of numerical results, signed by Steinhaus, dealing with the distribution of the number of matches in a box! After the start of war between Germany and Russia, the city was occupied by German troops that same summer and the inscriptions ceased.

The fate of the Scottish Book during the remaining years of the war is not known to me. According to Steinhaus, this document was brought back to the city of Wroclaw by Banach's son, now a physician in Poland. (Many of the surviving mathematicians from Lwów continue their work in Wroclaw. The tradition of the Scottish Book continues. Since 1945, new problems have been formulated and inscribed and a new volume is in progress.)

A general word of explanation may be in order here. I left Poland late in 1935 but, before the war, visited Lwów every summer in 1936, '37, '38, and '39. The last visit was during the summer preceding the outbreak of World War II, and I remember just a few days before I left Poland, around August 15, the conversation with Mazur on the likelihood of war. It seems that in general people were expecting another crisis like that of Munich in the preceding year, but were not prepared for the imminent world war. Mazur, in a discussion concerning such possibilities, suddenly said to me, "A world war may break out. What shall we do with the Scottish Book and our joint unpublished papers? You are leaving for the United States shortly, and presumably will be safe. In case of a bombardment of the city, I shall put all the manuscripts and the Scottish Book into a case which I shall bury in the ground." We even decided upon a location of this secret hiding place; it was to be near the goal post of a football field outside the city. It is not known to me whether anything of the sort really happened. Apparently, the manuscript of the Scottish Book survived in good enough shape to have a typewritten copy made, which Professor Steinhaus sent to me last year (1956).

The existence of such a collection of problems was mentioned on several occasions, during the last 20 years, to mathematical friends in this country. I have received, since, many requests for copies of this document. It was in answer to such oral and written requests that the present translation was made. This spring in an article, "Can We Grow Geniuses in Science?", which appears in Harper's June 1957 issue, L. L. Whyte alluded to the existence of the Scottish Book. Apparently, the diffusion of this small mystery became somewhat widespread, and this provided another incentive for this translation.

Before deciding to make such an informal distribution, I consulted my teacher and friend (and senior member of the group of authors of the problems), Professor Steinhaus, about the propriety of circulating this collection. With his agreement, I have translated the original text (the original is mostly in Polish) in order to make it available through this private communication.

Even as an author or coauthor of some of the problems, I have felt that the only practical and proper thing to do was to translate them verbatim. No explanations or reformulations of the problems have been made.

Many of the problems have since found their solution, some in the form of published papers (I know of some of my own problems, solutions to which were published in periodicals, among them Problem 17.1, Z. Zahorski, Fund. Math., Vol. 34, pp. 183–245 and Problem 77(a), R. H. Fox, Fund. Math., Vol. 34, pp. 278–287).

The work of following the literature in the several fields with which the problems deal would have been prohibitive for me. The time necessary for supplying the definitions or explanations of terms, all very well understood among mathematicians in Lwów, but perhaps not in current use now, would also be considerable. Some of the authors of the problems are no longer living and since one could not treat uniformly all the material, I have decided to make no changes whatsoever.

Perhaps some of the problems will still present an actual interest to mathematicians. At least the collection gives some picture of the interests of a compact mathematical group, an illustration of the mode of their work and thought; and reflects informal features of life in a very vital mathematical center. I should be grateful if the recipients of this collection were willing to point out errors, supply information about solutions to problems, or indicate developments contained in recent literature in topics connected with the subjects discussed in the problems.

It is with great pleasure that I express thanks to Miss Marie Odell for help in editing the manuscript and to Dr. Milton Wing for looking over the translated manuscript.

S. Ulam
Los Alamos, NM
Fall 1957

20

ON THE NOTION OF
ANALOGY AND COMPLEXITY
IN SOME CONSTRUCTIVE
MATHEMATICAL SCHEMATA

(LA-9065-MS, October 1981)

This report is a study of some schemata which might be used in the processes of understanding and invention. (Author's note). *

ABSTRACT

Banach often remarked "Good mathematicians see analogies between theorems or theories; the very best ones see analogies between analogies." Mark Kac certainly belongs to the latter group. His work on problems in statistical mechanics and in number theory, disciplines so different from each other, exhibits a feeling for the role of the ideas of probability analogous in some way in these two domains which are so far apart.

In what follows I shall try to sketch an elementary approach to the notion of analogy and suggest a few mathematical problems that pose themselves once one tries to discuss this notion in a somewhat general way, namely, similarity or proximity of proofs and counting binary and unary operations at each stage with similar trees on the set of axioms.

---

* The report was published in *"Probability, Statistical Mechanics, and Number Theory,"* A volume dedicated to Mark Kac, G.-C. Rota editor, Academic Press, Inc., 1986 (Eds.)

I. Generalizations about Analogy

Throughout the development of mathematics and with the growth of new concepts and more complicated notions, a cohesive tendency and organic structure have been guided by a feeling of analogy between the old and new ideas.

Historically, problems posed by the development of a new mathematical discipline, which originally was only metamathematical, coalesced into new parts of mathematics itself. One could cite, as obvious examples, the study of transformations of a space as points of a new space of such transformations, and the study of algorithms for solving equations as entities per se (group theory, for instance).

The increasing proliferation of notions in pure mathematics may suggest that the idea of analogy itself is amenable to mathematical discussion. One finds that old and elementary formulations of this idea are, in special cases, present in the definitions of the similarity of geometrical figures, more generally in the equivalence of figures—sets, through the elements of a group of transformations, or, more generally yet, through the identity of proximity of such sets in spaces which encompass them.

Two abstract sets of elements may be felt to be "analogous" if the difference between their cardinalities (in the finite case) is small compared to the cardinalities themselves. Two classes of such sets may be deemed to be analogous if the numbers of sets in the two classes differ by "little" and if the cardinalities of the corresponding sets also do not differ by much. Obviously, one needs to attempt to formulate a quantitative criterion, and it is clear *a priori* that the notion of analogy will not be, in general, transitive.

In this report we merely want to discuss some of the salient features of analogy and exhibit them on a class of examples where we shall attempt to define it, at first, as proximity in the sense of a metric distance in suitably defined spaces.

Our first example dealing with a linear or one-dimensional array of symbols will illustrate the definitions and the role of possible metrics in determining the corresponding analogy. For simplicity's sake we consider finite sequences of 0's and 1's. One kind of distance between them was defined in a report of mine.[1] These sequences are to represent the DNA codes (which factually use four symbols, but the whole discussion and results are equally valid in that case). The distance ρ between sequences $x = \alpha_1, \alpha_2, \ldots, \alpha_n$ and $y = \beta_1, \beta_2, \ldots, \beta_m$, where α_i and β_i are 0 or 1, is defined as the smallest number of *steps* that, performed on one or both of these sequences, will render them identical. The steps are of two kinds: replacing 0 by 1 or vice-versa, or erasing (or inserting) a symbol. One can prove easily that the number just defined satisfies all the properties of a metric, in particular the triangle inequality holds. We may consider two such sequences or codes as analogous or related if the distance between them is small compared to their lengths.

Another type of distance between these sequences can be described as follows:

Given the two sequences $x = \{\alpha_1, \alpha_2, \ldots, \alpha_m\}$ and $y = \{\beta_1, \beta_2, \ldots, \beta_m\}$, α_i and β_i are 0's or 1's; we look for the difference in the number of each type in x and in y. The absolute value of these differences of two types we describe by K_1. Next, given a system of *two* successive symbols either $(0,0)$, $(0,1)$, $(1,0)$, $(1,1)$, we compute the difference of the number of occurrences of each type in x with the corresponding number in y. Their sum is K_2. Similarly for triplets and so on. The sum of the absolute values of all these differences may serve as a "statistical" type of distance between x and y.

Still another distance could be the following:

Again we consider two types of steps. The first one is an erasure of any symbol (in either x or y). The second consists of choosing any element in x (or y) and placing it in between in another position. The smallest number of such steps making the two sequences identical is a metric distance.

We could modify this definition by "weighing" the transposition by its length. The distance would be the minimal "work" sufficient to make the two sequences identical.

One may define a distance for infinite sequences of this type by suitable passages to the limit. We shall discuss this later. Needless to say, one-dimensional (linear) sequences of symbols may represent a succession of sounds or other signals such as encephalographic scans. The metric or a distance between two such sequences should obviously depend on the interpretation of these sequences.

One may consider two-dimensional arrays of points or of more general numerical symbols.

To begin, we imagine that the sets or pictures are subsets of a bounded part of a Euclidean space with, say, the Euclidean metric in it (or a similar Minkowski-type metric). A widely used metric for closed subsets of this space is the Hausdorff distance. It is defined as follows.

$$\rho_H(A,B) = \underset{y \epsilon A}{\text{Max}} \; \underset{x \epsilon B}{\text{Min}} \; \rho_E(x,y) + \underset{y \epsilon A}{\text{Max}} \; \underset{x \epsilon B}{\text{Min}} \; \rho_E(x,y)$$

where A, B are sets (closed) in E; ρ_E is the metric in E. We want to define a metric in a space whose elements are classes of sets in E^2. Given a set or a picture, we consider sets which are translations or diminutions or enlargements of the given set; more generally, sets obtainable from the given one by "small" deformations. "Small" means that we have an ϵ-neighborhood of a group or a semigroup of transformations of E^2 and consider sets obtained from a given one as similar to it. This way we have a collection of classes of sets and one may define a distance function between any two such classes by iterating the Hausdorff procedure.

$$\rho(X, Y) = \underset{A \epsilon X}{\text{Max}} \ \underset{B \epsilon Y}{\text{Min}} \ \rho_H(A, B) + \underset{B \epsilon Y}{\text{Max}} \ \underset{A \epsilon X}{\text{Min}} \ \rho_H(A, B)$$

where X, Y are (compact) classes of sets and ρ_H is the Hausdorff distance as defined above.

In particular this definition applies to subsets of a finite set of points (that is, a grid or lattice of points—a "screen"). Other distances are possible in analogy to the different metrics defined earlier for the one-dimensional case. We may, given two sets A and B, add or subtract individual points to any of them. A "step" involving k points will contribute k units of distance. A metric between the two sets can be defined as the minimum of the Hausdorff distance added to k. In other words, the Hausdorff difference between two fixed sets is modified by the optimum (smallest) k. Given two classes of sets we again iterate the Hausdorff procedure as above. One can prove again the triangle property of this distance.

A more general procedure of this type can be defined by starting with a more general distance in the original grid of points, for example starting with a distance of the type proposed by A. Bednarek.[2] Again, it is a minimum number of steps that was used for the distance formula.*

In the case where sets A, B are infinite, the changes affected by adding or subtracting a finite number of points have to be generalized to adding or removing sets of "small" *measure* (or length).

The analogy between sets or classes of sets is thus representable, in one way at least, by the smallness of a distance in a suitable metric

---

* The minimum number of steps could be thought of as the minimum work of transition from one configuration to the other. A memoir of Appell[3] deals with the problem of déblais and remblais. This involves the calculation of the minimum work sufficient to move a pile of "sand" of a given shape into another prescribed one.

defined for them. Problems on the stability of properties of sets or transformations or their invariance with respect to some changes have been considered in many special cases. Thus, for example, the problem of transformations of a space, say the Euclidean or Hilbert spaces, that change the distances between points by less than a given $\epsilon > 0$, was studied in a paper by Hyers and myself.[4] Similarly, the problem of transformations that are almost linear is discussed in another paper with Hyers.[5]

The general problem of whether a transformation of a metric group into itself which is almost an isomorphism must be close to a strict isomorphism is still open. A recent result of D. Cenzer[6] brings a positive solution in the case of some Abelian groups, for example, the group of rotations on a circle or torus.

One could pose analogous (sic!) more general problems about other functional equations. For example, suppose a function of two complex variables satisfies up to an $\epsilon > 0$ an algebraic addition theorem, must there exist then a function satisfying an algebraic addition theorem exactly and within $c(\epsilon)$ of the given function?

Let us consider still another question of "stability" of more geometric nature. Given two surfaces which can be mapped into each other in such a way that the curvatures and the inverses of curvatures in corresponding points differ by less than $\epsilon > 0$, do there exist surfaces within $c(\epsilon)$ of the given ones that are strictly isometric in the sense of internal geometry?

Certain topological properties are also stable with respect to transformations that are not necessarily homeomorphisms but more general ones—continuous, and not collapsing two points if the distance between them exceeds a certain $\epsilon > 0$. There is, for example, a result of Borsuk and myself[7] which asserts that if a continuous transformation T of the surface of the unit sphere in n-dimensions is such that if the distance between p and q is $> (\sqrt{2}/2)\, r(S)$ where r is the radius of S, then $T(p) \neq T(q)$. Then the image separates the space. An interesting distance between topological types proposed by Borsuk[8] is defined as follows. Given two sets A, B in a metric space E, one considers all continuous mappings of A onto B and vice-versa. If A and B are homeomorphic, such mappings, if they exist at all, must collapse the images of a pair of points and either A or B or both. In the compact case the smallest sum of their distances is Borsuk's measure of the topological distinction between A and B.

One way to mathematize some ideas of analogy is then to consider a space 2^E with a metric ρ, ϵ. Two sets of elements of 2^E will belong to a cluster if their distance according to the metric ρ is $\leq \epsilon$. A cluster of analogous sets we define as a collection such that any two sets have a distance $\leq \epsilon$. (These clusters in general are, of course, not disjoint.) If T is a transformation of the set E into itself such that if two sets A, B, are analogous, then their images $T(A)$ and $T(B)$ are also analogous. One should want to find conditions on E, ρ, ϵ such that a transformation T of this sort preserves the analogy; that is, A is analogous to $T(A)$ for all A. Theorems of this sort would be generalizations of some of the stability theorems mentioned above on ϵ-isometries, ϵ-linear transformations, etc.

Another investigation should concern, so to speak, analogies between analogies. Considering now E, ρ, ϵ, we may, as the next order, define a metric between clusters of sets within which any two sets are analogous. Using the Hausdorff formula as above, we will define a metric between such clusters. Given this metric we then have super clusters consisting of analogous classes.

Given E, ρ, ϵ, and another set F, σ, η, where σ is a metric in F and $\eta > 0$, we shall call the F systems of analogies a representation or image of the E systems if the combinatorial structure of the clusters in 2^E can be mapped into a combinatorial structure of 2^F with a monotone correspondence of $\epsilon \rightarrow \eta$, and preserving the Boolean relations between clusters of 2^E homomorphically into some clusters in 2^F.

An interesting case is if there should exist a *finite* representation of E, ρ, ϵ, that is to say, the set F consists of a finite number of points.

II. Complexity

In recent years, a great number of studies have dealt with the mathematization of the idea of complexity in mathematical schemata. In this report, we want to study some special definitions of complexity, especially relative complexity of mathematical constructs, and consider the equality or approximate equality of complexity in two different mathematical setups as one characteristic of analogy.

We shall start with the account of work contained in a Los Alamos Laboratory report.[9]

1. Consider the set of positive integers Z and the operations addition, multiplication, and exponentiation. The complexity of an integer n will be the smallest number of steps using the above operations and starting with 1, adding the complexities of the numbers used to obtain N plus the complexities of the operations addition, multiplication, and exponentiation that in our first exercise we took as all $= 1$. The complexity of 1 we assumed to be $= 1$. We denoted the complexity of n

by $c(n)$. In our report we prepared a table of complexities of the first thousand integers, indicated an algorithm on a computer to calculate it, and gave some asymptotic expressions on the behavior of $c(n)$. We have also stated a number of problems. One of them is the following.

Call a number k complicated if $c(k) < c(\ell)$ for all $\ell < k$. One of the conjectures was that beginning with a certain n_0, all complicated numbers are primes.

We could define a relative complexity of an integer n relative to m by the smallest number of operations sufficient to obtain n not counting the complexity of m. (We will discuss this notion later in a more general context.)

The complexity $c(n)$ may be considered more generally as follows. We have a number of symbols a_1, a_2, $\ldots$, a_n whose complexities are given at the start. We have also a number of binary or unary operations R_1, R_2, $\ldots$, R_t. The complexity of any new symbol $c(x)$ is defined by the following: $c(a_1), c(a_2), \ldots, c(a_n)$; $c(R_1), c(R_2), \ldots, c(R_t)$ are given initially. If $z = R_i(x,y)$ then $c(z) \le c(x) + c(y) + c(R_i)$ and the sign of equality obtains for at least one such representation.

2. One could discuss, in the same spirit as above, the complexities of integers allowing, for example, the operation of subtraction $a - b$. It is obvious that the complexity values will change radically. We have indicated in our report a complexity within a system of integers modulo p.

Analogously one could define complexities within the system of Gaussian integers $a + bi$. More generally yet, in a finite or countable infinite ring, it is obvious how to define a complexity of its elements assuming the complexity of some starting elements and of the ring operations.

3. Another "complex system" could be a group of permutations, say the symmetric group S_n of permutations or the semigroup T_n of all transformations of the set of n integers into itself. Here one would again start with a number of algebraic elements, that is, permutations with given complexities, and assign a value of complexity to the operation of composition and/or to the operation of taking an inverse of a permutation. The complexity of a permutation would be the smallest number of steps starting with the given elements sufficient to obtain a given one. Again an interesting question would be: what is the average complexity of a permutation; what is the dependence of this value on the initial "generating" elements; what are the asymptotic bounds on these for n tending to infinity, etc.?

4. We may consider a set E finite or consisting of all the integers, its direct product E^n and the Boolean algebra on its subsets. We also

519

allow the arithmetic operations, say of addition and/or multiplication of numbers on the subsets. For example, given two subsets A, B, we define $C = A + B$ as the set of all elements c representable by sum or multiplication of one element from each of A, B, $c = a + b$, $c = a \cdot b$, $a \in A$, $b \in B$. Starting with some "elementary" subsets $S_1 \ldots S_k$ with assigned complexities, we define the complexity of sets they generate in the same way as in the above examples, by counting the number and complexity of steps in the shortest way.

5. One may consider the same as in 4, that is, subsets of E^n by allowing the operators of *projection* of a set in E^n on $E^m (m < n)$, which is an unary operation, also the binary operation of direct product of sets in the "lower dimensions." These, of course, allow a treatment of quantifiers on a given class of sets obtainable from an initial base of sets. For the case of $n = 2$, one has a way to define complexity in the projective algebras[10] or cylindrical algebras.[11]

6. One may consider in the case of $n = 2$ the operation of composition of sets treated as relations, together with Boolean algebra operations. Starting with some elementary sets, one will arrive at complexity definitions for sets in the relation algebra generated by them.[12]

7. Genealogical systems:[13] A branching process of a multiplicative system[14] constitutes possibly the simplest example of a complexity structure. In this process, starting with one or more elements, one considers the generation of progeny by mitosis, that is to say, each particle by itself gives rise to 0, 1, ... other particles and the process continues. The complexity of an element would be, of course, its number of ancestors leading to the original ones.

A much more complicated situation arises when the population is endowed with "sex" and the generation proceeds by pairing and the production of off-springs with the process continuing. Compared to the rather well-developed theory of branching processes, results on such "pair trees" are still meager and the analogs of the theorems known for branching processes remain unproved.

In some simple cases metrics have been defined for such genealogical trees,[15] and a probabilistic study of the degree of relationship has been initiated.

The complexity of an element could be again the shortest path leading to the original ancestor (in our general schema both direct mitosis and meiosis are allowed in the procreation). A relative relationship, or relative complexity, would be defined as above as an asymmetric distance between two elements.

8. Formal systems more generally: A most comprehensive example would be given by a formal system of mathematics, for example, one due to Gödel or a system of expressions obtainable by a Turing machine operating, say, in lexicographic order on all possible continuations and combinations of operations. Meaningful results could be obtained only with a "natural or economic" Gödel numbering system to mirror our intuitive and historical feeling about the development of logic and mathematics itself.

We might construct such trees of formal systems by discrete generations: given an n-th collection of expressions, we operate with all possible unary and binary combinations of the ones so far obtained to construct the $(n + 1)^{th}$ collection.

This gives rise to the "meaningful" expressions. If we have a system of axioms to start with, we may establish an identity between very many of these expressions. The identity of many pairs of statements, or that certain expressions equal 0, will not be obtainable from the axioms and rules of identification; these are the undecidable propositions or statements independent of the axioms. The number of meaningful expressions grows, roughly speaking, as 2^n asymptotically, but with "pruning" due to the identities provable from the axioms will be much smaller. The number of essentially different ones will be enormously reduced.

III. Comparisons Between the Complexity of Constructive Systems

Given two systems such as those above, for example, among the ones enumerated earlier, we may consider the question of *homomorphism* of one with a subsystem of the other. Our systems are examples of partially ordered sets and differ from genealogical systems proper in that a single element or "expression" may be the offspring of many different pairs of parents. Homomorphic mapping of one into the other would be one where relative complexities between pairs of elements are either preserved or decreased. (We repeat that a relative complexity of an element b with respect to a is the length of the shortest path leading from a terminating in b where we add up the complexities of all expressions and the complexities of operations linking the intervening pairs without adding the complexity of a.)

A number of problems arise concerning the comparison between our examples as homomorphic images transforming one complexity system onto parts of another.

A question of combinatorial interest deals with an asymptotic expression for the complexity of elements of the n-th generation as a function of n; for example, the average such complexity and, if we have a system which is finite, the average complexity for all elements and the average relative complexity of pairs (a, b) where $a \prec b$; a precedes b in the construction.

In any of our systems, given an element, that is, an expression, we may look for the first time or the earliest generation where its equivalent appeared. This occurs, say, in the k-th generation. We denote by ℓ the length of the expression (this is the total number of symbols and operations involved in it together). One may be interested in a comparison of ℓ with k. Expressions or statements where k is very large compared to ℓ may be "interesting." In a system of integers with only addition and multiplication allowed (starting with 1 as in the above example) certain primes (for example, those of the form $a^2 + 1$) will be "interesting."

In a development of a "constructive system," one often introduces new operations which are abbreviations, in addition to the original R_i. For example, an n-th iteration or repetition of a binary operation may be denoted by a new symbol R whose complexity is defined as the complexity of R plus the complexity of n *without* adding the complexity of R n times.

A genetic development of a mathematical discipline in some way resembles the evolution of organisms and perhaps of matter itself. (See, for example, a mathematical schema for the development of physical patterns through a process of transmutations.[16])

In the development by recursion of geometrical figures[17] and in the development of organisms or in the evolution of species, one may study a principle of "minimum total complexity" of the intervening stages between the initial and final positions. A "variational" principle of this kind would single out some out of all possible histories of the process between two given states.

Coming back to mathematical schemata, we can define a complexity of a rational number starting with the definition of complexity for integers by considering the operation of division with a given complexity and defining the complexity of the fraction a/b as the sum of the complexity of a and b plus the complexity of the operation of division. One could even attempt to define a complexity of a real number as the inferior limit of the complexities of the rational numbers converging to it normed by the complexity of the numerators and denominators suitably scaled, for example, by the logarithm of the two complexities.

Given two finite metric spaces we can consider a mapping of one into the other that minimizes the sum of the differences in the distances between corresponding pairs of points. Again, if the two spaces are, say, bounded, and the two metrics are defined on a topological measure space, we can consider as the measure of analogy between the two metrics an integral of these differences taken over the space of all pairs of points.

Much of mathematics consists in ascertaining, in the developing and increasing formal systems, identities between different expressions, for example, in a set theoretical framework, showing some of them to be equal to zero.

Another endeavor or exercise could be an attempt to define, perhaps in a manner similar to the above, the complexity of *proofs* starting with a number of identities, that is, axioms. The similarity or analogy of this process continuing indefinitely suggests that mathematics itself exhibits the behavior of "*eadem mutata resurgit.*"

In our search for properties of the notion of analogy, we may consider in a given system analogy-preserving transformations. We mentioned earlier theorems asserting that a transformation satisfying a linear function equation up to ϵ must be close to one satisfying it exactly: a linear transformation that is almost isometric is close to a strictly isometric one.

In a space of elements which we metrize to define analogy by an ϵ-proximity we may consider transformations of the space on itself that preserve analogy, that is, any two analogous elements are transformed into analogous ones. One would like to know in some cases that the transformation of the Euclidean space that has the property that a pair of ϵ-congruent sets goes over into ϵ-congruent ones is close to a congruence. One can easily show that the transformation itself is either a congruence or a change of scale in the distance. Similarly, one may try to prove analogous statements for more general definitions of analogy.

References

1. S. M. Ulam, "Some Ideas and Prospects in Biomathematics," Annual Review of Biophysics and Bioengineering **1**, 277–291 (1972).
2. A. R. Bednarek and Temple F. Smith, "A Taxonomic Distance Applicable to Paleontology," Mathematical Biosciences **50**, 285–295 (1980).
3. Paul Appell, "Le Problème Géométrique des Déblais et Remblais," Mémorial des Sciences Mathématiques de l'Académie des Sciences, fascicule **27**, (1928) Gauthier-Villars, Paris.
4. D. H. Hyers and S. Ulam, "On Approximate Isometries," Bull. Am. Math. Soc. **51**, 208–216 (1945).

5. D. H. Hyers and S. Ulam, "Approximate Isometries of the Space of Continuous Functions," Ann. Math. ser. 2, **49**, 285–289 (1947).

6. D. Cenzer, "The Stability Problem for Transformations of the Circle," Proc. Royal Soc. of Edinburgh **84A**, 279–281 (1979).

7. K. Borsuk and S. Ulam, "Uber gewisse Invarianten der ε-Abbildungen," Math. Annalen, 312–318 (1933).

8. K. Borsuk, "On Some Metrizations of the Hyperspace of Compact Sets," Fund. Math. **41**, 168–201 (1955).

9. W. A. Beyer, M. L. Stein, and S. M. Ulam, "The Notion of Complexity," Los Alamos Scientific Laboratory Report LA-4822 (December 1971).

10. C. J. Everett and S. Ulam, "Projective Algebra I," Amer. Jour. Math. **68**, 77–88 (1946).

11. L. Henkin, D. Monk, and A. Tarski, *Cylindrical Algebras* (North Holland Publishing Co., 1971).

12. A. R. Bednarek and S. Ulam, "Projective Algebra and the Calculus of Relations," Jour. Symb. Logic **43**, 54–64 (1978).

13. J. Mycielski and S. Ulam, "On the Pairing Process and the Notion of Genealogical Distance," Journ. Comb. Theory **6**, 227–234 (1969).

14. C. J. Everett and S. Ulam, "Multiplicative Systems I," Proc. Nat. Acad. Sci. USA **34**, 403–405 (1948). Also, C. J. Everett and S. Ulam, "Multiplicative Systems in Several Variables, I, II," Los Alamos Scientific Laboratory report LA-683 (1948), LA-690 (June 1948).

15. J. Kahane and R. Marr, "On a Class of Stochastic Pairing Processes and the Mycielski-Ulam Notion of Genealogical Distance," Jour. Comb. Theory, A, **13**, 33–40 (1972).

16. S. Ulam, "On the Operations of Pair Production, Transmutations and Generalized Random Walks," Adv. Appl. Math., **1**, 7–12 (1980).

17. R. G. Schrandt and S. M. Ulam, "On Recursively Defined Geometrical Objects and Patterns of Growth," Los Alamos Scientific Laboratory Report LA-3762 (November 1967).

21

SPECULATIONS ABOUT THE MECHANISM OF RECOGNITION AND DISCRIMINATION

(LA-UR-82-62, 1982)

This report is a preprint of a talk I gave at Los Alamos in 1981, speculating about some of the methods which may be used in some processes occurring in the nervous system. (Author's note.)

Let me first say that this title is not quite exact. I may want to speculate, but it won't be about the physiological or anatomical nature of memory, about which I know nothing! At the end I may venture my own private little questions. I don't think anybody knows really what the true physiological elements of recognition are.

The talk will be about more abstract mathematical schemata, some of which may perhaps have a physical basis. I'll try to talk first about various ways in which a visual picture is recognized. Towards the end, I'll talk about how, with suitable changes, this may apply to other sets of objects—for example linear arrays like DNA codes, or to auditory experiences which are more or less linear too, as for example a sequence of musical notes.

I don't dare speculate too seriously about three-dimensional objects; that indeed is the domain of immunology, for example, or about olfactory recognition which refers to the recognition of molecules by something in our nose. But could all this be coded up linearly perhaps? Or could the shape of molecules have something to do with infrared radiation? I have looked at a book about olfactory problems, but it is ten years old and I suppose people know more now.

Let me now talk about some purely mathematical attempts to give the combinatorial schemata for what we call recognition.

Recognition is already more ambitious than what I would call discernment or discrimination, namely the finding of a difference between two signals. I have not seen that discussed *per se* in the literature, but it seems to me that it is more elementary to distinguish two different letters, for example, than to recognize an object stored in the memory. Discernment or discrimination—I don't know what to call it—is something we have experimented on long ago on computers in this lab.

The main discussion will be about distinguishing between two-dimensional objects or pictures. The mathematical tool, or at least notion—for I don't think it deserves yet the name of tool—the mechanism I will talk about is the idea of a distance between objects such as, for instance, pictures on a screen.

First let me explain the properties of a distance:

Suppose we have a set E of objects we will call a, b, c,.... A distance $\rho(a,b)$ is a real valued function of pairs of elements of E. It should be ≥ 0. When it is 0, this means that $a = b$. It is symmetric, $\rho(a,b) = \rho(b,a)$. It also has a very important property for all applications called the triangle inequality, $\rho(a,c) \geq \rho(a,b) + \rho(a,c)$ for all a, b, c. If you have such a function on a set E, the set is called a metric space.

Mathematicians and physicists are familiar, of course, with metric spaces such as Euclidean space, Hilbert space and all kinds of function spaces, manifolds with "curved" geometry, with distance measured on geodesic lines, and so on.

I would like to give examples of differently constructed metric spaces. These may have something to do with at least a language for certain biological phenomena different from metric spaces in physics.

Here are various fifteen-year-old attempts to define a distance between sequences of symbols of a set of DNA codes, for example. The sequences consist of long arrays of four letters, A, C, T, G or U, looking for instance like ACTTGGA ....

For simplicity's sake, instead of using four letters, I'll take a sequence of just two letters and will call them 0 and 1. One such sequence may be $A = 011110101001...01...$ the other B may be $B = 101101....$ The sequences are quite long—they may have a thousand or two thousand letters, and perhaps they form the code for some definite molecule.

How can we compare them? The idea is to define a quantitative measure of this comparison, or a distance between them. Walter Goad*

---

* Los Alamos physicists who have become interested in biology. (Eds.)

and George Bell have occasionally defined various distances. I myself have played with this ten or twelve years ago and considered distances differently from the line of pure mathematics.

One of the simplest ways to define a metric for sequences $\alpha_1 \ldots \alpha_N$, of symbols 0's and 1's is, for example:

$$x = \alpha_1 \ldots \alpha_N, \; y = \beta_1 \ldots \beta_N, \; \rho(x,y) = \sum_{i=1}^{N} \mid \alpha_i - \beta_i \mid ,$$

sometimes called the Hamming distance. Clearly this distance is not what one would want in biology for a distance between codes. In "pure" mathematics, as in most physical situations, the description of objects, i.e., their positions are fixed. They are so to say rigid, and their beginning and end are fixed, whereas in biological situations the objects are pliable. The above distance between the two sequences $x = 0101\ldots0101$, $y = 1010\ldots1010$ would have a large distance N because they differ in every place. But if we erase just one letter from each we see that the sequences are the same. If they were written on a circle they would be exactly the same.

The distance between two linear biological arrays could be defined as the minimum number of steps which will change one sequence into the other, or more generally, will work on both to bring them into the same form.

What are these steps? Given two sequences of 0's and 1's, $x = \alpha_1 \ldots \alpha_2 \ldots \alpha_N$, $y = \beta_1 \ldots \beta_2 \ldots \beta_M$ (M could be different from N), we define the distance between x and y as the minimum number of *steps* which operating on one or the other or both of the sequences will bring x and y into the same form. One can prove that this minimum number satisfies the properties of a distance.

The steps are of the following types:

1. A change of a 0 into a 1 or vice versa.
2. An erasure of a symbol and a contraction of the remaining ones or an insertion of a symbol anywhere in the sequence.

It turns out that in order to find out about this distance, a rather efficient algorithm devised by Peter Sellers determines it in less than N^3 steps.

A definition of the distance given above was presented in an article I wrote in the Annual Review of Biophysics.[1]

Distances of this sort can be used to compare sequences of DNA defining proteins, for instance. The biologist Margoliash had the idea of trying to reconstruct the evolutionary tree for organisms and animals based on the codes for cytochrome C present in essentially every living organism and seemingly constant within a species but different from species to species. His idea was to find species whose codes for cytochrome C are less different from each other, or related more closely in the evolutionary tree, and species whose codes for this protein differ by more. To reconstruct a possible evolutionary tree led mathematically to the problem of a binary graph for the species now existing and perhaps some non-extant ones which have disappeared, starting with some very primitive organisms and in such a way that the sum of the distances in this binary graph of "descent" should have the sum of all the distances as small as possible. This postulate of Margoliash translates mathematically into an assumption that the collection of all mutations which occurred giving rise to the now existing variety of species was the least improbable among the possible ones.

In the space of the DNA codes defining the various cytochrome C sequences as a finite metric space we have a generalization of a problem by the nineteenth century German geometer Steiner who considered a finite system of points in the plane. His problem was to draw a graph through all the points, perhaps adding new auxiliary ones, so that the sum of all the edges would be minimal. Here, we have this kind of problem in a much more general combinatorial setting for a finite metric space.

One can think of many other types of distances for the sets of sequences of 0's and 1's, or more generally for sequences with k symbols, k being a fixed number. One such definition, again involving the smallest number of steps which allow passage from one sequence to another, with the nature of steps defined ahead of time, is, for example, the following:

Suppose that we compare two sequences of 0's and 1's with the same number of 1's and 0's in each and allow a step which consists of moving a 0 or a 1 from one position to another, the cost of this step being the length through which we move it. The "minimum" work to effect this is a possible distance between two such sequences.

Another definition can be obtained by comparing the number of 1's in the two sequences, noting the difference between these numbers, then considering the 0's and 1's in the two sequences, again noting the difference in these configurations and then the difference between symbols 1 followed by 1, etc. After this survey of pairs of succeeding symbols, counting triplets and so on, the sum total of such differences suitably normalized can serve as a distance.

We will now present a larger variety of possible distances between objects as sequences of two-dimensional "pictures." We shall concentrate on visual or two-dimensional impressions and on ways to quantify to a degree the similarity or lack of it between two-dimensional objects or "pictures." A variety of possible distances between two sets in a

plane, or more generally between two classes of sets in a plane, will
be discussed. Just as in the case of one dimension, the possibility
of "recognition" of a sequence of symbols involves the smallness of a
distance between impressions (auditory or tactile, for instance), and
the strings of symbols coding them which reside in the memory from
previous impressions, a two-dimensional visual impression is compared
with the picture or pictures stored in the memory of an organism.

Without at the moment going into possible physiological or anatomical ways to evaluate such distances, we shall discuss abstractly
various ways of considering a metric space whose elements are sets in
the plane. For our purpose it is sufficient to consider them finite, or if
infinite, closed and bounded.

One can consider a topological distance due to Hausdorff. It is
defined for closed subsets of a metric space E as follows: Let E be a
space with metric $\rho_E(x, y)$. For any two closed sets A and B one can
write as a distance $\rho_H(A, B)$ where

$$\rho_H(A, B) \;=\; \underset{x \epsilon A \;\; y \epsilon B}{\text{Max Min}} \; \rho_E(x, y) \;+\; \underset{y \epsilon B \;\; x \epsilon A}{\text{Max Min}} \; \rho_E(x, y) \;.$$

But again this is not quite what one wants for a distance between
two impressions or two sets given separately since the distance above
depends on the position or mutual relation of two subsets A, B, in the
plane (screen).

One can obtain a more satisfactory distance by *iterating* the Hausdorff formula as follows: Instead of a fixed set A consider a whole class
of sets "like" A by slightly deforming, shifting, turning the given set
A, and more generally by applying to A a number of transformations
forming a neighborhood of the identity of a whole group of transformations. In this way we obtain a whole class $\mathcal{A}$ of sets. Proceeding
analogously with the set B we obtain a class $\mathcal{B}$. Assuming that these
classes are finite, or compact in the case of an infinity of transformations, we may now consider a distance between $\mathcal{A}$ and $\mathcal{B}$ as:

$$\rho(\mathcal{A}, \mathcal{B}) \;=\; \underset{\mathcal{B} \epsilon \mathcal{B} \;\; \mathcal{A} \epsilon \mathcal{A}}{\text{Max Min}} \; \rho_H(A, B) \;+\; \underset{\mathcal{A} \epsilon \mathcal{A} \;\; \mathcal{B} \epsilon \mathcal{B}}{\text{Max Min}} \; \rho_H(A, B) \;.$$

One of our contentions is that in problems of reaction to impressions
(visual, auditory, tactile, or chemical), the organism produces a number
of small variations of the impressions stored in the memory. Perhaps
one is allowed to speculate that the memory could reside not only in
the central nervous system, but could exist in the immunological or
other autonomous parts of the organism.

There are other distances, in addition to the Hausdorff distance, which might actually be more suitable for the arrangements in the visual systems. The distance as defined above has the drawback that its value for a pair of sets which are almost identical except for a few points added to one of the sets may be considerable.

One can generalize the Hausdorff idea still further by considering in the notion of the class A variations in the sets A by looking at them "modulo" a small number of points, or in the infinite case, "modulo" sets of small linear measure.

Another distance between two sets, each of which has its own metric between points in them (e.g., if they are both subsets of a plane with Euclidean metric), can be obtained by trying to map the set A into B and vice versa with the smallest number of errors. If both sets are finite, we consider all mappings of one into the other trying to achieve an isometry as much as possible, that is to say, a transformation such that a pair of points x, y in A should go into a pair of points x', y' whose distance is equal to the distance between x and y.

Given a mapping, we can calculate the sum of the errors under an optimal mapping. The distance between two sets A and B can be defined as the minimum of the sum of errors under all possible mappings. In practice, of course, the number of all possible trial mappings is enormous, even for sets A and B consisting of a small number of points, and trying all mappings is impractical. Instead one can take recourse to a Monte Carlo type assay by looking at very small subsets of A and mapping them into subsets of B and vice versa.

Even if the number of such subsets is large, say hundreds or thousands, the total computation will be vastly shorter than the exponentially increasing (factorial n) number of all mappings.

The above definition bears a resemblance to the one involving a problem considered by Appell in his study of "Déblais and Remblais"[2]: the minimum work necessary for transforming a given pile of sand ("points" of a set) into a given different configuration.

What such a definition suggests is that, given a set on a screen providing a new impression, there may be a mechanism of attempting to map the points which form it onto a set of points of a set residing in the memory—this with a small number of errors in the distances between pairs of corresponding points. If this is possible, we consider the new impression as recognition of a previous one. If this is not possible, we might put it into the memory as a new object.

Here are a few more possible distances between sets:

Given a set of points in the unit square, we may imagine it transformed homothetically so that it would touch the boundary of the square. We now consider a successive division of the square, first into four squares of size 1/2, then a division into sixteen squares of size 1/4, then sixty-four, etc... We examine the *characteristic function* of the given set in the subdividing squares. We allot weights each normalized so that the squares of the first subdivision have weight 1, in the second subdivision 1/4, then 1/16, etc... The set will then be coded by a sequence of these numbers.

Given two sets A and B, we may define their distance as the norm of the absolute value of the difference between the two coding sequences. It is distances of this sort that were used by Schrandt and myself in experiments on the computer which we performed in Los Alamos around 1965 to attempt recognition via computer of hand written letters of the alphabet.

We proceeded as follows: We wanted to discriminate between two handwritten letters A and B. We stored in the memory 256 examples of each in the following way: Obviously it would have been laborious, time consuming, and slow to do it by changing the styles of these letters by handwriting. Instead we produced varied examples of a handwritten letter on the computer rather quickly. It is well known that there exists on an interval and similarly on the unit square two continuous transformations whose composition will produce a set dense in the space of all such continuous transformations.

Remembering this fact, we chose two transformations of the unit square S and T, each letter different from identity (small deformations). If we consider transformations of the form STTST...SSTSTS... etc., of say 7 letters, we have $2^7 - 128$ different transformations, each still not too violently distorting the geometry of the square. Thus we obtain 128 examples of each of the letters A and B initially written once by hand into the machine.

Given a new letter also handwritten, the problem was for the machine to decide whether it was an A or a B.

We computed the distance between the problem letter and the 128 examples of each which were now stored in the memory. Whichever sum of distances was smaller determined the answer. As it turned out, the deformed examples of each letter produced by the computer looked like different handwritten styles, some written by an old man with a shaking hand, some more rounded or pointed, etc., as if they had really been produced by people. The first computer trials gave more than 80% correct answers!

Is it possible that in the actual process of recognition—or discernment or distinction—between two visual impressions one does not need to have recourse to very many examples stored in the memory? That instead, by taking one of the examples in the memory, we might use internally some deformations and compare them with the given impression?

This would be a great saving of memory storage and it could be applied not only to single pictures, but to pairs or triplets or short "films" of pictures, thus enabling recognition or distinction between different new impressions.

A more sophisticated schema of recognition, in fact the beginning of more abstract reasoning, would involve a distance more general yet than the ones mentioned above. This could be based on the comparison of two given sets by decomposing them into pieces and considering a distance calculated from the sum of the comparisons of the parts.

A still more general distance leading to a beginning of "logic" would involve a comparison of classes of pictures. The *post hoc ergo propter hoc* (after that hence because of that) conclusion serving as an example.

About problems of recognition of shapes by comparison with a large storage of examples:

One question which occurs in auditory, tactile, or visual impressions concerns the taking of a decision that the new impression is "novel" and not to be considered as a variant of one of the examples stored in the memory. Thus, for example, antibodies are able to recognize a foreign or strange object.

We may postulate that there is a list of objects in the memory considered "familiar." This list might be, for example, arranged lexicographically or alphabetically or coded by numbers. As we will see, the way of arranging it in the memory is important.

For example, suppose we have listed 10^6 numbers, each of ten digits, say, so we have a rather sparse collection. A new number is presented having ten digits. The first question is: Is it equal to some number in the list? On a computer the answer can be obtained immediately. Similarly in a putative mechanism in the brain, since it suffices to go through the digits in succession, which is a fast and efficient process.

Suppose however that the question is: Is the new number if not equal to any number in the list, at least close to some such number, e.g., differs from it by 1 or 2 in some position? The search for such a close number would be time consuming if we tried to compare the given number with each of the 10^6 numbers in the memory. Obviously there is a better way.

From the given number we fabricate 20 numbers which differ from it by 1 in some one of the ten digits. Then we look for each of the 20 whether one of them might be in the list, as above, very quickly. Again we see that should there exist in the brain a mechanism to produce some small changes or deformations, then compare those with the contents of the memory, the recognition or discrimination would be much more efficient.

The above example refers to one particular distance which depends on the absolute value of the difference in the digits in the same position.

For other types of distances described above, analogous but combinatorially more complicated procedures are possible. For each of these distances the question of what is the most economical or practical clustering presents an interesting exercise.

The principal contention or conjecture is then of the existence of a mechanism in the nervous system capable of producing a number of ϵ-modifications of the impressions that affect distinction or recognition.

Astronomers trying to find a new star on a photograph of a portion of the sky quickly flip a number of pictures of this region and the new object jumps out visually from the collection of the others which are constantly present. A parade of coded pictures in the memory compared with the new impression might serve a similar purpose. Our supposition is that there exist ways of sensing quantitatively a number of different distances between the impression and the memory data, distances that are perhaps not unlike the ones we have enumerated above—an "averted" memory may, like averted vision, aid in the search.

Going beyond, we may have the beginning of a "logical" or "reasoning" process by considering *sequences* of impressions of pictures and measuring their analogy by metric distances.

References

1. S. Ulam, "Some Ideas and Prospects in Biomathematics," Ann. Rev. Biophys. and Bioeng. **1**, 277–292, (1972).
2. Paul Appell, "Le Problème Géométrique des Déblais et Remblais," Mémorial des Sciences Mathématiques de l'Académie des Sciences, fasc. **27** (1928), Gauthier-Villars, Paris.

APPENDIX A

PUBLICATIONS OF
STANISLAW M. ULAM

by Barbara Hendry
(LA-3923-MS, 1968)

*The material contained in this report was revised, expanded and brought
up to date in 1987 with the help of Nancy Shera and Dixie MacDonald. (Eds.)*

Remark on the generalized Bernstein's theorem. * *Fundamenta Math-
ematicae* 13(1929): 281-3.
Concerning functions of sets.* *Fundamenta Mathematicae* 14(1929):
231-3.
Zur Masstheorie in der allgemeinen Mengenlehre.* *Fundamenta Mathe-
maticae* 16(1930): 140-50. Also in *Mengenlehre*, edited by U. Felgner,
223-33. Darmstadt: Wissenschaftliche Gesellschaft, 1979.
On symmetric products of topological spaces (with Karol Borsuk).*
Bulletin of the American Mathematical Society 37(1931): 875-82.
Sur une propriété de la mesure de M. Lebesgue (with J. Schreier).*
Comptes Rendus de l'Académie des Sciences 192(1931): 539-42.
Zum Massbegriff in Produkträumen.* In *Verhandlungen, Interna-
tionaler Mathematikerkongress Zürich 1932*, volume 2, 118-9.
Zürich: Orell Füssli Verlag, 1932.

This publication appears in *Stanislaw Ulam: Sets, Numbers, and Universes*,
edited by W. A. Beyer, J. Mycielski, and G.-C. Rota. Cambridge, Massachusetts: The MIT Press, 1974.

Quelques propriétés topologiques du produit combinatoire (with C. Kuratowski).* *Fundamenta Mathematicae* 19(1932): 247–51.

Sur les transformations isométriques d'espaces vectoriels normés (with S. Mazur).* *Comptes Rendus de l'Académie des Sciences* 194(1932): 946–8.

Über gewisse Zerlegungen von Mengen.* *Fundamenta Mathematicae* 20(1933): 221–3.

Problème 56. *Fundamenta Mathematicae* 20(1933): 285.

Über gewisse Invarianten der ϵ-Abbildungen (with Karol Borsuk).* *Mathematische Annalen* 108(1933): 311–8.

Sur un coefficient lié aux transformations continues d'ensembles (with C. Kuratowski).* *Fundamenta Mathematicae* 20(1933): 244–53.

Sur le groupe des permutations de la suite des nombres naturels (with J. Schreier).* *Comptes Rendus de l'Académie des Sciences* 197(1933): 737–8.

Sur les transformations continues des sphères euclidiennes (with J. Schreier).* *Comptes Rendus de l'Académie des Sciences* 197(1933): 967–8.

Über die Permutationsgruppe der natürlichen Zahlenfolge (with J. Schreier).* *Studia Mathematica* 4(1933): 134–41.

Sur la théorie de la mesure dans les espaces combinatoires et son application au calcul des probabilités: I. Variables indépendantes (with Z. Lomnicki).* *Fundamenta Mathematicae* 23(1934): 237–78.

Über topologische Abbildungen der euklidischen Sphären (with J. Schreier).* *Fundamenta Mathematicae* 23(1934): 102–18.

Eine Bemerkung über die Gruppe der topologischen Abbildungen der Kreislinie auf sich selbst (with J. Schreier).* *Studia Mathametica* 5(1934): 155–9.

Sur le nombre de générateurs d'un groupe semi-simple (with H. Auerbach).* *Comptes Rendus de l'Académie des Sciences* 201(1935): 117–9.

Sur une propriété caractéristique de l'ellipsoïde (with H. Auerbach and S. Mazur).* *Monatsheften für Mathematik und Physik* 42(1935): 45–8.

Sur le nombre des générateurs d'un groupe topologique compact et connexe (with J. Schreier).* *Fundamenta Mathematicae* 24(1935): 302–4.

Über die Automorphismen der Permutationsgruppe der natürlichen Zahlenfolge (with J. Schreier).* *Fundamenta Mathematicae* 28(1937): 258–60.

Problème 74. *Fundamenta Mathematicae* 30(1938): 365.

On the equivalence of any set of first category to a set of measure zero (with J. C. Oxtoby).* *Fundamenta Mathematicae* 31(1938): 201–6.

On the existence of a measure invariant·under a transformation (with J. C. Oxtoby).* *Annals of Mathematics, Second Series* 40(1939): 560–6.

Measure-preserving homeomorphisms and metrical transitivity (with J. C. Oxtoby).* *Annals of Mathematics, Second Series* 42(1941): 874–920.

What is measure?* *The American Mathematical Monthly* 50(1943): 597–602.

Theory of multiplicative processes. I (with D. Hawkins). Los Alamos Scientific Laboratory report LA–171, 1944.

On ordered groups (with C. J. Everett).* *Transactions of the American Mathematical Society* 57(1945): 208–16.

On approximate isometries (with D. H. Hyers).* *Bulletin of the American Mathematical Society* 51(1945): 288–92.

Stefan Banach, 1892–1945. ** *Bulletin of the American Mathematical Society* 52(1946): 600–3.

Projective algebra I (with C. J. Everett).* *American Journal of Mathematics* 68(1946): 77–88.

Problèmes P34; P35; P35,R1 (with S. Banach). *Colloquium Mathematicum* 1(1947): 152–3.

Approximate isometries of the space of continuous functions (with D. H. Hyers).* *Annals of Mathematics, Second Series* 48(1947): 285–9.

Statistical methods in neutron diffusion (with J. von Neumann). Report written by R. D. Richtmyer and J. von Neumann. Los Alamos Scientific Laboratory report LAMS–551, 1947. Also in *Von Neumann: Collected Works, 1903–1957*, edited by A. H. Taub, volume 5. Oxford: Pergamon Press, 1963.

Multiplicative systems, I (with C. J. Everett).* *Proceedings of the National Academy of Sciences of the United States of America* 34(1948): 403–5.

Multiplicative systems in several variables. Parts I, II, and III (with C. J. Everett). Los Alamos Scientific Laboratory reports LA–683, LA–690, and LA–707, 1948.

The Monte Carlo method (with Nicholas Metropolis).* *Journal of the American Statistical Association* 44(1949): 335–41.

On the Monte Carlo method. In *Proceedings of the 1949 Symposium on Large-Scale Digital Calculating Machines*, 207–12. Cambridge, Massachusetts: Harvard University Press, 1951.

---

* This publication appears in *Science, Computers and People: From the Tree of Mathematics*, edited by Mark C. Reynolds and Gian-Carlo Rota. Boston: Birkhaüser, 1986.

Random processes and transformations. In *Proceedings of the International Congress of Mathematicians (Cambridge, Massachusetts, August 30–September 6, 1950)*, volume 2, 264–75. Providence, Rhode Island: American Mathematical Society, 1952.

Approximately convex functions (with D. H. Hyers).* *Proceedings of the American Mathematical Society* 3(1952): 821–8.

A property of randomness of an arithmetical function (with N. Metropolis).* *The American Mathematical Monthly* 60(1953): 252–3.

Heuristic studies in problems of mathematical physics on high speed computing machines (with J. Pasta). Los Alamos Scientific Laboratory report LA–1557, 1953.

On the stability of differential expressions (with D. H. Hyers).* *Mathematics Magazine* 28(1954): 59–64.

Homage to Fermi. *Santa Fe New Mexican*, January 6, 1955.

On a method of propulsion of projectiles by means of external nuclear explosions (with C. J. Everett). Los Alamos Scientific Laboratory report LAMS-1955, 1955.

Studies of nonlinear problems. I (with E. Fermi, J. Pasta, and M. Tsingou).* Los Alamos Scientific Laboratory report LA–1940, 1955. Also in *Enrico Fermi: Collected Papers*, volume 2, edited by E. Amaldi, H. L. Anderson, E. Persico, E. Segrè, and A. Wattenberg. Chicago: University of Chicago Press, 1965.

Study of certain combinatorial problems through experiments on computing machines (with P. R. Stein). In *Proceedings of the 1955 High-Speed Computer Conference (Louisiana State University, Baton Rouge, Louisiana, February 14–16, 1955)*, 101–6.

On the ergodic behavior of dynamical systems. In "Series of lectures on physics of ionized gases." Los Alamos Scientific Laboratory report LA–2055, 1956.

On certain sequences of integers defined by sieves (with Verna Gardiner, R. Lazarus, and N. Metropolis).* *Mathematics Magazine* 29(1956): 117–22.

Infinite models in physics.* In *Proceedings of the Seventh Symposium in Applied Mathematics (Brooklyn Polytechnic Institute, April 14–15, 1955)*. American Mathematical Society Symposia in Applied Mathematics, volume 7, 87–95. New York: McGraw-Hill Book Company, Inc., 1957.

Marian Smoluchowski and the theory of probabilities in physics.** *American Journal of Physics* 25(1957): 475–81.

The Scottish Book: A Collection of Problems. An edited translation of a notebook kept at the Scottish Café for the Lwów Section of the Société Polonaise de Mathématiques. Privately mimeographed and distributed by S. M. Ulam in 1957. Reprinted as Los Alamos Scientific Laboratory report LA–6832, 1967.

On some new possibilities in the organization and use of computing machines. IBM research report RC-86, 1957.

Experiments in chess (with J. Kister, P. Stein, W. Walden, and M. Wells).* *Journal of the Association for Computing Machinery* 4(1957): 174–7.

Experiments in chess on electronic computing machines (with P. R. Stein).** *Chess Review*, January 1957, 13–5. Also in *Computers and Automation*, September 1957.

John von Neumann, 1903–1957.** *Bulletin of the American Mathematical Society* 64(1958): 1–49.

The late John von Neumann on computers and the brain.** *Scientific American*, June 1958, 127.

On the possibility of extracting energy from gravitational systems by navigating space vehicles. Los Alamos Scientific Laboratory report LAMS–2219, 1958.

Statement before the U.S. House of Representatives. *Hearings on Astronautics and Space Exploration.* 85th Congress, 2nd session, April 15–May 12, 1958.

Review of *Funkcje Rzeczywiste* by Roman Sikorski. *Bulletin of the American Mathematical Society* 65(1959): 305–6.

Quadratic transformations. Part I (with M. T. Menzel and P. R. Stein). Los Alamos Scientific Laboratory report LA–2305, 1959.

Heuristic numerical work in some problems of hydrodynamics (with John R. Pasta).* In *Mathematical Tables and Other Aids to Computation* 13(1959): 1–12.

A Collection of Mathematical Problems. New York: Interscience Publishers, 1960. Reprinted as *Problems in Modern Mathematics.* John Wiley & Sons, Inc., 1964. Translated into Russian (1964).

Statement before the Joint Committee on Atomic Energy. In *Frontiers in Atomic Energy Research: Hearings before the Subcommittee on Research and Development of the Joint Committee on Atomic Energy, Eighty-sixth Congress, Second Session, on Frontiers in Atomic Energy Research, March 22–25, 1960,* 282–5. Washington, D.C.: U.S. Government Printing Office, 1960.

Monte Carlo calculations in problems of mathematical physics. In *Modern Mathematics for the Engineer, Second Series,* edited by Edwin F. Beckenbach, 95–108. New York: McGraw-Hill Book Company, Inc., 1961.

Nuclear propelled vehicle, such as a rocket (with C. J. Everett). British Patent 877,392, 1961.

How to formulate mathematically problems of the rate of evolution?** In *Proceedings of the Symposium on Mathematical Challenges to the*

Neo-Darwinian Interpretation of Evolution (New York, April 5–8, 1961), edited by Paul S. Moorhead and Martin M. Kaplan. Providence, Rhode Island: American Mathematical Society. Wistar Institute Monograph 5: (1967) 21–33, April 25-26, 1966 New York: A. Liss, 1985.

On some statistical properties of dynamical systems.* In *Proceedings of the Fourth Berkeley Symposium on Mathematical Statistics and Probability (University of California, Berkeley, June 20–July 30, 1960)*, edited by Lucian M. Le Cam, Jerzy Neyman, and Elizabeth Scott, volume 3, 315–20. Berkeley: University of California Press, 1961. Translated into Russian (1963).

Electronic computers and scientific research. In *The Age of Electronics*, edited by Carl F. J. Overhage, 95–108. New York: McGraw-Hill Book Company, Inc., 1962. Also in *Computers and Automation*, August 1963 and September 1963.

An open problem. In *Recent Advances in Game Theory (Papers Delivered at a Meeting of the Princeton University Conference, October 4-6,1961)*, 223. Princeton, New Jersey: Princeton University Conference, 1962.

On some mathematical problems connected with patterns of growth of figures.** *Applied Mathematics* 14(1962): 215–24. Also in *Essays on Cellular Automata*, edited by Arthur W. Burks. Urbana, Illinois: University of Illinois Press, 1970.

Stability of many-body computations. In *Hydrodynamic Instability*, edited by Garrett Birkhoff, Richard Bellman, and C. C. Lin, 247-58. American Mathematical Society Symposia in Applied Mathematics, volume 13. Providence, Rhode Island: American Mathematical Society, 1962.

Communication to the U.S. Senate Committee on Foreign Relations. In *Nuclear Test Ban Treaty: Hearings before the Committee on Foreign Relations, United States Senate, Eighty-eighth Congress, First Session, on Executive M, August 12–15, 19–23, 26–27, 1963*, 505–6 and 993. Washington, D.C.: U.S. Goverment Printing Office, 1963.

Problems 110, 111, and 112. In *Proceedings of the 1963 Number Theory Conference (University of Colorado, Boulder, Colorado, August 5–24, 1963)*, 114-5.

Some properties of certain non-linear transformations.* In *Mathematical Models in Physical Sciences: Proceedings of the Conference at the University of Notre Dame, 1962*, edited by Stefan Drobot, 85–95. Englewood Cliffs, New Jersey: Prentice-Hall, Inc., 1963.

Non-linear transformation studies on electronic computers (with P. R. Stein).* *Rozprawy Matematyczne* 39(1963): 1–66. The Introduction and Part I are also in *Essays on Cellular Automata*, edited by Arthur W. Burks. Urbana, Illinois: University of Illinois Press, 1970.

The future of nuclear energy in space: A panel discussion sponsored by the Aerospace Division, American Nuclear Society, at the 1963 winter meeting in New York City, N. Y. on November 20, 1963 (with F. deLuzio, W. von Braun, M. Hunter, and I. Asimov), edited by R. F. Trapp. Also in *Nuclear News*, July 1964.

Combinatorial analysis in infinite sets and some physical theories. *SIAM Review* 6(1964): 343–55.

Computers.** *Scientific American*, September 1964, 202–216.

A visual display of some properties of the distribution of primes (with M. L. Stein and M. B. Wells).* *The American Mathematical Monthly* 71(1964): 516–20.

Possibility of an accelerated process of collapse of stars in a very dense centre of a cluster or a galaxy (with W. E. Walden). *Nature* 201(1964): 1202.

The Orion project.** *Nuclear News*, January 1965, 25–7.

Collapse of stellar systems. In *Proceedings of the 25th International Astronomical Union Symposium (Thessaloniki, Greece, August 16–22, 1964)*, 76–7. International Astronomical Union, 1966.

La machine créatrice. In *Rencontres Internationales de Genève "Le robot, la bête et l'homme (1965)*, 31–42. Neuchatel: Éditions de la Baconnière, 1966.

On some possibilities of generalizing the Lorentz group in the special relativity theory (with C. J. Everett).* *Journal of Combinatorial Theory* 1(1966): 248–70.

Thermonuclear devices.** In *Perspectives in Modern Physics: Essays in Honor of Hans A. Bethe*, edited by R. E. Marshak and J. Warren Blaker, 593–601. New York: Interscience Publishers, 1966.

An education in applied math. In *Proceedings of May 24–27, 1966 SIAM Conference (Aspen, Colorado)*, edited by James Ortega, Paul I. Richards, and Frank W. Sinden. *SIAM Review* 9(1967): 343–4.

On general formulations of simulation and model construction. In *Prospects for Simulation and Simulators of Dynamic Systems*, edited by George Shapiro and Milton Rogers, 3–8. New York: Spartan Books, 1967.

On recursively defined geometrical objects and patterns of growth (with R. G. Schrandt).** Los Alamos Scientific Laboratory report LA–3762, 1967. Also in *Essays on Cellular Automata*, edited by Arthur W. Burks. Urbana, Illinois: University of Illinois Press, 1970.

On visual hulls of sets (with G. H. Meisters).* *Proceedings of the National Academy of Sciences of the United States of America* 57(1967): 1172–4.

An observation on the distribution of primes (with M. Stein).* *The American Mathematical Monthly* 74(1967): 43–4.

Computations on certain binary branching processes. In *Computers in Mathematical Research*, edited by R. F. Churchhouse and J.-C. Herz, 168–171. Amsterdam: North-Holland Publishing Company, 1968.

Numerical studies of star systems. In *Colloque sur le Problème des N Corps*, 265–7. Éditions du Centre National de la Recherche, 1968.

Philosophical implications of some recent scientific discoveries.** In *Science, Philosophy and Religion*. Proceeding Symposium Kirtland Air Force Laboratory, Albuquerque, New Mexico 44–48. (1968).

Note on the visual hull of a set (with W. A. Beyer).* *Journal of Combinatorial Theory* 4(1968): 240–5.

On equations with sets as unknowns (with Pal Erdös).* *Proceedings of the National Academy of Sciences of the United States of America* 60(1968): 1189–95.

Mathematics and Logic: Retrospect and Prospects (with Mark Kac). New York: Frederick A. Praeger, Inc., 1968. The text of this book first appeared as the article entitled "Mathematics and logic: Retrospect and prospects" in Britannica Perspectives, volume 1, 557–732. Chicago: Encyclopædia Britannica, Inc., 1968. Translated into French (1973), into Serbo-Croatian (1977), and into Spanish (1979).

The applicability of mathematics.** In *The Mathematical Sciences: A Collection of Essays*, edited by the National Research Council's Committee on Support of Research in the Mathematical Sciences, 1–6. Cambridge, Massachusetts: The M.I.T. Press, 1969.

Wspomnienia Kawiarni Szkockiej (Reminiscences of the Scottish Café). *Wiadomosci Matematyczne* 12(1969): 49–58. Available in English only in manuscript form.

Computer studies of some history-dependent random processes (with W. A. Beyer and R. G. Schrandt). Los Alamos Scientific Laboratory report LA–4246, 1969.

The entropy of interacting populations (with C. J. Everett). Los Alamos Scientific Laboratory report LA–4256, 1969.

On the pairing process and the notion of genealogical distance (with Jan Mycielski).* *Journal of Combinatorial Theory* 6(1969): 227–34.

Foreword to *My World Line: An Informal Autobiography* by G. Gamow. New York: Viking Press, 1970.

Generalizations of product isomorphisms. In *Recent Trends in Graph Theory*, 215. Lecture Notes in Mathematics, volume 186. Berlin: Springer-Verlag, 1971.

Testimony before the United States District Court, District of Minnesota, Minneapolis, Minnesota, September 17, 1971, in the case of Honeywell Incorporated versus Sperry Rand Corporation, 7342–438.

The notion of complexity (with W. A. Beyer and M. L. Stein). Los Alamos Scientific Laboratory report LA–4822, 1971.

Some probabilistic remarks on Fermat's last theorem (with P. Erdös).* *Rocky Mountain Journal of Mathematics* 1(1971): 613–6.

Some elementary attempts at numerical modeling of problems concerning rates of evolutionary processes (with R. Schrandt). Los Alamos Scientific Laboratory report LAMS–4573, 1971.

Gamow—and mathematics.** In *Cosmology, Fusion & Other Matters: George Gamow Memorial Volume*, edited by Frederick Reines, 272–9. Boulder, Colorado: Colorado Associated University Press, 1972.

Ideas of space and space-time.** *Rehovot Magazine*, Winter 1972–73, 29–33.

Some combinatorial problems studied experimentally on computing machines. In *Applications of Number Theory to Numerical Analysis*, edited by S. K. Zaremba, 1–10. New York: Academic Press, Inc., 1972.

Some ideas and prospects in biomathematics.** *Annual Review of Biophysics and Bioengineering* 1(1972): 277– 91.

Metrics in biology, an introduction (with W. A. Beyer, M. L. Stein, and Temple Smith). Los Alamos Scientific Laboratory report LA–4973, 1972.

Lectures in nonlinear algebraic transformations (with P. R. Stein). In *Studies in Mathematical Physics (Lectures Presented at the NATO Advanced Study Institute on Mathematical Physics held in Istanbul, August, 1970)*, edited by A. O. Barut, 263–314. Dordrecht, The Netherlands: D. Reidel Publishing Company, 1973.

Infinities. In *The Heritage of Copernicus*, edited by J. Neyman, 378–92. Cambridge, Massachusetts: The MIT Press, 1974.

New rules and old games. *Outlook*, Spring 1974, 32–3.

Stanislaw Ulam: Sets, Numbers, and Universes, edited by W. A. Beyer, J. Mycielski, and G.-C. Rota. Cambridge, Massachusetts: The MIT Press, 1974.

A molecular sequence metric and evolutionary trees (with William A. Beyer, Myron L. Stein, and Temple F. Smith). *Mathematical Biosciences* 19(1974): 9–25.

Arthur Koestler et le défi du hazard: Entretien avec Stan Ulam de Pierre Debray-Ritzen. In *Arthur Koestler*, 428–32. Cahiers de l'Herne. Paris: Èdition de L'Herne, 1975.

Adventures of a Mathematician. New York: Charles Scribner's Sons, 1976. Paperback editions published in 1977 and 1983. Translated into Japanese (1979).

Physics for mathematicians.** In *Physics and Our World: A Symposium in Honor of Victor F. Weisskopf (Massachusetts Institute of*

Technology, 1974), edited by Kerson Huang, 113–21. AIP Conference Proceedings, number 28. New York: American Institute of Physics, Inc., 1976.

Generators for algebras of relations (with A. R. Bednarek). *Bulletin of the American Mathematical Society* 82(1976): 781–2.

On the theory of relational structures and schemata for parallel computation (with A. R. Bednarek). Los Alamos Scientific Laboratory report LAMS–6734, 1977.

Some remarks on relational composition in computational theory and practice (with A. R. Bednarek). In *Fundamentals of Computation Theory: Proceedings of the 1977 International FCT-Conference (Poznań-Kórnik, Poland, September 19–23, 1977)*, edited by Marek Karpiński, 22–32. Lecture Notes in Computer Science, volume 56. Berlin: Springer-Verlag, 1977.

Przygody matematyka. *Kultura* 9(30 Lipca 1978). Translated into Polish by Jerzy Jaruzelski.

Banach i inni. *Kultura* 10(6 Sierpnia 1978). Translated into Polish by Jerzy Jaruzelski.

Narodziny "Księgi Szkockiej." *Kultura* 10(13 Sierpnia 1978). Translated into Polish by Jerzy Jaruzelski.

Projective algebra and the calculus of relations (with A. R. Bednarek). *Journal of Symbolic Logic* 43(1978): 56–64.

The role of abstract mathematical ideas in possible conceptual advances in natural sciences, more specifically biology. In *Proceedings of International Colloquium on the Role of Mathematical Physics in the Development of Science (Collège de France, Paris, June 13–15, 1977)*, edited by Dominique Akl, Moshe Flato, and Daniel Sternheimer, 12–25. UNESCO, 1978.

An integer-valued metric for patterns (with A. R. Bednarek). In *Fundamentals of Computation Theory*, 52–7. Berlin: Akademie-Verlag, 1979.

Minimal decomposition of two graphs into pairwise isomorphic subgraphs (with F. R. K. Chung, P. Erdös, R. L. Graham, and F. F. Yao). In *Proceedings of the Tenth Southeastern Conference on Combinatorics, Graph Theory, and Computing (Florida Atlantic University, Boca Raton, Florida, April 2–6, 1979)*, volume 1, 3–18. Congressus Numerantium, volume 23. Winnipeg, Manitoba: Utilitas Mathematica Publishing Incorporated, 1979.

A mathematical physicist looks at computing.** *Rehovot Magazine*, volume 9, number 1, 1980, 47–50.

On the operations of pair production, transmutations, and generalized random walks. *Advances in Applied Mathematics* 1(1980): 7–21.

Preface to *A Half Century of Polish Mathematics: Remembrances and Reflections* by Kazimierz Kuratowski. International Series in Pure and Applied Mathematics, volume 108. Oxford: Pergamon Press Ltd., 1980.

Von Neumann: The interaction of mathematics and computing.** In *A History of Computing in the Twentieth Century: A Collection of Essays,* edited by N. Metropolis, J. Howlett, and Gian-Carlo Rota, 93–9. New York: Academic Press, Inc., 1980.

Further applications of mathematics in the natural sciences.** In *American Mathematical Heritage: Algebra and Applied Mathematics,* edited by J. Dalton Tarwater, 101–14. Texas Tech University Mathematics Series, volume 13. Lubbock, Texas: Texas Technological University Press, 1981.

Kazimierz Kuratowski, 1896–1980.** *Polish Review* 26(1981): 62–6.

On the notion of analogy and complexity in some constructive mathematical schemata. Los Alamos National Laboratory report LA–9065, 1981. Also in *Probability, Statistical Mechanics, and Number Theory,* edited by Gian-Carlo Rota. Advances in Mathematics: Supplementary Studies: volume 9. New York: Academic Press, Inc., 1986.

An anecdotal history of the Scottish Book. In *The Scottish Book: Mathematics from the Scottish Café,* edited by R. Daniel Mauldin. Boston: Birkhäuser, 1982.

Introduction** to *Selected Studies: Physics-Astrophysics, Mathematics, History of Science. A Volume Dedicated to the Memory of Albert Einstein,* edited by Themistocles M. Rassias and George M. Rassias. Amsterdam: North-Holland Publishing Company, 1982.

Reflections of the Polish masters: An interview with Stan Ulam and Mark Kac. *Los Alamos Science,* volume 3, number 3, 1982, 54–65.

Speculations about the mechanism of recognition and discrimination. Los Alamos National Laboratory unclassified release LAUR 82–62, 1982.

Transformations, iterations and mixing flows. In *Dynamical Systems II,* edited by A. R. Bednarek and L. Cesari, 419–26. New York: Academic Press, 1982.

Kazimierz Kuratowski, Wspomnienia (Kazimierz Kuratowski: A reminiscence). *Wiadomosci Matematyczne,* 1983. Translated into Polish by R. Engelking. Also in *Kazimierz Kuratowski, Selected Papers,*Polish Academy of Sciences, K. Borsuk, editor, PWN, Warsaw, 1988.

Speculations on some possible mathematical frameworks for the foundations of certain physical theories. *Letters in Mathematical Physics* 10(1985): 101–6.

Science, Computers, and People: From the Tree of Mathematics, edited by Mark C. Reynolds and Gian-Carlo Rota. Boston: Birkhaüser, 1986.

Mathematical problems and games (with R. Daniel Mauldin). *Advances in Applied Mathematics* 8(1987): 281–344.

Reflections on the brain's attempts to understand itself. *Los Alamos Science*, number 15, 1987, 283–7.

Analogies between Analogies: The Mathematical Reports of S. Ulam and his Los Alamos Collaborators, edited by D. Sharp and M. Simmons. University of California Press.

Eleven weapons-related reports written by Ulam and his collaborators between 1944 and 1958 are still classified. These are listed in LAMS–3923, 1968 and in *Stanislaw Ulam: Sets, Numbers, and Universes*, edited by W. A. Beyer, J. Mycielski, and G.-C. Rota (The MIT Press, 1974).

Abstracts

Über unendliche Abelsche Gruppen (with S. Mazur). *Annales de la Société Polonaise de Mathématique*, 9(1930): 204.

Ein Betrag zum Massproblem. *Annales de la Société Polonaise de Mathématique*, 9(1930): 198.

Über die Eindeutigkeit des Masses von Gerardenmengen. *Annales de la Société Polonaise de Mathématique*, 9(1930): 200.

Über vollstandig additive Massfunktionen in abstrakten Räumen. *Annales de la Société Polonaise de Mathématique*, 9(1930): 195.

Zur Theorie des Fixpunktes. *Annales de la Société Polonaise de Mathématique*, 9(1930): 201–2.

Einige Sätze über Mengen II-er Kategorie. *Annales de la Société Polonaise de Mathématique*, 10(1931): 123–4.

Über eine charakteristische Eigenschaft des Ellipsoides (with H. Auerbach, S. Mazur). *Annales de la Société Polonaise de Mathématique*, 10(1931): 128.

Über eine neue topologische Operation (with K. Borsuk). *Annales de la Société Polonaise de Mathématique*, 10(1931): 125–6.

Über isometrische Abbildungen von normierten Vektorräumen (with S. Mazur). *Annales de la Société Polonaise de Mathématique*, 10(1931): 127.

Über die Grundlagen der Wahrscheinlichkeitsrechnung (with Z. Lomnicki). *Annales de la Société Polonaise de Mathématique*, 12(1933) 115.

Über die Gesetze der grossen Zahlen (with Z. Lomnicki). *Annales de la Société Polonaise de Mathématique*, 12(1933): 118.

Bemerkungen über die stetigen Abbildungen von Topologischen Räumen (with J. Schreier). *Annales de la Société Polonaise de Mathématique*, 13(1934): 142.

Über stetige Abbildungen von Mannigfaltigkeiten. *Annales de la Société Polonaise de Mathématique*, 13(1934): 141.

Existence of metrically transitive transformations. Preliminary report (with J. C. Oxtoby). *Bulletin of the American Mathematical Society*, 44(1938): 347.

On bounded transformations of space. Preliminary report. *Bulletin of the American Mathematical Society*, 44(1938): 195.

On the distribution of a general measure in any complete metric separable space. *Bulletin of the American Mathematical Society*, 44(1938): 786.

Set-theoretical invariants of the product operation. *Bulletin of the American Mathematical Society*, 44(1938): 195.

Sur les transformations ergodiques. *Annales de la Société Polonaise de Mathématique*, 17(1938): 112.

On the abstract theory of measure. *Bulletin of the American Mathematical Society*, 45(1939): 83.

ε-isomorphic transformations. Preliminary report. *Bulletin of the American Mathematical Society*, 45(1939): 232.

On approximate isometries. Preliminary report (with D. H. Hyers). *Bulletin of the American Mathematical Society*, 47(1941): 708.

On measures for subsets of sets of measure zero. *Bulletin of the American Mathematical Society*, 47(1941): 702.

Theory of operation of products of sets. I. Preliminary report. *Bulletin of the American Mathematical Society*, 47(1941): 702.

Approximate isometries of the space of continuous functions (with D. H. Hyers). *Bulletin of the American Mathematical Society*, 48(1942): 368.

Geometrical approach to the theory of representations of topological groups. Preliminary report. *Bulletin of the American Mathematical Society*, 48(1942): 44.

On the problem of completely additive measure in classes of sets with a general equivalence relation (with D. L. Bernstein). *Bulletin of the American Mathematical Society*, 48(1942): 361-2.

On the equivalence of functions. *Bulletin of the American Mathematical Society*, 49(1943): 49.

On the length of curves, the surface area and the isoperimetric problem under a general Minkowski metric. Preliminary report. *Bulletin of the American Mathematical Society*, 49(1943): 57.

Theory of the operation of products of sets. II. Preliminary report. *Bulletin of the American Mathematical Society*, 49(1943): 367-8.

On ordered groups (with J. C. Everett). *Bulletin of the American Mathematical Society*, 50(1944): 496.

On the algebra of systems of vectors and some problems in kinematics (with L. Cohen). *Bulletin of the American Mathematical Society*, 50(1944): 61.

Some combinatorial problems in set theory. Preliminary report (with P. Erdős). *Bulletin of the American Mathematical Society*, 50(1944): 57.

Theory of the operation of product of sets. III. Preliminary report. *Bulletin of the American Mathematical Society*, 50(1944): 60–1.

Projective algebra, I (with C. J. Everett). *Bulletin of the American Mathematical Society*, 51(1945): 59.

Random ergodic theorems (with J. von Neumann). *Bulletin of the American Mathematical Society*, 51(1945): 660.

On combination of stochastic and deterministic processes. Preliminary report (with J. von Neumann). *Bulletin of the American Mathematical Society*, 53(1947): 1120.

On quasi-fixed points for transformations in function spaces. *Bulletin of the American Mathematical Society*, 53(1947): 1120.

On the group of homeomorphisms of the surface of the sphere (with J. von Neumann). *Bulletin of the American Mathematical Society*, 53(1947): 506.

On the problem of determination of mathematical structures by their endomorphisms (with C. J. Everett). *Bulletin of the American Mathematical Society*, 54(1948): 646.

Statistical methods for problems involving equations of the diffusion type, (Monte Carlo). A.E.C. Information Meeting, Brookhaven Natl. Lab., Apr. 26–8, 1948, BNL-17, Special,(1948):27.

Multiplicative systems, I (with C. J. Everett). *Bulletin of the American Mathematical Society*, 55(1949): 51.

Multiplicative systems, II (with C. J. Everett). *Bulletin of the American Mathematical Society*, 55(1949): 51.

Multiplicative systems, III (with C. J. Everett). *Bulletin of the American Mathematical Society*, 55(1949): 51–2.

On motions of systems of mass points randomly distributed on the infinite line (with N. C. Metropolis). *Bulletin of the American Mathematical Society*, 55(1949): 670–1.

On an application of a correspondence between matrices over real algebras and matrices of positive real numbers (with C. J. Everett). *Bulletin of the American Mathematical Society*, 56(1950): 63.

Random walk and the Hamilton-Jacobi equation (with C. J. Everett). *Bulletin of the American Mathematical Society*, 56(1950): 63–4.

Approximately convex functions (with D. H. Hyers). *Bulletin of the American Mathematical Society*, 59(1951): 300–1.

Applications of Monte Carlo methods to tactical games. in Proceedings of March 16–17, 1954 Symposium on Monte Carlo Methods, Univ. of Fla., Herbert A. Meyer ed., Wiley (1956) 63.

Some mathematical problems investigated through computations on electronic machines. *The American Mathematical Monthly,* 63(1956): 607.

Future uses of future computers. *American Chemical Society Abstracts of Papers,* 133(1958): 33–4K.

On certain binary reaction systems (with P. R. Stein). *American Mathematical Society Notices,* 6(1959): 68–9.

On patterns of growth of figures in two dimensions (with R. G. Schrandt). *American Mathematical Society Notices,* 7(1960): 642.

On some combinatorial problems in patterns of growth, I (with J. C. Holladay). *American Mathematical Society Notices,* 7(1960): 234.

On a statistical method of solving multiplication and diffusion problems. Monsanto Chemical Company Meeting, Clinton National Laboratory, Oct. 13–5, 1947. *Monsanto Chemical Company Abstracts of Papers,* Mon-411, #14, 1961.

On some possibility of generalizing the Lorentz group in special relativity theory (with C. J. Everett). *American Mathematical Society Notices,* 12(1965): 614.

Recursive definitions of static changing patterns, in *Biomathematics and Computer Science in the Life Sciences,* Monograph of Proceedings of 3d Annual Symposium–Houston, Texas, (1965): IX.*

* As an amusing commentary on Ulam's personality the editors include the following remark from the foreword to the original 1968 Publication's Report:. . ."The most distinguishing personal traits of Stan Ulam are friendliness, simplicity, tenacity, and a certain disregard of formality or other mundane impertinences. . .One of our functions is to monitor all material published from this Laboratory. . .Everyone else sends his proposed paper. . .to us before mailing it. Most of our files on Ulam's publications were opened when he sent us a reprint and an invoice—yet there was never a problem about any of the papers.". . .

Leslie M. Redman
Technical Information Group
April 1968

APPENDIX B

Vita of Stanislaw M. Ulam

Born:
   April 13, 1909, Lwów, Poland

Studies:
   M.A. & D.Sc., 1933, Polytechnic Institute, Lwów. Post-doctoral studies in Vienna, Zürich & Cambridge (England), 1934

Positions:
   Came to U.S. on invitation of J. von Neumann, to Institute for Advanced Studies, Princeton, 1935
   Junior Fellow at Harvard Society of Fellows, then lecturer in mathematics, Harvard, 1936–40
   Assistant Professor, University of Wisconsin, 1941–43
   Staff member, then research advisor, Los Alamos Scientific Laboratory, 1944–1967 During that period visited:
   University of Southern California, Los Angeles 1945–46, Harvard University, 1951–52, M.I.T., 1956–57, University of Colorado, 1961, University of California, La Jolla, 1962
   Mathematics professor and chairman of department, University of Colorado, Boulder, 1965–77
   Consultant, Los Alamos National Laboratory, 1967–84
   Visiting Professor, M.I.T. and University of Paris, 1972
   Graduate research professor, University of Florida, Gainesville, 1974–84
   Professor of biomathematics, University of Colorado medical school, Denver, 1979–84
   Visiting professor, University of California, Davis, 1982

Member of:
   American Academy of Arts and Sciences, National Academy of Sciences, American Philosophical Society, Mathematical and Physical Societies, Board of Governors and Scientific Advisory Committee, Weizmann Institute of Science, Rehovot, Israel; Board of Governors, Jurzykowski Foundation, New York

Honorary Degrees and Awards:
   University of New Mexico, University of Wisconsin, University of Pittsburgh, Polish Millenium, AC.P.C.C. Scientific, Polish Heritage awards.

Committees:
   NAS Committee on Innovations, NRS Committee on Applications of Mathematics, Harvard Visiting Committees for Mathematics, and Applied Mathematics and Physics, General Twining's Air Force Committee

Consultant:
   President Kennedy's Science Advisory Committee, also IBM, General Atomic, North American Aviation, Hycon

Died:
   May 13, 1984, Santa Fe, New Mexico

INDEX

Abelian groups, 517

Absorption, neutron, 2, 3, 13, 21, 22, 24, 25

Acceleration, 167, 168-171, 174

Active material, 18, 21, 22, 24, 25, 34

ADAM, 431, 432-433, 436, 444

Additive processes, 1

Adiabatic equation, 124

Algebra, 144; Boolean, 348-349, 446, 479, 480, 481, 482, 491, 494, 498, 518, 520; cylindric, 480, 496, 520; pattern, 492, 494, 496 (*see also* Relational structures); polyadic, 480; projective, 480-481, 496, 520; relation, 492-496

American Philosophical Society, xii

Amino acids, 466, 469, 470, 471, 472

Analogy, ix, x, 514-518; analogies between, 513, 518; and complexity, 518; criteria for, 514; distance measures, 483, 485, 489; transformations preserve, 523

Appell, Paul-Emile, 516n, 530

Arithmetic: complexity, 445-463; multiprecision, 353n, 357n; significance, 483

Artificial intelligence, 466

Automatic plotting devices, 301, 303. *See also* Oscilloscope

Automorphism, 492

Autonomous systems, 344

Baire categories, 350

Banach, Stefan, ix, x, 510, 513

Bednarek, Alexander R., xii, xiv, 477, 482, 485, 516

Behavior: chaotic, 192; convergence, 192, 197, 199, 200, 201-202, 211, 213-217, 218, 224-225, 227, 233, 272, 305-306; ergodic, xvi, 13, 132, 140, 143, 155-162, 192, 293, 294, 348; of gas, 123-129; limiting, 189, 190, 192, 193, 198, 201, 207-209, 221, 227, 296 (*see also* Oscillation); pathological, 358; qualitative, 131; topological, 131, 133

Bell, George, 527

Bendixson, I., 344

Bernouillian, formulas, 2, 183, 345, 349

Beyer, William A., xiii, 399, 445, 465, 469, 482

Billowing, 126-129

Binary reaction systems, 194-286, 294-296; as commutative, 196; convergence behavior of, 199, 200, 201-202, 211, 213-217, 218, 224-225,

Evans, Trevor, 495
EVE, 434-442, 443; PQ, 437, 438, 441, 444; PM, 438-440; POS, 440-442, 443, 444
Everett, C.J., xi, xii, 1, 37, 163, 188, 417
Evolution: Darwinian, 430; distance in, 466, 467; in mathematics, 522; via mutations, 287, 430-431, 432-433, 434-442, 443-444; rate/development of, 429-430, 432, 434-442, 443, 468
Evolutionary trees, 466, 469, 470, 471, 472, 473
Explosion: external, 163-177; history-dependent, 402-405; nuclear, 163-177, 179-180, 182-183; velocity of, 165-166

Faber, Vance, 480
Feller, W., 400
Fermi, Enrico, x, xii, 139, 156
Feynman, Richard P., 2, 9
Fine-structure, 357
Fission, xi, 19, 21, 22, 24-25; bombs, 164-165; tamper, 34, 35
Fitch, W.M., 466, 472
Fixed points, 6, 8, 37, 40, 42-45, 198, 337, 338, 361, 368, 372, 481; attractive, 202, 294, 299, 308, 321, 330-333, 345, 348; boundary (*see* Boundary points); Brouwer on, 201, 202, 330n, 491; for continuous function, 4; convergence to, 4, 305, 308-309, 310; death, 43, 44, 85, 105; equation, 234; inside gap, 358; interior, 199, 200, 201, 202-207, 208, 210-211, 212, 213, 215, 222, 223, 224, 225, 226, 314, 328; invariant points as, 294-299, iteration behavior of, 321, 327; limit points as, 43; limit sets as, 330; nodal, 199, 200, 201, 213, 222; non-attractive, 225; repellent, 202, 294, 299, 316, 328, 330-333, 342, 345, 354, 359; in supercritical case, 38
Flow, 481-482; ergodic, 159; Liouville, 186; volume-preserving, 156-157, 159
Flux, 176
Ford, Kenneth, W., 188
Formal systems, 521
Form stability, 210-213
Fourier, Baron Joseph, 159; series of, 122, 139, 141, 143, 160, 161, 490
Fox, R.H., 500
France, research in, xvi
Frankel, Stanley P., 2, 9
Fréchet spaces, 497
Frisch, Otto, 12
Frobenius, G., 144, 218
Fusion, 155

Gases, 123-129

Moments: calculated, 5, 11-12, 14; combinatorial, 6; first, 2, 5, 7, 37, 38-39, 64, 65, 69, 70, 72, 74, 105; and generating function, 5, 11-12; properties of, 123, 129-130, 131; second, 2, 5, 15, 38, 40-42;

Montague, J.S., 492

Monte Carlo method, ix, xi, xvi, 17, 18-36, 402, 406, 477-478, 530

Morphism, 492. *See also* Homeomorphism; Homomorphism; Isomorphism

Motion, 130, 142, 144, 159

Multiplicative processes, xi, 1-15, 37-119; branching and, 520; as continuous, 2; fluctuations in, 8, 15; genealogies of, 85, 92-105

Mutation: evolution via, 287, 430-431, 432-433, 434-442, 443-444; as transformation, 471, 472

Mycielski, Jan, 466, 498

Neighborhoods, 98, 99-100, 102, 103, 491

Nervous system, 478, 484-485, 486, 489, 490, 532

Neumann, John von, ix, x, xii, xv, 17, 18-33, 125, 127, 157, 409

Neutron: absorption, 2, 3, 13, 21, 22, 24, 25; active material of, 18, 21, 22, 24, 25, 34; collision. 17, 18, 19, 20, 22, 24-25, 190, 195-196, 418; cross-section of, 19; density distribution of, 127, 130; diffusion, xi-xii, 17-36; fission, xi, 19, 21, 22, 24-25, 35; flux, 176; heating, 183; leakage of, 2, 13; linearly extrapolated path of, 22-24; mean free path of, 15, 20, 130; parent, 13; scattering. ix, 21, 22, 24, 25, 35, 400; slower-down material of, 18, 21, 22, 24, 25, 34, 35; sojourn time of, 143, 157, 350, 352; tamper material of, 18, 21, 22, 24, 25, 34; velocity, 18-19, 21-22, 34, 35

Nonlinearity/nonlinear transformations, 293-377; broken-linear (*see* Broken-linear transformations); computer study of, xii, xvi, 139-154, 297-377; cubic (*see* Cubic transformations); difference equations, 191-192, 344-345; differential equations, 300, 344-345; displacement in, 139, 140, 145, 151, 153, 160, 161; ergodic behavior of, 140; iterations of, 189; polynomial, 345; quadratic (*see* Quadratic transformations); time in study of, 140

Normalization, 192-193

Norris, E.N., 495

Nuclear: constant, 2; explosion, 163-177, 179-180, 182-183; propulsion, 163-177, 179-184

Numbers: complicated, 445, 460, 519; large, 8-9; p-adic, 468; prime, 409, 445, 460, 461, 519; random, 355n; rational/real, 522-523

Oscillation, 198, 199, 200, 201, 207, 208, 209, 211, 222, 224, 225, 227, 296, 343

Oscilloscope, 301, 307, 310, 319, 326

Project Orion, xiii, 163
Propellant/propulsion: acceleration in, 167, 168-171, 174; air as, 176; chemical, 174, 176; distance, 165, 167, 177; external, 163-177; gravity as, xiii; heating by, 164, 176; hydrogen as, 164; internal, 179-181; kinetic energy in, 166; magnetic field in, 164-165, 176; mass in, 165-166, 172-174; nuclear, 163-177, 179-184; positioning of, 176-177; temperature, 165; velocity in, 165-166, 168, 173, 182, 183
Protein, 466, 469, 470, 471, 472
Pseudo-periods. *See* Limit sets, Class III

Quadratic function, 190
Quadratic transformations, 139, 142, 144, 147, 149, 161, 189-291; four-variable, 298, 299, 302, 311, 313, 323-326; homogeneous, 191-192, 218, 225, 286-291, 294-296 (*see also* Binary reaction systems); iteration in, 191-192, 197-198; limit sets for, 323-326; three-variable, 189, 218-221, 300n, 304-305. *See also* Broken-linear transformations
Quasi-states, 144, 162

Rademacher, Hans Adolph, 122, 490
Radiation, 417, 418, 422
Random: ergodic theorem, 399, 409-410; history-dependent processes, 399-410; mating, 434, 436, 438, 439, 440, 444; pairing, 194, 286; processes, xi, 129, 399-410; procreation, 2; walk, 399-405
Rationals, binary, 351
Ratios, 61-64; strong, 106-118
Rayleigh, John W., 142, 161
Reaction systems. *See* Binary reaction systems
Recognition, 525; discrimination compared to, 526; distance as tool for, 526-533; memory and, 483, 485-490, 526, 529, 530, 531-532, 533; pattern, xiii, 465, 466, 478, 483, 485-490, 531; of two-dimensional objects, 526, 528
Recursion, 5-6, 11, 79, 379-397
Reichert, T.A., 471
Reines, F., 164
Relation: algebras, 492-496; of neighbors, 124-125, 126; theory, 496 elational structures, 480-492; semigroups of, 481, 492-496; topological, 494
Richtmyer, Robert D., xii, 17, 34-36
RNA, 471, 472
Rockets. *See* Propellant/propulsion; Space vehicles
Rota, Gian-Carlo, 482, 483
Rotations, irrational, 192

Scattering, ix, 21, 22, 24, 25, 35, 400. *See also* Monte Carlo method

Valuation theory, 468
Velocity, 18-19, 21-22, 34, 35; of explosion, 165-166; gravity affects, 174; of propellant, 165-166, 168, 173, 182, 183; time needed for, 186, 187
Vertex, geometric, 60
Vibrating string calculations, xii, 141, 142, 144, 146-154, 160-161
Viking/V-2, 176-177
Volume, 156-157, 159
von Neumann. *See* Neumann, John von

Walk, random, 399-405; self-avoiding, 399, 400, 404
Walsh. J.L., 490
Wave equation, 140, 160
Westinghouse, 163
Weyl, Herman, 158, 192, 218
Whyte, L. L., 511
Wilks, S., 20
Wistar Institute, 429, 430
Wolfram, S., 379
Wong, A.K.C., 471
Wyler, A., 446

Zarecki, K.A., 492

Lightning Source UK Ltd.
Milton Keynes UK
UKHW012021170622
404590UK00002B/230